EQUILIBRIUM STAGED SEPARATIONS

Separations in Chemical Engineering

EQUILIBRIUM STAGED SEPARATIONS

Phillip C. Wankat
Professor of Chemical Engineering
Purdue University
West Lafayette, Indiana

PRENTICE HALL, UPPER SADDLE RIVER, NEW JERSEY 07458

Library of Congress Cataloging-in-Publication Data:

Wankat, Phillip C., 1944–
  Separations in chemical engineering : equilibrium-staged separations /
Phillip C. Wankat.
  p. cm.
  Includes bibliographies and index.
  1. Separation (Technology) I. Title.
TP156.S45W36 1988
660.2'842–dc19   87-32949
                                                                    CIP

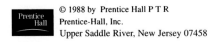
© 1988 by Prentice Hall P T R
Prentice-Hall, Inc.
Upper Saddle River, New Jersey 07458

Printed in the United States of America
20   19   18   17   16   15   14

ISBN   0-13-500968-5

Prentice-Hall International (UK) Limited, *London*
Prentice-Hall of Australia Pty. Limited, *Sydney*
Prentice-Hall Canada Inc., *Toronto*
Prentice-Hall Hispanoamericana, S. A., *Mexico*
Prentice-Hall of India Private Limited, *New Delhi*
Prentice-Hall of Japan, Inc., *Tokyo*
Prentice-Hall Asia Pte.Ltd., *Singapore*
Editora Prentice-Hall do Brasil, Ltda., *Rio de Janeiro*

To Dot and Charles

CONTENTS

PREFACE

Separations have always been very important in chemical engineering. This importance has recently escalated with the imminent emergence of new industries in biotechnology and high-performance materials. Separations will continue to remain important in bulk chemical manufacturing, petroleum processing, and the other standard areas of chemical engineering interest.

The development of new industries requiring the expertise of chemical engineers leads to problems and opportunities for chemical engineering education. Chemical engineering students need to be prepared for both the "known future" and the "unknown future." The known future includes the use of standard chemical engineering separation methods such as distillation and absorption which will remain important for many years. The unknown future involves the use of many relatively new separation methods such as adsorption, chromatography, electrophoresis, membrane separations, and zone melting. Up to now, these methods have not been included in the education of most undergraduate and graduate students.

In writing this book I have tried to satisfy the need of students to learn about the standard separation methods. *Equilibrium-Staged Separations,* covers the classical separations such as distillation, absorption and extraction. The equilibrium staged analysis procedure is developed in a rigorous way. Problem solving is emphasized throughout the text. There are detailed examples in each chapter and many homework problems which are solved in the solution manual.

Starting with flash distillation, the text leads the student into binary multistage distillation and then into multicomponent distillation. Both rigorous and short-cut methods for multicomponent distillation are explored. Azeotropic and extractive distillation are introduced, and binary batch distillation is discussed in detail. Then the design of staged and packed columns is discussed in depth. The section on distillation is finished with a chapter on economics and methods of coupling columns.

The principles developed for distillation are then used to study absorption, stripping, immiscible extraction, and washing. Since the basic tools are the same as for distillation, these subjects are covered in less depth. Partially miscible extraction and leaching are analyzed using triangular diagrams. The last chapter discusses mass transfer analyses.

This book has been extensively tested in a junior class at Purdue University. This class is taken before the students have had mass transfer. This book could also be used for sophmores or seniors. The order of prerequisite material is illustrated in Figure 1-2; this should help in arranging courses.

A second book, *Rate-Controlled Separations*, will include separation processes which require a mass transfer analysis for complete understanding. This includes most of the newer separation methods such as crystallization, adsorption, chromatography, and membrane separations. The style is similiar to the style of this book and problem solving is emphasized throughout. However, a higher level of mathematical analysis is required, and the second book is aimed for seniors and graduate students.

Many people have helped me with the writing of this book. Professor Joe Calo got me started writing the book, and Dr. Marjan Bace strongly supported this effort. My students have been most helpful in helping me develop clear methods to explain the separation methods. My teaching assistants over the last several years: Magdiel Agosto, Chris Buehler, Margret Shay, Sung-Sup Suh, and Narasimhan Sundaram, have been very helpful in solving problems and finding errors. Professor Ron Andres used the book in class. His comments were very helpful and have been incorporated into the text. Professors Karl T. Chuang and David P. Kessler, and Mr. Charles F. Gillard were very helpful in reviewing Chapter 12. Professors Alden Emery and James Caruthers were helpful in reviewing Chapter 19. A.P.V. Inc., Glitsch Inc., and The Norton Co. were very kind in providing photographs, as were Chris Roesel and Barb Naugle-Hildebrand in providing the art work.

Because of the length of this book many secretaries have been involved in the typing and production. Most of the work was done by Carolyn Blue, Debra Bowman, Jan Gray, and Becky Weston. I am grateful for their efforts. The copy editors and proof readers at Elsevier were very helpful in polishing the manuscript. The assistance of Philip Schafer is greatly appreciated.

A much more indirect but perhaps more important type of help came from the professors who helped me learn about the topics in this book. Professors Lowell B. Koppel and William R. Schowalter taught me my undergraduate and graduate courses in separations. They awakened my interest in separations. Dean C. Judson King through his book, his articles, and his personal example has helped keep that interest alive. My interest in problem solving has been sparked by Professors Richard Noble and Donald Woods.

Finally, my wife Dot has supported me when I thought I would never finish, and my son Charles has provided light to my life.

NOMENCLATURE

a	interfacial area per volume, ft^2/ft^3 or m^2/m^3
a_{P_1}, a_{P_2}, a_{P_3}, a_{T_1}, a_{T_2}, a_{T_6}	constants in Eq. (2-12) and Table 2-4
A,B,C	constants in Antoine Eq. (2-18)
A,B,C,D,E	constants in Eq. (3-51)
A_{active}	active area of tray, ft^2 or m^2
A_c	cross sectional area of column, ft^2 or m^2
A_d	downcomer area, ft^2 or m^2
A_{du}	flow area under downcomer apron, Eq. (12-28), ft^2
A_{hole}	area of holes in column, ft^2
A_I	interfacial area between two phases, ft^2 or m^2
A_{net}	net area, Eq. (12-13), ft^2 or m^2
b	equilibrium constant for linear equilibrium, $y = mx + b$
B	bottoms flow rate, kg mole/hr or lb mole/hr
C	number of components
C_{fL}	vapor load coefficient, Eq. (19-38)
C_p	heat capacity, Btu/lb °F or Btu/lbmole °F or cal/g °C or cal/g mole °C, etc.
C_{pH}	humid heat capacity, Eq. (19-89a)
C_{pW}	water heat capacity
C_{py}	vapor phase mass heat capacity
C_o	orifice coefficient, Eq. (12-25)

C_{sb}	capacity factor, Eq. (12-8)
d	dampening factor, Eq. (3-45)
D	diffusivity, cm^2/s or ft^2/hr
D, Dia	diameter of column, ft or m
D'_{col}	column diameter, see Table 19-2, ft
D_{total}	total amount of distillate (Chapt. 11), moles or kg
D	distillate flow rate, kg mole/hr or lb mole/hr
e	absolute entrainment, moles/hr
E	extract flow rate (Chapters 16 and 18), kg/hr
E_k	value of energy function for trial k, Eq. (3-38)
E_{ML}, E_{MV}	Murphree liquid and vapor efficiencies
E_o	Overall efficiency
E_{pt}	point efficiency, Eq. (12-3) or (19-76a)
$f = V/F$	fraction vaporized
f	friction factor, Eq. (19-86a)
$f_k(V/F)$	Rachford-Rice function for trial k, Eq. (3-29)
F	packing factor, Figure 13-4 and Tables 13-1 and 13-2
F	degrees of freedom, Eq. (2-5)
F	charge to still pot (Chapt. 11), moles or kg
F	feed flow rate, kgmole/hr or lbmole/hr or kg/hr etc.
F_D	diluent flow rate (Chapt. 16), kg/hr
F_{lv}	$\dfrac{W_L}{W_V} \sqrt{\dfrac{\rho_V}{\rho_L}} = \dfrac{L'}{G'} \sqrt{\dfrac{\rho_V}{\rho_L}}$, flow parameter
F_s, F_{solv}	flow rate solvent (Chapts. 16 and 17), kg/hr
F_{solid}	solids flow rate in leaching, kg insoluble solid/hr
F_w	modification factor, Eq. (12-26) and Figure 12-20

gap	gap from downcomer apron to tray, Eq. (12-28), ft
g	32.2 ft/s^2
G	flow rate carrier gas, kgmole/hr or kg/hr
G'	gas flux, lb/s,ft^2
h	pressure drop in head of clear liquid, inches liquid
h	height of liquid on stage (Chapter 19), ft.
h	height, m or ft
h	liquid enthalpy, kcal/kg, Btu/lb mole, etc.
$\tilde{h}$	pure component enthalphy
h_o	hole diameter, inches
h_p	packing height, ft or m
h_y	vapor phase heat transfer coefficient, $\text{Btu/(hr)}(\text{ft}^2)(°F)$
H	Henry's law constant, Eqs. (10-12) and (15-1)
H	vapor enthalpy, kcal/kg, Btu/lbmole, etc.
H_G	height of gas phase transfer unit, ft or m
$\overline{H}_G$	height of a transfer unit, Eq. (19-85b)
H_L	height of liquid phase transfer unit, ft or m
H_{OG}	height of overall gas phase transfer unit, ft or m
H_{OL}	height of overall liquid phase transfer unit, ft or m
H_{Ty}	gas phase height transfer unit for heat transfer, ft
H_y	enthalpy wet air, Eq. (19-89b), Btu/lb
HETP	height equivalent to a theoretical plate, ft or m
HTU	height of a transfer unit, ft or m
J_A	flux with respect to molar average velocity of fluid

k	thermal conductivity, Btu/(hr) (ft)($°$F)
k_x, k_y	individual mass transfer coefficients in liquid and vapor phases, see Table 19-2
k_y'	mass transfer coefficient in concentrated solutions, Eq. (19-43)
$\bar{k}_x$, $\bar{k}_y$	individual mass transfer coefficient in weight units
K	parameter to calculate column diameter, Eqs. (12-7), (12-8)
K_D	y/x, distribution coefficient for dilute extraction
K, K_i	y_i/x_i, equilibrium vapor-liquid ratio
K_{drum}	parameter to calculate u_{perm} for flash drums, Eq. (3-50)
K_x, K_y	overall mass transfer coefficient in liquid or vapor, lbmoles/ft^2 hr
l_w	weir length, ft
L	liquid flow rate, kgmoles/hr or lbmoles/hr
$\bar{L}$	mass liquid flow rate, lb/hr (Chapt. 19)
L'	liquid flux, lb/(s) (ft^2)
Le	Lewis number = $k/\rho\, D_{AB}\, C_{py}$
L_g	liquid flow rate in gal/min, Chapt. 12
m	linear equilibrium constant, $y = mx + b$
m	local slope of equilibrium curve, Eq. (19-5b)
M	flow rate of mixed stream (Chapt. 18), kg/hr
MW	molecular weight
n	moles
n_G	number of gas phase transfer units
$\bar{n}_G$	number of transfer units calculated from mass fractions, Eq. (19-85c)
n_{Hy}	number of transfer units for humidification, Eq. (19-92b)
n_L	number of liquid phase transfer units

n_{OG}	number of overall gas phase transfer units
n_{OL}	number of overall liquid phase transfer units
n_{Ty}	number of transfer units for heat transfer, Eq. (19-83b)
N	number of stages
N_A	flux of A, lbmoles/(hr)(ft^2)
N_f, N_{feed}	feed stage
N_{min}	number of stages at total reflux
$N_{feed,min}$	estimated feed stage location at total reflux
N_T	energy flux, Eqs. (19-80), Btu/(hr)(ft^2)
$\overline{N}_w$	mass flux of water, lb/(hr)(ft^2)
NTU	number of transfer units
O	total overflow rate in washing, kg/hr
p	pressure, atm, kPa, psi etc.
$\bar{p}, p_B$	partial pressure
P	Number of phases
q	$L_F/F = (\overline{L} - L)/F$, feed quality
Q	amount of energy transferred, Btu/hr, kcal/hr etc
Q_c	condenser heat load
Q_{flash}	heat loss from flash drum
Q_R	reboiler heat load
r	radius of column, ft or m
R	gas constant
R	raffinate flow rate (Chapts. 16 and 18), kg/hr
S	solvent flow rate (Chapt 10) kgmoles/hr or lbmoles/hr
S	solvent flow rate (Chapt. 18), kg/hr
Sc_L	Schmidt number for liquid $= \mu/\rho D$
Sc_v	Schmidt number for vapor $= \mu/\rho D$
t	time, s, min, or hr

t_{batch}	period for batch distillation, Eq. (11-25)
t_{down}	down time in batch distillation
$t_{operating}$	operating time in batch distillation
t_{res}	residence time in downcomer, Eq. (12-30), s
t_{tray}	tray thickness, inches
T	temperature, $^\circ$C, $^\circ$F, K or $^\circ$R
T_I	interfacial temperature
T_{ref}	reference temperature
T_w	water temperature
T_y	vapor phase temperature
u	vapor velocity, cm/s or ft/s
u_{flood}	flooding velocity, Eq. (12-7)
u_{op}	operating velocity, Eq. (12-11)
u_{perm}	permissible vapor velocity, Eq. (3-50)
U	underflow liquid rate, (Chapt. 17), kg/hr
v	vapor velocity, Eq. (19-86a)
v_o	vapor velocity through holes, Eq. (12-29), ft/s
$v_{o,bal}$	velocity where valve is balanced, Eq. (12-36)
V	vapor flow rate, kgmoles/hr or lbmoles/hr
V_{max}	maximum vapor flow rate
V_{surge}	surge volume in flash drum, Figure 3-4, ft^3
W	flow of phase co-current to L in 3 phase contactor (Chapt. 17)
W_L	liquid flow rate, kg/hr or lb/hr
W_L	liquid mass flux, lb/s ft^2 or lb/hr ft^2, (Chapt. 19)
W_V	vapor flow rate, kg/hr or lb/hr
x	weight or mole fraction in liquid
x	$[L/D - (L/D)_{min}]/(L/D + 1)$ in Eqs. (9-42)
x^*	equilibrium mole fraction in liquid

x_I	interfacial mole fraction in liquid
x_{out}^*	liquid mole fraction in equilibrium with inlet gas, Eq. (19-35b)
x'	pseudo-equilibrium, Eq. (17-15)
X	weight or mole ratio in liquid
y	weight or mole fraction in vapor
y^*	equilibrium mole fraction in vapor
y_{out}^*	vapor mole fraction in equilibrium with inlet liquid in counter current system, Eq. (19-35a) or in equilibrium with outlet liquid in cocurrent contactor, Eq. (19-71)
y_I	interfacial mole fraction in vapor
$\bar{y}$	mass fraction in vapor
Y	weight or mole ratio in vapor
Y_w	humidity, Eq. (19-88)
z	weight or mole fraction in feed
z	axial distance in bed (Chapt. 19)

Greek

α_{AB}	K_A/K_B, relative volatility
β	A_{hole}/A_{active}
γ	activity coefficient
Δ	change in variable
ϵ	limit for convergence
η	fraction of column available for vapor flow
θ	angle of downcomer, Figure 12-18B
λ	latent heat of vaporization, kcal/kg, Btu/lb, Btu/lbmole etc.
λ	mG/L in Eqs. (12-5) and (12-6)
λ_{ref}	latent heat of vaporization of water at 32°F
μ	viscosity, cp
μ_w	viscosity of water, cp

ρ_L	liquid density, g/cm^3 or lb/ft^3
ρ_V	vapor density
ρ_w	water density
σ	surface tension, dynes/cm
σ_w	surface tension of water, dynes/cm
ϕ	liquid phase packing parameter, Eq. (19-38)
ϕ_{dc}	relative froth density in downcomer, Eq. (12-29)
ψ	ρ_{water}/ρ_L, Chapter 13
ψ	$e/(e+L)$, fractional entrainment, Chapter 12
ψ	packing parameter for gas phase, Eq. (19-37)

EQUILIBRIUM STAGED SEPARATIONS

chapter 1
INTRODUCTION TO EQUILIBRIUM STAGED SEPARATIONS

1.1. IMPORTANCE OF SEPARATIONS

Why does chemical engineering require the study of separation techniques? Because separations are crucial in chemical engineering. A typical chemical plant is a chemical reactor surrounded by separators, as diagramed in the schematic flow sheet of Figure 1-1. Raw materials are prepurified in separation devices and fed to the chemical reactor; unreacted feed is separated from the reaction products and recycled back to the reactor. Products must be further separated and purified before they can be sold. This type of arrangement is very common. Examples for a variety of traditional processes are illustrated by Shreve and Brink (1977), Cavaseno *et al.* (1979), and Hatch and Matar (1981), whereas recent processes are shown in the Process Technology section of *Chemical Engineering* magazine. Chemical plants commonly have from 50 to 90% of their capital invested in separations equipment.

Since separations are ubiquitous in chemical plants and petroleum refineries, chemical engineers must be familiar with a variety of separation methods. We will focus on some of the most common chemical engineering separation methods: flash distillation, continuous column distillation, batch distillation, absorption, stripping, and extraction. These separations all contact two phases and can be designed and analyzed as equilibrium stage processes. Several other separation methods that can also be considered equilibrium stage processes will be briefly discussed. In a separate book (Wankat, 1988) important separations that do not operate as equilibrium stage systems will be explored.

The *equilibrium stage* concept is applicable when the process can be constructed as a series of discrete stages in which the two phases are contacted and then separated. The two separated phases are assumed to be in equilibrium with each other. For example, in distillation, a vapor and a liquid are commonly contacted on a metal plate with holes in it. Because of the intimate contact between the two phases, solute can transfer from one phase to another. Above the plate the vapor disengages from the liquid. Both liquid and vapor can be sent to additional

1

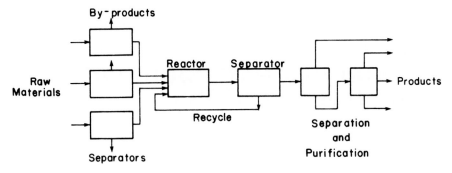

Figure 1-1. Typical chemical plant layout.

stages for further separation. Assuming that the stages are equilibrium stages, the engineer can calculate concentrations and temperatures without detailed knowledge of flow patterns and heat and mass transfer rates. Although this example shows the applicability of the equilibrium stage method for equipment built with a series of discrete stages, we will see that the staged design method can also be used for packed columns where there are no discrete stages. This method is a major simplification in the design and analysis of chemical engineering separations that is used throughout this book.

A second useful concept is that of a *unit operation*. The idea here is that although the specific design may vary depending on what chemicals are being separated, the basic design principles for a given separation method are always the same. For example, the basic principles of distillation are the same whether we are separating ethanol from water, or separating several hydrocarbons, or separating liquid metals. Consequently, distillation is often called a unit operation, as are absorption, extraction, etc.

A more general idea is that design methods for related unit operations are similar. Since distillation and absorption are both liquid-vapor contacting systems, the design is much the same for both. This similarity is useful because it allows us to apply a very few design tools to a variety of separation methods. We will focus on *stage-by-stage* methods where calculation is completed for one stage and then the results are used for calculation of the next stage.

1.2. PROBLEM-SOLVING METHODS

To help develop your problem-solving abilities, an explicit strategy, which is a modification of the strategy developed at McMaster University (Woods *et al.*, 1975), will be used throughout this book. The seven stages of this strategy are:

0. I want to and I can

1. Define the problem

2. Explore or think about it

3. Plan

4. Do it

5. Check

6. Generalize

Step 0 is a motivation and confidence step. It is a reminder that you got this far in chemical engineering because you can solve problems. The more different problems you solve, the better problem solver you will become. Remind yourself that you *want* to learn how to solve chemical engineering problems and you *can* do it.

In step 1 you want to define the problem. Make sure that you clearly understand all the words. Draw the system and label its parts. List all the known variables and constraints. Describe what you are asked to do. If you cannot define the problem clearly, you will probably be unable to solve it.

In step 2 you *explore* and *think about* the problem. What are you *really* being asked to do? What basic principles should be applied? Can you find a simple limiting solution that gives you bounds to the actual solution? Is the problem over- or underspecified? Let your mind play with the problem and chew on it. Then go back to the Define step to make sure that you are still looking at the problem in the same way. If not, revise the problem statement and continue. Experienced problem solvers always include an Explore step even if they don't explicitly state it.

In step 3 the problem solver *plans* how to subdivide the problem and decides what parts to attack first. The appropriate theory and principles must be selected, and mathematical methods chosen. The problem solver assembles required resources such as data, paper, and calculator. While doing this, new subproblems may arise; you may find there are not enough data to solve the problem. Recycle through the problem-solving sequence to solve these subproblems.

Step 4, *Do it,* is often the first step that inexperienced problem solvers try. In this step the mathematical manipulations are done, the numbers are plugged in, and an answer is generated. If your plan was incomplete, you may be unable to carry out this step. In that case, return to the Explore or Plan steps and recycle through the process.

In step 5, *check* your answer. Is it the right order of magnitude? For instance, commercial distillation columns are neither 12 inches nor 12 miles high. Does the answer seem reasonable? Have you avoided blunders such as plugging in the wrong number or incorrectly punching the calculator? Is there an alternative solution method that can serve as an independent check on the answer? If you find errors or inconsistencies, recycle to the appropriate step and solve the problem again.

The last step, *Generalize,* is important but is usually neglected. In this step you try to learn as much as possible from the problem. What have you learned about the physical situation? Did including a particular phenomenon have an important effect, or could you have ignored it? Generalizing allows you to learn and become a better problem solver.

At first these steps will not "feel" right. You will want to get on with it and start calculating instead of carefully defining the problem and working your way through the procedure. Stick with a systematic approach. It works much better on difficult problems than a "start calculating, maybe something will work" method. The more you use this or any other strategy the more familiar and less artificial it will become.

In this book, example problems are solved with this strategy. To avoid repeating myself I will not list step 0, but it is always there. The other six steps will usually be explicitly listed and developed. On the simpler examples some of the steps may be very short, but they are always present.

I strongly encourage you to use this strategy and write down each step as you do homework problems. In the long run this method will improve your problem-solving ability.

1.3. PREREQUISITE MATERIAL

No engineering book exists in a vacuum, and some preparatory material is always required. The first prerequisite, which is often overlooked, is that you must be able to read well. If you don't read well, get help immediately.

A second set of prerequisites involves certain mathematical abilities. You need to be comfortable with algebra and the manipulation of equations, as these skills are used throughout the text. Another required mathematical skill is graphical analysis, since many of the design methods are graphical methods. You need to be competent and to feel comfortable plotting curves and straight lines and solving simultaneous algebraic equations graphically. Familiarity with exponential and logarithmic manipulations is required for Chapter 9 on Short-Cut Methods for Distillation. The only chapters requiring calculus are Chapter 11, Batch Distillation, and Chapter 19, Mass Transfer.

The third area of prerequisites concerns mass balances, energy balances, and phase equilibria. Although the basics of mass and energy balances can be learned in a very short time, facility with their use requires practice. Thus this book will normally be preceded by a course on mass and energy balances. A knowledge of the basic ideas of phase equilibrium, including the concept of equilibrium, Gibbs's phase rule, distribution coefficients, and some familiarity with graphical representations of equilibrium data, will be helpful.

A fourth area of prerequisites are problem-solving skills. Because the chemical engineer must be a good problem solver, it is important to develop skills in this area. The ability to solve problems is a prerequisite for all chemical engineering courses.

Finally, you should have some skill in the use of programmable calculators and computers. The material in this text can be learned and the problems can be solved without computers, but real-life problems require the computer. It is a good idea to use computers and programmable calculators while you are learning the material and to do some of the computer homework problems.

In general, later chapters depend upon the earlier chapters, as shown schematically in Figure 1-2. Some chapters (10, 11, 14, 17, and 18) are not required for the understanding of later chapters and can be skipped if time is short. Figure 1-2 should be useful in planning the order in which to cover topics and for adapting this book for special purposes.

1.4. SUMMARY - OBJECTIVES

We have explored some of the reasons for studying separations and some of the methods we will use. At this point you should be able to satisfy the following objectives:

1. Explain how separations are used in a typical chemical plant.

2. Define the concepts of equilibrium stages and unit operations.

3. List the steps in the structured problem-solving approach, and start to use this approach.

4. Have some familiarity with the prerequisites.

6

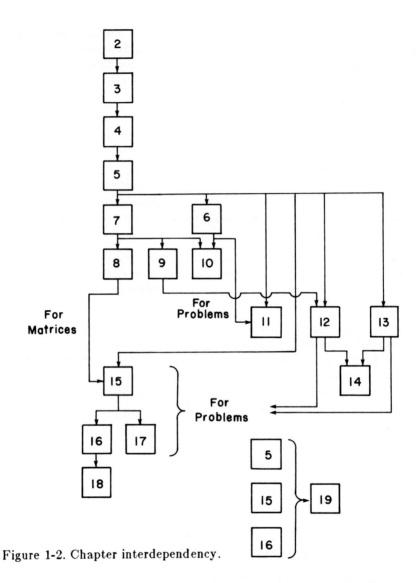

Figure 1-2. Chapter interdependency.

Note: In later chapters you may want to turn to the Summary - Objectives section first to help you see where you are going. Then when you've finished the chapter the Summary - Objectives section can help you decide if you got there.

REFERENCES

Cavaseno, V. and Staff of *Chemical Engineering* (eds.), *Process Technology and Flow Sheets*, McGraw-Hill, New York, 1979.

Hatch, L.F., and S. Matar, *From Hydrocarbons to Petrochemicals,* Gulf Publ. Co., Houston, TX, 1981.

Shreve, R.N., and J.A. Brink, *Chemical Process Industries,* 4th ed., McGraw-Hill, New York, 1977.

Wankat, P.C., *Mass Transfer Limited Separations,* Elsevier, New York (in press, 1988).

Woods, D.R., J.D. Wright, T.W. Hoffman, R.K. Swartman, and I.D. Doig, "Teaching Problem Solving Skills," *Engineering Education, 66* (3), 238 (Dec. 1975).

HOMEWORK

A.   *Discussion Problems*

A1.   Return to your successful solution of a fairly difficult problem in one of your previous technical courses (preferably chemical engineering). Look at this solution, but from the point of view of the *process* used to solve the problem instead of the technical details. Did you follow a structured method? Most people don't at first. Did you eventually do most of the steps listed? Usually, the define, explore, plan, and do it steps are done sometime during the solution. Rearrange your solution so that these steps are in order. Did you check your solution? If not, do that now. Finally, try generalizing your solution.

A2.   Without returning to the book, answer the following:

   a. Define a unit operation. Give a few examples.

   b. What is the equilibrium stage concept?

   c. What are the steps in the systematic problem solving approach?
   Explain each step in your own words.

A3.   Do you satisfy the prerequisites? If not, how can you remedy this situation?

A4.   Develop a key relations chart (one page or less) for this chapter. A key relations chart is a summary of everything you need to solve problems or answer questions from the chapter. In general, it will include equations, sketches, and key words. Organize it in your own way. The purpose of developing a key relations chart is to force your brain to actively organize the material. This will greatly aid you in remembering the material.

B.   *Generation of Alternatives*

B1.   List as many products and how they are purified or separated as you can. Go to a large supermarket and look at some of the household products. How many of these could you separate? At the end of this course you will know how to purify most of the liquid products.

C.   *Derivations*

C1.   Write the mass and energy balances (in general form) for the separator shown in Figure 1-1. If you have difficulty with this, review a book on mass and energy balances.

D.   *Problems* and E. *Complex Problems*
None for this chapter.

F.   *Problems Using Other Resources*

F1.   Look up the Process Technology section in a recent issue of *Chemical Engineering* magazine. Read the article and write a short (less than one page) critique. Explicitly comment on whether the flow sheet for the process fits (at least approximately) the general flow sheet shown in Figure 1-1.

chapter **2**
VAPOR-LIQUID PHASE EQUILIBRIUM

2.1. CONCEPT OF EQUILIBRIUM

The separation processes we are studying are based on the equilibrium stage concept, which states that streams leaving a stage are in equilibrium. What do we mean by equilibrium?

Consider a vapor and liquid that are in contact with each other as shown in Figure 2-1. Liquid molecules are continually vaporizing, while vapor molecules are continually condensing. If two chemical species are present, they will, in general, condense and vaporize at different rates. When not at equilibrium, the liquid and vapor can be at different pressures and temperatures and be present in different mole fractions. At equilibrium the temperatures, pressures, and fractions of the two phases cease to change. Although molecules continue to evaporate and condense, the rate at which each species condenses is equal to the rate at which it evaporates. Although on a molecular scale nothing has stopped, on the macroscopic scale, where we usually observe processes, there are no further changes in temperature, pressure, or composition.

Equilibrium conditions can be conveniently subdivided into thermal, mechanical, and chemical potential equilibrium. In thermal equilibrium, heat transfer stops and the temperatures of the two phases are equal.

$$T_{liquid} = T_{vapor} \qquad \text{(at equilibrium)} \qquad (2\text{-}1)$$

In mechanical equilibrium, the forces between vapor and liquid balance. In the staged separation processes we will study, this usually implies that the pressures are equal. Thus for the cases in this book,

$$P_{liquid} = P_{vapor} \qquad \text{(at equilibrium)} \qquad (2\text{-}2)$$

If the interface between liquid and vapor is curved, equal forces does not imply equal pressures. In this case the Laplace equation can be derived (e.g., see Levich, 1962).

9

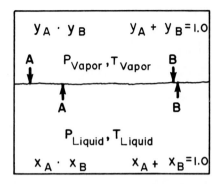

Figure 2-1.   Vapor-liquid
                contacting system.

In phase equilibrium, the rate at which each species is vaporizing is just equal to the rate at which it is condensing. Thus there is no change in composition (mole fraction in Figure 2-1). However, in general, the compositions of liquid and vapor are *not* equal. If the compositions were equal, no separation could be achieved in any equilibrium process. If temperature and pressure are constant, equal rates of vaporization and condensation require a minimum in the free energy of the system. The resulting condition for phase equilibrium is

$$\text{(chemical potential i)}_{\text{liquid}} = \text{(chemical potential i)}_{\text{vapor}} \qquad (2\text{-}3)$$

The development of Eq. (2-3), including the necessary definitions and concepts, is the subject of a large portion of many books on thermodynamics (e.g., Van Ness and Abbott, 1982; Smith and Van Ness, 1975; Balzhiser *et al.*, 1972; Denbigh, 1966, Walas, 1985) but is beyond the scope of this book. However, Eq. (2-3) does require that there be some relationship between liquid and vapor compositions. In real systems this relationship may be very complex and experimental data may be required. We will assume that the equilibrium data or appropriate equations are known, and will confine our discussion to the *use* of the equilibrium data in the design of separation equipment.

2.2. FORM AND SOURCES OF EQUILIBRIUM DATA

In principle, we can always experimentally determine the vapor-liquid equilibrium data we require. For a simple experiment we could take a chamber similar to Figure 2-1 and fill it with the chemicals of interest. Then, at different pressures and temperatures, we allow the liquid and vapor sufficient time to come to equilibrium and then take samples of liquid and vapor and analyze them. If we are very careful we can obtain reliable equilibrium data. In practice, the measurement is fairly difficult

Table 2-1. Vapor-Liquid Equilibrium Data for Ethanol and Water at 1 atm. y and x in Mole Fractions

x_{EtOH}	x_w	y_{EtOH}	y_w	$T, °C$
0	1.0	0	1.0	100
0.019	0.981	0.170	0.830	95.5
0.0721	0.9279	0.3891	0.6109	89.0
0.0966	0.9034	0.4375	0.5625	86.7
0.1238	0.8762	0.4704	0.5296	85.3
0.1661	0.8339	0.5089	0.4911	84.1
0.2377	0.7663	0.5445	0.4555	82.7
0.2608	0.7392	0.5580	0.4420	82.3
0.3273	0.6727	0.5826	0.4174	81.5
0.3965	0.6035	0.6122	0.3878	80.7
0.5079	0.4921	0.6564	0.3436	79.8
0.5198	0.4802	0.6599	0.3401	79.7
0.5732	0.4268	0.6841	0.3159	79.3
0.6763	0.3237	0.7385	0.2615	78.74
0.7472	0.2528	0.7815	0.2185	78.41
0.8943	0.1057	0.8943	0.1057	78.15
1.00	0	1.00	0	78.30

[R.H. Perry, C.H. Chilton and S.O. Kirkpatrick (Eds.), *Chemical Engineers Handbook*, 4th ed., New York, McGraw-Hill, p. 13-5, 1963.]

and a variety of special equilibrium stills have been developed (e.g., Hala *et al.*, 1967; Parker *et al.*, 1973; Peiffer *et al.*, 1972). Marsh (1978) and Van Ness and Abbott (1982, section 6-7) briefly review methods of determining equilibrium. With a static equilibrium cell, concentration measurements are not required for binary systems. Concentrations can be calculated from pressure and temperature data, but the calculation is complex.

If we obtained equilibrium measurements for a binary mixture of ethanol and water at 1 atm, we would generate data similar to those shown in Table 2-1. The mole fractions in each phase must sum to 1.0. Thus for this binary system,

$$x_1 + x_2 = 1.0 , \quad y_1 + y_2 = 1.0 \tag{2-4}$$

where x is mole fraction in the liquid and y is mole fraction in the vapor. Very often only the composition of the most volatile component (ethanol in this case) will be given. The mole fraction of the less volatile component can be found from Eqs. (2-4). Equilibrium depends on pressure. (Data in Table 2-1 are specified for a pressure of 1 atm.) Table 2-1 is only one source of equilibrium data for the ethanol-water system. Other sources are listed in Table 2-2, and data are contained in the more general sources listed in Table 2-3. The data in different references do not agree perfectly, and care must be taken in choosing good data.

We see in Table 2-1 that if pressure and temperature are set, then there is only one possible vapor composition for ethanol, y_{EtOH}, and one possible liquid composition, x_{EtOH}. Thus we cannot arbitrarily set as many variables as we might wish. For example, at 1 atm we cannot arbitrarily decide that we want a vapor-liquid equilibrium at 95°C and $x_{EtOH} = 0.1$.

The number of variables that we can arbitrarily specify, known as the degrees of freedom, is determined by subtracting the number of thermodynamic equilibrium equations from the number of variables. For nonreacting systems the resulting *Gibbs phase rule* is

$$F = C - P + 2 \qquad (2\text{-}5)$$

where F = degrees of freedom, C = number of components, and P = number of phases. For the binary system in Table 2-1, C = 2 (ethanol and water) and P = 2 (vapor and liquid). Thus,

$$F = 2 - 2 + 2 = 2$$

When pressure and temperature are set, all the degrees of freedom are used, and *at equilibrium* all compositions are determined from the experiment. Alternatively, we could set pressure and x_{EtOH} or x_w and determine temperature and the other mole fractions.

The amount of material and its flow rate are not controlled by the Gibbs phase rule. The phase rule refers to *intensive variables* such as pressure, temperature, or mole fraction, which do not depend on the total amount of material present. The *extensive variables,* such as number of moles, flow rate, and volume, do depend on the amount of material and are not included in the degrees of freedom. Thus a mixture in equilibrium must follow Table 2-1 whether there are 0.1, 1.0, 10, 100, or 1000 moles present.

Binary systems with only two degrees of freedom can be conveniently represented in tabular or graphical form by setting one variable (usually pressure) constant. Vapor-liquid equilibrium data have been determined for many binary systems. Sources for these data are listed in Table 2-3; you should become familiar with several of these sources. Note that the

Table 2-2.   Sources of Vapor-Liquid Equilibrium Data for
Ethanol-Water System

Altsheler, W.B., E.D. Unger, and P. Kolachov, *Ind. Eng. Chem., 43,* 2559 (1951).

Baker, E.H., *et al., Ind. Eng. Chem., 31,* 1260 (1939).

Beebe, H., K.E. Coulter, A. Lindsay, and E.M. Baker, *Ind. Eng. Chem., 34,* 1501 (1942).

Carey, J.S. and W.K. Lewis, *Ind. Eng. Chem., 24,* 882 (1932).

Cornell, L.W. and R.E. Montanna, *Ind. Eng. Chem., 25,* 1331 (1933).

Dalager, P., *J. Chem. Eng. Data, 14* (3), 198 (1969).

Evans, D.N., *Ind. Eng. Chem., 8,* 260 (1916).

Hughes, H.E. and J.O. Maloney, *Chem. Eng. Prog., 48* (4), 192 (1952).

Jones, C.A., E.M. Shoenborn, and A.P. Colburn, *Ind. Eng. Chem., 35,* 666 (1943).

Langdon, W.M. and D.B. Keyes, *Ind. Eng. Chem., 34,* 938 (1942).

Otsuki, H. and F.C. Williams, *Chem. Eng. Prog. Symp. Ser.,* No. *49* (6), 55 (1953).

Reider, R.M. and A.R. Thompson, *Ind. Eng. Chem., 41,* 2905 (1949).

Seader, J.D., "Distillation," in R.H. Perry and D.W. Green (Eds.), *Perry's Chemical Engineer's Handbook,* 6th ed., McGraw-Hill, New York, 1984, p. 13-61.

Vostrikova, W.N., M.E. Aerov, R.E Gurovich, and R.M. Solomatina, *Zh. Prikl. Khimii* (English Translation), *40* (3), 638 (1966).

Ethyl Alcohol Handbook, U.S.I. Chemicals, Division of National Distillers and Chemical Corporation, New York, 1969.

Table 2-3. Sources of Vapor-Liquid Equilibrium Data

Chu, J.C., R.J. Getty, L.F. Brennecke, and R. Paul, *Distillation Equilibrium Data,* Reinhold, New York, 1950.

Engineering Data Book, Natural Gasoline Supply Men's Association, 421 Kennedy Bldg., Tulsa, OK, 1953.

Gmehling, J. and U. Onken, *Vapor-Liquid Equilibrium Data Collection,* Chemical Data Series, 10 vols., DECHEMA, 1977.

Hala, E., I. Wichterle, J. Polak, and T. Boublik, *Vapor-Liquid Equilibrium Data at Normal Pressures,* Pergamon, New York, 1968.

Hala, E., J. Pick, V. Fried, and O. Vilim, *Vapor-Liquid Equilibrium,* 3rd ed., 2nd Engl. ed., Pergamon, New York, 1967.

Horsely, L.H., *Azeotropic Data,* ACS Advances in Chemistry, No. 6, American Chemical Society, Washington, DC, 1952.

Horsely, L.H., *Azeotropic Data (II),* ACS Advances in Chemistry, No. 35, American Chemical Society, Washington, DC, 1952.

Maxwell, J.B., *Data Book on Hydrocarbons,* Van Nostrand, Princeton, NJ, 1950.

Perry, R.H. and C.H. Chilton (Eds.), *Chemical Engineer's Handbook,* 5th ed., McGraw-Hill, New York, 1973.

Perry, R.H., C.H. Chilton, and S.D. Kirkpatrick (Eds.), *Chemical Engineer's Handbook,* 4th ed., McGraw-Hill, New York, 1963.

Perry, R.H. and Green, D. (Eds.), *Perry's Chemical Engineer's Handbook,* 6th ed., McGraw-Hill, New York, 1984.

Prausnitz, J.M., Anderson, T.F., Grens, E.A., Eckert, C.A., Hsieh, R., and O'Connell, J.P., *Computer Calculations for Multicomponent Vapor-Liquid and Liquid-Liquid Equilibria,* Prentice-Hall, Englewood Cliffs, NJ 1980. (See Table 1 for an extensive bibliography.)

Timmermans, J., *The Physico-Chemical Constants of Binary Systems in Concentrated Solutions,* 5 vols., Interscience, New York, 1959-1960.

Wichterle, I., J. Linek, and E. Hala, *Vapor-Liquid Equilibrium Data Bibliography,* Elsevier, Amsterdam, 1973.

data are not of equal quality. Methods for testing the thermodynamic consistency of equilibrium data are discussed in great detail by Van Ness and Abbott (1982, pp. 56-64, 301-348). Errors in the equilibrium data can have a profound effect on the design of the separation method (e.g., see Nelson *et al.*, 1983).

2.3. GRAPHICAL REPRESENTATION OF BINARY EQUILI-BRIUM DATA

Binary vapor-liquid equilibrium data can be represented graphically in several ways. The most convenient forms are temperature-composition, y-x, and enthalpy-composition diagrams. These figures all represent the same data and can be converted from one form to another.

Table 2-1 gives the equilibrium data for ethanol and water at 1 atmosphere. With pressure set, there is only one degree of freedom remaining. Thus we can select any of the intensive variables as the independent variable and plot any other intensive variable as the dependent variable. The simplest such graph is the y versus x graph shown in Figure 2-2. Typically, we plot the mole fraction of the more volatile component (the component that has y > x; ethanol in this case). This diagram is also called a McCabe-Thiele diagram when it is used for calculations. Pressure is constant, but the temperature is

Figure 2-2.   y versus x diagram for ethanol-water.

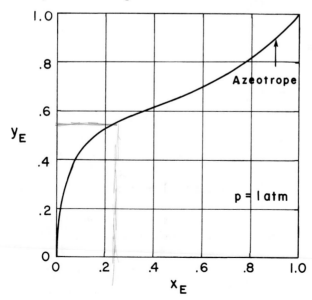

different at each point on the equilibrium curve. Points on the equilibrium curve represent two phases in equilibrium. Any point not on the equilibrium curve represents a system that may have both liquid and vapor, but they are not in equilibrium. As we will discover later, y-x diagrams are extremely convenient for calculation.

The data in Table 2-1 can also be plotted on a temperature-composition diagram as shown in Figure 2-3. The result is actually two graphs: one is liquid temperature versus x_{EtOH}, and the other is vapor temperature versus y_{EtOH}. These curves are called *saturated liquid* and *saturated vapor* lines, because they represent all possible liquid and vapor systems that can be in equilibrium at a pressure of 1 atm. Any point below the saturated liquid curve represents a subcooled liquid, whereas any point above the curve would be a superheated vapor. Points between the two saturation curves represent streams consisting of both liquid and vapor. If allowed to separate, these streams will give a liquid and vapor in equilibrium. Liquid and vapor in equilibrium must be at the same temperature; therefore, these streams will be connected by a horizontal isotherm as shown in Figure 2-3 for $x_{EtOH} = 0.2$.

Even more information can be shown on an enthalpy-composition or Ponchon-Savarit diagram, as illustrated for ethanol and water in Figure 2-4. Note that the units in Figure 2-4 differ from those in Figure 2-3.

Figure 2-3.   Temperature-composition diagram for ethanol-water.

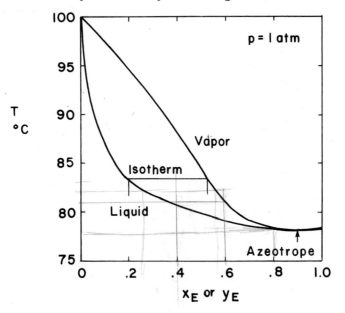

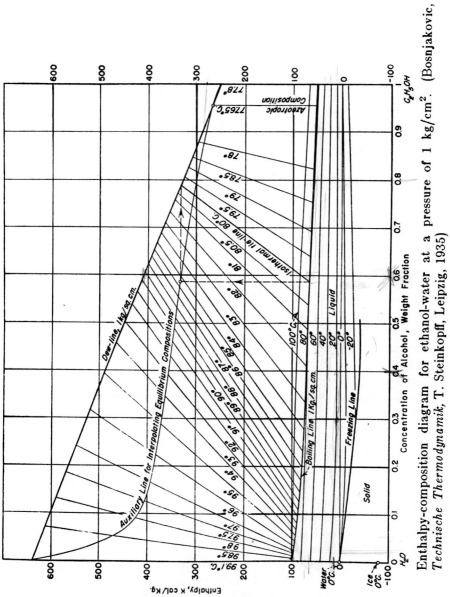

Figure 2-4. Enthalpy-composition diagram for ethanol-water at a pressure of 1 kg/cm². (Bosnjakovic, *Technische Thermodynamik*, T. Steinkopff, Leipzig, 1935)

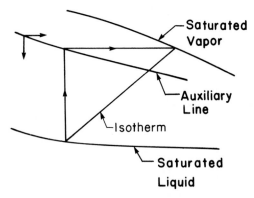

Figure 2-5.   Use of auxiliary line. To find an isotherm go vertically
from the saturated liquid curve to the auxiliary line.
Then go horizontally to the saturated vapor line. The
line connecting the points on the saturated vapor and
saturated liquid curves is the isotherm.

Again, there are really two plots: one for liquid and one for vapor. The
isotherms shown in Figure 2-4 show the change in enthalpy at constant
temperature as weight fraction varies. Because liquid and vapor in
equilibrium must be at the same temperature, these points are connected
by an isotherm. Points between the saturated vapor and liquid curves
represent two-phase systems. An isotherm through any point can be
generated using the auxiliary line with the construction shown in Figure
2-5. If an isotherm is desired through a point in the two-phase region, a
simple trial-and-error procedure is required.

   Isotherms on the enthalpy-composition diagram can also be generated from the y-x and temperature-composition diagrams. Since
these diagrams represent the same data, the vapor composition in equilibrium with a given liquid composition can be found from either the y-x
or temperature-composition graph, and the value transferred to the
enthalpy-composition diagram. This procedure can also be done graphically as shown in Figure 2-6 if the units are the same in all figures. In
Figure 2-6a we can start at point A and draw a vertical line to point A'
(constant x value). At constant temperature, we can find the equilibrium vapor composition (point B'). Following the vertical line (constant y), we proceed to point B. The isotherm connects points A and
B. A similar procedure is used in Figure 2-6b, except now the y-x line
must be used on the McCabe-Thiele graph. This is necessary because
points A and B in equilibrium appear as a single point, A'/B', on the
y-x graph. The y = x line allows us to convert the ordinate value (y)

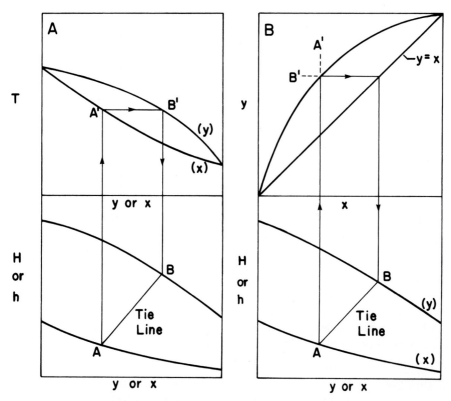

Figure 2-6.   Drawing isotherms on the enthalpy-composition diagram. A. from the temperature-composition diagram; B. from the y-x diagram.

on the y-x diagram to an abscissa value (also y) on the enthalpy-composition diagram. Thus the procedure is to start at point A and go up to point A′/B′ on the y-x graph. Then go horizontally to the y=x line and finally drop vertically to point B on the vapor curve. The isotherm now connects points A and B.

The data presented in Table 2-1 and illustrated in Figures 2-2, 2-3, and 2-4 show a minimum-boiling *azeotrope,* i.e., the liquid and vapor are of exactly the same composition at a mole fraction ethanol of 0.8943. This can be found from Figure 2-2 by drawing the y=x line and finding the intersection with the equilibrium curve. In Figure 2-3 the saturated liquid and vapor curves touch, while in Figure 2-4 the isotherm is vertical at the azeotrope. Note that the azeotrope composition is numerically different in Figure 2-4, but actually it is the same, since Figure 2-4 is in weight fractions, whereas the other figures are in mole fractions.

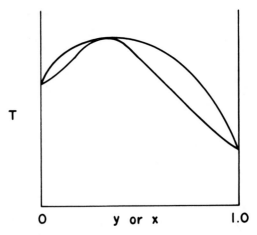

Figure 2-7.   Maximum boiling azeotrope system.

Below the azeotrope composition, ethanol is the more volatile component; above it, ethanol is the less volatile component. The system is called a minimum-boiling azeotrope because the azeotrope boils at 78.15° C, which is less than that of either pure ethanol or pure water. The azeotrope location is a function of pressure. Below 70 mmHg no

Figure 2-8.   Heterogeneous azeotrope system, n-butanol and water at 1 atmosphere (Perry *et al.*, 1963, p. 13-4).

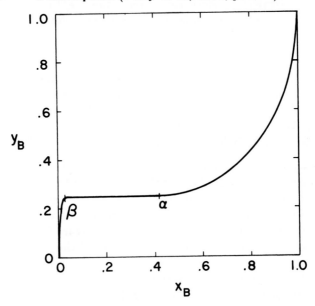

azeotrope exists for ethanol-water (Seader, 1984). Maximum-boiling azeotropes also occur (see Figure 2-7). Only the temperature-composition diagram will look significantly different.

The usual azeotrope system cannot be completely separated by distillation and requires the use of an additional separation method. An exception to this occurs when the two components are only partially miscible as liquid. Now when the azeotrope vapor is condensed it will separate into two separate liquid phases (points α and β on Figure 2-8, which shows the n-butanol-water system. Equilibrium for partially miscible systems is considered in Chapter 10.

2.4. GRAPHICAL MASS BALANCES FOR EQUILIBRIUM SYSTEMS

To determine the amounts of liquid and vapor formed when a two-phase mixture separates, mass balances are required. These balances can be obtained graphically on temperature-composition or enthalpy-composition diagrams.

The method is easily illustrated in a temperature-composition diagram. In Figure 2-9 a two-phase mixture at point A is allowed to separate into liquid at B and vapor at C. Because the mixture is in equilibrium, points A, B, and C are all on the same isotherm. If we let F be the moles of feed of composition z_A (we use z since this is a two-phase mixture at point A), the mass balances before and after the mixture separates are

$$F z_A = L x_B + V y_C \tag{2-6}$$

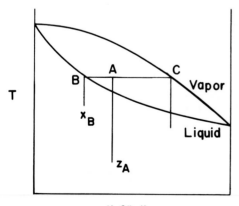

Figure 2-9. Mass balances on temperature-composition diagram.

$$F = L + V \qquad (2\text{-}7)$$

Substituting Eq. (2-7) into (2-6),

$$L z_A + V z_A = L x_B + V y_C$$

Rearranging this, we have

$$L(z_A - x_B) = V(y_C - z_A)$$

or

$$\frac{L}{V} = \frac{y_C - z_A}{z_A - x_B} \qquad (2\text{-}8)$$

From Figure 2-9 we see that $(y_C - z_A)$ is equal to the distance from point A to C, that is $\overline{AC}$. Also $(z_A - x_B)$ is equal to distance $\overline{BA}$. Thus

$$\frac{L}{V} = \frac{\overline{AC}}{\overline{BA}} \qquad (2\text{-}9)$$

Equation (2-9) can be used to find L/V from distances on Figure 2-9. This value of L/V and Eq. (2-7) can be used to determine the amounts of liquid and vapor.

Equation (2-9) is called the lever-arm rule because the same result is obtained when a moment-arm balance is done on a seesaw. Thus if we set moment arms of the seesaw in Figure 2-10 equal, we obtain

$$(\text{wt B})\,(\overline{BA}) = (\text{wt C})\,(\overline{AC})$$

or

$$\frac{\text{wt B}}{\text{wt C}} = \frac{\overline{AC}}{\overline{BA}}$$

Figure 2-10.   Illustration of lever-arm rule.

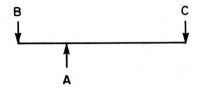

which gives the same result as Eq. (2-9). The seesaw is a convenient way to remember the form of the lever-arm rule.

The lever-arm rule can also be applied on enthalpy-composition diagrams and on ternary diagrams for extraction. In these cases the lever-arm rule has several other uses (see Chapter 18).

2.5. OTHER REPRESENTATIONS OF EQUILIBRIUM DATA

Graphical representations of equilibrium data are difficult to adapt to computers or programmed calculators and are not convenient for multicomponent systems. In these cases equations are very useful.

One way to represent equilibrium data is to define a distribution coefficient or K value as

$$K_A = y_A/x_A \tag{2-10}$$

In general, the K values depend on temperature, pressure, and composition. These nonideal K values are discussed in detail by Smith (1963) and Walas (1985) and in thermodynamics textbooks.

Fortunately, for many systems the K values are approximately independent of composition. Thus,

$$K = K(T,p) \quad \text{(approximate)} \tag{2-11}$$

For light hydrocarbons, the approximate K values can be determined from the monographs prepared by DePriester. These are shown in Figures 2-11 and 2-12, which cover different temperature ranges. If temperature and/or pressure of the equilibrium mixture are unknown, a trial-and-error procedure is required. DePriester charts in other temperature and pressure units are given by Perry and Chilton (1973), Perry et al. (1963), and Smith and Van Ness (1975). The DePriester charts have been fit to the following equation (McWilliams, 1973):

$$\ln K = \frac{a_{T1}}{T^2} + \frac{a_{T2}}{T} + a_{T6} + a_{p1} \ln p + \frac{a_{p2}}{p^2} + \frac{a_{p3}}{p} \tag{2-12}$$

Note that T is in °R and p is in psia. The constants a_{T1}, a_{T2}, a_{T6}, a_{p1}, a_{p2}, and a_{p3} are given in Table 2-4. The last line gives the mean errors in the K values compared to the values from the DePriester charts. This equation is valid from $-70°C$ ($365.7°R$) to

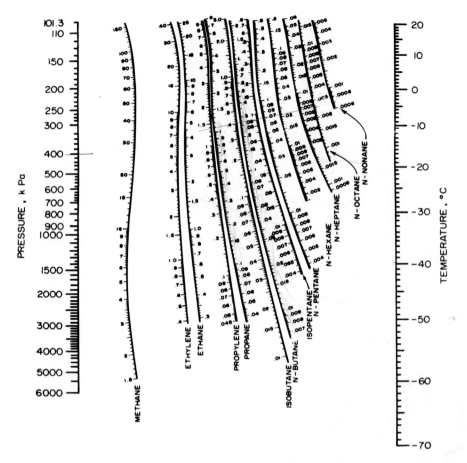

Figure 2-11.   Modified DePriester chart (in S.I. units) at low temperatures. From D.B. Dadyburjor, *Chem. Eng. Prog.*, 85,
April 1978. Copyright 1978, AIChE. Reproduced by
permission of the American Institute of Chemical
Engineers.

200° C (851.7° R) and for pressures from 101.3 kPa (14.69 psia) to 6000
kPa (870.1 psia). If K and p are known, then Eq. (2-12) can be solved
for T. The obvious advantage of an equation compared to the charts is
that it can be programmed into a computer or calculator.

The K values are used along with the stoichiometric equations which
state the mole fractions in liquid and vapor phases must sum to 1.0.

$$\sum_{i=1}^{N} y_i = 1.0 , \qquad \sum_{i=1}^{N} x_i = 1.0 \qquad (2-13)$$

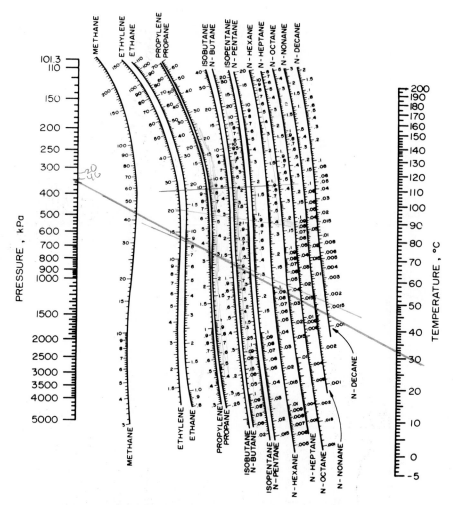

Figure 2-12. Modified DePriester chart at high temperatures. From D.B. Dadyburjor, *Chem. Eng. Prog.*, 85, April 1978. Copyright 1978, AIChE. Reproduced by permission of the American Institute of Chemical Engineers.

If only one component is present, then y = 1.0 and x = 1.0. This implies that $K_i = y/x = 1.0$. This gives a simple way of determining the boiling temperature of a pure compound at any pressure. For example, if we wish to find the boiling point of isobutane at p = 150 kPa, we set our straightedge on p = 150 and at 1.0 on the isobutane scale on Figure 2-11. Then read T = −1.5°C as the boiling point. Alternatively, Eq. (2-12) with values from Table 2-4 can be solved for T. This gives T = 488.68°R or −1.6°C.

Table 2-4. Constants for Fit to K Values Using Eq. (2-12). Note: T in °R and p in psia.

Compound	a_{T1}	a_{T2}	a_{T6}	a_{p1}	a_{p2}	a_{p3}	Mean Error
Methane	−292,860	0	8.2445	−.8951	59.8465	0	1.66
Ethylene	−600,076.875	0	7.90595	−.84677	42.94594	0	2.65
Ethane	−687,248.25	0	7.90694	−.88600	49.02654	0	1.95
Propylene	−923,484.6875	0	7.71725	−.87871	47.67624	0	1.90
Propane	−970,688.5625	0	7.15059	−.76984	0	6.90224	2.35
Isobutane	−1,166,846	0	7.72668	−.92213	0	0	2.52
n-Butane	−1,280,557	0	7.94986	−.96455	0	0	3.61
Isopentane	−1,481,583	0	7.58071	−.93159	0	0	4.56
n-Pentane	−1,524,891	0	7.33129	−.89143	0	0	4.30
n-Hexane	−1,778,901	0	6.96783	−.84634	0	0	4.90
n-Heptane	−2,013,803	0	6.52914	−.79543	0	0	6.34
n-Octane	0	−7646.81641	12.48457	−.73152	0	0	7.58
n-Nonane	−255,104	0	5.69313	−.67818	0	0	9.40
n-Decane	0	−9760.45703	13.80354	−.71470	0	0	5.69

Source: McWilliams (1973)

For ideal systems Raoult's law holds. Raoult's law states that the partial pressure of a component is equal to its vapor pressure multiplied by its mole fraction in the liquid. Thus,

$$p_A = x_A (VP)_A \qquad (2\text{-}14)$$

where vapor pressure (VP) depends on temperature. By Dalton's law of partial pressures,

$$y_A = \frac{p_A}{p} \qquad (2\text{-}15)$$

Combining these equations,

$$y_A = \frac{(VP)_A x_A}{p} \qquad (2\text{-}16)$$

Comparing Eqs. (2-16) and (2-10), the Raoult's law K value is

$$K_A = \frac{(VP)_A}{p} \qquad (2\text{-}17)$$

This is handy, since extensive tables of vapor pressures are available (e.g., Boublik *et al.*, 1984; Perry and Chilton, 1973; Perry and Green, 1984). Vapor pressure is often correlated in terms of the Antoine equation

$$\ln(VP) = A - \frac{B}{T+C} \qquad (2\text{-}18)$$

where A, B, and C are constants for each pure compound. These constants are tabulated in various data sources (Boublik *et al.*, 1984; Gmehling and Onken, 1977; and Yaws, 1977). The equations based on Raoult's law should be used with great care, since deviations from Raoult's law are extremely common.

Nonidealities in the liquid phase are taken into account with a liquid-phase activity coefficient, γ_i. Then Eq. (2-17) becomes

$$K_A = \frac{\gamma_A(VP_A)}{p_{total}} \qquad (2\text{-}19)$$

The activity coefficient depends on temperature, pressure and concentration. Excellent correlation procedures for activity coefficients such as the Margules, Van Laar, Wilson, NRTL and UNIQUAC methods have been developed (Balzhiser *et al.*, 1972; Fredenslund *et al.*, 1977; Prausnitz *et al.*, 1980; Reid *et al.*, 1977; Van Ness and Abbott, 1982; and Wales, 1985). The coefficients for these equations for a wide variety of mixtures have been tabulated by Gmehling and Onken (1977) along with the experimental data. When the binary data are not available, the UNIFAC method (Fredenslund *et al.*, 1977; Prausnitz *et al.*, 1980) can be used to predict the missing data. Most large companies that use distillation will use Eqs. (2-18) and (2-19) along with one of the equations for activity coefficients. A detailed description of these methods is beyond the scope of this book.

The K values are strongly dependent on temperature and thus vary greatly in a distillation column. Much of the temperature dependence is removed if we take the ratio of K values for different components. Thus the *relative volatility*, α_{AB}, is defined as

$$\alpha_{AB} = \frac{K_A}{K_B} = \frac{y_A/x_A}{y_B/x_B} \qquad (2\text{-}20)$$

If K_A and K_B have exactly the same temperature dependence, α_{AB} will be independent of temperature. However, the relative volatility is usually a weak function of temperature. Equation (2-20) is used in conjunction with the summation equations (2-13). If Raoult's law is valid, then we can determine relative volatility as

$$\alpha_{AB} = \frac{(VP)_A}{(VP)_B} \qquad (2\text{-}21)$$

For binary systems, Eqs. (2-13) or (2-4) become

$$y_B = 1 - y_A, \quad x_B = 1 - x_A$$

and the relative volatility is

$$\alpha_{AB} = \frac{y_A(1 - x_A)}{(1 - y_A)x_A} \qquad \text{(binary)} \qquad (2\text{-}22)$$

Solving Eq. (2-22) for y_A, we obtain

$$y_A = \frac{\alpha_{AB}\, x_A}{1 + (\alpha_{AB} - 1)x_A} \qquad \text{(binary)} \qquad (2\text{-}23)$$

Equation (2-23) is a simple way to represent binary vapor-liquid equilibrium. It is used later to develop analytical solutions for distillation columns.

2.6. CALCULATION OF BUBBLE-POINT AND DEW-POINT TEMPERATURES

The bubble-point temperature is the temperature at which a liquid mixture begins to boil. The dew-point temperature is the temperature at which a vapor mixture first begins to condense. Obviously, both temperatures depend on the system pressure.

For a bubble-point calculation, the pressure, p, and the mole fractions of the liquid, x_i, will be specified. We wish to find the temperature, T_{BP}, at which $\sum y_i = 1.0$ where the y_i are calculated as $y_i = K_i(T_{BP})x_i$. If a simple expression for K_i as a function of temperature is available, we may be able to solve the resulting equation for temperature. Otherwise a root-finding technique or a trial-and-error procedure will be required. When graphs or charts are used, a trial-and-error procedure is always needed. This procesure is shown in Figure 2-13. When we are finished with the calculation, the y_i calculated are the mole fractions of the first bubble of gas formed.

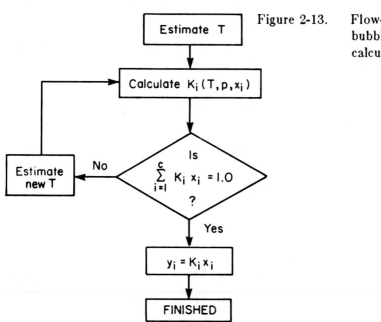

Figure 2-13. Flow-sheet for bubble-point calculation.

How can we make a good guess for the initial temperature?

Note that $\Sigma x_i = 1.0$ and $\Sigma K_i x_i = 1.0$. If all $K_i > 1.0$, then we must have $\Sigma K_i x_i > 1.0$. If all $K_i < 1.0$, then $\Sigma K_i x_i < 1.0$. Therefore, we should choose a temperature so that some K_i are greater than 1.0 and some are less than 1.0. One way to do this is to find the boiling temperature of each component ($K_i = 1.0$) and estimate T as $T = \sum z_i T_i(K_i = 1.0)$.

How do we pick a new temperature when the previous trial was not accurate?

If we look at the DePriester charts or Eq. (2-12) we see that K_i are complex functions of temperature. However, the function is quite similar for all K's. Thus the variation in K for one component (the reference component) and the variation in the value of $\Sigma(K_i x_i)$ will be quite similiar. We can estimate the appropriate K value for the reference component as

$$K_{ref}(T_{new}) = \frac{K_{ref}(T_{old})}{\sum_{i=1}^{N} (K_i x_i)_{calc}} \qquad (2\text{-}24)$$

The temperature for the next trial is determined from the new value, $K_{ref}(T_{new})$. This procedure should converge quite rapidly. As an alternative, a Newtonian convergence scheme can be used [see Eqs. (3-30) to (3-33)].

A similar calculation of the bubble point can be done with relative volatilities. For any component i, we can calculate y_i as

$$y_i = x_i \left(\frac{y_i/x_i}{y_{ref}/x_{ref}}\right) \left(\frac{y_{ref}}{x_{ref}}\right) = x_i \, \alpha_i \, K_{ref} \qquad (2\text{-}25)$$

Now apply the summation condition

$$\sum_{i=1}^{N} y_i = \sum_{i=1}^{N} x_i \, \alpha_i \, K_{ref} = 1.0$$

Solving for K_{ref},

$$K_{ref} = \frac{1}{\sum_{i=1}^{N} (x_i \alpha_i)} \qquad (2\text{-}26)$$

The bubble-point temperature can be calculated from K_{ref} and the y_i from Eq. (2-25).

The dew-point calculation will be similar, except that the y_i are given and we want $\Sigma x_i = \Sigma \left(\dfrac{y_i}{K_i}\right) = 1.0$. The next temperature can be estimated by calculating K_{ref} from

$$K_{ref}(T_{new}) = K_{ref}(T_{old}) \sum_{i=1}^{N} \left(\frac{y_i}{K_i}\right)_{calc} \qquad (2\text{-}27)$$

The new temperature is then determined from $K_{ref}(T_{new})$. The development of a convergence routine is straightforward (see Problem 2-C1).

Example 2-1. Calculation of Bubble-Point Temperature

What is the bubble-point temperature of a mixture that is 15 mole % isopentane, 30 mole % n-pentane, and 55 mole % n-hexane? Pressure is 1.0 atm.

Solution

A. Define. We want T for which $\Sigma y_i = \Sigma K_i x_i = 1.0$.

B. Explore. We will want to use a trial-and-error procedure. If we use the DePriester charts, Figures 2-11 and 2-12, convert atm to kPa.

$$p = 1.0 \text{ atm} \left[\frac{101.3 \text{ kPA}}{1.0 \text{ atm}} \right] = 101.3 \text{ kPa}$$

Equation (2-12) can also be used, and we will use it for the Check step.

C. Plan. For First T use a T for which $K_{ic5} > K_{nc5} > 1.0 > K_{nc6}$. (Use a DePriester chart.) Then use Eq. (2-24) to calculate K_{ref} and from the DePriester chart find T. Pick nC_6 = reference component (arbitrary).

D. Do It. First guess: Using Figure 2-12 at 50°C: $K_{ic5} = 2.02$, $K_{nc5} = 1.55$, $K_{nc6} = 0.56$.

Thus 50°C (and many other temperatures) satisfies our first guess criteria.

Check:
$$\sum_{i=1}^{3} y_i = \sum_{i=1}^{3} (K_i)(x_i)$$

$$= (2.02)(0.15) + (1.55)(0.30) + (0.56)(0.55) = 1.076$$

This is too high, so the temperature of 50° C is too high.

From Eq. (2-24), calculate

$$K_{c6_{New}} = \frac{K_{c6\ old}}{\Sigma K_i x_i} = \frac{0.56}{1.076} = 0.52$$

From the DePriester chart, the corresponding temperature is $T = 47.5°$ C.

At this temperature $K_{ic5} = 1.92$, $K_{nc5} = 1.50$. Note that all the K's are lower, so the summation will be lower. Now

$$\Sigma y_i = \Sigma K_i x_i = (1.92)(0.15) + (1.50)(0.30) + (0.52)(0.55) = 1.024$$

The next $K_{c6} = \dfrac{0.52}{1.024} = 0.508$, which corresponds to $T = 47°$ C.
At this temperature, $K_{ic5} = 1.89$, $K_{nc5} = 1.44$.

Check: $\Sigma K_i x_i = (1.89)(0.15) + (1.44)(0.30) + (0.508)(0.55) = 0.9949$.

This is about as close as we can get with the DePriester chart. Thus the bubble-point temperature is 47° C.

E.   Check. The y_i values are equal to $K_i x_i$. Thus, $y_{ic5} = (1.89)(0.15) = 0.2835$, $y_{nc5} = (1.44)(0.30) = 0.4320$, $y_{nc6} = (0.508)(0.55) = 0.2794$, and $\sum_{i=1}^{3} y_i = 0.9949$. The y values should be rounded off to two significant figures when they are reported. An alternative solution can be obtained using Eq. (2-12). This procedure is sketched briefly below.

lst guess: $T = 50°$ C $= 122°$ F $= 122 + 459.58 = 581.58$ °R
$p = 14.7$ psia
For iC_5, nC_5, nC_6, Eq. (2-12) simplifies to

$$\ln K = a_{T1}\left(\frac{1}{T^2}\right) + a_{T6} + a_{p1} \ln p \tag{2-28}$$

which gives $K_{iC_5} = 2.0065$, $K_{nC_5} = 1.5325$, $K_{nC_6} = 0.5676$. Then

$$\sum y_i = \sum K_i x_i$$

$$= (2.0065)(0.15) + (1.5325)(0.3) + (0.5676)(0.55) = 1.0729.$$

This is too high. To find next temperature, use Eq. (2-24).

$$K_{C_6 \text{ new}} = \frac{K_{C_6 \text{ old}}}{\sum K_i x_i} = \frac{0.5676}{1.0729} = 0.5290$$

Solving Eq. (2-28) for T,

$$T = \left(\frac{a_{T1}}{\ln K - a_{T6} - a_{p1} \ln p - a_{p2}/p^2 - a_{p3}/p}\right)^{1/2} \qquad (2\text{-}29)$$

and we obtain $T = 577.73°\,R$. Using this for the new guess we can continue. The final result is $T = 576.9°\,R = 47.4°\,C$. This result is within the error of Eq. (2-12) when compared to the $47.0°\,C$ found from the DePriester charts. Equation (2-29) is valid for all the hydrocarbons covered by the DePriester charts except n-octane and n-decane (see Problem 2-C3).

F. Generalize. If K values depend on composition, then an extra loop in the trial-and-error procedure will be required. When K values are in equation form such as Eq. (2-12), bubble-point calculations are easy to program on calculators or computers. With K in equation form, we can also write $\sum K_i x_i = 1$ as a single equation and solve for T numerically (see Problem 2-C2).

2.7. SUMMARY - OBJECTIVES

In this chapter we discussed vapor-liquid equilibrium data. At this point you should be able to satisfy the following objectives.

1. Explain what is meant by phase equilibrium.

2. Find desired vapor-liquid equilibrium data in the literature.

3. Plot and use y-x, temperature-composition, and enthalpy-composition diagrams. Explain the relationship between these three types of diagrams.

4. Derive and use the lever-arm rule.

5. Define and use K values, Raoult's law, and relative volatility.

6. Use trial-and-error methods to calculate bubble-point and dew-point temperatures.

REFERENCES

Balzhiser, R.E., M.R. Samuels, and J.D. Eliassen, *Chemical Engineering Thermodynamics: The Study of Energy, Entropy, and Equilibrium*, Prentice-Hall, Englewood Cliffs, NJ, 1972.

Beebe, H., K.E. Coulter, A. Lindsay, and E.M. Barker, "Equilibria in Ethanol-Water Systems at Pressures Less Than Atmospheric," *Ind. Eng. Chem., 34*, 1501 (1942).

Boublik, T., V. Fried, and E. Hala, *Vapour Pressures of Pure Substances*, Elsevier, Amsterdam, 1984.

Dadyburjor, D.B., "SI Units for Distribution Coefficients," *Chem. Eng. Prog., 74* (4), 85, (April, 1978).

Denbigh, K., *The Principles of Chemical Equilibrium*, 2nd ed., Cambridge University Press, Cambridge, England, 1966.

Fredenslund, A., J. Bmehling, and P. Rasmussen, *Vapor-Liquid Equilibria Using UNIFAC: A Group-Contribution Method*, Elsevier, Amsterdam, 1977.

Gmehling, J. and U. Onken, *Vapor-Liquid Equilibrium Data Collection*, Chemical Data Series, DECHEMA, 1977.

Hala, E., J. Pick, V. Fried, and O. Vilim, *Vapor-Liquid Equilibrium*, 3rd ed., 2nd English ed., Pergamon, New York, 1967.

Kojima, K. and T. Tochigi, *Prediction of Vapor-Liquid Equilibria by the ASOG Method*, Elsevier, Amsterdam, 1979.

Levich, V.G., *Physiochemical Hydrodynamics*, Prentice-Hall, Englewood Cliffs, New Jersey, 1962.

McWilliams, M.L., "An Equation to Relate K-factors to Pressure and Temperature," *Chem. Eng., 80* (25), 138 (Oct. 29, 1973).

Marsh, K.N., "The Measurement of Thermodynamic Excess Functions of Binary Liquid Mixtures," *Chemical Thermodynamics,* vol. 2, The Chemical Society, London, 1978, pp. 1-45.

Nelson, A.R., J.H. Olson, and S.I. Sandler, "Sensitivity of Distillation Process Design and Operation to VLE Data," *Ind. Eng. Chem. Process Des. Develop., 22,* 547 (1983).

Packer, L.G., S.R.M. Ellis, and L. De. J. Soares, "Miniature Equilibrium Still," *Chem. Eng. Sci., 28,* 597 (1973).

Pfeiffer, C.C., R.S. Metcalfe, R.J. Kopke, and R.H. McCormick, "A 20-Stage Cocurrent Contacting Vapor-Liquid Equilibrium Unit," *Ind. Eng. Chem. Process Des. Develop., 11,* 525 (1972).

Perry, R.H. and C.H. Chilton (Eds.), *Chemical Engineer's Handbook,* 5th ed., McGraw-Hill, New York, 1973.

Perry, R.H., C.H. Chilton, and S.D. Kirkpatrick (Eds.), *Chemical Engineer's Handbook,* 4th ed., McGraw-Hill, New York, 1963.

Perry, R.H. and D. Green (Eds.), *Perry's Chemical Engineer's Handbook,* 6th ed., McGraw-Hill, New York, 1984.

Prausnitz, J.M., T.F. Anderson, E.A. Grens, C.A. Eckert, R. Hsieh, and J.P. O'Connell, *Computer Calculations for Multicomponent Vapor-Liquid and Liquid-Liquid Equilibria,* Prentice-Hall, Englewood Cliffs, NJ, 1980.

Reid, R.C., J.M. Prausnitz, and T.K. Sherwood, *The Properties of Gases and Liquids,* 3rd ed., McGraw-Hill, New York, 1977.

Smith, J.M. and H.C. VanNess, *Introduction to Chemical Engineering Thermodynamics,* 3rd ed., McGraw-Hill, New York, 1975.

Van Ness, H.C. and M.M. Abbott, *Classical Thermodynamics of Non-Electrolyte Solutions. With Applications to Phase Equilibria,* McGraw-Hill, New York, 1982.

Walas, S.M., *Phase Equilibria in Chemical Engineering,* Butterworth, Boston, 1985.

HOMEWORK

A. *Discussion Problems.*

A1. Without looking in the text, define K value, relative volatility, equilibrium, azeotrope, Antoine equation, DePriester chart, dew-point temperature, Gibbs phase rule, lever-arm rule, auxiliary line. Check your answers in the text.

A2. What would Figure 2-2 look like if we plotted y_2 vs x_2 (i.e., plot less volatile component mole fractions)?

A3. What would Figure 2-3 look like if we plotted T vs x_2 or y_2 (less volatile component)?

A4. What would Figure 2-4 look like if we plotted H or h versus y_2 or x_2 (less volatile component)?

A5. Why is a dew-point calculation trial-and-error? Under what conditions might it not be trial-and-error?

A6. If a liquid mixture of n-butanol and water that is 20 mole % n-butanol is vaporized, what is the vapor composition? (See Figure 2-8.) Repeat for mixtures that are 10, 30, and 40 mole % n-butanol. Explain what is happening.

A7. For a typical straight-chain hydrocarbon, does:

   a. K increase, decrease, or stay the same when temperature is increased?

   b. K increase, decrease, or stay the same when pressure is increased?

   c. K increase, decrease, or stay the same when mole fraction in the liquid phase is increased?

   d. K increase, decrease, or stay the same when the molecular weight of the hydrocarbon is increased within a homologous series?

   Note: it will help to visualize the DePriester chart in answering this question.

A8. Why do the values of the azeotrope composition appear to be different in Figures 2-3 and 2-4?

A9. Develop your own key relations chart for this chapter. See Problem 1-A4 for a description of a key relations chart.

A10. In Example 2-1 part C, we decided to pick a temperature so that $K_{iC5} > K_{nC5} > K_{nC8}$. Why does this criterion give a good first guess for temperature?

B. *Generation of Alternatives*

B1. We have an equilibrium mixture of isobutane, n-butane, isopentane, and n-pentane. List as many ways as possible in which a bubble-point or dew-point problem could be specified. (Use the Gibbs phase rule first.)

B2. In principle, measuring vapor-liquid equilibrium (VLE) data is straightforward. In practice, actual measurement may be very difficult. Think of how you might do this. How would you take samples without perturbing the system? How would you analyze for the concentrations? What could go wrong? Look in your thermodynamics textbook for ideas.

C. *Derivations*

C1. Develop the convergence routine for a dew-point calculation.

C2. Repeat Example 2-1, except use Eq. (2-28) to write $\sum K_i x_i = 1.0$ as a single equation with T as the unknown. Solve numerically for T or use a canned root-finding method.

C3. Derive the solution for T from Eq. (2-12) for n-octane and n-decane.

D. *Problems*

D1. Generate the y-x diagram that would correspond to Figure 2-7.

D2. Use the DePriester chart to generate the temperature-composition diagram for isobutane and propane at 1000 kPa.

D3. Use the DePriester chart to generate a pressure-composition diagram for isobutane and propane to 60° C.

D4. If a 40 mole % ethanol, 60 mole % water mixture at 60° C and 1 atm is heated:

a. At what temperature does it first begin to boil? What is the composition of the first bubble of vapor?

b. At what temperature would it stop boiling (assume no material is removed)? What is the composition of the last droplet of liquid?

c. At 82° C, what fraction is liquid?

d. When 80% has been vaporized, what is the temperature, and what are the liquid and vapor compositions?

D5.   a. Use the DePriester charts to find the relative volatility of propane with respect to n-butane at a pressure of 400 kPa and at temperatures of −40° C, 0° C, 40° C, 80° C and 120° C. Compare the change in α with temperature to the changes in the K values of propane and n-butane.

b. Repeat part a using Eq. (2-12).

D6.   We have a mixture of n-hexane, n-heptane, and n-octane at 300 kPa. The composition of the mixture can be changed to any desired mole fractions of these three compounds.

a. What is the highest possible bubble-point temperature?

b. What is the lowest possible bubble-point temperature?

D7.   a. What is the boiling temperature of isopentane at 101.3 kPa? 200 kPa? 500 kPa? 2500 kPa?

b. What is the boiling pressure of isopentane at 100° and 150° C?

D8.   Find the pressure that a liquid and vapor equilibrium mixture of propane and n-butane must be at if the temperature is 100° F and the liquid is 0.40 mole fraction n-butane. Do with both a DePriester chart and Eq. (2-12).

D9.   A mixture of n-butane, n-pentane, and n-hexane is at 120° F and 20 psia. Liquid and vapor are in equilibrium. If the liquid is 0.10 mole fraction n-butane, find the compositions of liquid and vapor. Note: Watch your units. This problem is *not* trial-and-error.

D10.   Using Figure 2-4, find:

a. The enthalpy of a 40 wt % ethanol solution at −20° C and 40° C. What is the heat capacity of this liquid?

b. The enthalpy of a saturated liquid that is 20 wt % ethanol.

   c. The enthalpy of a saturated vapor that is 85 wt % ethanol.

   d. The liquid composition in equilibrium with a vapor that is 30 wt % alcohol.

   e. If a 60 wt % alcohol mixture is heated to 83°C and then allowed to separate into liquid and vapor, what are the compositions of the two phases at equilibrium? Use the lever-arm rule (along an isotherm) to determine the relative amounts of liquid and vapor.

D11. At 1 atm, the ethylene dibromide-propylene dibromide system has a constant relative volatility of $\alpha_{EP} = 1.30$ (Perry *et al.*, 1963, p. 13-3). Use the α_{EP} value to generate the y-x equilibrium diagram.

D12.   a. At 1 atm, we have a vapor that is 15 mole % n-butanol and 85 mole % water. If this vapor is entirely condensed, what are the compositions of the two liquid phases that form? If there were initially 1 kg mole of vapor, how many moles of each liquid phase are there? Use Figure 2-8 or the data in Table 10-2.

        b. Repeat this problem for an initial vapor of 30 mole % n-butanol.

E.   *More Complex Problems*

E1. Find the dew-point and bubble-point temperatures for a mixture that is 10 mole % ethylene, 40 mole % ethane, and 50 mole % propane at 690 kPa.

     a. Use the DePriester charts. b. Use Eq. (2-12).

E2. Find the dew-point and bubble-point temperatures for a mixture that is 20 mole % n-butane, 50 mole % n-pentane, and 30 mole % n-hexane. Pressure is 1 atm.

     a. Use the DePriester charts. b. Use Eq. (2-12).

E3. Write a computer or calculator program for calculation of the bubble-point temperature. Use Eq. (2-12) or the simplified forms shown in Eqs. (2-28) and (2-29) for K values. Test your program on Example 2-1.

F. *Problems Requiring Other Resources*

F1. Benzene-toluene equilibrium is often approximated as $\alpha_{BT} = 2.5$. Generate the y-x diagram for this relative volatility. Compare your results with data in the literature (see references in Table 2-3). Also, generate the equilibrium data using Raoult's law, and compare your results to these.

F2. Isopropanol and water form an azeotrope. At 1 atm:

    a. What is the azeotrope composition?

    b. Is the azeotrope homogeneous or heterogeneous?

    c. At 10 mole % water, which compound is more volatile? Which is more volatile at 90% water?

    d. If the liquid is 10 mole % water, what is the vapor composition? What is the temperature of the mixture?

chapter 3
FLASH DISTILLATION

3.1. BASIC METHOD

One of the simplest separation processes commonly employed is flash distillation. In this process, part of a feed stream vaporizes in a flash chamber, and the vapor and liquid in equilibrium with each other are separated. The more volatile component will be more concentrated in the vapor. Usually a large degree of separation is not achieved; however, in some cases, such as the desalination of seawater, complete separation results.

The equipment needed for flash distillation is shown in Figure 3-1. The fluid is pressurized and heated and is then passed through a throttling valve or nozzle into the flash drum. Because of the large drop in pressure, part of the fluid vaporizes. The vapor is taken off overhead, while the liquid drains to the bottom of the drum, where it is withdrawn. A demister or entrainment eliminator is often employed to prevent liquid droplets from being entrained in the vapor. The system is called "flash" distillation because the vaporization is extremely rapid after the feed enters the drum. Because of the intimate contact between liquid and vapor, the system in the flash chamber is very close to an equilibrium stage. Figure 3-1 shows a vertical flash drum, but horizontal drums are also common.

The designer of a flash system needs to know the pressure and temperature of the flash drum, the size of the drum, and the liquid and vapor compositions and flow rates. He or she also wishes to know the pressure, temperature, and flow rate of the feed entering the drum. In addition, he or she will need to know how much the original feed has to be pressurized and heated. The pressures must be chosen so that at the feed pressure, p_F, the feed is below its boiling point and remains liquid, while at the pressure of the flash drum, p_{drum}, the feed is above its boiling point and some of it vaporizes. If the feed is already hot and/or the pressure of the flash drum is quite low, the pump and heater shown in Figure 3-1 may not be needed.

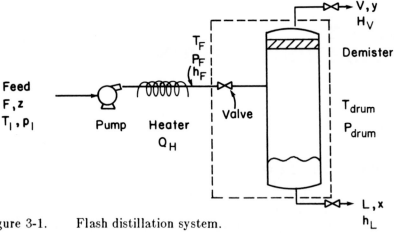

Figure 3-1.   Flash distillation system.

The designer has six degrees of freedom to work with for a binary separation. Usually, the original feed specifications take up four of these degrees of freedom:

Feed flow rate, F

Feed composition, z (mole fraction of the more volatile component)

Temperature, T_1

Pressure, p_1

Of the remaining, the designer will usually select first

Drum pressure, p_{drum}

A number of other variables are available to fulfill the last degree of freedom.

As is true in the design of many separation techniques, the choice of specified design variables controls the choice of the design method. For the flash chamber, we can use either a sequential solution method or a simultaneous solution method. In the sequential procedure, we solve the mass balances and equilibrium relationships first and then solve the energy balances and enthalpy equations. In the simultaneous solution method, all equations must be solved at the same time. In both cases we solve for flow rates, compositions, and temperatures before we size the flash drum.

3.2. BINARY FLASH DISTILLATION

We will assume that the flash drum shown in Figure 3-1 acts as an equilibrium stage. Then vapor and liquid are in equilibrium, and y, x, and T_{drum} can be calculated:

$$y = y_v(x, p_{drum}) \tag{3-1}$$

$$T_{drum} = T(x, p_{drum}) \tag{3-2}$$

Equations (3-1) and (3-2) may be represented by K values, by a relative volatility, or graphically. For binary systems it is usually convenient to plot equilibrium data on y-x or enthalpy-composition diagrams.

Mass and energy balances are written for the balance envelope shown as a dashed line in Figure 3-1. For a binary system there are two independent mass balances. The standard procedure is to use the overall mass balance,

$$F = V + L \tag{3-3}$$

and the component balance for the more volatile component,

$$Fz = Vy + Lx \tag{3-4}$$

The energy balance is

$$Fh_F + Q_{flash} = VH_v + Lh_L \tag{3-5}$$

where h_F, H_v and h_L are the enthalpies of the feed, vapor, and liquid streams. Usually $Q_{flash} = 0$, since the flash drum is insulated and the flash is considered to be adiabatic.

To utilize the energy balance equations, we need to know the enthalpies. Their general form is

$$h_F = h_F(T_F, z), \quad H_v = H_v(T_{drum}, y), \quad h_L = h_L(T_{drum}, x) \tag{3-6}$$

For binary systems it is often convenient to represent the enthalpy functions graphically on an enthalpy-composition diagram such as Figure 2-4. For ideal mixtures the enthalpies can be calculated from heat capacities and latent heats. Then,

$$h_L(T, x) = x_A C_{PL,A} (T - T_{ref}) + x_B C_{PL,B} (T - T_{ref}) \tag{3-7a}$$

$$h_F(T_F, z) = z_A C_{pL,A}(T_F - T_{ref}) + z_B C_{PL,B} (T_F - T_{ref}) \tag{3-7b}$$

$$H_V(T, y) = y_A[\lambda_A + C_{PV,A}(T - T_{ref})]$$

$$+ y_B[\lambda_B + C_{PV,B}(T - T_{ref})] \qquad (3\text{-}8)$$

where x_A and y_A are mole fractions of component A in liquid and vapor, respectively. C_p is the molar heat capacity, T_{ref} is the chosen reference temperature, T_{BP} is the boiling point of the mixture, and λ is the latent heat of vaporization at T_{ref}. For binary systems, $x_B = 1 - x_A$, and $y_B = 1 - y_A$.

3.2.1. Sequential Solution Procedure

In the sequential solution procedure, we first solve the mass balance and equilibrium relationships, and then we solve the energy balance and enthalpy equations. In other words, the two sets of equations are uncoupled. The sequential solution procedure is applicable when the last degree of freedom is used to specify a variable that relates to the conditions in the flash drum. Possible choices are:

Vapor mole fraction, y
Liquid mole fraction, x
Fraction feed vaporized, $f = V/F$
Fraction feed remaining liquid, $q = L/F$
Temperature of flash drum, T_{drum}

If one of the equilibrium conditions, y, x, or T_{drum}, is specified, then the other two can be found from Eqs. (3-1) and (3-2) or the graphical representation of equilibrium data. For example, if y is specified, x is obtained from Eq. (3-1) and T_{drum} from Eq. (3-2). In the mass balances, Eqs. (3-3) and (3-4), the only unknowns are L and V, and the two equations can be solved simultaneously.

If either the fraction vaporized or fraction remaining liquid is specified, Eqs. (3-1), (3-3), and (3-4) must be solved simultaneously. The most convenient way to do this is to combine the mass balances. Solving Eq. (3-4) for y, we obtain

$$y = -\frac{L}{V}x + \frac{F}{V}z \qquad (3\text{-}9)$$

Equation (3-9) is the *operating equation*, which for a single-stage system relates the compositions of the two streams leaving the stage. Equation (3-9) can be rewritten in terms of either the fraction vaporized, $f = V/F$, or the fraction remaining liquid, $q = L/F$.

From the overall mass balance, Eq. (3-3),

$$\frac{L}{V} = \frac{F - V}{V} = \frac{1 - V/F}{V/F} = \frac{1 - f}{f} \qquad (3\text{-}10)$$

Then the operating equation becomes

$$y = -\frac{1 - f}{f} x + \frac{z}{f} \qquad (3\text{-}11)$$

The alternative in terms of L/F is

$$\frac{L}{V} = \frac{L}{F - L} = \frac{L/F}{1 - L/F} = \frac{q}{1 - q} \qquad (3\text{-}12)$$

and the operating equation becomes

$$y = -\frac{q}{1 - q} x + \left(\frac{1}{1 - q}\right)z \qquad (3\text{-}13)$$

Although they have different forms, Eqs. (3-9), (3-11), and (3-13) are equivalent means of obtaining y, x, or z. We will use whichever operating equation is most convenient.

Now the equilibrium equation (3-1) and the operating equation (3-9, 3-11, or 3-13) must be solved simultaneously. The exact way to do this depends on the form of the equilibrium data. For binary systems a graphical solution is very convenient. Equations (3-9), (3-11), and (3-13) represent a single straight line, called the *operating line,* on a graph of y versus x. This straight line will have

$$\text{Slope} = -\frac{L}{V} = -\frac{1 - f}{f} = -\frac{q}{1 - q} \qquad (3\text{-}14)$$

and

$$y \text{ intercept } (x = 0) = \frac{F}{V} z = \frac{1}{f} z = \frac{1}{1 - q} z \qquad (3\text{-}15)$$

The equilibrium data at pressure p_{drum} can also be plotted on the y-x diagram. The intersection of the equilibrium curve and the operating line is the simultaneous solution of the mass balances and equilibrium. This plot of y vs x showing both equilibrium and operating lines is called a *McCabe-Thiele diagram* and is shown in Figure 3-2 for an

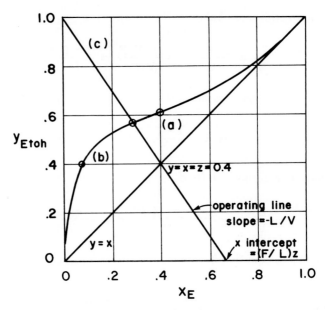

Figure 3-2.   McCabe-Thiele diagram for binary flash distillation. Illustrated for Example 3-1.

ethanol-water separation. The solution point gives the vapor and liquid compositions leaving the flash drum. Figure 3-2 shows three different operating lines as V/F varies from 0 to 0.4 to 1.0 (see Example 3-1). T_{drum} can be found from Eq. (3-2) or from a temperature-composition diagram.

Two other points shown on the McCabe-Thiele diagram are the x intercept (y = 0) of the operating line and its intersection with the y = x line. Either of these points can also be located algebraically and then used to plot the operating line.

The intersection of the operating line and the y = x line is often used because it is simple to plot. This point can be determined by simultaneously solving Eq. (3-9) and the equation y = x. Substituting y = x into Eq. (3-9), we have

$$y = -\frac{L}{V} y + \frac{F}{V} z$$

or $$y\left(1 + \frac{L}{V}\right) = \frac{F}{V} z$$

$$\text{or} \qquad y\left(\frac{V+L}{V}\right) = \frac{F}{V} z$$

since $V + L = F$, the result is $y = z$ and therefore

$$x = y = z$$

The intersection is at the feed composition.

It is important to realize that the $y = x$ line has no fundamental significance. It is often used in graphical solution methods because it simplifies the calculation. However, do *not* use it blindly.

Obviously, the graphical technique can be used if y, x, or T_{drum} is specified. The order in which you find points on the diagram will depend on what information you have to begin with.

Example 3-1. Flash Separator for Ethanol and Water

A flash distillation chamber operating at 101.3 kPa is separating an ethanol-water mixture. The feed mixture is 40 mole % ethanol. (a) What is the maximum vapor composition and (b) what is the minimum liquid composition that can be obtained if V/F is allowed to vary? (c) If $V/F = 0.4$, what are the liquid and vapor compositions? (d) Repeat step c, given that F is specified as 1000 kg moles/hr.

Solution

A. Define. We wish to analyze the performance of a flash separator at 1 atm.

a. Find y_{max}.

b. Find x_{min}.

c. and d. Find y and x for $V/F = 0.4$.

B. Explore. Note that $p_{drum} = 101.3$ kPa $= 1$ atm. Thus we must use data at this pressure. These data are conveniently available in Table 2-1 and Figure 2-2. Since p_{drum} and V/F for part c are given, a sequential solution procedure will be used. For parts a and b we will look at limiting values of V/F.

C. Plan. We will use the y-x diagram as illustrated in Figure 3-2. For all cases we will do a mass balance to derive an operating line [we could use Eqs. (3-9), (3-11), or (3-13), but I wish to illustrate how easy it is to derive an operating line]. Note that $0 \leq V/F \leq 1.0$. Thus our maximum and minimum values for V/F must lie within this range.

D. Do It. Sketch is shown.

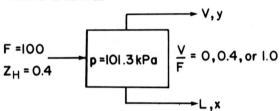

Mass Balances:
$$F = V + L$$
$$Fz = Vy + Lx$$

Solve for y:
$$y = -\frac{L}{V} x + \frac{F}{V} z$$

From the overall balance, $L = F - V$. Thus

when $\dfrac{V}{F} = 0.0$, $V = 0$, $L = F$, and $\dfrac{L}{V} = \dfrac{F}{0} = \infty$

when $\dfrac{V}{F} = 0.4$, $V = 0.4F$, $L = 0.6F$, and $\dfrac{L}{V} = \dfrac{0.6F}{0.4F} = 1.5$

when $\dfrac{V}{F} = 1.0$, $V = F$, $L = 0$, and $\dfrac{L}{V} = \dfrac{0}{F} = 0$

Thus the slopes $(-L/V)$ are $-\infty$, -1.5, and -0.

If we solve for the $y = x$ interception, we find it at $y = x = z = 0.4$ for all cases. Thus can plot three operating lines through $y = x = z = 0.4$, with slopes of $-\infty$, -1.5 and -0. These operating lines were shown in Figure 3-2.

a. Highest y is for V/F = 0: $y = 0.61$ [x = 0.4]

b. Lowest x is for V/F = 1.0: $x = 0.075$ [y = 0.4]

c. When V/F is 0.4 , $y = 0.57$ and $x = 0.29$

d. When $F = 1000$ with $V/F = 0.4$, the answer is exactly the same as in part c. The feed rate will affect the drum diameter and the energy needed in the preheater.

E. Check. We can check the solutions with the mass balance,
$Fz = Vy + Lx$.

a. $(100)(0.4) = 0(0.61) + (100)(0.4)$ checks

b. $(100)(0.4) = 100(0.4) + (0)(0.075)$ checks

c. $100(0.4) = (40)(0.57) + (60)(0.29)$
Note $V = 0.4F$ and $L = 0.6F$
This is $40 = 40.2$, which checks within the accuracy of the graph.

d. Check is similar to c : $400 = 402$

F. Generalization. The method for obtaining bounds for the answer (setting the V/F equation to its extreme values of 0.0 and 1.0) can be used in a variety of other situations. In general, the feed rate will not affect the compositions obtained in the design of staged separators. Feed rate does affect heat requirement and equipment diameters.

Once the conditions within the flash drum have been calculated, we proceed to the energy balance. With y, x, and T_{drum} known, the enthalpies H_v and h_L are easily calculated from Eqs. (3-6) or (3-7a) and (3-8). Then the only unknown in Eq. (3-5) is the feed enthalpy h_F. Once h_F is known, the inlet feed temperature T_F can be obtained from Eq. (3-6) or (3-7b).

The amount of heat required in the heater, Q_h, can be determined from an energy balance around the heater.

$$Q_h + Fh_1(T_1, z) = Fh_F(T_F, z) \qquad (3\text{-}17)$$

Since enthalpy h_1 can be calculated from T_1 and z, the only unknown is Q_h, which controls the size of the heater.

The feed pressure, p_F, required is semi-arbitrary. Any pressure high enough to prevent boiling at temperature T_F can be used.

Except for sizing the flash drum, which is covered later, this completes the sequential procedure. Note that the advantages of this procedure are that mass and energy balances are uncoupled and can be solved independently. Thus trial-and-error is not required.

3.2.2. Simultaneous Solution Procedure

If the temperature of the feed to the drum, T_F, is the specified variable, the mass and energy balances and the equilibrium equations must be solved simultaneously. You can see from the energy balance, Eq. (3-5), why this is true. The feed enthalpy, h_F, can be calculated, but the vapor and liquid enthalpies, H_v and h_L, depend upon T_{drum}, y, and x, which are unknown. Thus a sequential solution is not possible.

We could write Eqs. (3-1) to (3-6) and solve seven equations simultaneously for the seven unknowns y, x, L, V, H_v, h_L, and T_{drum}. This is feasible but rather difficult, particularly since Eqs. (3-1) and (3-2) and often Eqs. (3-6) are nonlinear, so we resort to a trial-and-error procedure. This method is: Guess the value of one of the variables, calculate the other variables, and then check the guessed value of the trial variable. For a binary system, we can select any one of several trial variables, such as y, x, T_{drum}, V/F, or L/F. For example, if we select the temperature of the drum, T_{drum}, as the trial variable, the calculation procedure is:

1. Calculate $h_F(T_F, z)$ [e.g., use Eq. (3-7b)].

2. Guess value of T_{drum}.

3. Calculate x and y from the equilibrium equations (3-1) and (3-2) or graphically (use temperature-composition diagram).

4. Find L and V by solving the mass balance equations (3-3) and (3-4), or find L/V from Figure 3-2 and use the overall mass balance, Eq. (3-3).

5. Calculate $h_L(T_{drum}, x)$ and $H_v(T_{drum}, y)$ from Eqs. (3-6) or (3-7a) and (3-8) or from the enthalpy-composition diagram.

6. Check: Is the energy balance equation (3-5) satisfied? If it is satisfied we are finished. Otherwise, return to step 2.

The procedures are similar for other trial variables.

For binary flash distillation, the simultaneous procedure can be conveniently carried out on an enthalpy-composition diagram. First calculate the feed enthalpy, h_F, from Eq. (3-6); then plot the feed point on Figure 3-3. In the flash drum the feed separates into liquid and vapor in equilibrium. Thus the isotherm through the feed point, which must be the T_{drum} isotherm, gives the correct values for x and y. The flow rates, L and V, can be determined from the mass balances, Eqs. (3-3) and (3-4), or from the lever-arm rule [see Eqs. (2-6) to (2-9)].

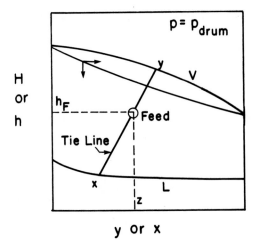

Figure 3-3.   Binary   flash   calculation   on   enthalpy-composition
diagram.

Determining the isotherm through the feed point requires a minor
trial-and-error procedure. Pick a y (or x), draw the isotherm, and check
whether it goes through the feed point. If not, repeat with a new y (or
x).

3.3. MULTICOMPONENT FLASH DISTILLATION

If there are more than two components, an analytical procedure is
needed. The basic equipment configuration is the same as in Figure 3-1.

The equations used are equilibrium, mass and energy balances, and
stoichiometric relations. The mass and energy balances are very similar
to those used in the binary case, but the equilibrium equations are usually written in terms of K values. The equilibrium form is

$$y_i = K_i x_i \tag{3-18}$$

where in general

$$K_i = K_i (T_{drum}, P_{drum}, \text{all } x_i) \tag{3-19}$$

Equations (3-18) and (3-19) are written once for each component. For
ideal systems the K values do not depend on mole fraction.

Equations (3-18) and (3-19) are solved along with the stoichiometric equations,

$$\sum_{i=1}^{C} y_i = 1.0, \qquad \sum_{i=1}^{C} x_i = 1.0 \tag{3-20}$$

the mass balances,

$$Fz_i = Lx_i + Vy_i \tag{3-21}$$

$$F = L + V \tag{3-22}$$

and the energy balance,

$$Fh_F + Q_{flash} = VH_V + Lh_L \tag{3-23}$$

In Eqs. (3-20), C is the number of components. Equations (3-21) and (3-23) are very similar to the binary mass balance, Eq. (3-4), and energy balance, Eq. (3-5). Equation (3-22) is the same as Eq. (3-3).

Usually the feed flow rate, F, and the feed mole fractions z_i for $C - 1$ of the components will be specified. If p_{drum} and T_{drum} or one liquid or vapor composition are also specified, then a sequential procedure can be used. That is, the mass balances, stoichiometric equations, and equilibrium equations are solved simultaneously, and then the energy balances are solved.

Now consider for a minute what this means. Suppose we have 10 components (C=10). Then we must find 10 K's, 10 x's, 10 y's, one L, and one V, or 32 variables. To do this we must solve 32 equations [10 Eq. (3-18), 10 Eq. (3-19), 2 Eq. (3-20), and 10 independent mass balances] simultaneously. And this is the simpler sequential solution for a relatively simple problem.

How does one solve 32 simultaneous equations? In general, the K value relations could be nonlinear functions of composition. However, we will restrict ourselves to ideal solutions where

$$K_i = K_i(T_{drum}, p_{drum})$$

Since T_{drum} and p_{drum} are known, the 10 K_i can be determined easily [say, from the DePriester charts or Eq. (2-12)]. Now there are only 22 linear equations to solve simultaneously. This can be done, but trial-and-error procedures are simpler.

To simplify the solution procedure, we first use equilibrium, $y_i = K_i x_i$, to remove y_i from Eq. (3-21):

$$Fz_i = Lx_i + VK_i x_i \qquad i = 1, C$$

Solving for x_i, we have

$$x_i = \frac{Fz_i}{L + VK_i} \qquad i = 1, C$$

If we solve Eq. (3-22) for L, $L = F - V$, and substitute this into the last equation we have

$$x_i = \frac{Fz_i}{F - V + K_i V} \qquad i = 1, C \qquad (3\text{-}24)$$

Now if the unknown V is determined, all of the x_i can be determined. It is usual to divide the numerator and denominator of Eq. (3-24) by the feed rate F and work in terms of the variable V/F. Then upon rearrangement we have

$$x_i = \frac{z_i}{1 + (K_i - 1)\dfrac{V}{F}} \qquad i = 1, C \qquad (3\text{-}25)$$

The reason for using V/F, the fraction vaporized, is that it is bounded between 0 and 1.0 for all possible problems. Since $y_i = K_i x_i$, we obtain

$$y_i = \frac{K_i z_i}{1 + (K_i - 1)\dfrac{V}{F}} \qquad i = 1, C \qquad (3\text{-}26)$$

Once V/F is determined, x_i and y_i are easily found from Eqs. (3-25) and (3-26).

How can we derive an equation that allows us to calculate V/F?

To answer this, first consider what equations have not been used. These are the two stoichiometric equations, $\Sigma x_i = 1.0$ and $\Sigma y_i = 1.0$. If we substitute Eqs. (3-25) and (3-26) into these equations, we obtain

$$\sum_{i=1}^{C} \frac{z_i}{1 + (K_i - 1)\dfrac{V}{F}} = 1.0 \tag{3-27}$$

and

$$\sum_{i=1}^{C} \frac{K_i \, z_i}{1 + (K_i - 1)\dfrac{V}{F}} = 1.0 \tag{3-28}$$

Either of these equations can be used to solve for V/F. If we clear fractions, these are Cth-order polynomials. Thus, if C is greater than 3, a trial-and-error procedure or root-finding technique must be used to find V/F. Although Eqs. (3-27) and (3-28) are both valid, they do not have good convergence properties. That is, if the wrong V/F is chosen, the V/F that is chosen next may not be better.

Fortunately, an equation that does have good convergence properties is easy to derive. To do this, subtract Eq. (3-27) from (3-28).

$$\sum \frac{K_i \, z_i}{1 + (K_i - 1)\dfrac{V}{F}} - \sum \frac{z_i}{1 + (K_i - 1)\dfrac{V}{F}} = 0$$

Subtracting the sums term by term, we have

$$f\left(\frac{V}{F}\right) = \sum_{i=1}^{C} \frac{(K_i - 1) \, z_i}{1 + (K_i - 1)\dfrac{V}{F}} = 0 \tag{3-29}$$

Equation (3-29), which is known as the Rachford-Rice equation, has excellent convergence properties.

Since the feed compositions, z_i, are specified and K_i can be calculated when T_{drum} and p_{drum} are given, the only variable in Eq. (3-29) is the fraction vaporized, V/F. This equation can be solved by many different convergence procedures. For instance, the secant method can be used by selecting two values of V/F and calculating the values of the summation (it will be zero only at the correct value of V/F). Then a linear interpolation is done to find the V/F that would make the summation zero if f(V/F) were linear. This calculation is the subject of Problems 3-C1 and 3-D16.

The Newtonian convergence procedure will converge faster. Since

$f(V/F)$ in Eq. (3-29) is a function of V/F that should have a zero value, the equation for the Newtonian convergence procedure is

$$f_{k+1} - f_k = \frac{df_k}{d(V/F)}\Delta(V/F) \qquad (3\text{-}30)$$

where f_k is the value of the function for trial k and $df_k/d(V/F)$ is the value of the derivative of the function for trial k. We desire to have f_{k+1} equal zero, so we set $f_{k+1} = 0$ and solve for $\Delta (V/F)$:

$$\Delta(V/F) = (V/F)_{k+1} - (V/F)_k = \frac{-f_k}{\left[\dfrac{df_k}{d(V/F)}\right]} \qquad (3\text{-}31)$$

This equation gives us the best next guess for the fraction vaporized. To use it, however, we need equations for both the function and the derivative. For f_k, use the Rachford-Rice equation, (3-29). Then the derivative is

$$\frac{df_k}{d(V/F)} = -\sum_{i=1}^{C} \frac{(K_i - 1)^2\, z_i}{[1 + (K_i - 1)V/F]^2} \qquad (3\text{-}32)$$

Substituting Eqs. (3-29) and (3-32) into (3-31) and solving for $(V/F)_{k+1}$, we obtain

$$\left(\frac{V}{F}\right)_{k+1} = \left(\frac{V}{F}\right)_k + \frac{\displaystyle\sum_{i=1}^{C}\left[\frac{(K_i - 1)\, z_i}{1 + (K_i - 1)V/F}\right]}{\displaystyle\sum_{i=1}^{C}\left[\frac{z_i(K_i - 1)^2}{[1 + (K_i - 1)V/F]^2}\right]} \qquad (3\text{-}33)$$

Equation (3-33) gives a good estimate for the next trial. Once $(V/F)_{k+1}$ is calculated the value of the Rachford-Rice function can be determined. If it is close enough to zero, the calculation is finished; otherwise, repeat the Newtonian convergence for the next trial.

Newtonian convergence procedures do not always converge. One advantage of using the Rachford-Rice equation with the Newtonian convergence procedure is that there is always rapid convergence. This is illustrated in Example 3-2.

Once V/F has been found, x_i and y_i are calculated from Eqs. (3-25) and (3-26). L and V are determined from the overall mass balance, Eq. (3-22). The enthalpies h_L and H_v can now be calculated. For ideal solutions the enthalpies can be determined from the sum of the pure component enthalpies multiplied by the corresponding mole fractions:

$$H_v = \sum_{i=1}^{C} y_i \, \tilde{H}_{v_i} \, (T_{drum}, p_{drum}) \tag{3-34}$$

$$h_L = \sum_{i=1}^{C} x_i \, \tilde{h}_{L_i} \, (T_{drum}, p_{drum}) \tag{3-35}$$

where $\tilde{H}_{v_i}$ and $\tilde{h}_{L_i}$ are enthalpies of the pure components. If the solutions are not ideal, heats of mixing are required. Then the energy balance, Eq. (3-23), is solved for h_F, and T_F is determined.

If V/F and p_{drum} are specified, then T_{drum} must be determined. This can be done by picking a value for T_{drum}, calculating K_i, and checking with the Rachford-Rice equation, (3-29). A plot of $f(V/F)$ versus T_{drum} will help us select the temperature value for the next trial. Alternatively, an approximate convergence procedure similar to that employed for bubble- and dew-point calculations can be used. The new K_{ref} can be determined from

$$K_{ref}(T_{new}) = \frac{K_{ref}(T_{old})}{1 + (d)f(T_{old})} \tag{3-36}$$

where the damping factor $d \leq 1.0$. In some cases this may overcorrect unless the initial guess is close to the correct answer.

Example 3-2. Multicomponent Flash Distillation

A flash chamber operating at 50 °C and 200 kPa is separating 1000 kg moles/hr of a feed that is 30 mole % propane, 10 mole % n-butane, 15 mole % n-pentane and 45 mole % n-hexane. Find the product compositions and flow rates.

Solution

A. Define. We want to calculate y_i, x_i, V, and L for the equilibrium flash chamber shown in the diagram.

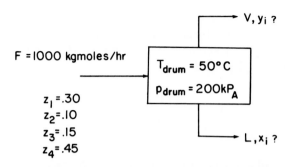

F = 1000 kgmoles/hr

$z_1 = .30$
$z_2 = .10$
$z_3 = .15$
$z_4 = .45$

V, y_i ?

$T_{drum} = 50°C$
$P_{drum} = 200 kP_A$

L, x_i ?

B. Explore. Since T_{drum} and p_{drum} are given, a sequential solution can be used. We can use the Rachford-Rice equation to solve for V/F and then find x_i, y_i, L, and V.

C. Plan. Calculate K_i from DePriester charts or from Eq. (2-12). Use Newtonian convergence with the Rachford-Rice equation, Eq. (3-33), to converge on the correct V/F value. Once the correct V/F has been found, calculate x_i from Eq. (3-25) and y_i from Eq. (3-26). Calculate V from V/F and L from overall mass balance, Eq. (3-22).

D. Do It. From the DePriester chart (Fig. 2-12), at 50°C and 200 kPa we find

$$
\begin{array}{ll}
K_1 = 7.0 & C_3 \\
K_2 = 2.4 & n\text{-}C_4 \\
K_3 = 0.80 & n\text{-}C_5 \\
K_4 = 0.30 & n\text{-}C_6
\end{array}
$$

Calculate f (V/F) from the Rachford-Rice equation:

$$
f\left(\frac{V}{F}\right) = \sum_{i=1}^{4} \frac{(K_i - 1)z_i}{1 + (K_i - 1)\, V/F}
$$

Pick V/F = 0.1 as first guess (this illustrates convergence for a poor first guess).

$$
f(0.1) = \frac{(7.0-1)(0.3)}{1+(7.0-1)(0.1)} + \frac{(2.4-1)(0.1)}{1+(2.4-1)(0.1)}
$$

$$
+ \frac{(0.8-1)(0.15)}{1+(0.8-1)(0.1)} + \frac{(0.1-3)(0.45)}{1+(0.3-1)(0.1)}
$$

$$
= 1.125 + 0.1228 + (-0.0306) + (-0.3387) = 0.8785
$$

Since f(0.1) is positive, a higher value for V/F is required. Note that only one term in the denominator of each term changes. Thus we can set up the equation so that only V/F will change. Then f(V/F) equals

$$\frac{1.8}{1 + 6(\frac{V}{F})} + \frac{0.14}{1 + 1.4(\frac{V}{F})} + \frac{-0.03}{1 - 0.2(\frac{V}{F})} + \frac{-0.315}{1 - 0.7 (\frac{V}{F})}$$

Now all subsequent calculations will be easier.

The derivative of the R-R equation can be calculated for this first guess

$$(\frac{df}{d(\frac{V}{F})})_1 = - \left\{ \frac{(K_1-1)^2 z_1}{[1+(K_1-1)\frac{V}{F}]^2} + \frac{(K_2-1)^2 z_2}{[1+(K_2-1)\frac{V}{F}]^2} \right.$$

$$+ \frac{(K_3-1)^2 z_3}{[1+(K_3-1)\frac{V}{F}]^2} + \left. \frac{(K_4-1)^2 z_4}{[1+(K_4-1)\frac{V}{F}]^2} \right\}$$

$$= - \left\{ \frac{10.8}{[1+(6.0)\frac{V}{F}]^2} + \frac{0.196}{[1+1.4\frac{V}{F}]^2} \right.$$

$$+ \frac{0.006}{[1-0.2\frac{V}{F}]^2} + \left. \frac{0.2205}{[1-0.7\frac{V}{F}]^2} \right\}$$

With V/F = 0.1 this is $(\frac{df}{d(V/F)})_1$ = 4.631. From Eq. (3-33) the next guess for V/F is $(V/F)_2 = 0.1 + \frac{0.8785}{4.631}$ = 0.29. Calculating the value of the Rachford-Rice equation, we have f(0.29) = 0.329. This is still positive and V/F is still too low. *2nd Trial:* $(\frac{df}{d(V/F)})|_{0.29}$ = 1.891 which gives

$$(V/F)_3 = 0.29 + 0.329/1.891 = 0.46$$

and the Rachford-Rice equation is $f(0.46) = 0.066$. This is closer, but V/F is still too low. Continue convergence.

3rd Trial: $\dfrac{df}{d(V/F)}\Big|_{0.46} = 1.32$ which gives

$$(V/F)_4 = 0.46 + 0.066/1.32 = 0.51$$

We calculate that $f(0.51) = 0.00173$ which is very close and is within the accuracy of the DePriester charts. Thus V/F = 0.51.

Now we calculate x_i from Eq. (3-25) and y_i from $K_i x_i$,

$$x_1 = \frac{z_1}{1+(K_1-1)\dfrac{V}{F}} = \frac{0.30}{1+(7.0-1)(0.51)} = 0.0739$$

$$y_1 = K_1\, x_1 = (7.0)\,(0.739) = 0.5172$$

By similar calculations,
$$x_2 = 0.0583, \quad y_2 = 0.1400$$
$$x_3 = 0.1670, \quad y_3 = 0.1336$$
$$x_4 = 0.6998 \quad y_4 = 0.2099$$

Since F = 1000 and V/F = 0.51, V = 0.51F = 510 kg moles/hr, and L = F − V = 1000 − 510 = 490 kg moles/hr.

E. Check. We can check $\Sigma\, y_i$ and $\Sigma\, x_i$.

$$\sum_{i=1}^{4} x_i = 0.999\ , \qquad \sum_{i=1}^{4} y_i = 1.0007$$

These are close enough. They aren't perfect, because V/F wasn't exact. Essentially the same answer is obtained if Eq. (2-12) is used for the K values.

F. Generalize. Since the Rachford-Rice equation is almost linear, the Newtonian convergence routine gives rapid convergence. Note that the convergence was monotonic and did not oscillate. Faster convergence would be achieved with a better first guess of V/F. This type of trial-and-error problem is easy to program.

If the specified variables are F, z_i, p_{drum}, and either x or y for one component, we can follow a sequential convergence procedure using Eq.

(3-25) or (3-26) to relate the specified composition (the reference component) to either K_{ref} or V/F. We can do this in either of two ways. The first is to guess T_{drum} and use Eq. (3-25) or (3-26) to solve for V/F. The Rachford-Rice equation is then the check equation on T_{drum}. If the Rachford-Rice equation is not satisfied we select a new temperature [by a secant method or Eq. (3-36)] and repeat the procedure. In the second approach, we guess V/F and calculate K_{ref} from Eq. (3-25) or (3-26). We then determine the drum temperature from this K_{ref}. The Rachford-Rice equation is again the check. If it is not satisfied, we select a new V/F and continue the process.

3.4. SIMULTANEOUS MULTICOMPONENT CONVERGENCE

If the feed rate F, the feed composition consisting of $(C - 1)$ z_i values, the flash drum pressure p_{drum}, and the feed temperature T_F are specified, then we must use a simultaneous solution procedure. First, we choose a feed pressure such that the feed will be liquid. Then we can calculate the feed enthalpy in the same way as Eqs. (3-34) and (3-35):

$$h_F = \sum_{i=1}^{C} z_i \, \tilde{h}_{F_i} \, (T_F, \, p_F) \qquad (3-37)$$

Although the mass and energy balances, equilibrium relations, and stoichiometric relations could all be solved simultaneously, it is again easier to use a trial-and-error procedure. This problem is now a double trial-and-error.

The first question to ask in setting up a trial-and-error procedure is: What are the possible trial variables and which ones shall we use? Here we first pick T_{drum}, since it is required to calculate all K_i, h_{L_i}, and H_{v_i} and since it is difficult to solve for. The second trial variable is V/F, because then we can use the Rachford-Rice approach with Newtonian convergence.

The second question to ask is: Should we converge on both variables simultaneously (that is change both T_{drum} and V/F at the same time), or should we converge sequentially? Both techniques will work, but, if applied properly, sequential convergence tends to be more stable. If we use sequential convergence, then a third question is: Which variable should we converge on first, V/F or T_{drum}? To answer this question we need to consider the chemical system we are separating. If the mixture is wide-boiling, that is, if the dew point and bubble point are far apart (say more than 80 to 100 ° C), then a small change in T_{drum} cannot have much effect on V/F. In this case we wish to converge on V/F first. Then when T_{drum} is changed, we will be close to the correct answer for V/F. For a significant separation in a flash system, the volatilities must be very different, so this is the typical situation for flash distillation.

3.4.1. Wide-Boiling Feeds

The procedure for wide-boiling feeds is shown in Figure 3-4. Note that the energy balance is used last. This is standard procedure since accurate values of x_i and y_i are available to calculate enthalpies for the energy balance.

The fourth question is: How should we carry out the individual convergence steps? For the Rachford-Rice equation, linear interpolation or Newtonian convergence will be satisfactory. Several methods can be used to estimate the next flash drum temperature. One of the fastest and easiest is to use a Newtonian convergence procedure. To do this we rearrange the energy balance (Eq. 3-23) into the functional form,

$$E_k \left(T_{drum}\right) = VH_v + Lh_L - Fh_F - Q_{flash} = 0 \qquad (3\text{-}38)$$

The subscript k again refers to the trial number. When E_k is zero, the problem has been solved. The Newtonian procedure estimates $E_{k+1} \left(T_{drum}\right)$ from the derivative,

$$E_{k+1} - E_k = \frac{dE_k}{dT_{drum}} \left(\Delta T_{drum}\right) \qquad (3\text{-}39)$$

where ΔT_{drum} is the change in T_{drum} from trial to trial,

$$\Delta T_{drum} = T_{k+1} - T_k \qquad (3\text{-}40)$$

and dE_k/dT_{drum} is the variation of E_k as temperature changes. Since the last two terms in Eq. (3-38) do not depend on T_{drum}, this derivative can be calculated as

$$\frac{dE_k}{dT_{drum_k}} = V \frac{dH_v}{dT_{drum}} + L \frac{dh_L}{dT_{drum}} = VC_{P_v} + Lc_{P_L} \qquad (3\text{-}41)$$

where we have used the definition of the heat capacity. We want the energy balance to be satisfied after the next trial. Thus we set $E_{k+1} =$

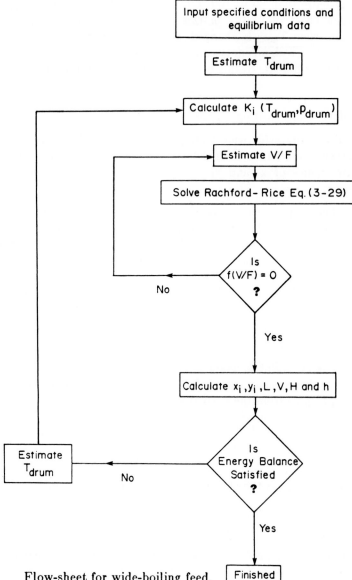

Figure 3-4.   Flow-sheet for wide-boiling feed.

0. Now Eq. (3-39) can be solved for ΔT_{drum}:

$$\Delta T_{drum} = \frac{-E_k(T_{drum_k})}{\dfrac{dE_k}{dT_{drum_k}}} \qquad (3\text{-}42)$$

Substituting the expression for ΔT_{drum} into this equation and solving for $T_{drum_{k+1}}$, we obtain the best guess for temperature for the next trial,

$$T_{drum_{k+1}} = T_{drum_k} - \frac{E_k(T_{drum_k})}{\dfrac{dE}{dT_{drum_k}}} \tag{3-43}$$

In this equation E_k is the calculated numerical value of the energy balance function from Eq. (3-38), and dE/dT_{drum} is the numerical value of the derivative calculated from Eq. (3-41).

The procedure has converged when

$$\left| \Delta T_{drum} \right| < \epsilon \tag{3-44}$$

For computer calculations, $\epsilon = 0.01\,^\circ C$ is a reasonable choice. For hand calculations, a less stringent limit such as $\epsilon = 0.2\,^\circ C$ would be used.

It is possible that this convergence scheme will predict values of ΔT_{drum} that are too large. When this occurs, the drum temperature may oscillate with a growing amplitude and not converge. To discourage this behavior, ΔT_{drum} can be damped.

$$\Delta T_{drum} = (d)(\Delta T_{drum, \text{ calc Eq. (3-42)}}) \tag{3-45}$$

where the damping factor d is about 0.5. Note than when d = 1.0 this is just the Newtonian approach.

The drum temperature should always lie between the bubble- and dew-point temperature of the feed. In addition, the temperature should converge toward some central value. If either of these criteria is violated, then the convergence scheme should be damped or an alternative convergence scheme should be used.

3.4.2. Narrow-Boiling Feeds

For flash distillation problems where the feed boils over a relatively narrow temperature range, a small change in temperature is a relatively large percentage of the total range. Thus V/F will be strongly affected by small temperature changes. In this case the flow chart used for wide-boiling feeds will not work well. We now want to converge on temperature first and then on V/F. The narrow-boiling feed procedure is shown in Figure 3-5. Again the energy balance is used last.

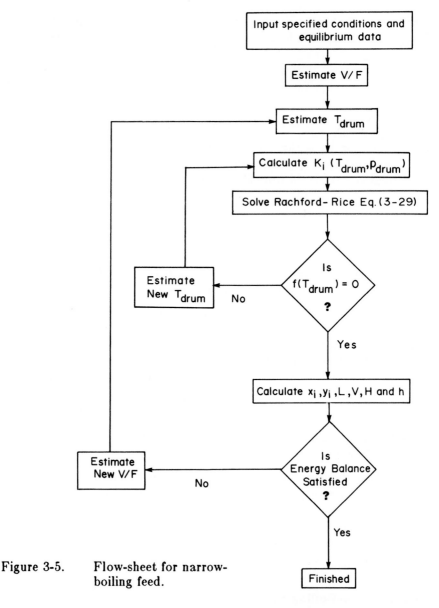

Figure 3-5.   Flow-sheet for narrow-
boiling feed.

The Rachford-Rice equation will again have good convergence properties even though T_{drum} is now the variable. For the energy balance
we can use either a substitution procedure or a Newtonian convergence
procedure. In the energy balance [Eq. (3-23) or (3-38)], replace L with F
− V, and then divide the equation by the feed rate, F. The result is

$$E_k\,(V/F) = (V/F)H_v + (1 - V/F)\,h_L - h_F - Q_{flash}/F = 0 \qquad (3\text{-}46)$$

In the direct substitution approach we solve Eq. (3-46) for V/F. That is,

$$(V/F)_{calc} = \frac{Q_{flash}/F + h_F - h_L}{H_v - h_L} \qquad (3\text{-}47)$$

If the V/F used was correct, then the calculated V/F from Eq. (3-47) will have the same value. If this is true, Eq. (3-46) is satisfied with that V/F, and we are finished. If the two V/F values differ, then we use the calculated V/F from Eq. (3-47) for the next trial.

This direct substitution method may oscillate, but the oscillation can be controlled by damping the new guess using

$$\left[\frac{V}{F}\right]_{new} = \left[\frac{V}{F}\right]_{old} + d\left[\left[\frac{V}{F}\right]_{calc} - \left[\frac{V}{F}\right]_{old}\right] \qquad (3\text{-}48)$$

where the damping factor d is about 0.5. Note that when d = 1.0 this is direct substitution.

If we apply the Newtonian approach, the result is that V/F for the next trial is also given by Eq. (3-47)! This development is left as an exercise (Problem 3-C4).

Example 3-3. Simultaneous Convergence

We have a feed that is 20 mole % methane, 45 mole % n-pentane, and 35 mole % n-hexane. Feed rate is 1500 kg moles/hr, and feed temperature is 42°C. The flash drum operates at 30 psia. Find: T_{drum}, V/F, x_i, y_i, L, V.

Solution

A. Define. The process is sketched in the diagram.

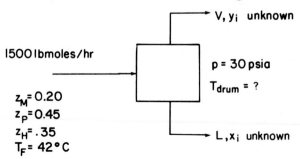

1500 lbmoles/hr

$z_M = 0.20$
$z_P = 0.45$
$z_H = .35$
$T_F = 42°C$

V, y_i unknown

p = 30 psia

T_{drum} = ?

L, x_i unknown

B. Explore. Since T_F is given, this will be a double trial and error. We must first determine if a wide-boiling or narrow-boiling procedure should be used. K values from the DePriester charts or from Eq. (2-12) can be used. For energy balances, enthalpies can be calculated from heat capacities and latent heats. The required data are listed in Table 3-1.

Table 3-1. Data for Methane, n-Pentane, and n-Hexane

Component	λ, kcal/g-mole,	Normal boiling pt., $^\circ$C	$C_{P,L}$, cal/g-mole-$^\circ$C
1. Methane	1.955	−161.48	11.0 (est.)
2. n-Pentane	6.160	36.08	39.66
3. n-Hexane	6.896	68.75	45.58

Vapor heat capacities in cal/g-mole-$^\circ$C; T in $^\circ$C:

$$C_{PV,1} = 8.20 + 0.01307\ T + 8.75 \times 10^{-7}T^2 - 2.63 \times 10^{-9}T^3$$

$$C_{PV,2} = 27.45 + 0.08148\ T - 4.538 \times 10^{-5}T^2 + 10.1 \times 10^{-9}T^3$$

$$C_{PV,3} = 32.85 + 0.09763\ T - 5.716 \times 10^{-5}T^2 + 13.78 \times 10^{-9}T^3$$

Source: Himmelblau (1974)

Since the determination of a wide- or narrow-boiling feed is so important, we will do that first. A glance at the normal boiling points indicates that this is probably a wide-boiling feed. The dew-point temperature was determined with a simple computer program as 68.1 $^\circ$C using Eq. (2-12) for K values. The bubble point is less than −70 $^\circ$C, which is the limit of validity of the DePriester charts. Thus this is clearly a wide-boiling feed.

C. Plan. For a wide-boiling feed the sum-rates type of procedure will be used. Since this is a double trial and error, all calculations will be done on the computer and summarized here. Newtonian convergence will be used for both the Rachford-Rice equation and the energy balance estimate of new drum temperature. $\epsilon = 0.02$ is used for energy convergence (Eq. 3-44). The Rachford-Rice equation is considered converged when

$$\left| (V/F)_{k+1} - (V/F)_k \right| < \epsilon_R \qquad (3\text{-}49)$$

$\epsilon_R = 0.005$ is used here.

D. Do It. The first guess is made by arbitrarily assuming that $T_{drum} = 15\,^{\circ}C$ and $V/F = 0.25$. Since convergence of the program is rapid, more effort on an accurate first guess is probably not justified. Using Eq. (3-33) as illustrated in Example 3-2, the following V/F values are obtained [K values from Eq. (2-12)]:

V/F = 0.25, 0.2485, 0.2470, 0.2457, 0.2445, 0.2434, 0.2424, 0.2414, 0.2405, 0.2397, 0.2390, 0.2383, 0.2377, 0.2371, 0.2366, 0.2361

Note that convergence is monotonic. With V/F known, x_i and y_i are found from Eqs. (3-25) and (3-26).

Compositions are: x_m = 0.0124, x_p = 0.5459, x_H = 0.4470, and y_m = 0.8072, y_p = 0.1398, y_H = 0.0362.

Flow rates L and V are found from the mass balance and V/F value. After determining enthalpies, Eq. (3-43) is used to determine $T_{drum,2}$ = 27.9 $^{\circ}$C. Obviously, this is still far from convergence.

The convergence procedure is continued, as summarized in Table 3-2. Note that the drum temperature oscillates, and because of this the converged V/F oscillates. Also, the number of trials to converge on V/F decreases as the calculation proceeds. The final compositions and flow rates are:

x_m = 0.0108, x_p = 0.5381, x_H = 0.4513
y_m = 0.7531, y_P = 0.1925, y_H = 0.0539
V = 382.3 kg-mole/hr and L = 1117.7 kg-mole/hr

Table 3-2. Iterations for Example 3-3

Iteration No.	Initial T_{drum}	Trials to find V/F	V/F	Calc T_{drum}
1	15.00	16	0.2361	27.903
2	27.903	13	0.2677	21.385
3	21.385	14	0.2496	25.149
4	25.149	7	0.2567	23.277
5	23.277	3	0.2551	24.128
6	24.128	2	0.2551	23.786
7	23.786	2	0.2550	23.930
8	23.930	2	0.2550	23.875
9	23.875	2	0.2549	23.900
10	23.900	2	0.2549	23.892

E. Check. The results are checked throughout the trial-and-error procedure. Naturally, they depend upon the validity of data used for the enthalpies and Ks. At least the results appear to be self-consistent (that is, $\Sigma \, x_i = 1.0$, $\Sigma \, y_i = 1.0$) and are of the right order of magnitude.

F. Generalization. Note that the bubble-point and dew-point calculations converged very rapidly. The preliminary bubble-point and dew-point calculations allowed us to use the correct convergence procedure. The narrow-boiling feed procedure does *not* converge for this problem. It is often a good idea to use a preliminary calculation to decide what procedure to follow. The use of the computer greatly reduces calculation time on this double trial-and-error problem.

3.5. SIZE CALCULATION

Once the vapor and liquid compositions and flow rates have been determined, the flash drum can be sized. This is an empirical procedure. We will discuss the specific procedure for vertical flash drums, like the one shown in Figure 3-1.

Step 1. Calculate the permissible vapor velocity, u_{perm},

$$u_{perm} = K_{drum} \sqrt{\frac{\rho_L - \rho_v}{\rho_v}} \tag{3-50}$$

u_{perm} is the maximum permissible vapor velocity in feet per second at the maximum cross-sectional area. ρ_L and ρ_v are the liquid and vapor densities.

K_{drum} is an empirical constant whose value has been correlated graphically by Watkins (1967) for 85% of flood with no demister. Approximately 5% liquid will be entrained with the vapor. Use of the same design with a demister will reduce entrainment to less than 1%. The demister traps small liquid droplets on fine wires and prevents them from exiting. The droplets then coalesce into larger droplets, which fall off the wire and through the rising vapor into the liquid pool at the bottom of the flash chamber. Blackwell (1984) fit Watkins' correlation to the equation

$$K_{drum} = \exp[A + B \ln F_{lv} + C(\ln F_{lv})^2$$

$$+ D(\ln F_{lv})^3 + E(\ln F_{lv})^4] \qquad (3\text{-}51)$$

where $F_{lv} = \dfrac{W_L}{W_v} \sqrt{\dfrac{\rho_v}{\rho_L}}$ with W_L and W_v being the liquid and vapor flow rates in weight units per hour (e.g., lb/hr). The constants are (Blackwell, 1984):

$A = -1.877478097 \qquad C = -0.1870744085 \qquad E = -0.0010148518$
$B = -0.8145804597 \qquad D = -0.0145228667$

Step 2. Using the known vapor rate, V, convert u_{perm} into a horizontal area. The vapor flow rate, V, in lb moles/hr is

$$V\left(\frac{\text{lb moles}}{\text{hr}}\right) = \frac{u_{perm}\left(\frac{\text{ft}}{\text{s}}\right)\left(\frac{3600\text{ s}}{\text{hr}}\right) A_c(\text{ft}^2)\, \rho_v\left(\frac{\text{lbm}}{\text{ft}^3}\right)}{MW_{vapor}\left(\frac{\text{lbm}}{\text{lb mole}}\right)}$$

Solving for the cross-sectional area,

$$A_c = \frac{V(MW_v)}{u_{perm}(3600)\rho_v} \qquad (3\text{-}52)$$

For a vertical drum, diameter D is

$$D = \sqrt{\frac{4A_c}{\pi}} \qquad (3\text{-}53)$$

Usually, the diameter is increased to the next largest 6-in. increment.

Step 3. Set the diameter/length ratio either by rule of thumb or by the required liquid surge volume. For vertical flash drums, the rule of thumb is that L/D ranges from 3.0 to 5.0. The appropriate value of L/D within this range can be found by minimizing the total vessel weight (which minimizes cost).

Flash drums are often used as liquid surge tanks in addition to separating liquid and vapor. The design procedure for this case is discussed by Watkins (1967) for petrochemical applications.

The height of the drum above the centerline of the feed nozzle, h_v, should be 36 in. plus one-half the diameter of the feed line (see Figure 3-6). The minimum of this distance is 48 in.

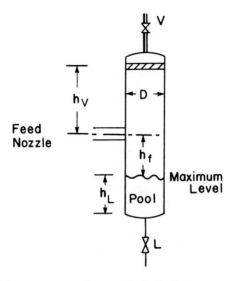

Figure 3-6.   Measurements for vertical flash drum.

The height of the center of the feed line above the maximum level of the liquid pool, h_f, should be 12 in. plus one-half the diameter of the feed line. The minimum distance for this free space is 18 in.

The depth of the liquid pool, h_L, can be determined from the desired surge volume, V_{surge}.

$$h_L = \frac{V_{surge}}{\pi D^2/4} \tag{3-54}$$

The geometry can now be checked, since

$$\frac{L}{D} = \frac{h_V + h_f + h_L}{D}$$

should be between 3 and 5. If $L/D < 3$, a larger liquid surge volume should be allowed. If $L/D > 5$, a horizontal flash drum should be used. Calculator programs for sizing both vertical and horizontal drums are available (Blackwell, 1984).

More detailed design procedures and methods for horizontal drums are presented by Evans (1980), Blackwell (1984), and Watkins (1967). Note that in industries other than petrochemicals the sizing may vary.

Example 3-4. Calculation of Drum Size

A vertical flash drum is to flash a feed of 1500 lb moles/hr that is 40 mole % n-hexane and 60 mole % n-octane at 101.3 kPa (1 atm). We wish to produce a vapor that is 60 mole % n-hexane. Solution of the flash equations with equilibrium data gives x_H = 0.19, T_{drum} = 378K, and V/F = 0.51. What size flash drum is required?

Solution

A. Define. We wish to find diameter and length of flash drum.

B. Explore. We want to use the empirical method developed in Eqs. (3-50) to (3-54). For this we need to estimate the following physical properties: ρ_L, ρ_v, MW_v. To do this we need to know something about the behavior of the gas and of the liquid.

C. Plan. Assume ideal gas and ideal mixtures for liquid. Calculate average ρ_L by assuming additive volumes. Calculate ρ_v from the ideal gas law. Then calculate u_{perm} from Eq. (3-50) and diameter from Eq. (3-54).

D. Do It.

1. Liquid Density

The average liquid molecular weight is

$$\overline{MW}_L = x_H MW_H + x_O MW_O$$

where subscript H is n-hexane and O is n-octane. Calculate or look up the molecular weights. MW_H = 86.17 and MW_O = 114.22. Then $\overline{MW}_L$ = (0.19)(86.17) + (0.81)(114.22) = 108.89.

The specific volume is the sum of mole fractions times pure component specific volumes (ideal mixture):

$$\overline{V}_L = x_H \overline{V}_H + x_O \overline{V}_O = x_H \frac{MW_H}{\rho_H} + \frac{x_O MW_O}{\rho_O}$$

From the *Handbook of Chemistry and Physics*, ρ_H = 0.659 g/mL and ρ_O = 0.703 g/mL at 20 °C. Thus,

$$\overline{V}_L = (0.19)\frac{86.17}{0.659} + (0.81)\frac{114.22}{0.703} = 156.45 \text{ mL/g-mole}$$

Then
$$\rho_L = \frac{MW_L}{\overline{V}_L} = \frac{108.89}{156.45} = 0.6960 \text{ g/mL}$$

2. Vapor Density

Density in moles per liter for ideal gas is $\rho'_v = \dfrac{n}{V} = \dfrac{p}{RT}$, which in grams per liter is $\rho_v = \dfrac{p\,\overline{MW}_v}{RT}$.

The average molecular weight of the vapor is

$$\overline{MW}_v = y_H\,MW_H + y_O\,MW_O$$

where $y_H = 0.60$ and $y_O = 0.40$, and thus $\overline{MW}_v = 97.39$ lb/lb-mole. This gives

$$\rho_v = \frac{(1.0\text{atm})(97.39 \text{ g/mole})}{(82.0575\,\frac{\text{mL atm}}{\text{mole K}})(378 \text{ K})} = 3.14 \times 10^{-3} \text{ g/mL}$$

3. K_{drum} Calculation.

Calculation of flow parameter F_{lv}:

$$V = (V/F)(F) = (0.51)(1500) = 765 \text{ lb moles/hr}$$

$$W_v = (V)(\overline{MW}_v) = (765)(97.39) = 74{,}503 \text{ lb/hr}$$

$$L = F - V = 735 \text{ lb moles/hr}$$

$$W_L = (L)(\overline{MW}_L) = (735)(108.89) = 80{,}034 \text{ lb/hr}$$

$$F_{lv} = \frac{W_L}{W_v}\sqrt{\frac{\rho_v}{\rho_L}} = \frac{80034}{74503}\sqrt{\frac{3.14\times 10^{-3}}{0.6960}} = 0.0722$$

K_{drum} from Eq. (3-51) gives $K_{drum} = 0.4433$, which seems a bit high but agrees with Watkins's (1967) chart.

4. $u_{perm} = K_{drum} \sqrt{\dfrac{\rho_L - \rho_v}{\rho_v}}$

$= 0.4433 \sqrt{\dfrac{0.6960 - 0.00314}{0.00314}} = 6.5849 \text{ft/s}$

5. $A_c = \dfrac{V(M_v)}{u_{perm}(3600)\rho_v}$

$= \dfrac{(765)(97.39)(454 \text{ g/lb})}{(6.5849)(3600)(0.00314 \text{ g/mL}) \ (28316.85 \text{ mL/ft}^2)}$

$= 16.047 \text{ ft}^2$

$D = \sqrt{\dfrac{4A_c}{\pi}}$ = 4.01 ft. Use a 4.0 ft diameter drum or 4.5 ft to be safe.

6. If use $L/D = 4$, $L = 4 \ (4.5 \text{ ft }) = 18.0 \text{ ft}$.

E. Check. This drum size is reasonable. Minimums for h_v and h_f are easily met. Note that units do work out in all calculations; however, one must be careful with units, particularly in step D5.

F. Generalization. If the ideal gas law is not valid, a compressibility factor could be inserted in the equation for ρ_v. Note that most of the work involved calculation of the physical properties. This is often true in designing equipment. In practice we pick a standard size drum (4.5 ft diameter) instead of custom building the drum.

3.6. UTILIZING EXISTING FLASH DRUMS

Individual pieces of equipment will often outlive the entire plant. This used equipment is then available either in the plant's salvage section or from used equipment dealers. As long as used equipment is clean and structurally sound (it pays to have an expert check it), it can be used instead of designing and building new equipment. Used equipment and off-the-shelf new equipment will often be cheaper and will have faster delivery than custom-designed new equipment; however, it may have been designed for a different separation. The challenge in using existing equipment is to adapt it with minimum cost to the new separation problem.

The existing flash drum already has its dimensions L and D specified. Solving Eqs. (3-52) and (3-53) for a vertical drum for V, we have

$$V_{max} = \frac{\pi(D)^2 \, u_{perm} \, (3600)\rho_v}{4 \, MW_v} \qquad (3\text{-}55)$$

This vapor velocity is the maximum for this existing drum, since it will give a linear vapor velocity equal to u_{perm}.

The maximum vapor capacity of the drum limits the product of (V/F) times F, since we must have

$$(V/F) \, F < V_{Max} \qquad (3\text{-}56)$$

If Eq. (3-56) is satisfied, then use of the drum is straightforward. If Eq. (3-56) is violated, something has to give. Some of the possible adjustments are:

a. Add a demister to increase V_{Max} or to reduce entrainment.

b. Reduce feed rate to the drum.

c. Reduce V/F. Less vapor product with more of the more volatile components will be produced.

d. Use existing drums in parallel. This reduces feed rate to each drum.

e. Use existing drums in series (see Problem 3-D2).

f. Buy a different flash drum or build a new one.

g. Use some combination of these alternatives.

The engineer can use ingenuity to solve the problem in the cheapest and quickest way.

3.7. SUMMARY - OBJECTIVES

This chapter has discussed the calculation procedures for binary and multicomponent flash distillation. At this point you should be able to satisfy the following objectives:

1. Explain and sketch the basic flash distillation process.

2. Derive and plot the operating equation for a binary flash distillation on a y-x diagram.

3. Solve both sequential and simultaneous binary flash distillation problems. Explain the differences between these types of problems.

4. Derive the Rachford-Rice equation for multicomponent flash distillation.

5. Solve sequential multicomponent flash distillation problems.

6. For simultaneous multicomponent flash distillation explain the difference between the solution methods for narrow- and wide-boiling feeds. Determine which solution method to use. Solve both types of simultaneous problems.

7. Determine the length and diameter of a flash drum.

8. Use existing flash drums for a new separation problem.

REFERENCES

Blackwell, W.W., *Chemical Process Design on a Programmable Calculator,* McGraw-Hill, New York, 1984, Chapt. 3.

Evans, Frank L., Jr., *Equipment Design Handbook for Refineries and Chemical Plants,* vol. 2, 2nd ed., Gulf Publishing Co., Houston, TX, 1980.

Himmelblau, D.M., *Basic Principles and Calculations in Chemical Engineering,* 3rd ed., Prentice-Hall, Englewood Cliffs, NJ, 1974.

King, C.J., *Separation Processes,* 2nd ed., McGraw-Hill, New York, 1981.

Maxwell, J.B., *Data Book on Hydrocarbons,* Van Nostrand, Princeton, NJ, 1950.

Perry, R.H., C.H. Chilton, and S.D. Kirkpatrick (Eds.), *Chemical Engineer's Handbook,* 4th ed., McGraw-Hill, New York, 1963.

Smith, B.D., *Design of Equilibrium Stage Processes,* McGraw-Hill, New York, 1963.

Watkins, R.N., "Sizing Separators and Accumulators," *Hydrocarbon Processing, 46* (1), 253 (Nov. 1967).

HOMEWORK

A. *Discussion Problems*

A1. In Figure 3-3 the feed plots as a two-phase mixture, whereas it is a liquid before introduction to the flash chamber. Explain why. Why can't the feed location be plotted directly from known values of T_F and z? In other words, why does h_F have to be calculated separately from an equation such as Eq. (3-7b)?

A2. Can weight units be used in the flash calculations instead of molar units?

A3. Explain why a sequential solution procedure cannot be used when T_{feed} is specified for a flash drum.

A4. Explain why graphical procedures will not be very useful for a four-component flash distillation calculation.

A5. In the flash distillation of salt water, the salt is totally nonvolatile (this is the equilibrium statement). Show a McCabe-Thiele diagram for a feed water containing 3.5 wt % salt. Be sure to plot weight fraction of more volatile component.

A6. Explain why damping factors [see Eq. (3-48)] can prevent oscillation of solutions in trial-and-error calculations.

A7. Develop your own key relations chart for this chapter. That is, on *one* page summarize everything you would want to know to solve problems in flash distillation. Include sketches, equations, and key words.

B. *Generation of Alternatives*

B1. Think of all the ways a binary flash distillation problem can be specified. For example, we have usually specified F, z, T_{drum}, P_{drum}. What other combinations of variables can be used? (I have over 20.) Then consider how you would solve the resulting problems.

B2. An existing flash drum is available. The vertical drum has a demister and is 4 ft in diameter and 12 ft tall. The feed is 30 mole % methanol and 70 mole % water. A vapor product that is 58 mole % methanol is desired. We have a feed rate of 25,000 lb moles/hr. Operation is at 1 atm pressure. Since this feed rate is too high for the existing drum, what can be done to produce a vapor of the desired composition? Design the new equipment for your new scheme. You should devise at least three alternatives. Data are given in Problem 3-D1.

C.   *Derivations*

C1.   Use the secant convergence method to derive the equation to estimate the next V/F value in the trial-and-error procedure for a sequential multicomponent flash distillation problem.

C2.   T.K. Serghides [*Chem. Eng.*, *89* (18), 107-110 (Sept. 6, 1982)] derives a direct substitution equation to determine V/F. This equation is

$$\frac{V}{F} = 1 - \sum_{i=1}^{C} \frac{z_i}{1 + \dfrac{K_i(V/F)}{1 - (V/F)}}$$

a.   Derive this equation starting with $\sum_{i=1}^{C} x_i = 1$.

b.   Derive a similar equation starting with $\sum_{i=1}^{C} y_i = 1$.

c.   Derive an equation using both $\sum_{i=1}^{C} x_i = 1$ and $\sum_{i=1}^{C} y_i = 1$.

d.   Which of the three equations is likely to have the best convergence properties?

C3.   Convert Eqs. (3-51) and (3-52) to SI units.

C4.   Show that for narrow-boiling feeds the Newtonian convergence procedure also results in Eq. (3-47). Generalize the meaning of this result. Be careful, since a similar result will not be true for Eq. (3-43).

C5.   Choosing to use V/F to develop the Rachford-Rice equation is conventional but arbitrary. We could also use L/F, the fraction remaining liquid, as the trial variable.

a. Develop the Rachford-Rice equation as f(L/F).
b. Develop the equation analogous to Eq. (3-47) to determine L/F instead of V/F.

C6.   In flash distillation a liquid mixture is partially vaporized. We could also take a vapor mixture and partially condense it. Draw a schematic diagram of partial condensation equipment. Derive the equations for this process. Are they different from flash distillation? If so, how?

C7. Without looking at the text, derive the Rachford-Rice equation.

C8. Plot Eq. (3-27) versus V/F for Example 3-2 to illustrate that convergence is not as linear as the Rachford-Rice equation.

D. *Problems*

D1. We are separating a mixture of methanol and water in a flash drum at 1 atm pressure. Equilibrium data are listed in Table 3-3.

Table 3-3. Vapor-Liquid Equilibrium Data for Methanol Water (p = 1 atm)

| Mole % Methanol | | |
Liquid	Vapor	Temp., °C
0	0	100
2.0	13.4	96.4
4.0	23.0	93.5
6.0	30.4	91.2
8.0	36.5	89.3
10.0	41.8	87.7
15.0	51.7	84.4
20.0	57.9	81.7
30.0	66.5	78.0
40.0	72.9	75.3
50.0	77.9	73.1
60.0	82.5	71.2
70.0	87.0	69.3
80.0	91.5	67.6
90.0	95.8	66.0
95.0	97.9	65.0
100.0	100.0	64.5

Source: Perry *et al.* (1963), p. 13-5.

a. Feed is 60 mole % methanol, and 40 % of the feed is vaporized. What are the vapor and liquid mole fractions and flow rates? Feed rate is 100 kg moles/hr.

b. Repeat part a for a feed rate of 1500 kg moles/hr.

c. If the feed is 30 mole % methanol and we desire a liquid product that is 20 mole % methanol, what V/F must be used? For a feed rate of 1000 lb moles/hr, find product flow rates and compositions.

d. We are operating the flash drum so that the liquid mole fraction is 45 mole % methanol. L = 1500 kg moles/hr, and V/F = 0.2. What must the flow rate and composition of the feed be?

e. Find the dimensions of a vertical flash drum for Problem D-1c.
Data: ρ_w = 1.00 g/cm³, $\rho_{m,L}$ = 0.7914 g/cm³, MW_w = 18.01, MW_m = 32.04. Assume vapors are ideal gas.

f. If z = 0.4, p = 1 atm, and T_{drum} = 77°C, find V/F, x_m, and y_m.

D2. Two flash distillation chambers are hooked together as shown in the diagram. Both are at 1 atm pressure. The feed to the first drum is a binary mixture of methanol and water that is 55 mole % methanol. Feed flow rate is 10,000 kg moles/hr. The second flash drum operates with $(V/F)_2$ = 0.7 and the liquid product composition is 25 mole % methanol. Equilibrium data are given in Table 3-3.

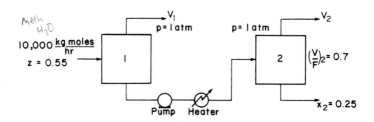

a. What is the fraction vaporized in the first flash drum?

b. What are y_1, y_2, x_1, T_1, and T_2?

D3. You want to flash a mixture with a drum pressure of 2 atm and a drum temperature of 25°C. The feed is 2000 kg moles/hr. The feed is 5 mole % methane, 10 mole % propane, and the rest n-hexane. Figures 2-11 and 2-12 or Eq. (2-12) can be used for equilibrium data. Find the fraction vaporized, vapor mole fractions, liquid mole fractions, and vapor and liquid flow rates.

D4. A mixture that is 40 mole % benzene and 60 mole % toluene is to be flashed in a flash distillation system. Feed is 100 kg

moles/day. We desire a liquid product that is 30 mole % benzene. The relative volatility is $\alpha_{BT} = 2.4$.
Find: (a) vapor composition, (b) liquid flow rate.

D5. We wish to flash distill an ethanol-water mixture that is 30 wt % ethanol and has a feed flow of 1000 kg/hr. Feed is at 200 °C. The flash drum operates at a pressure of 1 kg/cm². Find: T_{drum}, weight fractions of liquid and vapor products, and liquid and vapor flow rates.

Data: $C_{PL,EtOH} = 37.96$ at 100 °C, kcal/kg mole °C

$C_{PL,W} = 18.0$, kcal/kg mole °C

$C_{PV,EtOH} = 14.66 + 3.758 \times 10^{-2}T - 2.091 \times 10^{-5}T^2 + 4.74 \times 10^{-9}T^3$

$C_{PV,W} = 7.88 + 0.32 \times 10^{-2}T - 0.04833 \times 10^{-5}T^2$

Both C_{PV} values are in kcal/kg mole °C, with T in °C.

$\rho_{EtOH} = 0.789$ g/mL, $\rho_W = 1.0$ g/mL, $MW_{EtOH} = 46.07$, $MW_W = 18.016$, $\lambda_{EtOH} = 9.22$ kcal/g mole at 351.7 K, and $\lambda_W = 9.7171$ kcal/g mole at 373.16 K.

Enthalpy composition diagram at $p = 1$ kg/cm² is in Figure 2-4. Note: Be careful with units.

D6. We have a mixture that is 40 mole % propane and 60 mole % n-heptane. Feed flow rate is 1500 kg moles/hr. The drum pressure is 250 kPa. We wish to operate at V/F = 0.45.

  a. Use the DePriester charts (Figure 2-11 or 2-12) or Eq. (2-12) to generate temperature-composition and y-x diagrams. Then solve graphically for the product compositions and drum temperature.

  b. Solve by using the Rachford-Rice equation and a trial-and-error procedure.

  c. Compare the solution techniques you used in parts a and b. Which do you prefer? If you also wanted to determine the product compositions with V/F = 0.2, 0.4, 0.6 and 0.8, which solution procedure would you prefer and why?

D7. A flash drum is to treat 1150 kg moles/hr of a mixture of n-butane, n-hexane, and n-octane. The feed is 65 mole % n-butane, 15 mole % n-hexane, and the rest n-octane. The flash

drum operates at $100\,^\circ$ C, and 40% of the feed re
$(L/F = 0.40)$. What is the pressure of the flash drur
ure 2-11 or 2-12 or Eq. (2-12).

D8. We have a flash drum separating 50 kg moles/hr of a ... oi
ethane, isobutane and n-butane. The ratio of isobutane to n-
butane is held constant at 0.8 (that is, $z_{iC4}/z_{nC4} = 0.8$). The
mole fractions of all three components in the feed can change.
The flash drum operates at a pressure of 100 kPa and a temperature of $20\,^\circ$ C. If the drum is operating at $V/F = 0.4$, what
must the mole fractions of all three components in the feed be?
Use Figure 2-11 or 2-12 or Eq. (2-12).

D9. A flash drum is to flash 10,000 lb moles/hr of a feed that is 65
mole % n-hexane and 35 mole % n-octane at 1 atm pressure.
$V/F = 0.4$.

a. Find T_{drum}, liquid mole fraction, vapor mole fraction.

b. Find the size required for a vertical flash drum.

Note: y, x, T equilibrium data can be determined from the
DePriester charts or Eq. (2-12). Other required physical pro-
perty data are given in Example 3-4.

D10. An ethanol-water mixture is to be flash distilled at 1 atm. If the
feed is 25 mole % ethanol, what are the liquid and vapor compositions if:

a. All of the feed is vaporized.

b. None of the feed is vaporized.

c. One-third of the feed is vaporized.

d. Two-thirds of the feed is vaporized.

Solve this problem graphically. Data are in Table 2-1 and Figure 2-2.

D11. For the system given in Problem 3-D10, what fraction of the
liquid must be vaporized if:

a. A vapor product containing 50 mole % ethanol is desired.

b. A liquid product containing 7 mole % ethanol is desired.

D12. A flash chamber is being used to separate 1000 moles/hr of a
mixture of ethylene, ethane, n-butane, and n-pentane. Feed
mole fractions are ethylene, 0.5; ethane, 0.40; n-butane, 0.35;
and n-pentane, 0.20. The flash chamber operates at a tempera-

ture that will provide for vaporization of 30% of the feed. Find the temperature of the flash chamber and the compositions (in mole fractions) of the vapor and liquid products. Use the DePriester charts or Eq. (2-12) for equilibrium data. Pressure = 40 psia.

D13. We wish to flash distill a mixture that is 45 mole % benzene and 55 mole % toluene. Feed rate to the still is 700 moles/hr. Equilibrium data for the benzene-toluene system can be approximated with a constant relative volatility of 2.5, where benzene is more volatile. Operation of the still is at 1 atm.

   a. Plot a y-x diagram for benzene.

   b. If 60% of the feed is vaporized, find the liquid and vapor compositions.

   c. If we desire a vapor composition of 60 mole %, what is corresponding liquid composition, and what are the liquid and vapor flow rates?

   d. Repeat part b, but solve equilibrium and operating equations analytically.

D14. We wish to take the benzene-toluene feed given in Problem 3-D13 and flash 40% of the feed. The liquid product will then be sent to a second flash chamber where 30% is flashed. Find compositions and flow rates of all unknown streams. Both stills operate at 1 atm.

D15. We are flash distilling a mixture of ethanol and propanol. This mixture is ideal and has $\alpha_{EP} = 2.10$. The feed to the flash still is 100 kg moles/hr of 0.6 mole fraction ethanol. Your supervisor wants to know the effect of varying the ratio V/F. Find the composition of the liquid and vapor leaving the still for values of V/F = 0, 0.2, 0.4, 0.6, 0.8, 1.0.

   a. Do this solution analytically. No graphs allowed.

   b. Solve the problem graphically. This will require a y-x plot for $\alpha_{EP} = 2.10$.

   c. Plot compositions of vapor and liquid leaving the still versus V/F for part b.

   d. We desire an outlet liquid composition of 0.5. What value of V/F do we use to obtain this? Solve graphically.

D16. Repeat Example 3-2, but use the secant convergence procedure (see Problem 3-C1).

D17. A vapor stream is partially condensed in a heat exchanger, and then vapor and liquid are separated in a drum. The feed is 60 mole % methanol and 40 mole % water at 50 kg moles/hr. The drum operates at 101.3 kPa.

   a. If 80% of the stream is liquefied, what are outlet compositions?

   b. If the product vapor y = 0.64, what is the liquid flow rate?

   Equilibrium data are given in Problem 3 D1.

D18. We wish to flash distill a feed that is 10 mole % propane, 30 mole % n-butane, and 60 mole % n-hexane. Feed rate is 10 kg moles/hr, and drum pressure is 200 kPa. We desire a liquid that is 85 mole % n-hexane. Use DePriester charts (Figure 2-11 or 2-12) or Eq. (2-12) for K values. Find T_{drum} and V/F. Continue until your answer is within 0.5 ° C of the correct answer. Note: This is a single trial-and-error, NOT a simultaneous mass and energy balance convergence problem.

D19. A flash drum operating at 300 kPa is separating a mixture that is fed in as 40 mole % isobutane, 25% n-pentane, and 35% n-hexane. We wish a 90% recovery of n-hexane in the liquid (that is, 90% of the n-hexane in the feed exits in the liquid product). F = 1000 kg moles/hr. Find T_{drum}, x_j, y_j, V/F.

D20. We wish to remove methane from a mixture of heavier hydrocarbons. The feed is 15 mole % CH_4, 60% n-C_5H_{12} and 25% n-C_6H_{14}. Feed rate is 2500 kg moles/hr. Flash drum is at 80 ° F and 30 psia. Find: V/F, x_i, y_i, L, V.

E. *More Complex Problems*

E1. (Solve by hand or write a computer program.) A feed consisting of 16 mole % ethane, 67 mole % n-butane, and 17 mole % n-pentane is to be flash distilled. On the basis of 1 mole of feed, find L, V, and the compositions of the liquid and vapor products if the flash distillation is done at:

   a. 100 ° F and 50 psia

   b. 95 ° F and 50 psia

   c. 90 ° F and 50 psia

   The Σx and Σy should be 1.000 ± 0.001. Check one of your answers with hand calculation, using the DePriester chart for K value data.

E2.   (Solve by hand or write a computer program.) The same feed as in Problem 3-E1 is to be flash distilled at 500 psia and such a temperature that a specified amount of the feed will be vaporized. Find the operating temperature and compositions of the liquid and vapor if:

   a. 25% of feed is to be vaporized

   b. 50% of feed is to be vaporized

   c. 75% of feed is to be vaporized

   Σx and Σy should be 1.000 ± 0.001. Check one of your answers with a hand calculation using the DePriester chart for K value data.

F.   *Problems Requiring Other Resources*

F1.   Calculate T_F and p_F for Example 3-2.

F2.   (Difficult convergence.) We wish to flash distill a mixture that is 0.648 mole fraction ethane, 0.145 mole fraction n-butane, and 0.207 mole fraction n-pentane. The feed rate is 100 g moles/hr. The feed to the flash is at 96.6° C. The flash drum will operate at 1500 kPa. Find the drum temperature, the value of V/F, and the mole fractions of each component in the liquid and vapor products.

F3.   Ethylene glycol and water are to be flash distilled in a cascade of three stills connected as shown in the figure. All stills operate at 228 mm Hg. Feed is 40 mole % water. One-third of the feed is vaporized in the first still, two-thirds of feed to second still is vaporized, and one-half the feed to the third still is vaporized. What are the compositions of streams L_3 and V_3? Note that water is the more volatile component.

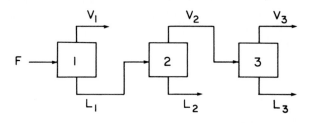

F4.   A feed that is 75 mole % chloroform and 25 mole % benzene is fed to a flash chamber at a rate of 153.5 moles/hr. 60% of the feed is vaporized in the flash chamber. The liquid product from the first chamber is sent to a second flash chamber, where it is reflashed. Find the fraction vaporized in the second flash chamber to obtain a liquid that is 55 mole % chloroform.

F5.   If you have a computer package available for solving multicomponent flash distillation, pick four or five components in a series (e.g., C_3, i C_4, n C_5 and n C_6) for which data are available.

   a.   What would you expect to happen if there was a very large amount of the most volatile component (say 78% C_3)? Qualitatively, what would you expect the concentration and temperature of the vapor product to look like as V/F increased? Try this: Pick V/F values of 0.1, 0.3, 0.5, 0.7 and 0.9 (more values can be run at interesting values - this first cut gives you an idea of whether more runs will tell you if any more information will be useful and where you want to obtain it). Alternatively, pick T_{drum} instead of V/F, and increase T_{drum}. What temperature must the feed be for these runs? For *this* feed composition, plot T_{feed} versus T_{drum}. Note how this graph could be used to give you a good first guess of T_{drum} if T_{feed} were specified.

   b.   Try another feed composition of the same components, but this time with larger quantities of the less volatile components. Repeat the runs you used before (but since you're smarter now perhaps you can do fewer of them).

   c.   Repeat part b but with a feed with large quantities of one of the middle components (C_4 or C_5).

   d.   Repeat part b but with a feed with approximately equimolar composition of all components.

   e.   Repeat part b but with a feed that is almost a binary but contains traces of the other two components.

   f.   Repeat part b but with a binary feed of two of the components. Compare with results from part e.

What general conclusions can you draw about the effect of feed composition on product composition, T_{drum}, and T_{feed} as V/F increases? This exercise should help you in making good first guesses for trial-and-error problems.

F6.   We wish to flash distill a mixture that is 0.517 mole fraction propane, 0.091 mole fraction n-pentane, and 0.392 mole fraction n-octane. The feed rate is 100 kg moles/hr. The feed to the flash drum is at 95° C. The flash drum will operate at 250 kPa. Find the drum temperature, the value of V/F, and the mole fractions of each component in the liquid and vapor products. Converge $| \Delta T_{drum} |$ to $< 0.20°$ C.

G.   *Open-Ended and Synthesis Problems*

G1.   We wish to flash 150,000 kg/hr of a mixture that is 25 wt % ethanol and 75 wt % water and get a vapor product that is 50 wt % ethanol. Flow rate may vary by 12%. Feed can range from 21 to 26 wt % ethanol. Feed is currently at ambient temperature and pressure (both of which vary during the year).

   a. Design a new flash system with one flash drum.

   b. Design a new flash system with two flash drums so that the final concentration of ethanol in the liquid wastes will be decreased and more 50 wt % ethanol vapor product will be produced.

chapter 4
INTRODUCTION TO COLUMN DISTILLATION

4.1. DEVELOPING A DISTILLATION CASCADE

In Chapter 3 we learned how to do the calculations for flash distillation. Flash distillation is a very simple unit operation, but in most cases it produces a limited amount of separation. In Problems 3-D2, 3-D14, and 3-F3 we saw that more separation could be obtained by adding on (or *cascading*) more flash separators. The cascading procedure can be extended into a process that produces one pure vapor and one pure liquid product. First, we could send the vapor streams to additional flash chambers at increasing pressures and the liquid streams to flash chambers with decreasing pressures, as shown in Figure 4-1. Stream V_1 will have a high concentration of the more volatile component and stream L_5 will have a low concentration of the more volatile component. Each flash chamber in Figure 4-1 can be analyzed by the methods developed previously.

One difficulty with the cascade shown in Figure 4-1 is that the intermediate product streams, L_1, L_2, V_4, and V_5, are of intermediate concentration and need further separation. Of course, each of these streams could be fed to another flash cascade, but then the intermediate products from those cascades would have to be sent to additional cascades, and so forth. A much cleverer solution is to use the intermediate product streams as additional feeds within the same cascade.

Consider stream L_2, which was generated by flashing part of the feed stream and then condensing part of the resulting vapor. Since the material in L_2 has been vaporized once and condensed once, it probably has a concentration close to that of the original feed stream. (To check this, you can do the appropriate flash calculation on a McCabe-Thiele diagram.) Thus, it is appropriate to use L_2 as an additional feed stream to stage 3. However, since $P_2 > P_3$, its pressure must first be decreased.

Stream L_1 is the liquid obtained by partially condensing V_2, the vapor obtained from partially condensing vapor stream V_3. After one vaporization and then a condensation, stream L_1 will have a concentra-

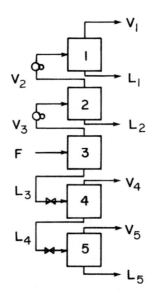

Figure 4-1   Cascade of flash chambers. $p_1 > p_2 > p_3 > p_4 > p_5$

tion close to that of stream V_3. Thus it is appropriate to use stream L_1 as an additional feed to stage 2 after pressure reduction.

A similar argument can be applied to the intermediate vapor products below the feed, V_4 and V_5. V_4 was obtained by partially condensing the feed stream and then partially vaporizing the resulting liquid. Since its concentration is approximately the same as the feed, stream V_4 can be used as an additional feed to stage 3 after compression to a higher pressure. By the same reasoning, stream V_5 can be fed to stage 4.

Figure 4-2 shows the resulting *countercurrent cascade,* so called because vapor and liquid streams go in opposite directions. The advantages of this cascade over the one shown in Figure 4-1 are that there are no intermediate products and the two end products can both be pure *and* obtained in high yield. Thus V_1 can be almost 100% of the more volatile components and contain almost all of the more volatile component of the feed stream.

Although a significant advance, this variable pressure (or isothermal distillation) system is seldom used commercially. Operation at different pressures requires a large number of compressors, which are expensive. It is much cheaper to operate at constant pressure and force the temperature to vary. Thus, in stage 1 of Figure 4-2 a relatively low temperature would be employed, since the concentration of the more vola-

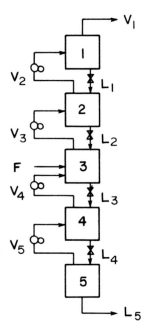

Figure 4-2.   Countercurrent cascade of flash chambers. $p_1 > p_2 > p_3 > p_4 > p_5$

tile component, which boils at a lower temperature, is high. For stage 5, where the less volatile component is concentrated, the temperature would be high. To achieve this temperature variation, we can use heat exchangers (condensers) to partially condense the vapor streams and other heat exchangers (reboilers) to partially vaporize the liquid streams. This is illustrated in Figure 4-3, where partial condensers and partial reboilers are used.

The cascade shown in Figure 4-3 has a decreasing vapor flow rate as we go from the feed stage to the top stage, and a decreasing liquid flow rate as we go from the feed stage to the bottom stage. Operation and design will be easier if part of the top vapor stream V_1 is condensed and returned to stage 1 and if part of the bottom liquid stream L_5 is vaporized and returned to stage 5, as illustrated in Figure 4-4. This allows us to control the internal liquid and vapor flow rates at any desired level. Stream D is the distillate product, while B is the bottom product. Stream L_0 is called the *reflux* while V_6 is the *boilup*.

The use of reflux and boilup allows for a further simplification. We can now apply all of the heat required for the distillation to the bottom reboiler, and we can do all of the required cooling in the top condenser.

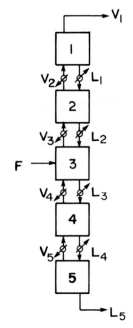

Figure 4-3. Countercurrent cascade of flash chambers with intermediate reboilers and condensers. p = constant. $T_1 < T_2 < T_3 < T_4 < T_5$.

Figure 4-4. Countercurrent cascade of flash chambers with reflux and boilup. p = constant. $T_1 < T_2 < T_3 < T_4 < T_5$

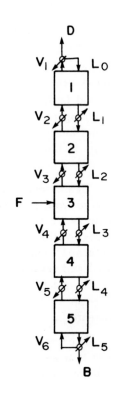

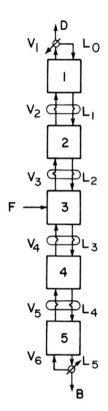

Figure 4-5.   Countercurrent   cascade   with
intermediate heat exchangers.

The required partial condensation of intermediate vapor streams and partial vaporization of liquid streams can both be done with the same heat exchangers as shown in Figure 4-5. Here stream V_2 is partially condensed by stream L_1 while L_1 is simultaneously partially vaporized. Since L_1 has a higher concentration of more volatile component, it will boil at a lower temperature and heat transfer is in the appropriate direction. Since the heat of vaporization per mole is usually approximately constant, condensation of 1 mole of vapor will vaporize 1 mole of liquid. Thus liquid and vapor flow rates tend to remain constant. Heat exchangers can be used for all other pairs of *passing streams:* L_2 and V_3, L_3 and V_4, L_4 and V_5.

Note that reflux and boilup are *not* the same as recycle. Recycle returns a stream to the feed to the process. Reflux (or boilup) first changes the phase of a stream and then returns the stream to the *same* stage the vapor (or liquid) was withdrawn from. This return at the same location helps increase the concentration of the component that is concentrated at that stage.

A

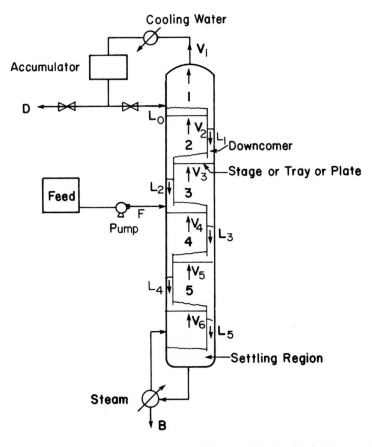

Figure 4-6.   Distillation column. (A) Schematic of five-stage column $(T_1 < T_2 < T_3 < T_4 < T_5 < T_6)$. (B) Photograph of distillation columns. Courtesy of APV Equipment, Inc. Tonawanda, NY.

The cascade shown in Figure 4-5 can be further simplified by building the entire system in a column instead of as a series of individual stages. The intermediate heat exchange can be done very efficiently with the liquid and vapor in direct contact on each stage. The result is a much simpler and cheaper device. A schematic of such a distillation column is shown in Figure 4-6.

The cascade shown in Figure 4-6 is the usual form in which distillation is done. Because of the repeated vaporizations and condensations as we go up the column, the top product (distillate) can be highly con-

B

centrated in the more volatile component. The section of the column above the feed stage is known as the *enriching* or *rectifying* section. The bottom product (bottoms) is highly concentrated in the less volatile component, since the more volatile component has been stripped out by the rising vapors. This section is called the *stripping* section.

The distillation separation works because every time we vaporize material the more volatile component tends to concentrate in the vapor, and the less volatile component in the liquid. As the relative volatility of the system decreases, distillation becomes more difficult. If $\alpha = 1.0$, the liquid and vapor will have the same composition, and no separation will occur. Liquid and vapor also have the same composition when an azeotrope occurs. In this case one can approach the azeotrope concentration at the top or bottom of the column but cannot get past it except with a heterogeneous azeotrope (see Chapter 10). The third limit to distillation is the presence of either chemical reactions between components or decomposition reactions. This problem can often be controlled by operating at lower temperatures and using vacuum or steam distillation (see Chapter 10).

While we are still thinking of flash distillation chambers, a simple but useful result can be developed. In a flash chamber a component will tend to exit in the vapor if $y_i V > x_i L$. Rearranging this, if $K_i V/L > 1$ a component tends to exit in the vapor. In a distillation column this means that components with $K_i V/L > 1$ tend to exit in the distillate, and components with $K_i V/L < 1$ tend to exit in the bottoms. This is only a qualitative guide, since the separation on each stage is far from perfect, and $K_i V$, and L all vary in the column; however, it is useful to remember.

4.2. DISTILLATION EQUIPMENT

It will be helpful for you to have a basic understanding of distillation equipment before studying the design methods. A detailed description of equipment will be delayed until Chapter 12. Figure 4-6A is a schematic of a distillation column, and Figure 4-6B is a photograph of several columns.

The column is usually metal and has a circular cross section. It contains trays (plates or stages) where liquid-vapor contact occurs. The simplest type of tray is a sieve tray, which is a sheet of metal with holes punched into it for vapor to pass through. This is illustrated in Figure 4-7. The liquid flows down from the tray above in a downcomer and then across the sieve tray, where it is intimately mixed with the vapor. The vapor flowing up through the holes prevents the liquid from dripping downward, and the weir acts as a dam to keep a sufficient level of liquid on the plate. The liquid that flows over the weir is a frothy mixture containing a lot of vapor. This vapor disengages in the downcomer so that clear liquid flows into the stage below. The space above the tray allows for disengagement of liquid from vapor and needs to be high enough to prevent excessive entrainment (carryover of liquid from one stage to the next). Distances between trays vary from 2 to 48 inches and tend to be greater the larger the diameter of the column.

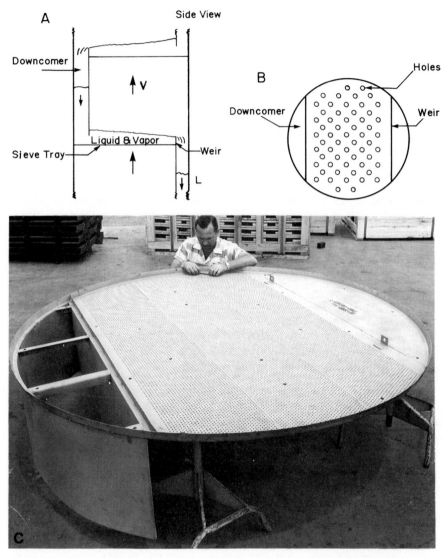

Figure 4-7.   Sieve trays. (A) Schematic side view; (B) schematic top view; (C) photograph, courtesy of Glitsch, Inc.

To say that there is a liquid pool on the tray is an oversimplification. In practice, any one of four distinct flow regimes can be observed on trays, depending on the gas flow rate. In the *bubble* regime the liquid is close to being a stagnant pool with distinct bubbles rising through it. This regime occurs at low gas flow rates; because of poor mixing there is poor liquid and vapor contact, which results in low stage efficiency. Because of the low gas flow rate and low efficiency, the bubble regime is undesirable in commercial applications.

At higher gas flow rates the stage will often be in a *foam* regime. In this regime, the liquid phase is continuous and has fairly distinct bubbles rapidly rising through it; there is a distinct foam like the head on a beer on top of the liquid. Because of the large surface area in a foam, the area for vapor-liquid mass transfer is large, and stage efficiency may be quite high. However, if the foam is too stable it can fill the entire region between stages. When this occurs, *entrainment* (the carryover of liquid to the tray above) becomes excessive, stage efficiency drops, and the column may flood (fill up with liquid and become inoperative). This may require use of a chemical antifoam agent. The foam regime is usually at vapor flow rates that are too low for most industrial applications.

At even higher vapor flow rates the *froth* regime occurs. In this regime the liquid is continuous and has large, pulsating voids of vapor rapidly passing through it. The surface of the liquid is boiling violently, and there is considerable splashing. The liquid phase is thoroughly mixed, but the vapor phase is not. In most distillation systems where the liquid-phase mass transfer controls, this regime has good efficiency. Because of the good efficiency and reasonable vapor capacity, this is usually the flow regime used in commercial operation.

At even higher gas flow rates the vapor-liquid contact on the stage changes markedly. In the *spray* regime the vapor is continuous and the liquid occurs as a discontinuous spray of droplets. The vapor is very well mixed, but the liquid droplets usually are not. Because of this poor liquid mixing, the mass transfer rate is usually low and stage efficiencies are low. The significance of this is that relatively small increases in vapor velocity can cause the column to go from the froth to the spray regime and cause a significant decrease in stage efficiency (for example, from 65% to 40%).

A variety of other configurations and modifications of the basic design shown in Figures 4-6 and 4-7 are possible. Both bubble cap and valve trays are popular. Downcomers can be chords of a circle as shown or circular pipes. Both partial and total condensers and a variety of reboilers are used. The column may have multiple feeds, sidestream withdrawals, intermediate reboilers or condensers, and so forth. The column also usually has a host of temperature, pressure, flow rate, and level measurement and control devices. Despite this variety, the operating principles are the same as for the simple distillation column shown in Figure 4-6.

4.3. SPECIFICATIONS

In the design or operation of a distillation column, a number of variables must be specified. For both design and simulation problems we usually specify column pressure (which sets the equilibrium data); feed composition, flow rate and feed temperature or feed enthalpy or feed quality; and temperature or enthalpy of the reflux liquid. The usual reflux condition set is a saturated liquid reflux. These variables are listed in Table 4-1. The other variables set depend upon the type of problem.

In *design* problems, the desired separation is set and a column is designed that will achieve this separation. For a binary distillation we would usually specify the mole fraction of the more volatile component in the distillate and bottoms products. In addition, the external reflux ratio, L_0/D in Figure 4-6, is usually specified. Finally, we usually specify that the *optimum feed location* be used, that is, the feed location that will result in the fewest total number of stages. The designer's job is to calculate distillate and bottoms flow rates, the heating and cooling requirements in the reboiler and condenser, the number of stages required and the optimum feed stage location, and finally the required column diameter. Alternative specifications such as the splits (fraction of a component recovered in the distillate or bottoms) or distillate or bottoms flow rates are common. Some of the possibilities are summarized in Table 4-2.

In *simulation* problems, the column has already been built and we wish to predict how much separation can be achieved for a given feed. Since the column has already been built, the number of stages and the feed stage location are already specified. In addition, the column diameter and the reboiler size, which usually control a maximum vapor flow rate, are set. There are a variety of ways to specify the remainder of the problem (see Table 4-3). The desired composition of more volatile component in the distillate and bottoms could be specified, and the engineer would then have to determine the external reflux ratio, L_0/D, that will produce this separation and check that the maximum vapor flow rate will not be exceeded. An alternative is to specify L_0/D and

Table 4-1. Usual Specified Variables for Binary Distillation

1.	Column pressure
2.	Feed flow rate
3.	Feed composition
4.	Feed temperature or enthalpy or quality
5.	Reflux temperature or enthalpy (usually saturated liquid)

Table 4-2. Specifications and Calculated Variables for Binary
Distillation for Design Problems

Specified Variables	Designer Calculates
A. 1. Mole fraction more volatile component in distillate, x_D 2. Mole fraction more volatile component in bottoms, x_B 3. External reflux ratio, L_0/D 4. Use optimum feed plate	A. Distillate and bottoms flow rates, D and B Heating and cooling loads, Q_R and Q_c Number of stages, N Optimum feed plate Column diameter
B. 1,2. Fractional recoveries of components in distillate and bottoms, $(FR_A)_{dist}, (Fr_B)_{bot}$ 3. External reflux ratio, L_0/D 4. Use optimum feed plate	B. x_B, x_D, D, B Q_R, Q_c N N_{feed} Column diameter
C. 1. D or B 2. x_D or x_B 3. External reflux ratio, L_0/D 4. Use optimum feed plate	C. B or D x_B or x_D Q_R, Q_c N and N_{Feed} Column diameter
D. 1,2. x_D and x_B 3. Boilup ratio, $\overline{V}/B$ 4. Use optimum feed plate	D. D and B, Q_R and Q_c N, N_{Feed} Column diameter

either distillate or bottoms composition, in which case the engineer
determines the unknown composition and checks the vapor flow rate.
Another alternative is to specify the heat load in the reboiler and the
distillate or bottoms composition. The engineer would then determine
the reflux ratio and unknown product composition and check the vapor
flow rate. The thread that runs through all these alternatives is that
since the column has been built, some method of specifying the separation must be used.

The engineer always specifies variables that can be controlled.
Several sets of possible specifications and calculated variables are outlined in Tables 4-1 to 4-3. Study these tables to determine the
difference between design-type and simulation-type problems. Note that
other combinations of specifications are possible.

Table 4-3. Specifications and Calculated Variables for Binary
Distillation for Simulation Problems

Specified Variables		Designer Calculates	
A.	1,2. N, N_{feed} 3,4. x_D and x_B Column diameter	A.	L_0/D B, D, Q_c, Q_R Check $V < V_{max}$
B.	1,2. N, N_{feed} 3,4. L_0/D, x_D (or x_B) Column diameter	B.	x_B (or x_D) B, D, Q_c, Q_R Check $V < V_{max}$
C.	1,2. N, N_{feed} 3. x_D(or x_B) 4. Column diameter (set $V = V_{max}$)	C.	L_0/D, x_B(or x_D) B, D, Q_c, Q_R
D.	1,2. N, N_{feed} 3. Q_R 4. x_D (or x_B) Column diameter	D.	B, D, Q_c, x_B (or x_D), L_0/D Check $V < V_{max}$

In Table 4-1 we find five specified variables common to both types of problems. For design problems (Table 4-2), four additional variables must be set. Note that whereas column diameter is a specified variable in simulation problem C, it serves as a constraint in simulation problems A, B, and D (Table 4-3). Column diameter will allow us to calculate V_{max} and then we can check that $V < V_{max}$. However, we have not specified a variable for simulation. In problem C, where we specify $V = V_{max}$, the column diameter serves as a variable for simulation.

Chapter 5 discusses the simple design problem. Simulation and other more complicated problems are considered in Chapter 6.

4.4. EXTERNAL COLUMN BALANCES

Once the problem has been specified, the engineer must calculate the unknown variables. Often it is not necessary to solve the entire problem, since only limited answers are required. The first step is to do mass and energy balances around the entire column. For binary design problems, these balances can usually be solved without doing stage-by-stage

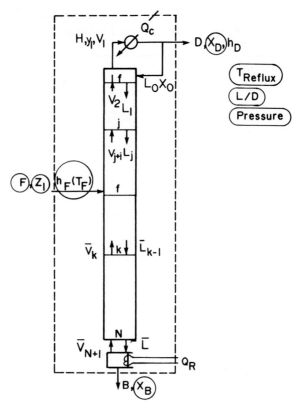

Figure 4-8.   Binary distillation column. Circled variables are typically specified in design problems.

calculations. Figure 4-8 shows the schematic of a distillation column.
The specified variables for a typical design problem are circled. We will
assume that the column is well insulated and can be considered adiabatic. All of the heat transfer takes place in the condenser and reboiler.
Column pressure is assumed to be constant.

From the balances around the entire column we wish to calculate
distillate and bottoms flow rates, D and B, and the heat loads in the
condenser and reboiler, Q_c and Q_R. We can start with mass balances
around the entire column using the balance envelope shown by the
dashed outline in the figure. The overall mass balance is

$$F = B + D \qquad (4\text{-}1)$$

and the more volatile component mass balance is

$$Fz = Bx_B + Dx_D \qquad (4\text{-}2)$$

For the design problem shown in Figure 4-8, Eqs. (4-1) and (4-2) can be solved immediately, since the only unknowns are B and D. Solving Eq. (4-1) for B, substituting this into Eq. (4-2), and solving for D, we obtain

$$D = (\frac{z - x_B}{x_D - x_B})F \qquad (4-3)$$

and

$$B = F - D = (\frac{x_D - z}{x_D - x_B})F \qquad (4-4)$$

Don't memorize equations like these; they can be derived as needed.

For the energy balance we will use the convention that all heat loads will be treated as inputs. If energy is removed, then the numerical value of the heat load will be negative. The steady-state energy balance around the entire column is

$$Fh_F + Q_c + Q_R = Dh_D + Bh_B \qquad (4-5)$$

where we have assumed that kinetic and potential energy and work terms are negligible. The column is assumed to be well insulated and adiabatic. Q_R will be positive and Q_c negative. The enthalpies in Eq. (4-5) can all be determined from an enthalpy-composition diagram (e.g., Figure 2-4) or from the heat capacities and latent heats of vaporization. In general,

$$h_F = h_F(z, T_F, p), \quad h_D = h_D(x_D, T_{Reflux}, p) \qquad (4\text{-}6a,b)$$

$$h_B = h_B(x_B, \text{saturated liquid}, p) \qquad (4\text{-}6c)$$

These three enthalpies can all be determined.

Since F was specified and D and B were just calculated, we are left with two unknowns, Q_R and Q_c, in Eq. (4-5). Obviously another equation is required.

For the total condenser shown in Figure 4-8 we can determine Q_c. The total condenser changes the phase of the entering vapor stream but does not affect the composition. The splitter after the condenser changes only flow rates. Thus composition is unchanged and

$$y_1 = x_D = x_0 \qquad (4-7)$$

The condenser mass balance is

$$V_1 = L_0 + D \tag{4-8}$$

Since the external reflux ratio, L_0/D, is specified, we can substitute its value into Eq. (4-8).

$$V_1 = \left(\frac{L_0}{D}\right)D + D = \left(1 + \frac{L_0}{D}\right)D \tag{4-9}$$

Then, since the terms on the right-hand side of Eq. (4-9) are known, we can calculate V_1. The condenser energy balance is

$$V_1 H_1 + Q_c = D h_D + L_0 h_0 \tag{4-10}$$

Since stream V_1 is a vapor leaving an equilibrium stage in the distillation column, it is a saturated vapor. Thus,

$$H_1 = H_1(y_1, \text{ saturated vapor}, p) \tag{4-11}$$

and the enthalpy can be determined. Since the reflux and distillate streams are at the same composition, temperature, and pressure, $h_0 = h_D$. Thus,

$$V_1 H_1 + Q_c = (D + L_0)h_D = V_1 h_D \tag{4-12}$$

Solving for Q_c we have

$$Q_c = V_1(h_D - H_1) \tag{4-13}$$

or, substituting in Eq. (4-9) and then Eq. (4-4),

$$Q_c = \left(1 + \frac{L_0}{D}\right) D(h_D - H_1)$$

$$= \left(1 + \frac{L_0}{D}\right)\left(\frac{z - x_B}{x_D - x_B}\right)F(h_D - H_1) \tag{4-14}$$

Note that $Q_c < 0$ because the liquid enthalpy, h_D, is less than the vapor enthalpy, H_1. This agrees with our convention. If the reflux is a saturated liquid, $H_1 - h_D = \lambda$, the latent heat of vaporization per mole.

With Q_c known we can solve the column energy balance, Eq. (4-5), for Q_R.

$$Q_R = Dh_D + Bh_B - Fh_F - Q_c \qquad (4\text{-}15a)$$

or

$$Q_R = Dh_D + Bh_B - Fh_F + (1 + \frac{L_0}{D})D(H_1 - h_D) \qquad (4\text{-}15b)$$

or

$$Q_R = (\frac{z - x_B}{x_D - x_B})Fh_D + (\frac{x_D - z}{x_D - x_B})Fh_B - Fh_F$$

$$+ (1 + \frac{L_0}{D})(\frac{z - x_B}{x_D - x_B})F(H_1 - h_D) \qquad (4\text{-}16)$$

Q_R will be a positive number.

Example 4-1. External Balances for Binary Distillation

A steady-state, countercurrent, staged distillation column is to be used to separate ethanol from water. The feed is a 30 wt % ethanol, 70 wt % water mixture at 40°C. Flow rate of feed is 10,000 kg/hr. The column operates at a pressure of 1 kg/cm². The reflux is returned as a saturated liquid. A reflux ratio of $L/D = 3.0$ is being used. We desire a bottoms composition of $x_B = 0.05$ (weight fraction ethanol) and a distillate composition of $x_D = 0.80$ (weight fraction ethanol). The system has a total condenser and a partial reboiler. Find D, B, Q_c, and Q_R.

Solution

A. Define. The column and known information are sketched in the following figure.

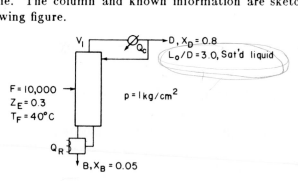

Find D, B, Q_c, Q_R.

B. Explore. Since there are only two unknowns in the mass balances, B and D, we can solve for these variables immediately. Either solve Eqs. (4-1) and (4-2) simultaneously or use Eqs. (4-3) and (4-4). For the energy balances, enthalpies must be determined. These can be read from the enthalpy-composition diagram (Figure 2-4). Then Q_c can be determined from the balance around the condenser and Q_R from the overall energy balance.

C. Plan. Use Eqs. (4-3) and (4-4) to find D and B, Eq. (4-14) to determine Q_c, and Eq. (4-15a) to determine Q_R.

D. Do It. From Eq. (4-3),

$$D = F\left(\frac{z - x_B}{x_D - x_B}\right) = 10,000\left[\frac{0.3 - 0.05}{0.8 - 0.05}\right] = 3333 \text{ kg/hr}$$

From Eq. (4-4), $B = F - D = 10,000 - 3333 = 667$ kg/hr
From Figure 2-4 the enthalpies are

$$h_D(x_D = 0.8, \text{ satd liquid}) = 60 \text{ kcal/kg}$$

$$h_B(x_B = 0.05, \text{ satd liquid}) = 90 \text{ kcal/kg}$$

$$h_f(z = 0.3, 40°C) = 30 \text{ kcal/kg}$$

$$H_1(y_1 = x_D = 0.8, \text{ satd vapor}) = 330 \text{ kcal/kg}$$

From Eq. (4-14),

$$Q_c = \left(1 + \frac{L_0}{D}\right) D(h_D - H_1)$$

$$= (1 + 3)(3333)(60 - 330) = -3,559,640 \text{ kcal/hr}$$

From Eq. (4-15a), $\qquad Q_R = Dh_D + Bh_B - Fh_F - Q_C$

$$Q_R = (3333)(60) + (6667)(90) - (10,000)(30) - (-3,599,640)$$

$$= 4,099,650 \text{ kcal/hr}$$

E. Check. The overall balances, Eqs. (4-1) and (4-5), are satisfied.

F. Generalize. In this case we could solve the mass and energy balances sequentially. This is not always the case. Sometimes the equations must be solved simultaneously (see Problem 4-D3). Also, the mass balances and energy balances derived in the text were for the specific case shown in Figure 4-8. When the column configuration is changed, the mass and energy balances change (see Problems 4-D2, 4-D3 and 4-D5). For binary distillation we can usually determine the external flows and energy requirements from the external balances. Exceptions will be discussed in Chapter 6.

4.5. SUMMARY - OBJECTIVES

In this chapter we have introduced the idea of distillation columns and have seen how to do external balances. At this point you should be able to satisfy the following objectives.

1. Explain physically how a countercurrent distillation column works.

2. Sketch and label the parts of a distillation system. Explain the operation of each part and the flow regime on the trays.

3. Explain the difference between design and simulation problems. List the specifications for typical problems.

4. Write and solve external mass and energy balances for binary distillation systems.

HOMEWORK

A. *Discussion Problems*

A1. Sketch a y-x equilibrium diagram (similar to methanol and water). Refer to Figure 4-1. For an arbitrary feed, determine how close L_2 and V_4 will be to the feed concentration. Explore this by varying the fractions vaporized.

A2. Without looking at the text, define the following.

   a. Isothermal distillation.

   b. The four flow regimes in a staged distillation column.

    c.   Reflux and reflux ratio.

    d.   Boilup and boilup ratio.

    e.   Rectifying (enriching) and stripping sections.

    f.   Simulation and design problems.

   Check the text for definitions you did not know.

A3.   Explain the reasons a constant pressure distillation column is preferable to:

    a.   An isothermal distillation system.

    b.   A cascade of flash separators at constant temperature.

    c.   A cascade of flash separators at constant pressure.

A4.   In a countercurrent distillation column at constant pressure, where is the temperature highest? Where is it lowest?

A5.   Develop your own key relations chart for this chapter. In one page or less draw sketches, write equations, and include all key words you would want for solving problems.

A6.   Explain the difference between sequential and simultaneous solution of the external mass and energy balances.

A7.   What type of specifications will lead to simultaneous solution of the mass and energy balances?

A8.   What are the purposes of reflux? How does it differ from recycle?

A9.   Without looking at the text, name the streams or column parts labeled A to H in the following figure.

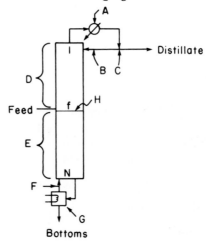

B.   *Generation of Alternatives*

B1.   There are ways in which columns can be specified other than those listed in Tables 4-1 to 4-3.

      a.   Develop alternative specifications for design problems.

      b.   Develop alternative specifications for simulation problems.

C.   *Derivations*

C1.   Derive Eq. (4-3) for the column shown in Figure 4-8.

C2.   Derive Eq. (4-4) for the column shown in Figure 4-8.

C3.   Derive Eq. (4-14) for the column shown in Figure 4-8.

C4.   Derive Eq. (4-16) for the column shown in Figure 4-8.

C5.   For the column shown in Problem 4-D2, derive equations for D, B, Q_c, and L/D.

C6.   For the column shown in Problem 4-D3, derive equations for D, B, $\overline{V}$, and Q_R.

D.   *Problems*

D1.   A steady-state, countercurrent, staged distillation column is to be used to separate ethanol from water. The feed is a mixture of 40 wt % ethanol and 60 wt % water at 20 °C. Flow rate of feed is 20,000 kg/hr. The column operates at a pressure of 1 kg/cm². The reflux is cooled to 40 °C. A reflux ratio of $L/D = 3.5$ is being used. We desire a bottoms composition of $x_B = 0.002$ (weight fraction ethanol) and a distillate composition of $x_D = 0.91$ (weight fraction ethanol). The system has a total condenser and a partial reboiler. Find D, B, Q_c, Q_R. Use Figure 2-4 for data.

D2.   A distillation column separating ethanol from water is shown. Pressure is 1 kg/cm². Instead of having a reboiler, steam (pure water) is injected directly into the bottom of the column to provide heat. The injected steam is a saturated vapor. The feed is 30 wt % ethanol and is at 20 °C. Feed flow rate is 100 kg/min. Reflux is a saturated liquid. We desire a distillate concentration of 60 wt % ethanol and a bottoms product that is 5 wt % ethanol. The steam is input at 100 kg/min. What is the external reflux ratio, L/D?

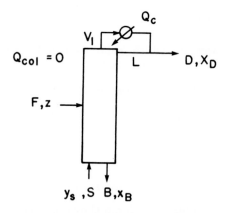

D3.   A distillation column separating ethanol from water is shown. Pressure is 1 kg/cm². Instead of having a condenser, a stream of pure ethanol is added directly to the column to serve as the reflux. This stream is a saturated liquid. The feed is 40 wt % ethanol and is at $-20\,^{\circ}$C. Feed flow rate is 2000 kg/hr. We desire a distillate concentration of 80 wt % ethanol and a bottoms composition of 5 wt % ethanol. A total reboiler is used, and the boilup is a saturated vapor. The cooling stream is input at $C = 1000$ kg/hr. Find the external boilup rate, $\overline{V}$. Note: Set up the equations, solve in equation form for $\overline{V}$ including explicit equations for all required terms, read off all required enthalpies from the enthalpy composition diagram (Figure 2-4), and then calculate a numerical answer.

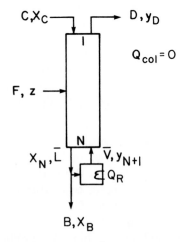

D4.   A mixture of ethanol and water is to be separated in a distillation column. The feed to the column is 1000 lb moles/hr of a 13 mole

% ethanol solution. We desire a distillate product of 80 mole % ethanol and a bottoms product of 0.1 mole % ethanol. The column has a total condenser and a partial reboiler. The column has an external reflux ratio L_0/D of 2.0. The feed is a two-phase mixture at 90 °C. Reflux leaves as a liquid at 60 °C. Pressure is 1 kg/cm².

   a. Find the flow rates of bottoms, B, distillate D, and vapor feed to condenser V_1, in lb moles/hr.

   b. Find the heat loads in the condenser, Q_C, and the reboiler, Q_R, in Btu/hr.

D5. A distillation column is separating ethanol from water at a pressure of 1 kg/cm². A two-phase feed of 20 wt% ethanol at 93 °C is input at 100 kg/min. The column has a total condenser and a partial reboiler. The distillate composition is 90 wt % ethanol. Distillate and reflux are at 20 °C. Bottoms composition is 1 wt % ethanol. Reflux ratio is $L_0/D = 3$. A liquid side stream is withdrawn above the feed stage. Side stream is 70 wt % ethanol, and side stream flow rate is 10 kg/min. Find D, B, q_c, and q_R. Data are in Figure 2-4.

D6. A distillation column receives a feed that is 40 mole % n-pentane and 60 mole % n-hexane. Feed flow rate is 2500 lb moles/hr and feed temperature is 30 °C. The column is at 1 atm. A distillate that is 99.9 mole % n-pentane is desired. A total condenser is used. Reflux is a saturated liquid. The external reflux ratio is $L_0/D = 3$. Bottoms from the partial reboiler is 99.8 mole % n-hexane. Find D, B, Q_R, Q_c. Note: Watch your units on temperature.

Data: λ_{c5} = 11,369 Btu/lb-mole
λ_{c6} = 13,572 Btu/lb-mole
$C_{PL,C5}$ — 39.7 Btu/lb-mole − °F (assume constant)
$C_{PL,C6}$ — 51.7 Btu/lb-mole − °F (assume constant)
$C_{PV,C5}$ — 27.45 + 0.08148 T − 4.538 × 10^{-5} T^2 + 10.1 × 10^{-9} T^3
$C_{PV,C6}$ — 32.85 + 0.09763 T − 5.716 × 10^{-5} T^2 + 13.78 × 10^{-9} T^3

where T is in °C and C_{PV} is $\dfrac{\text{cal}}{\text{g mole °C}}$ or $\dfrac{\text{Btu}}{\text{lb mole °F}}$

F. *Problems Requiring Other Resources*

F1. A mixture of oxygen and nitrogen is to be distilled at low temperature. The feed is 25,000 kg moles/hr and is 21 mole % oxygen and 79 mole % nitrogen. An ordinary column (as shown in Figure 4-8) will be used. Column pressure is 1 atm. The feed is a superheated vapor at 100 K. We desire a bottoms composition of 99.6 mole % oxygen and a distillate that is 99.7 mole % nitrogen. Reflux ratio is $L_0/D = 4$, and reflux is returned as a saturated liquid. Find D, B, Q_R, and Q_c.

F2. A mixture of water and ammonia is to be distilled in an ordinary distillation system (Figure 4-8) at a pressure of 6 kg/cm^2. The feed is 30 wt % ammonia and is at 20 °C. We desire a distillate product that is 98 wt % ammonia and a 95% recovery of the ammonia in the distillate. The external reflux ratio is $L_0/D = 2.0$. Reflux is returned at −20 °C. Find D, B, x_B, Q_R, and Q_c per mole of feed.

F3. A distillation column is separating ammonia from water at a pressure of 10 atm. Feed rate is 1500 lb/hr of a 37 wt % ammonia feed. Feed is at 10 atm pressure and 80 °F. We desire a distillate product that is 95 wt % ammonia and a bottoms product that is 5 wt % ammonia. Reflux is returned to the column from the total condenser as a saturated liquid. The column has a partial reboiler and will be designed so that the optimum feed plate is used. An external reflux ratio of $L_0/D = 1.6$ is employed. Find distillate and bottoms flow rates (in lb/hr) and the heat loads on the reboiler and condenser (in Btu/hr).

chapter 5
COLUMN DISTILLATION: INTERNAL STAGE-BY-STAGE BALANCES

5.1. INTERNAL BALANCES

In Chapter 4 we introduced column distillation and developed the external balance equations. In this chapter we start looking inside the column. For binary systems the number of stages required for the separation can conveniently be obtained by use of stage-by-stage balances. We start at the top of the column and write the balances and equilibrium relationship for the first stage, and then once we have determined the unknown variables for the first stage we write balances for the second stage. Utilizing the variables just calculated, we can again calculate the unknowns. We can now proceed down the column in this stage-by-stage fashion until we reach the bottom. We could also start at the bottom and proceed upwards. This procedure assumes that each stage is an equilibrium stage, but this assumption may not be true. Ways to handle nonequilibrium stages are discussed in Chapter 6.

In the enriching section of the column it is convenient to use a balance envelope that goes around the desired stage and around the condenser. This is shown in Figure 5-1. For the first stage the balance envelope is shown in Figure 5-1A. The overall mass balance is then

$$V_2 = L_1 + D \qquad \text{(5-1, stage 1)}$$

The more volatile component mass balance is

$$V_2 y_2 = L_1 x_1 + D x_D \qquad \text{(5-2, stage 1)}$$

For a well-insulated, adiabatic column, the energy balance is

$$V_2 H_2 + Q_c = L_1 h_1 + D h_D \qquad \text{(5-3, stage 1)}$$

Assuming that each stage is an equilibrium stage, we know that the

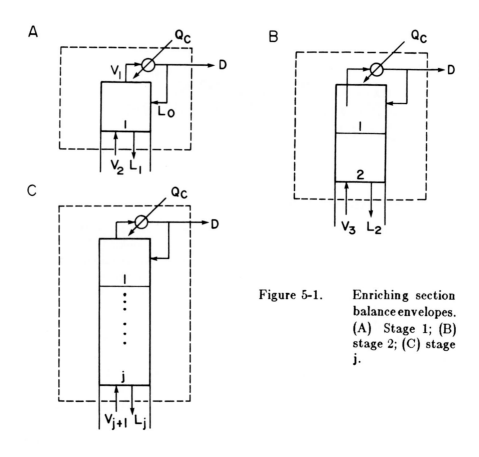

Figure 5-1.   Enriching section balance envelopes. (A) Stage 1; (B) stage 2; (C) stage j.

liquid and vapor leaving the stage are in equilibrium. For a binary system, the Gibbs phase rule becomes

$$\text{Degrees of freedom} = C - P + 2 = 2 - 2 + 2 = 2$$

Since pressure has been set, there is one remaining degree of freedom. Thus for the equilibrium stage the variables are all functions of a single variable. For the saturated liquid we can write

$$h_1 = h_1(x_1) \qquad \text{(5-4a, stage 1)}$$

and for the saturated vapor,

$$H_2 = H_2(y_2) \qquad \text{(5-4b, stage 1)}$$

The liquid and vapor mole fractions leaving a stage are also related:

$$x_1 = x_1(y_1) \qquad \text{(5-4c, stage 1)}$$

Equations (5-4) for stage 1 represent the equilibrium relationship. Their exact form depends on the chemical system being separated. Equations (5-1, stage 1) to (5-4c, stage 1) are six equations with six unknowns: L_1, V_2, x_1, y_2, H_2, and h_1.

Since we have six equations and six unknowns, we can solve for the six unknowns. The exact methods for doing this are the subjects of the remainder of this chapter and Chapter 6. For now we will just note that we can solve for the unknowns and then proceed to the second stage. For the second stage we use the balance envelope shown in Figure 5-1B. The mass balances are now

$$V_3 = L_2 + D \qquad \text{(5-1, stage 2)}$$

$$V_3 y_3 = L_2 x_2 + D x_D \qquad \text{(5-2, stage 2)}$$

while the energy balance is

$$Q_c + V_3 H_3 = L_2 h_2 + D h_D \qquad \text{(5-3, stage 2)}$$

The equilibrium relationships are

$$h_2 = h_2(x_2), \quad H_3 = H_3(y_3), \quad x_2 = x_2(y_2) \qquad \text{(5-4, stage 2)}$$

Again we have six equations with six unknowns. The unknowns are now L_2, V_3, x_2, y_3, H_3, and h_2.

We can now proceed to the third stage and utilize the same procedures. After that, we can go to the fourth stage and then the fifth stage and so forth. For a general stage j (j can be from 1 to f − 1) in the enriching section, the balance envelope is shown in Figure 1C. For this stage the mass and energy balances are

$$V_{j+1} = L_j + D \qquad \text{(5-1, stage j)}$$

$$V_{j+1} y_{j+1} = L_j x_j + D x_D \qquad \text{(5-2, stage j)}$$

and

$$Q_c + V_{j+1}H_{j+1} = L_jh_j + Dh_D \qquad \text{(5-3, stage j)}$$

while the equilibrium relationships are

$$h_j = h_j(x_j) , \quad H_{j+1} = H_{j+1}(y_{j+1}) , \quad x_j = x_j(y_j) \quad \text{(5-4, stage j)}$$

When we reach stage j, the values of y_j, Q_c, D, and h_D will be known, and the unknown variables will be L_j, V_{j+1}, x_j, y_{j+1}, H_{j+1}, and h_j. At the feed stage, the mass and energy balances will change because of the addition of the feed stream.

Before continuing, we will stop to note the symmetry of the mass and energy balances and the equilibrium relationships as we go from stage to stage. A look at Eqs. (5-1) for stages 1, 2, and j will show that these equations all have the same structure and differ only in subscripts. Equations (5-1, stage 1) or (5-1, stage 2) can be obtained from the general equation (5-1, stage j) by replacing j with 1 or 2, respectively. The same observations can be made for the other equations (5-2, 5-3, 5-4a, 5-4b, and 5-4c). The unknown variables as we go from stage to stage are also similar and differ in subscript only.

In addition to this symmetry from stage to stage, there is a symmetry between equations for the same stage. Thus Eqs. (5-1, stage j), (5-2, stage j), and (5-3, stage j) are all steady-state balances that state

$$\text{Input} = \text{output}$$

In all three equations the output (of overall mass, solute, or energy) is associated with streams L_j and D. The input is associated with stream V_{j+1} and (for energy) with the cooling load, Q_c.

Below the feed stage the balance equations must change, but the equilibrium relationships in Eqs. (5-4a, b, c) will be unchanged. The balance envelopes in the stripping section are shown in Figure 5-2 for a column with a partial reboiler. The bars over flow rates signify that they are in the stripping section. It is traditional and simplest to write the stripping section balances around the bottom of the column using the balance envelope shown in Figure 5-2. Then these balances around stage $f+1$ (immediately below the feed plate) are

$$\overline{V}_{f+1} = \overline{L}_f - B \qquad \text{(5-5, stage } f+1\text{)}$$

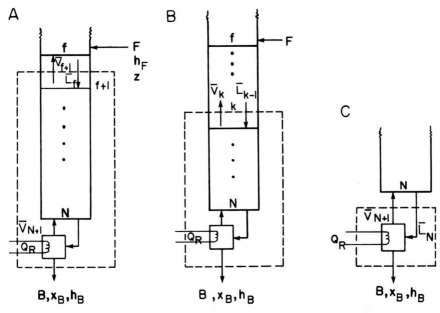

Figure 5-2.   Stripping section balance envelopes. (A) Below feed stage (stage $f + 1$); (B) Stage k; (C) Partial reboiler.

$$\overline{V}_{f+1}y_{f+1} = \overline{L}_f x_f - Bx_B \qquad \text{(5-6, stage } f+1)$$

$$\overline{V}_{f+1}H_{f+1} = \overline{L}_f h_F - Bh_B + Q_R \qquad \text{(5-7, stage } f+1)$$

The equilibrium relationships are Eqs. (5-4) written for stage $f+1$.

$$h_f = h_f(x_f), \quad H_{f+1} = H_{f+1}(y_{f+1}), \quad x_f = x_f(y_f) \qquad \text{(5-4, stage } f+1)$$

These six equations have six unknowns: $\overline{L}_f$, $\overline{V}_{f+1}$, x_f, y_{f+1}, H_{f+1}, and h_f. x_B was specified in the problem statement, B and Q_R were calculated from the column balances, and y_f (required for the last equation) was obtained from the solution of Eqs. (5-1, stage j) to (5-4c, stage j) with j $= f - 1$. At the feed stage we change from one set of balance envelopes to another.

Note that the same equations will be obtained if we write the balances above stage $f+1$ and around the top of the distillation column (use a different balance envelope). This is easily illustrated with the overall mass balance which is now

$$\overline{V}_{f+1} + F = D + \overline{L}_f$$

Rearranging, we have

$$\overline{V}_{f+1} = \overline{L}_f - (F - D)$$

However, since the external column mass balance says $F - D = B$, the last equation becomes

$$\overline{V}_{f+1} = \overline{L}_f - B$$

which is Eq. (5-5, stage $f+1$). Similar results are obtained for the other balance equations.

Once the six equations (5-4a) to (5-7) for stage $f+1$ have been solved, we can proceed down the column to the next stage, $f+2$. For a balance envelope around general stage k as shown in Figure 5-2B, the equations are

$$\overline{V}_k = \overline{L}_{k-1} - B \qquad \text{(5-5, stage k)}$$

$$\overline{V}_k y_k = \overline{L}_{k-1} x_{k-1} - B x_B \qquad \text{(5-6, stage k)}$$

$$\overline{V}_k H_k = \overline{L}_{k-1} h_{k-1} - B h_B + Q_R \qquad \text{(5-7, stage k)}$$

the equilibrium expressions will correspond to Eqs. (5-4, stage $f+1$) with $k - 1$ replacing f as a subscript. Thus,

$$h_{k-1} = h_{k-1}(x_{k-1}) , \quad H_k = H_k(y_k) , \quad x_{k-1} = x_{k-1}(y_{k-1}) \qquad \text{(5-4, stage k)}$$

A partial reboiler as shown in Figure 5-2C acts as an equilibrium contact. If we consider the reboiler as stage $N+1$, the balances for the envelope shown in Figure 5-2C can be obtained by setting $k = N+1$ and $k - 1 = N$ in Eqs. (5-5, stage k), (5-6, stage k) and (5-7, stage k).

If there were additional stages, the liquid leaving this stage, x_{N+1}, would be in equilibrium with y_{N+1}. If $x_{N+1} = x_B$, then $N+1$ equilibrium contacts gives us exactly the specified separation, and the problem is finished. If $x_{N+1} < x_B$ while $x_N > x_B$, then $N+1$ equilibrium contacts give slightly more separation than is required.

Just as the balance equations in the enriching section are symmetric from stage to stage, they are also symmetric in the stripping section.

5.2. BINARY STAGE-BY-STAGE SOLUTION METHODS

The challenge for any stage-by-stage solution method is to solve the three balance equations and the three equilibrium relationships simultaneously in an efficient manner. This problem was first solved by Sorel (1893), and graphical solutions of Sorel's method were developed independently by Ponchon (1921) and Savarit (1922). These methods all solve the complete mass and energy balance and equilibrium relationships stage by stage. Starting at the top of the column as shown in Figure 5-1a, we can find the liquid composition, x_1, in equilibrium with the leaving vapor composition, y_1, from Eq. (5-4c, stage 1). The liquid enthalpy, h_1, is easily found from Eqs. (5-4a, stage 1). The remaining four equations (5-1) to (5-3) and (5-4b) for stage 1 are coupled and must be solved simultaneously. The Ponchon-Savarit method does this graphically. The Sorel method uses a trial-and-error procedure on each stage.

The trial-and-error calculation on every stage of the Sorel method is obviously slow and laborious. Lewis (1922) noted that in many cases the molar vapor and liquid flow rates in each section (a region between input and output ports) were constant. Thus in Figures 5-1 and 5-2,

$$L_1 = L_2 = \cdots = L_j = \cdots = L_{j-1} = L$$

$$V_1 = V_2 = \cdots = V_{j+1} = \cdots = V_f = V \tag{5-8}$$

and

$$\overline{L}_f = \overline{L}_{f+1} = \cdots = \overline{L}_{k-1} = \cdots = \overline{L}_N = \overline{L}$$

$$\overline{V}_{f+1} = \overline{V}_{f+2} = \cdots = \overline{V}_k = \cdots = \overline{V}_{N+1} = \overline{V} \tag{5-9}$$

For each additional column section there will be another set of equations for constant flow rates. Note that in general $L \neq \overline{L}$ and $V \neq \overline{V}$. Equations (5-8) and (5-9) will be valid if every time a mole of vapor is condensed a mole of liquid is vaporized. This will occur if:

1. The column is adiabatic.

2. The specific heat changes are negligible compared to latent heat changes.

3. The heat of vaporization per mole, λ, is constant; that is, if λ does not depend on concentration.

Condition 3 is the most important criterion. Lewis called this set of conditions *constant molal overflow* (CMO). An alternative to conditions 2 and 3 is

4. The saturated liquid and vapor lines on an enthalpy-composition diagram (in molar units) are parallel.

For some systems, such as hydrocarbons, the latent heat of vaporization per kilogram is constant. Then the mass flow rates are constant, and constant *mass* overflow should be used.

The Lewis method assumes before the calculation is done that constant molal overflow is valid. Thus Eqs. (5-8) and (5-9) are valid. With this assumption, the energy balance, Eqs. (5-3) and (5-7), will be automatically satisfied. Then only Eqs. (5-1), (5-2), and (5-4c), or (5-5), (5-6), and (5-4c) need be solved. Equations (5-1, stage j) and (5-2, stage j) can be combined. Thus,

$$V_{j+1}y_{j+1} = L_jx_j + (V_{j+1} - L_j)x_D \qquad (5\text{-}10)$$

Solving for y_{j+1}, we have

$$y_{j+1} = \frac{L_j}{V_{j+1}}x_j + \left(1 - \frac{L_j}{V_{j+1}}\right)x_D \qquad (5\text{-}11)$$

Since L and V are constant, this equation becomes

$$y_{j+1} = \frac{L}{V}x_j + \left(1 - \frac{L}{V}\right)x_D \qquad (5\text{-}12)$$

Equation (5-12) is the *operating equation* in the enriching section. It relates the concentrations of two passing streams in the column and thus represents the mass balances in the enriching section. Equation (5-12) is solved sequentially with the equilibrium expression for x_j, which is Eq. (5-4c, stage j).

To start we first use the column balances to calculate D and B. Then $L_0 = (L_0/D)D$ and $V_1 = L_0 + D$. For a saturated liquid reflux, $L_0 = L_1 = L_2 = L$ and $V_1 = V_2 = V$. At the top of the column we know that $y_1 = x_D$. The vapor leaving the top stage is in equilibrium with the liquid leaving the bottom stage (see Figure 5-1a). Thus x_1 can be calculated from Eq. (5-4c, stage j) with $j = 1$. Then y_2 is found from Eq. (5-12) with $j = 1$. We then proceed to the second stage, set $j = 2$, and obtain x_2 from Eq. (5-4c, stage j) and y_3 from Eq. (5-12). We continue this procedure down to the feed stage.

In the stripping section, Eqs. (5-5, stage k) and (5-6, stage k) are combined to give

$$y_k = \frac{\overline{L}_{k-1}}{\overline{V}_k} x_{k-1} - (\frac{\overline{L}_{k-1}}{\overline{V}_k} - 1)x_B \tag{5-13}$$

With constant molal overflow, $\overline{L}$ and $\overline{V}$ are constant, and the resulting stripping section operating equation is

$$y_k = \frac{\overline{L}}{\overline{V}} x_{k-1} - (\frac{\overline{L}}{\overline{V}} - 1)x_B \tag{5-14}$$

Once we know $\overline{L}/\overline{V}$ we can obviously alternate between the operating equation (5-14) and the equilibrium equation (5-4c, stage k).

The phase and temperature of the feed obviously affect the vapor and liquid flow rates in the column. For instance, if the feed is liquid, the liquid flow rate below the feed stage must be greater than liquid flow above the feed stage, $\overline{L} > L$. If the feed is a vapor, $V > \overline{V}$. These effects can be quantified by writing mass and energy balances around the feed stage. The feed stage is shown schematically in Figure 5-3. The overall mass balance and the energy balance for the balance envelope shown in Figure 5-3 are

$$F + \overline{V} + L = \overline{L} + V \tag{5-15}$$

and

$$Fh_F + \overline{V}H_{f+1} + Lh_{f-1} = \overline{L}h_f + VH_f \tag{5-16}$$

(Despite the use of "h_F" as the symbol for the feed enthalpy, the feed can be a liquid or vapor or a two-phase mixture.) If we assume constant molal overflow neither the vapor enthalpies nor the liquid enthalpies vary much from stage to stage. Thus $H_{f+1} \sim H_f$ and $h_{f-1} \sim h_f$. Then Eq. (5-16) can be written as

$$Fh_F + (\overline{V} - V)H = (\overline{L} - L)h$$

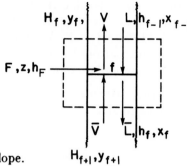

Figure 5-3. Feed-stage balance envelope.

The mass balance equation (5-15) can be conveniently solved for $\bar{V} - V$,

$$\bar{V} - V = \bar{L} - L - F$$

which can be substituted into the energy balance to give us

$$Fh_F + (\bar{L} - L)H - FH = (\bar{L} - L)h$$

Combining terms, this is

$$(\bar{L} - L)(H - h) = F(H - h_F)$$

or

$$q \equiv \frac{\bar{L} - L}{F} = \frac{H - h_F}{H - h} \tag{5-17}$$

In words, the "quality" q is

$$q \equiv \frac{\text{liquid flow rate below feed} - \text{liquid flow rate above feed}}{\text{feed rate}}$$

$$q = \frac{\text{vapor enthalpy on feed plate} - \text{feed enthalpy}}{\text{vapor enthalpy on feed plate} - \text{liquid enthalpy on feed plate}} \tag{5-18}$$

This result is analogous to the equation for an adiabatic flash distillation system [see Eq. (3-47)] but for the case where L/F is the variable, not V/F. Since the liquid and vapor enthalpies can be estimated, we can easily calculate q from Eq. (5-17). Then

$$\bar{L} = L + qF \tag{5-19}$$

The quality q is the fraction of feed that is liquid. For example, if the feed is a saturated liquid, $h_F = h$, $q = 1$, and $\bar{L} = L + F$. Once $\bar{L}$ has been determined, $\bar{V}$ is calculated from either Eq. (5-15) or Eq. (5-5, stage $f + 1$) or from

$$\bar{V} = V - (1 - q)F \tag{5-20}$$

which can be derived from Eqs. (5-15) and (5-19).

Example 5-1. Stage-by-Stage Calculation by the Lewis Method

A steady-state, countercurrent, staged distillation column is to be used to separate ethanol from water. The feed is a 30 wt % ethanol, 70 wt % water mixture that is a saturated liquid at 1 atm pressure. Flow rate of feed is 10,000 kg/hr. The column operates at a pressure of 1 atm. The reflux is returned as a saturated liquid. A reflux ratio of $L/D = 3.0$ is being used. We desire a bottoms composition of $x_B = 0.05$ (weight fraction ethanol) and a distillate composition of $x_D = 0.80$ (weight fraction ethanol). The system has a total condenser and a partial reboiler. The column is well insulated.

Use the Lewis method to find the number of equilibrium contacts required if the feed is input on the second stage from the top.

Solution

A. Define. The column and known information are shown in the following figure.

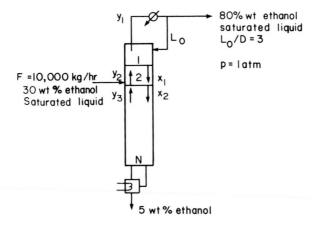

Find the number of equilibrium contacts required.

B. Explore. Except for some slight changes in the feed temperature and column pressure, this problem is very similar to Example 4-1. The solution for B and D obtained in that example is still correct. B = 6667 kg/hr, D = 3333 kg/hr. Equilibrium data are available in weight fractions in Figure 2-4 and in mole fraction units in Figure 2-2 and Table 2-1. To use the Lewis method we must have constant molal overflow. We can check this by comparing the latent heat per mole of pure ethanol and pure water. (This checks the third and most important criterion for CMO. Since the column is well insulated, the first criterion, adiabatic, will be satisfied.) The latent heats are (Himmelblau, 1974):

$$\lambda_E = 9.22 \text{ kcal/g-mole}, \qquad \lambda_W = 9.7171 \text{ kcal/g-mole}$$

This difference of roughly 5% is reasonable, particularly since we always use the ratio of L/V or $\bar{L}/\bar{V}$. (Using the ratio causes some of the change in L and V to divide out.) Thus we will assume CMO. Now we must convert flows and compositions to molar units.

C. Plan. First, convert to molar units. Carry out preliminary calculations to determine L/V and $\bar{L}/\bar{V}$. Then start at the top, alternating between equilibrium (Figure 2-2) and the top operating equation (5-12). Since stage 2 is the feed stage, calculate y_3 from the bottom operating equation (5-14).

D. Do It. *Preliminary Calculations:* To Convert to Molar Units:

$$MW_W = 18, \qquad MW_E = 46, \qquad z_E = \frac{0.3/46}{0.3/46 + 0.7/18} = 0.144$$

$$x_{D,E} = \frac{0.8/46}{0.8/46 + 0.2/18} = 0.61, \qquad x_{B,E} = 0.02$$

For distillate, the average molecular weight is

$$\overline{MW} = (0.61)(46) + (0.39)(18) = 35.08$$

which is also the average for the reflux liquid and vapor stream V.

then $\quad$ D = (3333 kg/hr)/35.08 = 95.23 kg moles/hr

and $\quad$ L = $(\frac{L}{D})$D = (3)(95.23) = 285.69 kg moles/hr

while $\quad$ V = L + D = 380.92

$$\frac{L}{V} = \frac{285.69}{380.92} = 0.75$$

Because of CMO, L/V is constant in the rectifying section.

Since the feed is a saturated liquid,

$\overline{L}$ = L + F = 285.69 + 454.1 = 739.79 kg moles/hr

where we have converted F to kg moles/hr. Since a saturated liquid feed does not affect the vapor, $\quad$ V = $\overline{V}$ = 380.92. Thus,

$$\frac{\overline{L}}{\overline{V}} = \frac{739.79}{380.92} = 1.942$$

An internal check on consistency is L/V < 1 and $\overline{L}/\overline{V}$ > 1.

Stage-by-Stage Calculations: At the top of the column, $y_1 = x_D = 0.61$. Liquid stream L_1 of concentration x_1 is in equilibrium with the vapor stream y_1. From Figure 2-2, $x_1 = 0.4$. (Note that $y_1 > x_1$ since ethanol is the more volatile component.) Vapor stream y_2 is a passing stream relative to x_1 and can be determined from the operating equation (5-12).

$$y_2 = \frac{L}{V} x_1 + (1 - \frac{L}{V}) x_D = (0.75)(0.4) + (0.25)(0.61) = 0.453$$

Stream x_2 is in equilibrium with y_2. From Figure 2-2 we obtain $x_2 = 0.11$.

Since stage 2 is the feed stage, use bottom operating equation (5-14) for y_3.

$$y_3 = \frac{\overline{L}}{\overline{V}} x_2 + (1 - \frac{\overline{L}}{\overline{V}}) x_B = (1.942)(0.11) + (-0.942)(0.02) = 0.195$$

Stream x_3 is in equilibrium with y_3. From Figure 2-2, this is $x_3 = 0.02$. Since $x_3 = x_B$ (in mole fractions), we are finished.

The third equilibrium contact would be the partial reboiler. Thus the column has two equilibrium stages plus the partial reboiler.

E. Check. This is a small number of stages. However, not much separation is required, the external reflux ratio is large, and the separation of ethanol from water is easy in this concentration range. Thus the answer is reasonable. We can check the calculation of L/V with mass balances.

Since $V_1 = L_0 + D$,

$$\frac{L_0}{V_1} = \frac{L_0}{D+L_0} = \frac{L_0/D}{\dfrac{D}{D} + \dfrac{L_0}{D}} = \frac{L_0/D}{1 + \dfrac{L_0}{D}} = \frac{3}{4}$$

Since L_0, V_1, and D, are the same composition, L_0/D and L_0/V_1 have the same values in mass and molar units.

F. Generalizations. We should always check that constant molal overflow is valid. Then convert all flows and compositions into molar units. The procedure for stepping off stages is easily programmed on a calculator or computer as long as the equilibrium data are in equation form. We could also have started at the bottom and worked our way up the column stage by stage. Going up the column we calculate y values from equilibrium and x values from the operating equations.

Note that $L/V < 1$ and $\overline{L}/\overline{V} > 1$. This makes sense, since we must have a net flow of material upwards in the rectifying section (to obtain a distillate product) and a net flow downwards in the stripping section. We must also have a net upward flow of ethanol in the rectifying section ($Lx_j < Vy_{j+1}$) and in the stripping section ($\overline{L}x_j < \overline{V}y_{j+1}$). These conditions are satisifed by all pairs of passing streams.

The Lewis method is obviously much faster and more convenient than the Sorel method. It is also easier to program on a computer or programmable calculator. In addition, it is easier to understand the physical reasons why separation occurs instead of becoming lost in the algebraic details. However, remember that the Lewis method is based on the assumption of constant molal overflow. If CMO is not valid, the answers will be incorrect.

If the calculation procedure in the Lewis method is confusing to you, continue on to the next section. The graphical McCabe-Thiele procedure explained there is easier for many students to understand. After completing the McCabe-Thiele procedure, return to this section and study the Lewis method again.

5.3. INTRODUCTION TO THE McCABE-THIELE METHOD

McCabe and Thiele (1925) developed a graphical solution method based on Lewis's method and the observation that the operating equations (5-12) and (5-14) plot as straight lines (the *operating lines)* on a y-x diagram. On this graph the equilibrium relationship can be solved from the y-x equilibrium curve and the mass balances from the operating lines.

To illustrate, consider a typical design problem for a binary distillation column such as the one illustrated in Figure 4-8. We will assume that equilibrium data are available at the operating pressure of the column. These data are plotted as shown in Figure 5-4. At the top of the column is a total condenser. As noted in Chapter 4 in Eq. (4-7), this means that $y_1 = x_D = x_0$. The vapor leaving the first stage is in equilibrium with the liquid leaving the first stage. This liquid composition, x_1, can be determined from the equilibrium curve at $y = y_1$. This is illustrated in Figure 5-4.

Liquid stream L_1 of composition x_1 passes vapor stream V_2 of composition y_2 inside the column. When the mass balances are written around

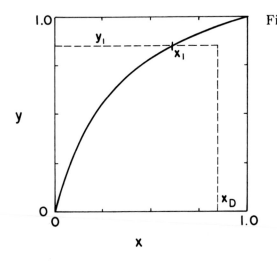

Figure 5-4. Equilibrium for top stage on McCabe-Thiele diagram.

stage 1 and the top of the column (see balance envelope in Figure 5-1a), the result after assuming constant molal overflow and doing some algebraic manipulations is Eq. (5-12) with j = 1. This equation can be plotted as a straight line on the y-x diagram. Suppressing the subscripts j+1 and j, we write Eq. (5-12) as

$$y = \frac{L}{V} x + \left(1 - \frac{L}{V}\right) x_D \qquad (5\text{-}21)$$

which is understood to apply to passing streams. Equation (5-21) plots as a straight line (the *top operating line*) with a slope of L/V and a y intercept (x = 0) of $(1 - L/V)x_D$. Once Eq. (5-21) has been plotted, y_2 is easily found from the y value at x = x_1. This is illustrated in Figure 5-5. Note that the top operating line goes through the point (y_1, x_D) since these coordinates satisfy Eq. (5-21).

With y_2 known we can proceed down the column. Since x_2 and y_2 are in equilibrium, we easily obtain x_2 from the equilibrium curve. Then we obtain y_3 from the operating line (mass balances), since x_2 and y_3 are the compositions of passing streams. This procedure of *stepping off stages* is shown in Figure 5-6. It can be continued as long as we are in the rectifying section. Note that this produces a staircase on the y-x, or McCabe-Thiele, diagram. Instead of memorizing this procedure, you should follow the points on the diagram and compare them to the schematics of a distillation column (Figures 4-8 and 5-1). Note that the horizontal and vertical lines have no physical meaning. The points on

Figure 5-5.   Stage 1 calculation on McCabe-Thiele diagram.

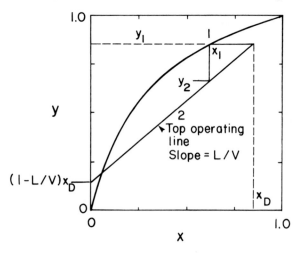

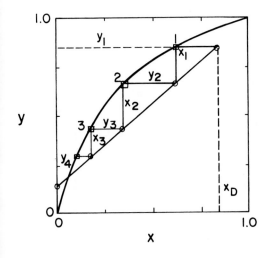

Figure 5-6. Stepping off stages in rectifying section.

the equilibrium curve (squares) represent liquid and vapor streams leaving an equilibrium stage. The points on the operating line (circles) represent the liquid and vapor streams passing each other in the column.

In the stripping section the top operating line is no longer valid, since different mass balances and hence a different operating equation are required. The stripping section operating equation was given in Eq. (5-14). When the subscripts k and k − 1 are suppressed, this equation becomes

$$y = \frac{\overline{L}}{\overline{V}} x - (\frac{\overline{L}}{\overline{V}} - 1) x_B \qquad (5-22)$$

Equation (5-22) plots as a straight line with slope $\overline{L}/\overline{V}$ and y intercept $-(\overline{L}/\overline{V} - 1)x_B$, as shown in Figure 5-7. This *bottom operating line* applies to passing streams in the stripping section. Starting with the liquid leaving the partial reboiler, of mole fraction $x_B = x_{N+1}$, we know that the vapor leaving the partial reboiler is in equilibrium with x_B. Thus we can find y_{N+1} from the equilibrium curve. x_N is easily found from the bottom operating line, since liquid of composition x_N is a passing stream to vapor of composition y_{N+1} (compare Figures 5-2 and 5-7). We can continue alternating between the equilibrium curve and the bottom operating line as long as we are in the stripping section.

If we are stepping off stages down the column, at the feed stage f we switch from the top operating line to the bottom operating line (refer to Figure 5-3, a schematic of the feed stage). Above the feed stage, we calculate x_{f-1} from equilibrium and y_f from the top operating line. Since

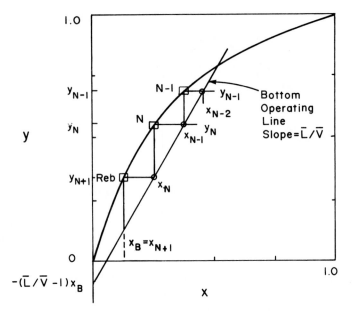

Figure 5-7.   Stepping off stages in stripping section.

liquid and vapor leaving the feed stage are assumed to be in equilibrium, we can determine x_f from the equilibrium curve at $y = y_f$ and then find y_{f+1} from the bottom operating line. This procedure is illustrated in Figure 5-8A, where stage 3 is the feed stage. The separation shown in Figure 5-8A would require 5 equilibrium stages plus an equilibrium partial reboiler, or 6 equilibrium contacts, when stage 3 is used as the feed stage. In this problem, stage 3 is the *optimum feed stage*. That is, a separation will require the fewest total number of stages when feed stage 3 is used. Note in Figure 5-8B and 5-8C that if stage 2 or stage 5 is used, more total stages are required. For binary distillation the optimum feed plate is easy to determine; it will always be the stage where the step in the staircase includes the point of intersection of the two operating lines (compare Figure 5-8A to Figures 5-8B and C). A mathematical analysis of the optimum feed plate location suitable for computer calculation with the Lewis method is developed later.

The fractional number of stages can be calculated as

$$\text{Fraction} = \frac{\text{distance from operating line to } x_B}{\text{distance from operating line to equilibrium curve}} \qquad (5\text{-}23)$$

Now that we have seen how to do the stage-by-stage calculations on

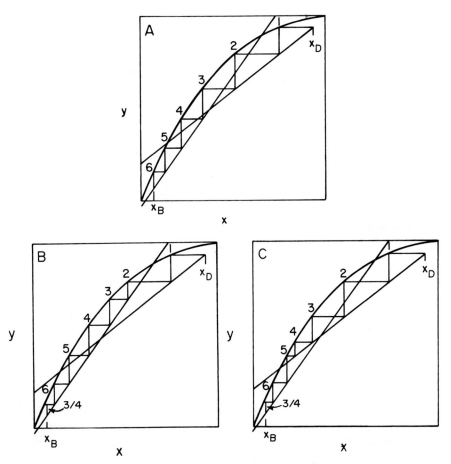

Figure 5-8.   McCabe-Thiele diagram for entire column. (A) Optimum feed stage (stage 3); (B) Feed stage too high (stage 2); (C) Feed stage too low (stage 5).

a McCabe-Thiele diagram, let us consider how to start with the design problem given in Figure 4-8 and Tables 4-1 and 4-2. The known variables are F, z, q, x_D, x_B, L_0/D, p, there is saturated liquid reflux, and we use the optimum feed location. Since the reflux is a saturated liquid, then there will be no change in the liquid or vapor flow rates on stage 1. Thus $L_0 = L_1$ and $V_1 = V_2$. This allows us to calculate the internal reflux ratio, L/V, from the external reflux ratio, L_0/D, which is specified.

$$\frac{L}{V} = \frac{L}{L+D} = \frac{L/D}{L/D+1} \tag{5-24}$$

With L/V and x_D known, the top operating line is fully specified and can be plotted.

Since the boilup ratio, $\overline{V}/B$, was not specified, we cannot directly calculate $\overline{L}/\overline{V}$, which is the slope of the bottom operating line. Instead, we need to utilize the condition of the feed to determine flow rates in the stripping section. The same procedure used with the Lewis method can be used here. The feed quality, q, is calculated from Eq. (5-17), which is repeated below:

$$q \equiv \frac{\overline{L} - L}{F} = \frac{H_v - h_F}{H_v - h_L} \tag{5-17}$$

Then $\overline{L}$ is given by Eq. (5-19), $\overline{L} = L + qF$, and $\overline{V} = \overline{L} - B$. We can calculate L as $(L/D)D$, where D and B are found from mass balances around the entire column. Alternatively, for a simple column Eqs. (4-3) and (4-4) can be substituted into the equations for $\overline{L}$ and $\overline{V}$. When this is done, we obtain

$$\frac{\overline{L}}{\overline{V}} = \frac{\dfrac{L_0}{D}(z - x_B) + q(x_D - x_B)}{\dfrac{L_0}{D}(z - x_B) + q(x_D - x_B) - (x_D - z)} \tag{5-25}$$

With $\overline{L}/\overline{V}$ and x_B known, the bottom operating equation is fully specified, and the bottom operating line can be plotted. Equation (5-25) is convenient for computer calculations but is specific for the simple column shown in Figure 4-8. For graphical calculations the alternative procedure shown in the next section is usually employed.

5.4. FEED LINE

In any section of the column between feeds and/or product streams the mass balances are represented by the operating line. In general, the operating line can be derived by drawing a mass balance envelope through an arbitrary stage in the section and around the top or bottom of the column. When material is added or withdrawn from the column the mass balances will change and the operating lines will have different slopes and intercepts. In the previous section the effect of a feed on the operating lines was determined from the feed quality and mass balances around the entire column or from Eq. (5-24). Here we will develop a graphical method for determining the effect of a feed on the operating lines.

Consider the simple single-feed column with a total condenser and a partial reboiler shown in Figure 4-8. The mass balance in the rectifying section for the more volatile component is

$$yV = Lx + Dx_D \tag{5-26}$$

while the balance in the stripping section is

$$y\overline{V} = \overline{L}x - Bx_B \tag{5-27}$$

where we have assumed that constant molal overflow is valid. At the feed plate we switch from one mass balance to the other. We wish to find the point at which the top operating line [representing Eq. (5-26)] intersects the bottom operating line [representing Eq. (5-27)].

The intersection of these two lines means that

$$y_{\text{top op}} = y_{\text{bot op}}, \quad x_{\text{top op}} = x_{\text{bot op}} \tag{5-28}$$

Equations (5-28) are valid only at the point of intersection. Since the y's and x's are equal at the point of intersection, we can subtract Eq. (5-26) from Eq. (5-27) and obtain

$$y(\overline{V} - V) = (\overline{L} - L)x - (Dx_D + Bx_B) \tag{5-29}$$

From the overall mass balance around the entire column, we know that the last term is $-Fz_F$. Then, solving Eq. (5-29) for y,

$$y = -\left(\frac{\overline{L} - L}{V - \overline{V}}\right)x + \frac{Fz_F}{V - \overline{V}} \tag{5-30}$$

Equation (5-30) is one form of the feed equation. Since $\overline{L}$, L, $\overline{V}$, V, F, and z_F are constant, it represents a straight line (the feed line) on a McCabe-Thiele diagram. Every possible intersection point of the two operating lines *must* occur on the feed line.

For the special case of a feed that flashes in the column to form a vapor and a liquid phase, we can relate Eq. (5-30) to flash distillation. In this case we have the situation shown in Figure 5-9. Part of the feed, V_F, vaporizes, while the remainder is liquid, L_F. Looking at the terms in Eq. (5-30), we note that $\overline{L} - L$ is the change in liquid flow rates at the feed stage. In this case,

$$\overline{L} - L = L_F \tag{5-31}$$

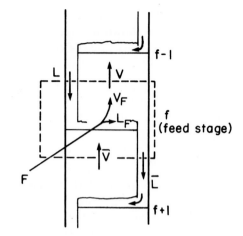

Figure 5-9. Two-phase feed.

The change in vapor flow rates is

$$V - \overline{V} = V_F \qquad (5\text{-}32)$$

Equation (5-30) then becomes

$$y = -\frac{L_F}{V_F}x + \frac{F}{V_F}z_F \qquad (5\text{-}33)$$

which is the same as Eq. (3-9), the operating equation for flash distillation. Thus the feed line represents the flashing of the feed into the column. Equation (5-33) can also be written in terms of the fraction vaporized, $f = V_F/F$, as [see Eqs. (3-10) and (3-11)]

$$y = -\frac{1-f}{f}x + \frac{1}{f}z_F \qquad (5\text{-}34)$$

In terms of the fraction remaining liquid, $q = L_F/F$ [see Eqs. (3-12) and (3-13)], Eq. (5-33) is

$$y = \frac{q}{q-1}x + \frac{1}{1-q}z_F \qquad (5\text{-}35)$$

Equations (5-33) through (5-35) were all derived for the special case where the feed is a two-phase mixture, but they can be used for any type of feed. For example, if we want to derive Eq. (5-35) for the general case we can start with Eq. (5-30). An overall mass balance around the feed stage (balance envelope shown in Figure 5-9) is

$$F + \overline{V} + L = V + \overline{L}$$

which can be rearranged to

$$V - \overline{V} = F - (\overline{L} - L)$$

Substituting this result into Eq. (5-30) gives

$$y = - \frac{\overline{L} - L}{F - (\overline{L} - L)} \, x + \frac{F z_F}{F - (\overline{L} - L)}$$

and dividing numerator and denominator of each term by the feed rate F, we get

$$y = - \frac{(\overline{L} - L)/F}{F/F - (\dfrac{\overline{L} - L}{F})} \, x + \frac{F/F \, z_F}{F/F - (\dfrac{\overline{L} - L}{F})}$$

which becomes Eq. (5-35), since q is defined to be $(\overline{L} - L)/F$. Equation (5-34) can be derived in a similar fashion (obviously, another homework problem).

Previously, we solved the mass and energy balances and found that

$$q \equiv \frac{\overline{L} - L}{F} = \frac{H - h_F}{H - h} \tag{5-17}$$

From Eq. (5-17) we can determine the value of q and hence the slope, $q/(q-1)$, of the feed line. For example, if the feed enters as a saturated liquid (that is, at the liquid boiling temperature at the column pressure), then $h_F = h$ and the numerator of Eq. (5-17) equals the denominator. Thus $q = 1.0$ and the slope of the feed line, $q/(q-1) = \infty$. The feed line is vertical.

The various types of feeds and the slopes of the feed line are illustrated in Table 5-1 and Figure 5-10. Note that all the feed lines intersect at one point, which is at $y = x$. If we set $y = x$ in Eq. (5-35), we obtain

$$y = x = z_F \tag{5-36}$$

Table 5-1. Feed Conditions

Type Feed	T^*	h_F	q	f	Slope
Subcooled liquid	$T_F < T_{BP}$	$h_F < h$	$q > 1$	$f < 0$	> 1.0
Saturated liquid	$T_F = T_{BP}$	h	1	0	∞
Two-phase mixture	$T_{DP} > T_F > T_{BP}$	$H > h_F > h$	$1 > q > 0$	$0 < f < 1$	Negative
Saturated vapor	$T_F = T_{DP}$	H	0	1	0
Superheated vapor	$T_F > T_{DP}$	$h_F > H$	$q < 0$	$f > 1$	$1 > \text{slope} > 0$

T_{BP} = bubble point of feed; T_{DP} = dew point of feed.

as the point of intersection. The feed line is easy to plot from the points $y = x = z_F$ or y intercept $(x = 0) = z_F/(1 - q)$ or x intercept $(y = 0) = z_F/q$, and the slope, which is $q/(q - 1)$. (This entire process of plotting the feed line should remind you of graphical binary flash distillation.)

The feed line was derived from the intersection of the top and bottom operating lines. It thus represents *all possible locations at which the two operating lines can intersect* for a given feed (z_F, q). Thus if we change the reflux ratio, we change the points of intersection, but they all lie on the feed line. This is illustrated in Figure 5-11A. If the reflux ratio is fixed (the top operating line is fixed) but q varies, the

Figure 5-10. Feed lines.

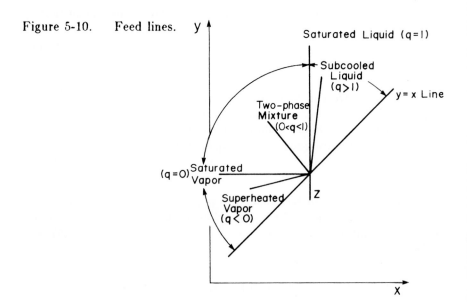

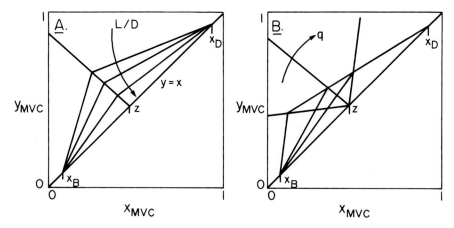

Figure 5-11.   Operating line intersections. (A) Changing reflux ratio
with constant q; (B) Changing q with fixed reflux ratio.

intersection point varies as shown in Figure 5-11b. The slope of the bottom operating line, $\overline{L}/\overline{V}$, depends upon L_0/D, x_D, x_B, and q as was shown in Eq. (5-25).

In Figure 5-8 we illustrated how to determine the optimum feed stage graphically. For computer applications an explicit test is required. If the point of intersection of the two operating lines, (y_I, x_I), is determined, then the optimum feed plate, f, is the one for which

$$y_{f-1} < y_I < y_f \tag{5-37a}$$

and

$$x_f < x_I < x_{f-1} \tag{5-37b}$$

This is illustrated in Figure 5-12. The intersection point can be determined by straightforward but tedious algebraic manipulation as

$$y_I = \frac{z_F + \dfrac{x_D q}{L/D}}{1 + q/(L/D)} \;,\; x_I = \frac{-(q-1)(1 - \dfrac{L}{V})x_D - z_F}{(q-1)(\dfrac{L}{V}) - q} \tag{5-38}$$

for the simple column shown in Figure 4-8.

The feed equations were developed for this simple column; however,

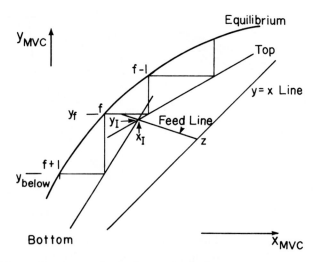

Figure 5-12.   Optimum feed plate calculation.

Eqs. (5-30), and (5-33) through (5-35) are valid for any column configuration if we generalize the definitions of f and q. In general,

$$q = \frac{L_{\text{below feed}} - L_{\text{above feed}}}{\text{feed flow rate}} \qquad (5\text{-}39a)$$

$$f = \frac{V_{\text{above feed}} - V_{\text{below feed}}}{\text{feed flow rate}} \qquad (5\text{-}39b)$$

Example 5-2. Feed Line

Calculate the feed line slope for the following cases.

a. A two-phase feed where 80% of the feed is vaporized under column conditions.

Solution.   The   slope   is $q/(q-1)$,   where $q = (L_{\text{below feed}} - L_{\text{above feed}})/F$ (other expressions could also be used). With a two-phase feed we have the situation shown.

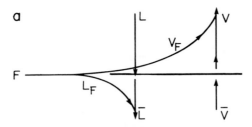

$\bar{L} = L + L_F$. Since 80% of the feed is vapor, 20% is liquid and $L_F = 0.2F$.

Then $q = \dfrac{\bar{L} - L}{F} = \dfrac{(L + 0.2F) - L}{F} = \dfrac{0.2F}{F} = 0.2$

Slope $= \dfrac{q}{q - 1} = \dfrac{0.2}{0.2 - 1} = -\dfrac{1}{4}$

This agrees with Figure 5-10.

b. A superheated vapor feed where 1 mole of liquid will vaporize on the feed stage for each 9 moles of feed input.

Solution. Now the situation is shown in the following figure.

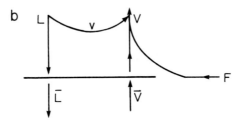

When the feed enters, some liquid must be boiled to cool the feed. Thus,

$\bar{L} = L -$ amount vaporized, v

and the amount vaporized is $v = \dfrac{1}{9} F$.

Thus, $q = \dfrac{\bar{L} - L}{F} = \dfrac{L - \dfrac{1}{9}F - L}{F} = -\dfrac{\dfrac{1}{9}F}{F} = -\dfrac{1}{9}$

$$\text{Slope} = \frac{q}{q-1} = \frac{-\dfrac{1}{9}}{-\dfrac{1}{9}-1} = \frac{-\dfrac{1}{9}}{-\dfrac{10}{9}} = \frac{1}{10}$$

which agrees with Figure 5-10.

c. A liquid feed subcooled by 35 ° F. Average liquid heat capacity is 30 Btu/lb-mole- ° F and λ = 15,000 Btu/lb-mole.

Solution. Here some vapor must be condensed by the entering feed. Thus the situation can be depicted as shown.

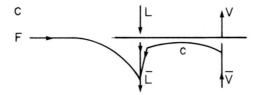

and $\bar{L} = L + F + c$, where c is the amount condensed.

Since the column is insulated, the source of energy to heat the feed to its boiling point is the condensing vapor.

$$F\, C_P(\Delta T) = c\lambda \quad \text{where } \Delta T = T_{BP} - T_F = 35\,°$$

or

$$c = \frac{C_P(\Delta T)}{\lambda}\, F = \frac{(30)(35)}{15,000}\, F = 0.07F$$

$$q = \frac{\bar{L} - L}{F} = \frac{L + F + 0.07F - L}{F} = 1.07$$

$$\text{Slope} = \frac{q}{q-1} = \frac{1.07}{1.07 - 1} = 15.29$$

This agrees with Figure 5-10. Despite the large amount of subcooling, the feed line is fairly close to vertical, and the results will be similar to a saturated liquid feed.

d. A mixture of ethanol and water that is 40 mole % ethanol. Feed is at 40 ° C. Pressure is 1 kg/cm².

Solution. We can now use Eq. (5-17):

$$q = \frac{H - h_F}{H - h}$$

The enthalpy data are available in Figure 2-4. To use that figure we must convert to weight fraction. 0.4 mole fraction is 0.63 wt frac. Then from Figure 2-4 we have

$$h_F(0.63, 40°C) = 20 \text{ kcal/kg}$$

The H and h terms should be in equilibrium at the feed stage, but the concentrations of the feed stage are not known. However, since CMO is valid, H and h in molal units will be constant. We can calculate all enthalpies at a weight fraction of 0.63, convert the enthalpies to enthalpies per kilogram mole, and estimate q. Thus, H (0.63, satd vapor) = 395, h (0.63, satd liquid) = 65, and

$$q = \frac{H(MW) - h_F(MW)}{H(MW) - h(MW)}$$

Since all the molecular weights are at the same concentration, they divide out.

$$q = \frac{H - h_F}{H - h} = \frac{395 - 20}{395 - 65} = 1.136$$

$$\text{Slope} = \frac{q}{q - 1} = \frac{1.136}{0.136} = 8.35$$

This agrees with Figure 5-10.

Note that feed rate was not needed for any of these calculations.

5.5. COMPLETE McCABE-THIELE METHOD

We are now ready to put all the pieces together and solve a design distillation problem by the McCabe-Thiele method. We will do this in the following example.

Example 5-3. McCabe-Thiele Method

A distillation column with a total condenser and a partial reboiler is separating an ethanol-water mixture. The feed is 20 mole % ethanol, feed rate is 1000 kg moles/hr, and feed temperature is 80 °F. A distillate composition of 80 mole % ethanol and a bottoms composition of 2 mole % ethanol are desired. The external reflux ratio is 5/3. The reflux is returned as a saturated liquid and constant molal overflow can be assumed. Find the optimum feed plate location and the total number of equilibrium stages required. Pressure is 1 atm.

Solution

A. Define. The column is sketched in the figure.

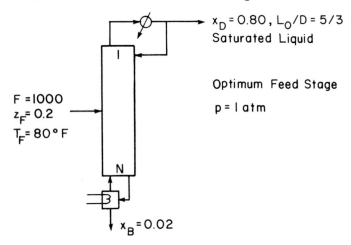

$x_D = 0.80$, $L_0/D = 5/3$
Saturated Liquid

Optimum Feed Stage

$p = 1$ atm

F = 1000
z_F = 0.2
T_F = 80° F

$x_B = 0.02$

Find the optimum feed plate location and the total number of equilibrium stages.

B. Explore. Equilibrium data at 1 atm is given in Figure 2-2. An enthalpy-composition diagram at 1 atm will be helpful to estimate q. These are available in other sources (e.g., Brown *et al.* (1950) or Foust *et al.* (1980) p.36, or a good estimate of q could be made from Figure 2-4 despite the pressure difference. In Example 5-1 we showed that CMO is valid. Thus we can apply the McCabe-Thiele method.

C. Plan. Determine q from Eq. (5-17) and the enthalpy-composition diagram at 1 atm. Plot the feed line. Calculate

L/V. Plot the top operating line; then plot the bottom operating line and step off stages.

D. Do It. *Feed Line:* To find q, first convert feed concentration, 20 mole %, to wt % ethanol = 39 wt %. Two calculations are shown.

Exact Calculation	Approx Calc.
Data at p = 1 atm	$(p = 1 \text{ kg/cm}^2)$
from Brown *et al.* (1950)	using Figure 2-4

$h_F = 25$ Btu/lb $(80\,^\circ F)$ $\qquad$ $h_F = 15$ kcal/kg $(30\,^\circ C)$

$H = 880$ Btu/lb (satd vapor) $\qquad$ $H = 485$ kcal/kg (satd vapor)

$h = 125$ Btu/lb (satd liquid) $\qquad$ $h = 70$ kcal/kg (satd liquid)

$$q = \frac{880 - 25}{880 - 125} = 1.13 \qquad\qquad q = \frac{485 - 15}{485 - 70} = 1.13$$

Thus small differences caused by pressure differences in the diagrams do not change the value of q. Note that molecular weight terms divide out as in Example 5-2d. Then

$$\text{Slope of feed line} = \frac{q}{q-1} = 8.7$$

Feed line intersects y=x line at feed concentration z=0.2. Feed line is plotted in Figure 5-13.

Top Operating Line:

$$y = \frac{L}{V}x + \left(1 - \frac{L}{V}\right) x_D$$

$$\text{Slope} = \frac{L}{V} = \frac{L/D}{1+L/D} = \frac{5/3}{1+5/3} = 5/8$$

$$y \text{ intercept} = \left(1 - \frac{L}{V}\right) x_D = (3/8)(0.8) = 0.3$$

Alternative solution: Intersection of top operating line and y = x (solve top operating line and y = x simultaneously) is at y = x = x_D. The top operating line is plotted in Figure 5-13.

Bottom Operating Line:

$$y = \frac{\overline{L}}{\overline{V}} x - \left(\frac{\overline{L}}{\overline{V}} - 1\right) x_B$$

We know that the bottom operating line intersects the top operating line at the feed line; this is one point. We could calculate $\overline{L}/\overline{V}$ from mass balances or from Eq. (5-25), but it is easier to find another point. The intersection of the bottom operating line and the $y=x$ line is at $y=x=x_B$ (see Problem 5-C6). This gives a second point.

The feed line, top operating line, and bottom operating line are shown in Figure 5-13. We stepped off stages from the bottom up (this is an arbitrary choice). The optimum feed stage is the second above the partial reboiler. 12 equilibrium stages plus a partial reboiler are required.

E. Check. We have a built-in check on the top operating line, since a slope and two points were calculated. The bottom operating line can be checked by calculating $\overline{L}/\overline{V}$ from mass balances and comparing it to the slope. The numbers are reasonable, since $L/V < 1$, $\overline{L}/\overline{V} > 1$, and $q > 1$ as expected. The most likely cause of error in Figure 5-13 (and the hardest to check) is the equilibrium data.

F. Generalization. If constructed carefully, the McCabe-Thiele diagram is quite accurate. Note that there is no need to plot parts of the equilibrium diagram that are greater than x_D or less than x_B. Specified parts of the diagram can be expanded to increase the accuracy.

We did not have to use external balances in this example, while in Example 5-1 we did. This is because we used the feed line as an aid in finding the bottom operating line. The $y=x$ intersection points are useful, but when the column configuration is changed their location may change.

5.6. PROFILES

Figure 5-13 essentially shows the complete solution of Example 5-3; however, it is useful to plot compositions, temperatures, and flow rates leaving each stage (these are known as *profiles*). From Figure 5-13 we

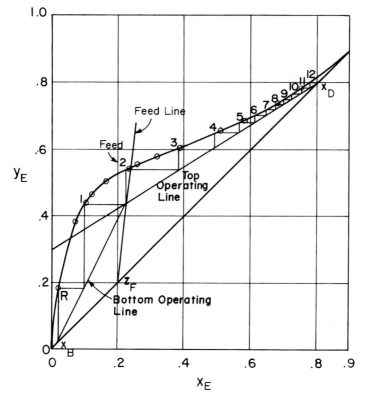

Figure 5-13.   Solution for Example 5-3.

can easily find the ethanol mole fractions in the liquid and vapor leaving each stage. Then $x_w = 1 - x_E$ and $y_w = 1 - y_E$. The temperature of each stage can be found from equilibrium data (Figure 2-3) because the stages are equilibrium stages. Since we assumed CMO, the flow rates of liquid and vapor will be constant in the enriching and stripping sections, and we can determine the changes in the flow rates at the feed stage from the calculated value of q.

The profiles are shown in Figure 5-14. As expected, the water concentration in both liquid and vapor streams decreases monotonically as we go up the column, while the ethanol concentration increases. Since the stages are discrete, the profiles are not smooth curves. Compare Figures 5-13 and 5-14. Note that the concentration and temperature changes from stage to stage become much less when the operating line and the equilibrium curve are close together. When these two lines almost touch, we have a *pinch point*. Then the composition profiles will

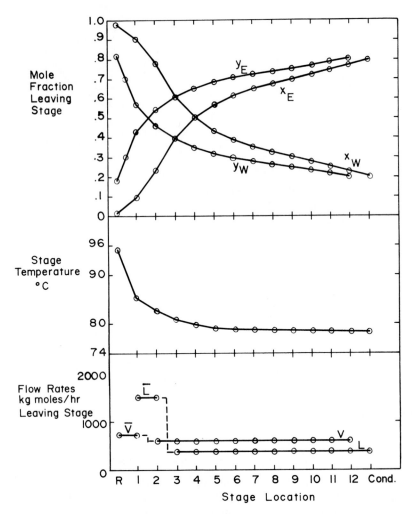

Figure 5-14. Profiles for Example 5-3.

become almost horizontal and there will be very little change in composition from stage to stage. The location of a pinch point within the column depends on the system and the operating conditions.

In this ethanol-water column the temperature decreases rapidly for the first few contacts above the reboiler but is almost constant for the last eight stages. This occurs mainly because of the shape of the temperature-composition diagram for ethanol-water (see Figure 2-3).

Since we assumed CMO, the flow profiles are flat in each section of the column. As expected, $\bar{L} > \bar{V}$ and $V > L$ (a convenient check to use). Since stage 2 is the feed stage, L_2 is in the stripping section while V_2 is in the enriching section (draw a sketch of the feed stage if this isn't clear). Different quality feeds will have different changes at the feed stage. Liquid and vapor flow rates can increase, decrease, or remain unchanged in passing from the stripping to the enriching section.

5.7. COMPARISONS BETWEEN ANALYTICAL LEWIS METHOD AND GRAPHICAL McCABE-THIELE DIAGRAM

Both the Lewis and McCabe-Thiele methods are based on the constant molal overflow assumption. If the latent heat of vaporization per mole varies but the specific heats are constant and the column is adiabatic, the CMO assumption can be modified. This will be explored in Chapter 6. When the CMO assumption is valid, the energy balances are automatically satisfied, which greatly simplifies the stage-by-stage calculations. The reboiler and condenser duties, Q_R and Q_c, are still determined from the balances around the entire column. If CMO is not valid, Q_R and Q_c will be unaffected if the same external reflux ratio can be used, but this may not be possible. The exact stage-by-stage calculation may show that a higher or lower L_0/D is required if CMO is invalid.

If calculations are to be done by hand, the graphical method is faster than alternating between the analytical forms of the equilibrium relationship and the operating equations. In the McCabe-Thiele method, solution of the equilibrium relationship is simply a matter of drawing a line to the equilibrium curve. In the analytical solution we must first either fit the equilibrium data to an analytical expression or develop an interpolation routine. Then we must solve this equation, which may be nonlinear, each time we do an equilibrium calculation. Since the operating equation is linear, it is easy to solve analytically. With a sharp pencil, large graph paper, and care, the McCabe-Thiele technique can easily be as accurate as the equilibrium data (two significant figures).

When a programmable calculator or a computer is to be used, the analytical method is more convenient. The computer calculations are obviously more convenient if a very large number of stages are required, if a trial-and-error solution is required, or if many cases are to be run to explore the effect of many variables. The computer calculations are *not* more accurate than the graphical method, however, since accuracy is limited by the equilibrium data.

One of the most convenient ways to discuss computer and calculator methods is by reference to a flow chart. The flow chart is fairly general while computer programs are very specific to the machine being used.

146

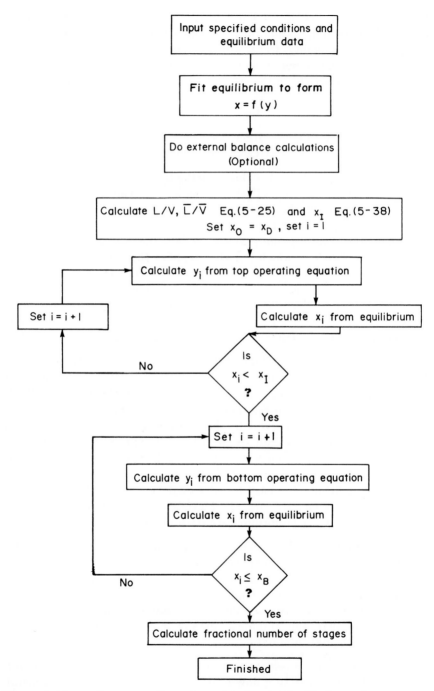

Figure 5-15.   Computer flow chart for simple distillation column.

Consider the specific design problem we solved in Example 5-3. Assume that we decide to step off stages from the top down. Now the computer or calculator program can proceed as shown in Figure 5-15. Other flow charts are possible. If we step off stages from the bottom up, we will calculate y_i from equilibrium and x_i from the operating equation. Note that a McCabe-Thiele diagram is very useful for following the logic of flow charts. Try this with Figure 5-15.

The development of digital computers has made the graphical McCabe-Thiele technique obsolete for most detailed design calculations. Engineers used to cover an entire wall with graph paper to do a McCabe-Thiele diagram when a very large number of stages were required. The method is still useful for one or two calculations, but its major uses are as a teaching tool and as a conceptual tool. The graphical procedure presents a very clear visual picture of the calculation that is easier to understand than the interactions of the equations. The graphs are also extremely useful as a tool to help determine what the effect of changing variables will be, as a diagnostic when the computer program appears to be malfunctioning, and as a diagnostic when the column appears to be malfunctioning. Because of its visual impact, we will use the McCabe-Thiele diagram extensively in the next chapter to explore a variety of distillation systems.

5.8. SUMMARY - OBJECTIVES

In this chapter we have developed the stage-by-stage balances for a distillation column and showed how to solve these equations when constant molar overflow is valid. You should now be able to satisfy the following objectives:

1. Write the mass and energy balances and equilibrium expressions for any stage in a column.

2. Explain what constant molal overflow (CMO) is, and determine if CMO is valid in a given situation.

3. Derive the operating equations for CMO systems.

4. Calculate the feed quality and determine its effect on flow rates. Plot the feed line on a y-x diagram.

5. Determine the number of stages required, using the Lewis method.

6. Determine the number of stages required, using the McCabe-Thiele method.

148

7. Develop and explain composition, temperature, and flow profiles.

8. Develop a computer flow chart and a computer program using the Lewis method.

REFERENCES

Brown, G.G., and associates, *Unit Operations,* John Wiley & Sons, New York, 1950.

Foust, A.S., L.A. Wenzel, C.W. Clump, L. Maus, and L.B. Andersen, *Principles of Unit Operations,* 2nd ed., Wiley, New York, NY, 1980.

Himmelblau, D.M., *Basic Principles and Calculations in Chemical Engineering,* 3rd ed., Prentice-Hall, Englewood Cliffs, NJ, 1974.

Lewis, W.K., "The Efficiency and Design of Rectifying Columns for Binary Mixtures," *Ind. Eng. Chem., 14,* 492 (1922).

Perry, R.H., C.H. Chilton and S.D. Kirkpatrick (Eds.), *Chemical Engineer's Handbook,* 4th ed., McGraw-Hill, New York, 1963.

Ponchon, M., *Tech. Moderne, 13,* 20, 53 (1921).

Savarit, R., *Arts et Metiers, 65,* 142, 178, 241, 266, 307 (1922).

Sorel, E., *La Retification de l'Alcohol,* Gauthier-Villars, Paris, 1893.

HOMEWORK

A. *Discussion Problems*

A1. In the figure shown, what streams are represented by point A? By point B? How would you detemine the temperature of stage 2? How about the temperature in the reboiler? If feed composition is as shown, how can the liquid composition on the optimum feed stage be so much less than z?

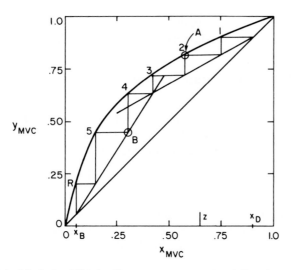

A2.   For this McCabe-Thiele diagram answer the following questions.

a. (1)   The actual feed tray is?

(2)   The mole fraction MVC in the feed is?

(3)   The vapor composition on the feed tray is?

(4)   The liquid composition on the feed tray is?

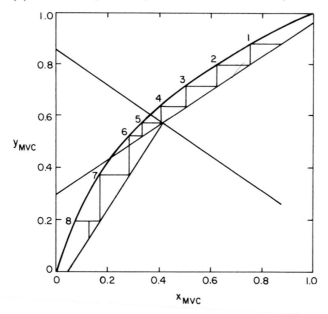

b. Is the feed a superheated vapor feed, saturated vapor feed, two-phase feed, saturated liquid feed, or subcooled liquid feed?

c. Is the temperature at stage 7 higher, lower, or the same as at stage 1?

A3. Suppose that constant mass overflow is valid instead of constant molal overflow. Explain how to carry out the Lewis and McCabe-Thiele procedures in this case.

A4. If you assume CMO and do all the calculations, but CMO is not true, what happens?

A5. Drawing the McCabe-Thiele graph as y_{MVC} versus x_{MVC} is traditional but not necessary. Repeat Example 5-3, but plot y_w versus x_w. Note the differences in the diagram. Do you expect to get the same answer?

A6. For distillation at constant molal overflow, show the flow profiles schematically (plot L_j and V_j versus stage location) for

a. Subcooled liquid feed

b. Two-phase feed

c. Superheated vapor feed

A7. Develop your key relations chart (maximum of one page) for this chapter. You will probably want to include sketches and equations.

C. *Derivations*

C1. Derive Eq. (5-20).

C2. Derive Eq. (5-25).

C3. Derive Eq. (5-30).

C4. Derive Eq. (5-34) for the general case.

C5. Derive Eq. (5-38).

C6. For Example 5-3 prove that:

a. The top operating line and the $y=x$ line intersect at $y=x=x_D$.

b. The bottom operating line and the $y=x$ line intersect at $y=x=x_B$.

C7. The boilup ratio $\overline{V}/B$ may be specified. Derive an expression for $\overline{L}/\overline{V}$ as a function of $\overline{V}/B$ for a partial reboiler.

C8. Derive an alternative flow chart (different from Figure 5-15) for a computer or calculator program for stepping off stages from the top down.

C9. Derive a flow chart for a computer or calculator program for stepping off stages from the bottom up.

D.   *Problems*

D1. Redo Example 5-1, starting at the bottom of the column. Use the Lewis method.

D2. Redo Example 5-1, using the McCabe-Thiele method.

D3.   a.   A feed mixture of ethanol and water is 40 mole % ethanol. The feed is at 200 °C and is at a high enough pressure that it is a liquid. It is input into the column through a valve, where it flashes. Column pressure is 1 kg/cm². Find the slope of the feed line.

   b.   We are separating ethanol and water in a distillation column at a pressure of 1 kg/cm². Feed is 50 wt % ethanol, and feed rate is 1 kg/hr. The feed is initially a liquid at 250 °C and then flashes when the pressure is dropped as it enters the column. Find q. Data are in Problem 4-D5 and in Figure 2-4. You may assume that CMO is valid.

D4.   a.   We have a superheated vapor feed of 60 mole % more volatile component at 350 °C. Feed flow rate is 1000 kg moles/hr. On the feed plate the temperature is 50 °C. For this mixture the average heat capacities are

$$C_{PL} = 50 \text{ cal/g-mole-}°C \, , \; C_{PV} = 25 \text{ cal/g-mole-}°C$$

   while the latent heat of vaporization is $\lambda = 5000$ cal/g-mole. Plot the feed line for this feed.

   b.   If a feed to a column is a two-phase feed that is 40 mole % vapor, find the value of q and the slope of the feed line.

   c.   If the feed to a column is a superheated vapor and 1 mole of liquid is vaporized on the feed plate to cool 5 moles of feed to a saturated vapor, what is the value of q? What is the slope of the feed line?

D5.   We desire to use a distillation column to separate an ethanol-water mixture. The column has a total condenser, a partial reboiler, and a saturated liquid reflux. The feed is a saturated liquid of composition 0.10 mole fraction ethanol and a flow rate of 250 moles/hr. A bottoms of 0.005 and a distillate of 0.75 mole fraction ethanol are desired. For an external reflux ratio of 2.0, find the liquid and vapor compositions leaving the fourth stage from the top of the column. Assume constant molal overflow, and solve the problem graphically. Pressure is 1 atm.

D6.   We desire to separate a feed of 75 mole % ethanol and 25 mole % propanol that is a saturated vapor at 556 moles/hr. Distillate composition should be 99 mole % ethanol and bottoms 10 mole % ethanol. A partial reboiler and a total condenser are used. Reflux is a saturated liquid, and constant molal overflow can be assumed. The system has a constant relative volatility of 2.1, and ethanol is more volatile. Boilup ratio $\overline{V}/B$ is 2.0. We desire to find the liquid composition leaving the second stage above the reboiler. Note that the reboiler acts as an equilibrium contact. Solve this problem both analytically and graphically.

D7.   a.   A distillation column with a total condenser is separating acetone from ethanol. A distillate concentration of $x_D = 0.90$ mole fraction acetone is desired. Since CMO is valid, L/V = constant. If L/V is equal to 0.8, find the composition of the liquid leaving the fifth stage below the total condenser.

      b.   A distillation column separating acetone and ethanol has a partial reboiler that acts as an equilibrium contact. If the bottoms composition is $x_B = 0.13$ mole fraction acetone and the boilup ratio $\overline{V}/B = 1.0$, find the vapor composition leaving the second stage above the partial reboiler.

      c.   The distillation column in parts a and b is separating acetone from ethanol and has $x_D = 0.9$, $x_B = 0.13$, $L/V = 0.8$, and $\overline{V}/B = 1.0$. If the feed composition is $z = 0.3$ (all concentrations are mole fraction of more volatile component), find the optimum feed plate location, total number of stages, and required q value of the feed. Equilibrium data for acetone and ethanol at 1 atm (Perry *et al.*, 1963, p. 13-4) are

x_A	.10	.15	.20	.25	.30	.35	.40	.50	.60	.70	.80	.90
y_A	.262	.348	.417	.478	.524	.566	.605	.674	.739	.802	.865	.929

D8.   A distillation column is separating acetone from ethanol. We are interested in the enriching section of the column. The distillate composition is 90 mole % acetone. Reflux is returned from the condenser to the column as a saturated liquid. Constant molal overflow is valid. If the external reflux ratio is $L/D = 3$, find the liquid and vapor compositions leaving the fifth stage below the total condenser. Data are given in Problem 5-D7.

D9.   A distillation column is separating ethylene dichloride from trichloroethane. We are interested in the stripping section. The column has a partial reboiler, and the bottoms composition is 0.10 mole fraction ethylene dichloride. CMO is valid. A boilup ratio of $\overline{V}/B = 2$ is used. The relative volatility is approximately constant at 2.40, and ethylene dichloride is the more volatile. Find the liquid and vapor compositions leaving the second stage above the partial reboiler (remember that the partial reboiler is an equilibrium contact). Do this:

   a.   Analytically: Use equation for equilibrium.

   b.   Graphically: Plot y-x diagram for $\alpha = 2.4$.

D10.   A distillation column is separating phenol from p-cresol at 1 atm pressure. The distillate composition desired is 0.96 mole fraction phenol. An external reflux ratio of $L/D = 4$ is used, and the reflux is returned to the column as a saturated liquid. The equilibrium data can be represented by a constant relative volatility, $\alpha_{phenol-cresol} = 1.76$ (Perry et al., 1963, p. 13-3). CMO can be assumed.

   a.   What is the vapor composition leaving the third equilibrium stage below the total condenser? Solve this by an analytical stage-by-stage calculation alternating between the operating equation and the equilibrium equation.

   b.   What is the liquid composition leaving the sixth equilibrium stage below the total condenser? Solve this problem graphically using a McCabe-Thiele diagram plotted for $\alpha_{p-c} = 1.76$.

D11.   A distillation column is separating methanol and water. The column has a partial reboiler, a total condenser, and a saturated liquid reflux. The feed rate is 152 kg moles/hr, and feed composition is 40 mole % methanol. The feed is a two-phase mixture that is 60% liquid and 40% vapor and will be put into the column at the optimum feed plate. The distillate composition is 0.92, and bottoms is 0.08 mole fraction methanol. CMO can be assumed. Use an external reflux ratio of $L/D = 0.9$. Data are in Table 3-3.

a.   Find the value of q, and plot the feed line.

b.   Plot the top and bottom operating lines.

c.   Find the optimum feed plate location and total number of equilibrium stages.

D12.   For the same feed, same distillate, and same bottoms composition as in problem 5-D11, we desire to use a boilup ratio $\overline{V}/B = 1$. If the feed is on the third stage above the reboiler, find the number of equilibrium stages needed for the separation.

D13.   Find the number of stages and the best feed location for a column separating ethanol and propanol, $\alpha = 2.1$. Feed composition $= 0.48$, $x_D = 0.96$, $x_B = 0.04$. Constant molal overflow can be assumed, and reflux is a saturated liquid. Column has a total condenser and a partial reboiler. Pressure is 101.3 kPa.

a. $L/D = 3$ for $q = 0.4$ and $q = 2.0$.

b. $q = 0.4$ for $L_0/D = 2.5$ and $L_0/D = 6$.

D14.   Write a computer or calculator program to find the number of equilibrium stages and the optimum feed plate location for a binary distillation with a constant relative volatility. System will have constant molal overflow, saturated liquid reflux, total condenser, and a partial reboiler. The given variables will be F, z_F, q, x_B, x_D, α, and L_0/D. Test your program by solving the following problems:

a.   Separation of phenol from p-cresol. $F = 100$, $z_F = 0.6$, $q = 0.4$, $x_B = 0.04$, $x_D = 0.98$, $\alpha = 1.76$, and $L/D = 4.00$.

b.   Separation of benzene from toluene. $F = 200$, $z_F = 0.4$, $q = 1.3$, $x_B = 0.0005$, $x_D = 0.98$, $\alpha = 2.5$, and $L/D = 2.00$.

c.   Since the relative volatility of benzene and toluene can vary from 2.61 for pure benzene to 2.315 for pure toluene, repeat part b for $\alpha = 2.315$, 2.4, and 2.61.

Write your program so that it will calculate the fractional number of stages required. Check your program by doing a hand calculation.

D15.   A distillation column is separating ethanol and water at a pressure of 1 atm. Feed is 40 mole % ethanol, and feed rate is 100 kg moles/hr. The feed is a superheated vapor, and 1 mole of liquid must vaporize in the column for each 10 moles of feed. The distillate product is 72 mole % ethanol. Reflux is a saturated liquid, and $L_0/D = 3.0$. The bottoms composition should be 2 mole % ethanol. The column has a total condenser and a partial reboiler. Find the optimum feed plate location and the total number of equilibrium stages required. Data are listed in Problems 4-D5 and 5-E2.

D16. Estimate q for Problem 4-D6. Estimate that the feed stage is at same composition as the feed.

E. *More Complex Problems*

E1. Repeat Problem 5-D7, but program it for a calculator or computer. The equilibrium data must be fit by an appropriate equation.

E2. A distillation column with a total condenser and a partial reboiler is separating ethanol and water at 1 kg/cm^2 pressure. Feed is 0.32 mole fraction ethanol and is at 30 C. Feed flow rate is 100 kg moles/hr. The distillate product is a saturated liquid, and the distillate is 80 mole % ethanol. The condenser removes 2,065,113 kcal/hr. The bottoms is 0.04 mole fraction ethanol. Assume that CMO is valid. Find the number of stages and the optimum feed stage location.

Data: $C_{PL\ EtOH} = 24.65$ cal/g-mole-$^\circ$C at $0\,^\circ$C.

Other data are in Problem 4-D5. The enthalpy-composition diagram is given in Figure 2-4, and the y-x diagram is in Figure 2-2. Note: Watch your units.

F. *Problems Requiring Other Resources*

F1. If we wish to separate the following systems by distillation, is CMO valid?

    a. Methanol and water

    b. Isopropanol and water

    c. Acetic acid and water

    d. n-Butane from n-pentane

    e. Benzene from toluene

F2. A distillation column with a total condenser and a partial reboiler is separating 250 moles/hr of a feed that is 80 mole % benzene and 20 mole % ethanol. The feed enters the column as a two-phase mixture that is 30% vapor. A distillate composition of 40 mole % ethanol is desired, and a bottoms of 2.5 mole % ethanol. Reflux is returned as a saturated liquid, and CMO can be assumed. An external reflux ratio of L/D = 1.0 is used. Pressure = 101.3 kPa.

a. What is the optimum feed location, and how many equilibrium stages are needed?

b. If the fourth stage below the condenser is used as the feed stage, how many equilibrium stages are needed for the separation?

F3. A distillation column with a total condenser and a partial reboiler is separating benzene and ethanol. The feed is a saturated vapor of concentration 25 mole % benzene. Feed flow rate is 105 moles/hr. We desire a distillate concentration of $x_D = 0.478$ mole fraction benzene and a bottoms concentration of $x_B = 0.025$ mole fraction benzene. A boilup ratio $\overline{V}/B = 1.0$ is used, and constant molal overflow can be assumed. Pressure = 101.3 kPa. Use the optimum feed plate. How many equilibrium stages are needed for this separation?

F4. For the feed given in Problem 5-F3, what is the maximum distillate concentration that can be obtained?

F5. A distillation column is being used to separate a chloroform-benzene mixture. The column has a total condenser, and a partial reboiler, and CMO can be assumed to be valid. The feed of 2000 kg moles/hr is a two-phase mixture that is 30% liquid, and its composition is 55 mole % chloroform. The reflux is a saturated liquid, the distillate should be 90 mole % chloroform, and the bottoms should be 20 mole % chloroform. Pressure = 101.3 kPa. If the external reflux ratio $L_0/D = 4$, find the optimum feed plate location and the total number of stages required.

F6. A distillation column with a total condenser and a partial reboiler is separating acetone and ethanol. The feed is 35 mole % acetone, has a flow rate of 1500 kg moles/hr, and is a two-phase mixture that is 45% vapor. CMO is valid, and the reflux is a saturated liquid. The distillate concentration is set at 85 mole % acetone, and an external reflux ratio of $L_0/D = 3.0$ is used. The reboiler is operated so that its temperature is 75.4 °C. Pressure = 101.3 kPa. Find the optimum feed plate location and the total number of equilibrium stages needed.

chapter 6
ADVANCED BINARY DISTILLATION: McCABE-THIELE AND LEWIS ANALYSES

In the previous chapter we considered the basic design approach for a continuous distillation column with a single feed, a partial reboiler, and a total condenser. A large variety of alternatives to this basic arrangement are used and can be analyzed by means of the McCabe-Thiele or Lewis approaches. In this chapter we will first develop the graphical McCabe-Thiele procedure further. Then we will use this tool to study a variety of distillation systems. Modifications in the McCabe-Thiele method to handle nonconstant molal overflow systems will be developed. Finally, the use of these equations in calculator or computer routines will be discussed.

6.1. EXAMPLE: OPEN STEAM DISTILLATION

To illustrate the application of these methods to other situations we will start with an example.

> Example 6-1. McCabe-Thiele Analysis Procedure - Open Steam Distillation
>
> We now have all the tools required to solve any binary distillation problem with the graphical McCabe-Thiele procedure. As a specific example, consider the separation of methanol from water in a staged distillation column. The feed is 60 mole % methanol and 40 mole % water and is input as a two-phase mixture that flashes so that $V_F/F = 0.3$. Feed flow rate is 350 kg moles/hr. The column is well insulated and has a total condenser. The reflux is returned to the column as a saturated liquid. An external reflux ratio of $L_0/D = 3.0$ is used. We desire a distillate concentration of 95 mole % methanol and a bottoms concentration of 8 mole % methanol. Instead of using a reboiler, saturated steam at 1 atm is sparged directly into the bottom of the column

to provide boilup. (This is called direct or open steam.) Column pressure is 1 atm. Calculate the number of equilibrium stages and the optimum feed plate location.

Solution

A. Define. It helps to draw a schematic diagram of the apparatus, particularly since a new type of distillation is involved. This is shown in Figure 6-1. We wish to find the optimum feed plate location, N_F, and the total number of equilibrium stages, N, required for this separation. We could also calculate Q_c, D, B, and steam rate S, but these were not asked for. We assume that the column is adiabatic since it is well insulated.

B. Explore. The first thing we need is equilibrium data. Fortunately, these are readily available (see Table 3-3).

Figure 6-1.   Distillation with direct steam heating, Example 6-1.

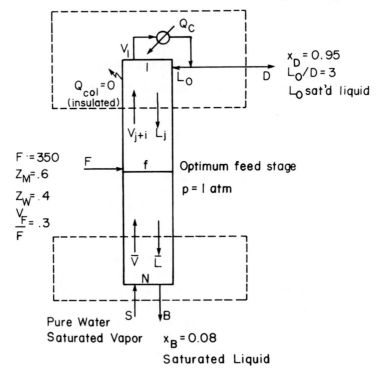

Second, we would like to assume constant molal overflow (CMO) so that we can use the McCabe-Thiele analysis procedure. An easy way to check this assumption is to compare the latent heats of vaporization per mole (Himmelblau, 1974).

$$\Delta H_{vap} \text{ methanol (at bp)} = 8.43 \text{ kcal/g-mole}$$

$$\Delta H_{vap} \text{ water (at bp)} = 9.72 \text{ kcal/g-mole}$$

These values are not equal, and in fact, water's latent heat is 15.3% higher than methanol's. Thus, CMO is not strictly valid. We will solve this problem assuming CMO and later in this chapter compare our result with a more exact solution.

A look at Figure 6-1 shows that the configuration at the bottom of the column is different than when a reboiler is present. Thus we should expect that the bottom operating equations will be different from those derived previously.

C. Plan. We will use a McCabe-Thiele analysis. Plot the equilibrium data on a y-x graph.

Top Operating Line: Mass balances in the rectifying section (see Fig. 6-1) are

$$V_{j+1} = L_j + D$$

$$y_{j+1}V_{j+1} = L_j x_j + Dx_D$$

Assume constant molal overflow and solve for y_{j+1}.

$$y_{j+1} = (L/V)x_j + (1 - L/V)x_D$$

Slope $= L/V$,   y intercept $(x = 0) = (1 - L/V) x_D$

Intersection $y = x = x_D$

Since the reflux is returned as a saturated liquid,

$$L/V = \frac{L_0}{L_0 + D} = \frac{L_0/D}{L_0/D + 1}$$

Enough information is available to plot the top operating line.

Feed Line: $y = \dfrac{q}{(q-1)} x + \dfrac{z}{(1-q)}$

$\quad$ Slope $= \dfrac{q}{q-1}$, $\qquad$ y intercept $(x = 0) = \dfrac{z}{q-1}$

Intersection: $y = x = z$

$$q = \frac{\bar{L}-L}{F} = \frac{\bar{V}-V+F}{F} = \frac{F}{F} - \frac{V_F}{F} = 1 - \frac{V_F}{F}$$

Now we can plot the feed line.

Bottom Operating Line: The mass balances are

$$\bar{V} + B = \bar{L} + S$$

$$\bar{V}y + Bx_B = \bar{L}x + Sy_s \quad \text{(for methanol)}$$

Solve for y:

$$y = (\bar{L}/\bar{V}) x + (S/\bar{V}) y_s - (B/\bar{V}) x_B$$

Simplifications: Since the steam is pure, $y_s = 0.0$ (contains no methanol). Since steam is saturated, $S = \bar{V}$ and $B = \bar{L}$ (constant molal overflow). Then

$$y = (\bar{L}/\bar{V}) x - (\bar{L}/\bar{V}) x_B \qquad (6\text{-}1)$$

Note this *is* different from the operating equation for the bottom section when a reboiler is present.

$\quad$ Slope $= \bar{L}/\bar{V}$ (unknown), y intercept $= -(\bar{L}/\bar{V})x_B$ (unknown)

$$y = x = \frac{-(\bar{L}/\bar{V})x_B}{(1 - \bar{L}/\bar{V})} \quad \text{(unknown)}$$

One known point is the intercept of the top operating line with the feed line. We still need a second point, and we can find it at the x intercept. When y is set to zero, $x = x_B$ (this is left as Problem 6-C1).

D. Do It. Equilibrium data are plotted on Figure 6-2.

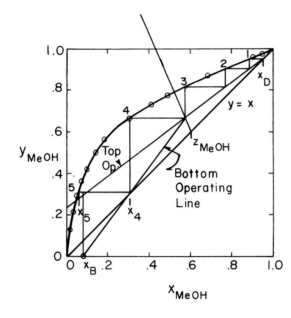

Figure 6-2.
Solution for Example 6-1.

Top Operating Line: $L/V = \dfrac{L/D}{1 + L/D} = 3/4 = $ slope

$$y = x = x_D = 0.95$$

$$y \text{ intercept} = (1 - L/V) x_D = 0.2375$$

We can plot this straight line as shown in Figure 6-2.

Feed Line: Slope $ = q/(q-1) = 0.7/(0.7 - 1) = -7/3.$

Intersects at $y = x = z = 0.6$. Plotted on Figure 6-2.

Bottom Operating Line: We can plot this line between two points, the intercept of top operating line and feed line, and

$$x \text{ intercept } (y = 0) = x_B = 0.08$$

This is also shown in Figure 6-2.

Step off stages, starting at the top. x_1 is in equilibrium with y_1 at x_D. Drawing a horizontal line to the equilibrium curve gives value x_1. y_2 and x_1 are related by the operating line. At constant y_2 (horizontal line), go to the equilibrium curve to find x_2. Continue this stage-by-stage procedure.

Optimum feed stage is determined as in Figure 5-8A. Optimum feed in Figure 6-2 is on stage 3 or 4 (since by accident x_3 is at intersection point of feed and operating lines). Since the feed is a two-phase feed, we would introduce it above stage 4 in this case.

Number of stages: 5 is more than enough. We can calculate a fractional number of stages.

$$\text{Frac} = \frac{\text{distance from operating line to product}}{\text{distance from operating line to equilibrium curve}} \qquad (6\text{-}2)$$

In Figure 6-2,

$$\text{Frac} = \frac{x_B - x_4}{x_5 - x_4} = \frac{0.08 - 0.305}{0.06 - 0.305} = 0.92$$

We need $4 + 0.92 = 4.92$ equilibrium contacts.

E. Check. There are a series of internal consistency checks that can be made. Equilibrium should be a smooth curve. This will pick up misplotted points. $L/V < 1$ (otherwise no distillate product), and $\bar{L}/\bar{V} > 1$ (otherwise no bottoms product). The feed line's slope is in the correct direction for a two-phase feed. A final check on the assumption of constant molal overflow would be advisable since the latent heats vary by 15% (see Example 6-3).

F. Generalize. Note that the $y = x$ line is not always useful. Don't memorize locations of points. Learn to derive what is needed. The total condenser does not change compositions and is not counted as an equilibrium stage. The total condenser appears in Figure 6-2 as the single point $y = x = x_D$. Think about why this is true. In general, all inputs to the column can change flow rates and hence slopes inside the column. The purpose of the feed line is to help determine this effect. The reflux stream and open steam are also inputs to the column. If they are not saturated streams the flow rates are calculated differently; this is discussed later.

6.2. GENERAL ANALYSIS PROCEDURE

The open steam example illustrated one specific case. It is useful to generalize this analysis procedure. A *section* of the column is the segment

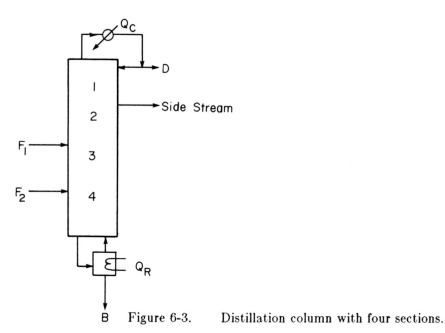

B　Figure 6-3.　Distillation column with four sections.

of stages between two input or exit streams. Thus in Figure 6-1 there are two sections: top and bottom. Figure 6-3 illustrates a column with four sections. Each section's operating equation can be derived independently. Thus, the secret (if that's what it is) is to treat each section as an independent subproblem connected to the other subproblems by the feed lines (which are also independent).

An algorithm for any problem is the following:

1.　Draw a figure of the column and label all known variables (e.g., as in Figure 6-1). Check to see if constant molal overflow is valid.

2.　For each section:

　a.　Draw a mass balance envelope. We desire this envelope to cut the unknown liquid and vapor streams in the section and known streams (feeds, specified products or specified side-streams). The fewer streams involved, the simpler the mass balances will be. This step is important, since it controls how easy the following steps will be.

　b.　Write the overall and most volatile component mass balances.

　c.　Derive the operating equation.

　d.　Simplify.

　e.　Calculate all known slopes, intercepts, and intersections.

3. Develop feed line equations. Calculate q values, slopes, and y = x intersections.

4. For operating and feed lines:

   a. Plot as many of the operating lines and feed lines as you can.

   b. If all operating lines cannot be plotted, step off stages if the stage location of any feed or side stream is specified.

   c. If needed, do external mass and energy balances (see Example 6-2). Use the values of D and B in step 2.

5. When all operating lines have been plotted, step off stages, determine optimum feed plate locations and the total number of stages. Calculate a fractional number of stages.

Not all of these general steps were illustrated in the previous examples, but they will be illustrated in the examples that follow.

This problem-solving algorithm should be used as a guide, not as a computer code to be followed exactly. The wide variety of possible configurations for distillation columns will allow a lot of practice in using the McCabe-Thiele method for solving problems.

Example 6-2. Distillation with Two Feeds

We wish to separate ethanol from water in a distillation column with a total condenser and a partial reboiler. We have 200 kg moles/hr of feed 1, which is 30 mole % ethanol and is a saturated vapor. We also have 300 kg moles/hr of feed 2, which is 40 mole % ethanol. Feed 2 is a subcooled liquid. One mole of vapor must condense inside the column to heat up 4 moles of feed to its boiling point. We desire a bottoms product that is 2 mole % ethanol and a distillate product that is 72 mole % ethanol. External reflux ratio is $L_0/D = 1.0$. The reflux is a saturated liquid. Column pressure is 101.3 kPa, and the column is well insulated. The feeds are to be input at their optimum feed locations. Find the optimum feed locations (reported as stages above the reboiler) and the total number of equilibrium stages required.

Solution

A. Define. Again a sketch will be helpful; see Figure 6-4. Since the two feed streams are already partially separated, it makes sense to input them separately to maintain the separation that

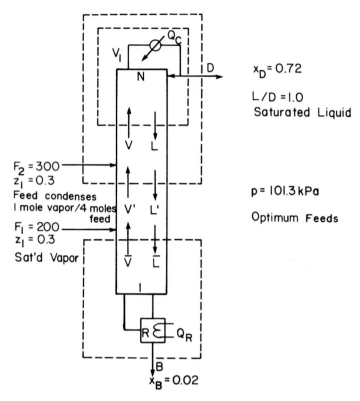

Figure 6-4.   Two feed distillation column for Example 6-2.

already exists. We have made an inherent assumption in Figure 6-4. That is, feed 2 of higher mole fraction ethanol should enter the column higher up than feed 1. This assumption will be checked when the optimum feed plate locations are calculated, but it will affect the way we do the preliminary calculations. Since the feed plate locations were asked for as stages above the reboiler, the stages have been numbered from the bottom up.

B. Explore. Obviously, equilibrium data are required, and they are available from Figure 2-2. We already checked (Example 5-1) that CMO is a reasonable assumption. A look at Figure 6-4 shows that the top section is the same as top sections used previously. The bottom section is also familiar. Thus the major new calculations problem concerns the middle section. There will be two feed lines and three operating lines.

C. Plan. We will look at the two feed lines, top operating line, bottom operating line, and middle operating line. The simple numerical calculations will also be done here.

Feeds: $y = \dfrac{q_i}{q_i - 1} x - \dfrac{z_i}{q_i - 1}$

Feed 1: Saturated vapor, $q_i = 0$, slope $= 0$, y intercept $= \dfrac{z_i}{q_i - 1} = 0.3$, intersection at $y = x = z_i = 0.3$.

Feed 2: $q_2 = \dfrac{L_{below} - L_{above}}{F_2}$

The feed stage for feed 2 looks schematically as shown in the figure:

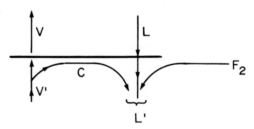

Then,

$$L_{below\ feed} = L' = L + F_2 + c$$

Amount condensed $= c = (1/4) F_2$,

$$q_2 = \frac{(L + F_2 + c) - L}{F_2} = \frac{F_2 + (1/4) F_2}{F_2} = 5/4$$

$$\text{slope} = \frac{q_2}{q_2 - 1} = \frac{5/4}{5/4 - 1} = 5$$

Intersection: $y = x = z_2 = 0.4$

Top Operating Line: We can derive the top operating equation:

$$y = \frac{L}{V} x + \left(1 - \frac{L}{V}\right) x_D$$

This is the usual top operating line. With saturated liquid reflux the slope is,

$$\frac{L}{V} = \frac{L}{L + D} = \frac{L/D}{L/D + 1} = 1/2$$

Intersection: $y = x = x_D = 0.72$

Bottom Operating Line: We can derive the bottom operating equation:

$$y = \frac{\overline{L}}{\overline{V}} x - (\frac{\overline{L}}{\overline{V}} - 1) x_B$$

This is the usual bottom operating line. Then, slope $= \overline{L}/\overline{V}$ is unknown.

$y = x$ intersection is at $y = x = x_B = 0.02$. One other point is the intersection of the F_1 feed line and the middle operating line, if we can find the middle operating line.

Middle Operating Line: To derive an operating equation for the middle section, we can write a mass balance around the top or the bottom of the column. The resulting equations will look different, but they are equivalent. We will use the mass balance envelope around the top of the column as shown in Figure 6-4. These mass balances are

$$F_2 + V' = L' + D$$

$$F_2 z_2 + V' y = L' x + D x_D \quad (MVC)$$

Solve this equation for y to develop the operating equation,

$$y = \frac{L'}{V'} x + \frac{D x_D - F_2 z_2}{V'} \tag{6-3}$$

One known point is the intersection of the F_2 feed line and the top operating line. A second point is needed. We can try

y intercept $= \dfrac{D x_D - F_2 z_2}{V'}$ but this is unknown.

The intersection of the top operating line with the $y = x$ line is found by setting $y = x$ in Eq. (6-3) and solving,

$$x = y = \frac{Dx_D - F_2z_2}{V' - L'}$$

From the overall balance equation,

$$V' - L' = D - F_2$$

Thus,

$$x = y = \frac{Dx_D - F_2z_2}{D - F_2} \qquad (6\text{-}4)$$

This point is not known, but it can be calculated once D is known.

Slope $= L'/V'$ is unknown, but it can be calculated from the feed-stage calculation. From the definition of q_2,

$$L' = F_2q_2 + L \qquad (6\text{-}5)$$

where $L = (L/D)D$. Once D is determined, L and then L' can be calculated. Then the mass balance gives

$$V' = L' + D - F_2$$

and the slope L'/V' can be calculated.

The conclusion from these calculations is, we have to calculate D. To do this we need external mass balances:

$$F_1 + F_2 = D + B$$

$$F_1z_1 + F_2z_2 = Dx_D + Bx_B \quad (\text{MVC})$$

These two equations have two unknowns, D and B, and we can solve for them. Rearranging the first, we have $B = F_1 + F_2 - D$. Then, we substitute this into the more volatile component (MVC) mass balance,

$$F_1z_1 + F_2z_2 = Dx_D + (F_1 + F_2)x_B - Dx_B$$

and we solve for D:

$$D = \frac{F_1 z_1 + F_2 z_2 - (F_1 + F_2)x_B}{x_D - x_B} \qquad (6\text{-}6)$$

D. Do It. The two feed lines and the top operating line can immediately be plotted on a y-x diagram. This is shown in Figure 6-5. Before plotting the middle operating line we must find D from Eq. (6-6).

$$D = \frac{(200)(0.3) + (300)(0.4) - (500)(0.02)}{0.72 - 0.02}$$

$$= 242.9 \text{ kg moles/hr}$$

Now the middle operating line y=x intercept can be determined from Eq. (6-4).

$$y = x = \frac{Dx_D - F_2 z_2}{D - F_2}$$

$$= \frac{(242.9)(0.72) - (300)(0.4)}{242.9 - 300} = -0.96$$

This intersection point could be used, but it is off of the graph in Figure 6-5. Instead of using a larger sheet of paper, we will calculate the slope L'/V'.

$$L' = F_2 q_2 + L = F_2 q_2 + (\frac{L}{D})(D)$$

$$L' = 300(5/4) + (1.0)(242.9) = 617.9 \text{ kg moles/hr}$$

$$V' = L' + D - F_2 = 617.9 + 242.9 - 300 = 560.8$$

$$L'/V' = 1.10$$

Now the middle operating line is plotted in Figure 6-5 from the intersection of feed line 2 and the top operating line, with a slope of 1.10. The bottom operating line then goes from $y=x=x_B$ to the intersection of the middle operating line and feed line 1 (see Figure 6-5).

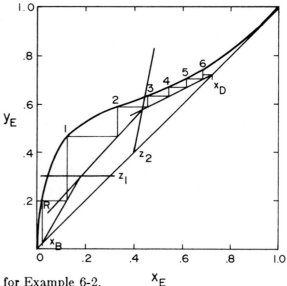

Figure 6-5.   Solution for Example 6-2.

Since feed locations were desired as stages above the reboiler, we step off stages from the bottom up, starting with the partial reboiler as the first equilibrium contact. The optimum feed stage for feed 1 is the first stage, while the optimum feed stage for feed 2 is the second stage. 6 stages + partial reboiler are more than sufficient.

$$\text{Frac} = \frac{x_D - y_5}{y_6 - y_5} = 1/2$$

We need 5 1/2 stages + partial reboiler.

E. Check. The internal consistency checks all make sense. Note that L'/V' can be greater than or less than 1.0. Since the latent heats of vaporization per mole are close, CMO is probably a good assumption. Our initial assumption that feed 2 enters the column higher up than feed 1 is shown to be valid by the McCabe-Thiele diagram. We could also calculate $\bar{L}$ and $\bar{V}$ and check that the slope of the bottom operating line is correct.

F. Generalize. The method of inserting the overall mass balance to simplify the intersection of the y = x line and middle operating line to derive Eq. (6-4) can be used in other cases. The method for calculating L'/V' can also be generalized to other situations. That is, we can calculate D (or B), find flow rate in section above (or below), and use feed conditions to find flow rates in the desired section. Since we stepped off stages from the bottom up, the fractional stage is calculated from the difference in y values (that is, vertical distances). Industrial problems use lower reflux ratios and have more stages. A relatively large reflux ratio is used in this example to keep the graph simple.

6.3. OTHER COLUMN SITUATIONS

A variety of modifications of the basic column are often used. In this section we will briefly consider the unique aspects of several of these. Constant molal overflow will be assumed. Detailed examples will not be given but will be left to serve as homework problems.

6.3.1. Partial Condensers

A partial condenser condenses only part of the overhead stream and returns this as reflux. The distillate product is removed as a vapor as shown in Figure 6-6A. If a vapor distillate is desired, then a partial condenser will be very convenient. The partial condenser acts as one equilibrium contact.

If a mass balance is done on the more volatile component using the mass balance envelope shown in Figure 6-6A, we obtain

$$Vy = Lx + Dy_D$$

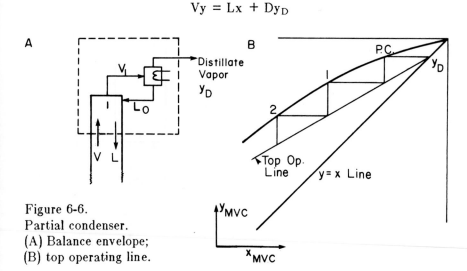

Figure 6-6.
Partial condenser.
(A) Balance envelope;
(B) top operating line.

Removing D and solving for y, we obtain the operating equation

$$y = \frac{L}{V} x + (1 - \frac{L}{V})y_D \qquad (6\text{-}7)$$

This is essentially the same as the equation for a top operating line with a total condenser except that y_D has replaced x_D. The top operating line will intersect the $y = x$ line at

$$y = x = y_D$$

The top operating line is shown in Figure 6-6B. The major difference between this case and that for a total condenser is that the partial condenser serves as the first equilibrium contact.

6.3.2. Total Reboilers

A total reboiler vaporizes the entire stream sent to it; thus, the vapor composition is the same as the liquid composition. This is illustrated in Figure 6-7. The mass balance and the bottom operating equation with a total reboiler are exactly the same as with a partial reboiler. The only difference is that a partial reboiler is an equilibrium contact and is labeled as such on the McCabe-Thiele diagram (e.g., Figure 6-5). The

Figure 6-7.   Total reboiler.

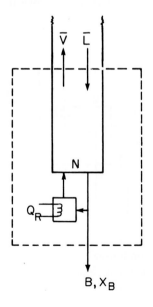

total reboiler is not an equilibrium contact and appears on the McCabe-Thiele diagram as the single point $y = x = x_B$.

Some types of partial reboilers (e.g., Perry and Chilton, 1973, Sec. 11) may act as more or less than one equilibrium contact. In these cases, exact details of the reboiler construction are required.

6.3.3. Side Streams

If a product of intermediate composition is required, a vapor or liquid side stream may be withdrawn. This is commonly done in petroleum refineries and is illustrated in Figure 6-8A for a liquid side stream. Three additional variables such as flow rate, S, type of side draw (liquid or vapor), and location or composition x_S or y_S, must be specified. The operating equation for the middle section can be derived from mass balances around the top or bottom of the column. For the situation shown in Figure 6-8A, the middle operating equation is

$$y = \frac{L'}{V'} x + \frac{Dx_D + Sx_S}{V'} \qquad (6\text{-}8)$$

Figure 6-8.   Liquid side stream. (A) Column; (B) McCabe-Thiele diagram.

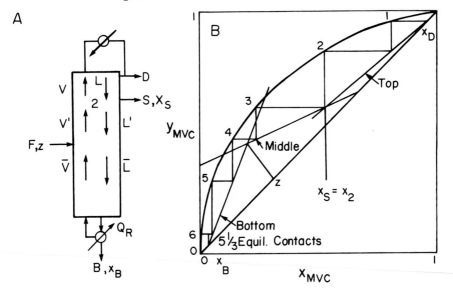

The $y = x$ intercept is

$$x = y = \frac{Dx_D + Sx_S}{D + S} \qquad (6\text{-}9)$$

This point can be plotted if S, x_S, D, and x_D are known. Derivation of Eqs. (6-8) and (6-9) is left as Problem 6-C3.

A second point can be found where the side stream is withdrawn. A saturated liquid withdrawal is equivalent to a negative feed of concentration x_S. Thus there must be a vertical feed line at $x = x_S$. The top and middle operating lines must intersect at this feed line.

Side-stream calculations have one difference that sets them apart from feed calculations. The stage must hit exactly at the point of intersection of the two operating lines. This is illustrated in Figure 6-8B. Since the liquid side stream is withdrawn from tray 2, we must have $x_S = x_2$. If the stage location is given, x_S can be found by stepping off the required number of stages.

For a liquid withdrawal, a balance on the liquid gives

$$L = L' + S \qquad (6\text{-}10)$$

while vapor flow rates are unchanged, $V = V'$. Thus slope, L'/V', of the middle operating line can be determined if L and V are known. L and V can be determined from L/D and D, where D can be found from external balances once x_S is known.

For a vapor side stream, the feed line is horizontal at $y = y_S$. A balance on vapor flow rates gives

$$V' = V + S_V \qquad (6\text{-}11)$$

while liquid flow rates are unchanged. Again L'/V' can be calculated if L and V are known.

If a specified x_S or y_s is desired, the problem is one of trial and error. The top operating line is adjusted (change L/D) until a stage ends exactly at x_S.

Calculations for side streams below the feed can be developed using similar principles (see Problem 6-C4).

6.3.4. Intermediate Reboilers and Intermediate Condensers

Another modification that is used occasionally is to have an intermediate reboiler or an intermediate condenser. The intermediate reboiler removes a liquid side stream from the column, vaporizes it, and reinjects the vapor into the column. An intermediate condenser removes a vapor side stream, condenses it, and reinjects it into the column. Figure 6-9A illustrates an intermediate reboiler.

An energy balance around the column will show that Q_R without an intermediate reboiler is equal to $Q_R + Q_I$ with the intermediate reboiler (F, z, q, x_D, x_B, p, L/D constant). Thus the amount of energy required is unchanged; what does change is the temperature at which it is required. Since $x_S > x_B$, the temperature of the intermediate reboiler is lower than that of the reboiler, and a cheaper heat source can be used. (Check this out with equilibrium data.)

Since the column shown in Figure 6-9A has four sections, there will be four operating lines. This is illustrated in the McCabe-Thiele diagram of Figure 6-9B. One would specify that the liquid be withdrawn at flow rate S at either a specified concentration x_S or a given

Figure 6-9.   Intermediate reboiler.   (A) Balance envelopes; (B) McCabe-Thiele diagram.

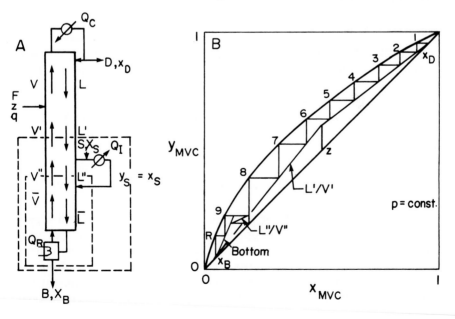

stage location. The saturated vapor is at concentration $y_S = x_S$. Thus there is a horizontal feed line at y_S. If the optimum location for inputting the vapor is immediately below the stage where the liquid is withdrawn, the L''/V'' line will be present, but no stages will be stepped off on it, as shown in Figure 6-9B. (The optimum location for vapor feed may be several stages below the liquid withdrawal point.) Development of the two middle operating lines is left as Problem 6-C5. Use the mass balance envelopes shown in Figure 6-9A.

Intermediate condensers are useful since the coolant can be at a higher temperature. See Problem 6-C6.

6.3.5. Stripping and Enriching Columns

Up to this point we have considered complete distillation columns with at least two sections. Columns with only a stripping section or only an enriching section are also commonly used. These are illustrated in Figure 6-10A. When only a stripping section is used, the feed must be a subcooled or saturated liquid. No reflux is used. A very pure bottoms product can be obtained but the vapor distillate will not be pure. In the enriching or rectifying column, on the other hand, the feed is a superheated vapor or a saturated vapor, and the distillate can be very pure but the bottoms will not be very pure. Stripping columns and enriching columns are used when a pure distillate or a pure bottoms, respectively, is not needed.

Analysis of stripping and enriching columns is similar. We will analyze the stripping column here and leave the analysis of the enriching column as a homework assignment (Problem 6-C8). The stripping column shown in Figure 6-10A can be thought of as a complete distillation column with zero liquid flow rate in the enriching section. Then the top operating line is $y = y_D$. The bottom operating line can be derived as

$$y = (\overline{L}/\overline{V})x - (\overline{L}/\overline{V} - 1)x_B$$

which is the usual equation for a bottom operating equation with a partial reboiler. Top and bottom operating lines intersect at the feed line. If the specified variables are F, q, z, p, x_B, and y_D, the feed line can be plotted and then the bottom operating line can be obtained from its intersection at $y = x = x_B$ and its intersection with the feed line at y_D. (Proof is left as Problem 6-C7.) If the boilup rate, $\overline{V}/B$, is specified, then

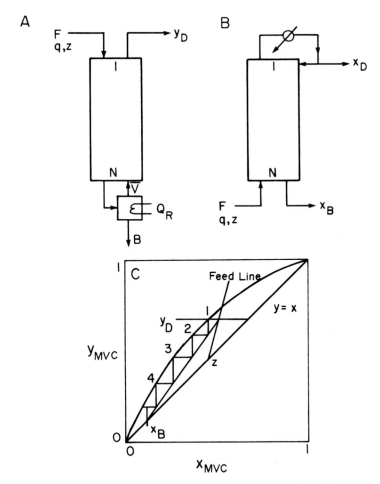

Figure 6-10.   Stripping and enriching columns. (A) Stripping; (B) enriching; (C) McCabe-Thiele diagram for stripping column.

y_D will not be specified and can be solved for. The McCabe-Thiele diagram for a stripping column is shown in Figure 6-10C.

6.4. LIMITING OPERATING CONDITIONS

It is always useful to look at limiting conditions. For distillation, two limiting conditions are total reflux and minimum reflux. In total reflux, all of the overhead vapor is returned to the column as reflux, and all of

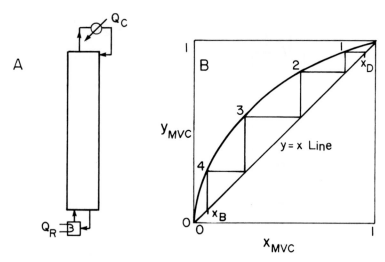

Figure 6-11.   Total reflux. (A) Column; (B) McCabe-Thiele diagram.

the underflow liquid is returned as boilup. Thus distillate and bottoms flow rates are zero. At steady state the feed rate must also be zero. Total reflux is used for starting up columns, for keeping a column operating when another part of the plant is shut down, and for testing column efficiency.

The analysis of total reflux is very simple. Since all of the vapor is refluxed, $L = V$ and $L/V = 1.0$. Also, $\bar{L} = \bar{V}$ and $\bar{L}/\bar{V} = 1.0$. Thus both operating lines become the $y = x$ line. This is illustrated in Figure 6-11. Total reflux represents the maximum separation that can be obtained with a given number of stages but zero throughput. Total reflux also gives the minimum number of stages required for a given separation.

Minimum reflux, $(L/D)_{min}$, is defined as the external reflux ratio at which the desired separation could just be obtained with an infinite number of stages. This is obviously not a real condition, but it is a useful hypothetical construct. To have an infinite number of stages, the operating and equilibrium lines must touch. In general, this can happen either at the feed or at a point tangent to the equilibrium curve. These two points are illustrated in Figures 6-12A and B. The point where the operating line touches the equilibrium curve is called the *pinch point*. At the pinch point the concentrations of liquid and vapor do not change from stage to stage. This is illustrated in Figure 6-12C for a pinch at the feed stage. If the reflux ratio is increased slightly, then the desired separation can be achieved with a finite number of stages.

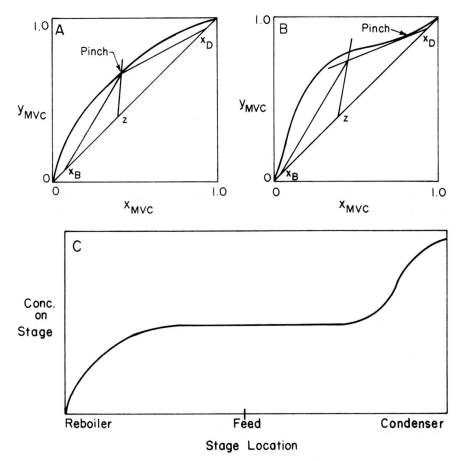

Figure 6-12.   Minimum reflux. (A) Pinch at feed stage; (B) tangent pinch; (C) concentration profile for $L/D \sim (L/D)_{min}$.

For binary systems the minimum reflux ratio is easily determined. The top operating line is drawn to a pinch point as in Figure 6-12A or B. Then $(L/V)_{min}$ is equal to the slope of this top operating line (which cannot be used for an actual column, since an infinite number of stages are needed), and

$$\left(\frac{L}{D}\right)_{min} = \left(\frac{L}{V-L}\right)_{min} = \frac{(L/V)_{min}}{1 - (L/V)_{min}} \qquad (6\text{-}12)$$

Note that the minimum reflux ratio depends on x_D, z, and q and can depend on x_B. The calculation of minimum reflux may be more complex when there are two feeds or a sidestream. This is explored in the homework problems.

The minimum reflux ratio is commonly used in specifying operating conditions. For example, we may specify the reflux ratio as $L/D = 1.2(L/D)_{min}$. Minimum reflux would use the minimum amount of reflux liquid and hence the minimum amount of heat in the reboiler, but the maximum (infinite) number of stages. Obviously, the best operating conditions lie somewhere between minimum and total reflux. As a rule of thumb, the optimum external reflux ratio is between 1.05 and 1.25 times $(L/D)_{min}$. (See Chapter 14 for more details.)

A maximum $\overline{L}/\overline{V}$ and hence a minimum boilup ratio $\overline{V}/B$ can also be defined. The pinch points will look the same as in Figures 6-12A or B. Problem 6-C12 looks at this situation further.

6.5. EFFICIENCIES

Up until now we have always assumed that the stages are equilibrium stages. Stages that are very close to equilibrium can be constructed, but they are only used for special purposes. To compare the performance of an actual stage to an equilibrium stage, we use a measure of efficiency.

Many different measures of efficiency have been defined. Two that are in common use are the overall efficiency and the Murphree efficiency. The overall efficiency, E_o, is defined as the number of equilibrium stages required for the separation divided by the actual number of stages required:

$$E_o = \frac{N_{equil}}{N_{actual}} \qquad (6\text{-}13)$$

Partial condensers and partial reboilers are not included in either the actual or equilibrium number of stages, since they will not have the same efficiency as the stages in the column.

The overall efficiency lumps together everything that happens in the column. What variables would we expect to affect column efficiency? The hydrodynamic flow properties such as viscosity and gas flow rate would affect the flow regime. The mass transfer rate, which is affected by the diffusivity, will in turn affect efficiency. Overall efficiency is usually smaller as the separation becomes easier (α_{AB} increases). The column size can also have an effect. Correlations for determining the overall efficiency will be discussed in Chapter 12. For now, we will con-

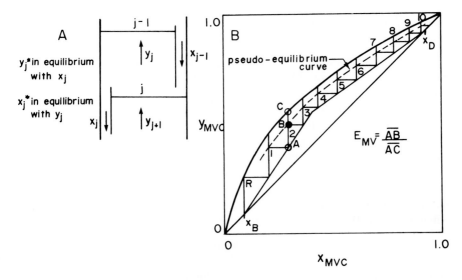

Figure 6-13. Murphree efficiency. (A) Stage nomenclature; (B) McCabe-Thiele diagram for E_{MV}.

sider that the overall efficiency is determined from operating experience with similar distillation columns.

The overall efficiency has the advantage of being easy to use but the disadvantage that it is very difficult to calculate from first principles. Stage efficiencies are defined for each stage and may vary from stage to stage. The stage efficiencies are easier to estimate from first principles or to correlate with operating data. The most commonly used stage efficiencies are the Murphree vapor and liquid efficiencies (Murphree, 1925). The Murphree vapor efficiency is defined as

$$E_{MV} = \frac{\text{actual change in vapor}}{\text{change in vapor for equilibrium stage}} \qquad (6\text{-}14)$$

Murphree postulated that the vapor between trays is well mixed, that the liquid in the downcomers is well mixed, and that the liquid on the tray is well mixed and is of the same composition as the liquid in the downcomer leaving the tray. For the nomenclature illustrated in Figure 6-13A, the Murphree vapor efficiency is

$$E_{MV} = \frac{y_j - y_{j+1}}{y_j^* - y_{j+1}} \qquad (6\text{-}15)$$

where y_j^* is vapor mole fraction in equilibrium with actual liquid mole fraction x_j.

Once the Murphree vapor efficiency is known for every stage, it can easily be used on a McCabe-Thiele diagram. (In fact, Murphree adjusted his paper to use the newly developed McCabe-Thiele diagram.) The denominator in Eq. (6-15) represents the vertical distance from the operating line to the equilibrium line. The numerator is the vertical distance from the operating line to the actual outlet concentration. Thus the Murphree vapor efficiency is the fractional amount of the total vertical distance to move from the operating line. If we step off stages from the bottom up, we get the result shown in Figure 6-13B. Note that the partial reboiler is treated separately since it will have a different efficiency than the remainder of the column.

The Murphree efficiency can be used as a ratio of distances as shown in Figure 6-13B. If the Murphree efficiencies are accurate, the locations labeled by the stage numbers represent the actual vapor and liquid compositions leaving a stage. These points can be connected to form a pseudo-equilibrium curve, but this curve depends on the operating lines used and thus has to be redrawn for each new set of operating lines. Figure 6-13B allows us to calculate the real optimum feed plate location and the real total number of stages.

A Murphree liquid efficiency can be defined as

$$E_{ML} = \frac{x_j - x_{j-1}}{x_j^* - x_{j-1}} \qquad (6\text{-}16)$$

which tells us that E_{ML} is the actual change divided by the change for an equilibrium stage. The Murphree liquid efficiency is similar to the Murphree vapor efficiency except that it uses horizontal distances. Note that $E_{ML} \neq E_{MV}$.

For binary mixtures the Murphree efficiencies are the same whether they are written in terms of the more volatile or least volatile component. For multicomponent mixtures they can be different for different components.

6.6. SIMULATION PROBLEMS

In a simulation problem the column has already been built, and we want to know how much separation can be obtained. As noted in Tables 4-1 and 4-3, the real number of stages, the real feed location, the column diameter, and the types and sizes of reboiler and condenser are known.

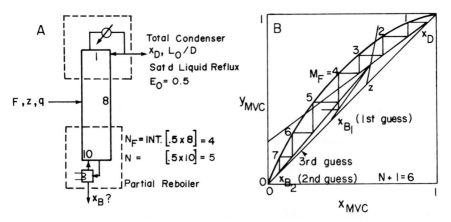

Figure 6-14.   Simulation   problems.   (A)   Existing   column;   (B) McCabe-Thiele diagram.

The engineer does the detailed stage-by-stage calculation and the detailed diameter calculation, and finally he or she checks that the operation is feasible.

To be specific, consider the situation where the known variables are F, z, q, x_D, T_{reflux} (saturated liquid), p, $Q_{col} = 0$, N_{actual}, $N_{feed\ actual}$, diameter, L_0/D, the overall efficiency E_o, and constant molal overflow. This column is illustrated in Figure 6-14A. The engineer wishes to determine the bottoms composition, x_B.

We start by deriving the top and bottom operating equations. These are the familiar forms,

$$y = \frac{L}{V} x + \left(1 - \frac{L}{V}\right)x_D$$

and

$$y = \frac{\overline{L}}{\overline{V}} x - \left(\frac{\overline{L}}{\overline{V}} - 1\right)x_B$$

The feed line equation is also unchanged

$$y = \frac{q}{q - 1} x + \frac{z_F}{1 - q}$$

Since the reflux is a saturated liquid and the external reflux ratio, L_0/D, is known, we calculate L/V and plot the top operating line (Figure 6-

14B). The feed line can also be plotted. The intersection of the top operating line and the feed line gives one point on the bottom operating line. Unfortunately, the bottom operating line cannot be plotted since neither $\overline{L}/\overline{V}$ nor x_B is known (why won't external balances give one of these variables?).

To proceed we must use the three items of information that have not been used yet. These are N_{actual}, $N_{feed\ actual}$, and E_o. We can estimate the equilibrium number of stages as

$$N = E_o\ N_{actual}, \quad N_F = \text{Integer}\ [N_{feed\ actual}\ E_o] \qquad (6\text{-}17)$$

The feed location in equilibrium stages, N_F, must be estimated as an integer, but the total number of equilibrium stages, N, could be a fractional number. Now we can step off equilibrium stages on the top operating line until we reach the feed stage N_F. At this point we need to switch to the bottom operating line, which is not known. To use the final bit of information, the value N, we must guess x_B, plot the bottom operating line, and check to see if the separation is achieved with $N+1$ (the $+1$ includes the partial reboiler) equilibrium contacts. Thus the simulation problem is one of trial and error when a stage-by-stage computation procedure is used. This procedure is illustrated in Figure 6-14B. Note that the actual feed stage may not be (and probably is not) optimum.

Once x_B has been determined the external balances can be completed, and we can determine B, D, Q_c, and Q_R. Now L, V, $\overline{L}$, and $\overline{V}$ can be calculated and we can proceed to an exact check that V and $\overline{V}$ are less than V_{max}. This is done by calculating a permissible vapor velocity. This calculation is similar to calculating u_{perm} for a flash drum and is shown in Chapter 12. The condenser and reboiler sizes can also be checked. If the flow rates are too large or the condenser and reboiler are too small, the existing column will not satisfactorily do the desired separation. Either the feed rate can be decreased or L/D can be decreased. This latter change obviously requires that the entire solution be repeated.

When other variables are specified, the stage-by-stage calculation is still trial and error. The basic procedure remains the same. That is, calculate and plot everything you can first, guess the needed variable, and then check whether the separation can be obtained with the existing number of stages. Murphree stage efficiencies are easily employed in these calculations.

6.7. NEW USES FOR OLD COLUMNS

Closely related to simulation is the use of existing or used distillation systems for new separations. The new use may be debottlenecking — that is, increasing capacity for the same separation. With increasing turnover of products, the problem of using equipment for new separations is becoming much more common.

Why would we want to use an existing column for a problem it wasn't designed for? First, it is usually cheaper to modify a column that has already been paid for than to buy a new one. Second, it is usually quicker to do minor modifications than to wait for a new column. Finally, for many engineers there is an esthetic appeal to solving the often knotty problems involved in adapting a column to a new separation.

The first thing to do when new chemicals are to be separated is clean the entire system and inspect it thoroughly. Is the system in good shape? If not, will minor maintenance and parts replacement put the equipment in working order? If there are major structural problems such as major corrosion, it may be cheaper and less of a long-term headache to buy new equipment.

Do simulation calculations to determine how close the column will come to meeting the new separation specifications. Rarely will the column provide a perfect answer to the new problem. Difficulties can be classified as problems with the separation required and problems with capacity.

What can be done if the existing column cannot produce the desired product purities? The following steps can be explored (they are listed roughly in the order of increasing cost).

1. Find out whether the product specifications can be relaxed. A purity of 99.5% is much easier to obtain than 99.99%.

2. See if a higher reflux ratio will do the separation. Remember to check if column vapor capacity and the reboiler and condenser are large enough. If they are, changes in L/D affect only operating costs.

3. Change the feed temperature. This may make a nonoptimum feed stage optimum.

4. Will a new feed stage at the optimum location (the existing feed stage is probably nonoptimum) allow you to meet product specifications?

5. Consider replacing the existing trays (or packing) with more efficient or more closely spaced trays (or new packing). This is relatively expensive but is cheaper than buying a completely new system.

6. Check to see if two existing columns can be hooked together in series to achieve the desired separation. Feed can be introduced at the feed tray of either column or in between the two columns. Since vapor loading requirements are different in different sections of the column (see Chapter 12), the columns do not have to be the same diameter.

What if the column produces product much purer than specifications? This problem is pleasant. Usually the reflux ratio can be decreased, which will decrease operating expenses.

Problems with vapor capacity are discussed in more detail in Chapter 12. Briefly, if the column diameter is not large enough, the engineer can consider:

1. Operating at a reduced L/D, which reduces V. This may make it difficult to meet the product specifications.

2. Operating at a higher pressure, which increases the vapor density. Note that the column must have been designed for these higher pressures.

3. Using two columns in parallel.

4. Replacing the downcomers with larger downcomers (see Chapter 12).

5. Replacing the trays or packing with trays or packing with a higher capacity. Major increases in capacity are unlikely.

If the column diameter is too large, vapor velocities will be low. The trays will operate at tray efficiencies lower than designed, and in severe cases they may not operate at all since liquid may dump through the holes. Possible solutions include:

1. Decrease column pressure to decrease vapor density. This increases the linear vapor velocity.

2. Increase L/D to increase V.

3. Recycle some distillate and bottoms product to effectively increase F.

Using existing columns for new uses often requires a creative solution. Thus these problems can be both challenging and fun; they are also often assigned to engineers just out of school.

6.8. SUBCOOLED REFLUX AND SUPERHEATED BOILUP

What happens if the reflux liquid is subcooled or the boilup vapor is superheated? We have already looked at two similar cases where we have a subcooled liquid or a superheated vapor *feed*. In those cases we found that a subcooled liquid would condense some vapor in the column, while a superheated vapor would vaporize some liquid. Since reflux and boilup are inputs to the column, we should expect exactly the same behavior if these streams are subcooled or superheated.

Subcooled reflux often occurs if the condenser is at ground level. Then a pump is required to return the reflux to the top of the column. A saturated liquid will cause cavitation and destroy the pump; thus, the liquid must be subcooled if it is to be pumped. To analyze the effect of subcooled reflux, consider the top of the column shown in Figure 6-15. The cold liquid stream, L_0, must be heated up to its boiling point. This energy must come from condensing vapor on the top stage, stream c in Figure 6-15. Thus the flow rates on the first stage are different from those in the rest of the rectifying section. Constant molal overflow is valid in the remainder of the column. The internal reflux ratio in the rectifying column is $L_1/V_2 = L/V$, and the top operating line is

$$y = \frac{L}{V} x + (1 - \frac{L}{V})x_D$$

Now, L/V cannot be directly calculated from the external reflux ratio L_0/D, since L and V change on the top stage.

Balances on vapor and liquid streams give

$$V_2 = V_1 + c \tag{6-18}$$

$$L_1 = L_0 + c \tag{6-19}$$

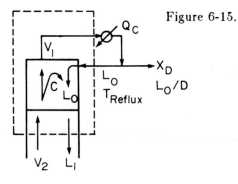

Figure 6-15.   Balance envelope for subcooled reflux.

An energy balance using the balance envelope in Figure 6-15 is

$$V_1 H_1 + L_1 h_1 = L_0 h_{reflux} + V_2 H_2 \qquad (6\text{-}20)$$

With constant molal overflow, $H_1 = H_2$. Then Eq. (6-20) becomes

$$(V_2 - V_1)H = L_1 h_1 - L_0 h_{reflux}$$

Using Eqs. (6-18) and (6-19), this becomes

$$cH = (L_0 + c)h_1 - L_0 h_{reflux}$$

Solving for amount condensed, c, we get

$$c = \left(\frac{h_{liq} - h_{reflux}}{H_{vap} - h_{liq}}\right) L_0 = f_c L_0 \qquad (6\text{-}21a)$$

which can also be written as

$$c = \frac{\overline{C}_{P_L}(T_{BP} - T_{reflux})}{\lambda} L_0 = f_c L_0 \qquad (6\text{-}21b)$$

or

$$c = \left(\frac{\text{energy to heat reflux to boiling}}{\text{latent heat of vaporization}}\right) L_0 = f_c L_0 \qquad (6\text{-}21c)$$

where f_c is the fraction condensed per mole of reflux. We can now calculate the internal reflux ratio, $L/V = L_1/V_2$. To do this, we start with the ratio we desire and use Eqs. (6-18), (6-19), (6-21), and (6-22). That is,

$$\frac{L}{V} = \frac{L_1}{V_2} = \frac{L_0 + c}{V_1 + c} = \frac{L_0 + f_c L_0}{V_1 + f_c L_0} = \frac{(1 + f_c)L_0/V_1}{1 + f_c L_0/V_1} \qquad (6\text{-}22)$$

The ratio L_0/V_1 is easily found from L_0/D as

$$\frac{L_0}{V_1} = \frac{L_0/D}{1 + L_0/D}$$

Using this expression in Eq. (6-22) we obtain

$$\frac{L}{V} = \frac{L_1}{V_2} = \frac{(1 + f_c)L_0/D}{1 + (1 + f_c)L_0/D} \tag{6-23}$$

Note that when $f_c=0$, Eqs. (6-22) and (6-23) both say $L_1/V_2 = L_0/V_1$. As the fraction condensed increases (reflux is subcooled more), the internal reflux ratio, L_1/V_2, becomes larger. Thus the net result of subcooled reflux is equivalent to increasing the reflux ratio. Numerical calculations (such as Problems 6-A9 or 6-D5) show that a large amount of subcooling is required to have a significant effect on L/V.

A superheated direct steam input or a superheated boilup from a total reboiler will cause vaporization of liquid inside the column. This is equivalent to a net increase in the boilup ratio, $\overline{V}/B$, and makes the slope of the stripping section operating line approach 1.0. Since superheated vapor inputs can be analyzed in the same fashion as the subcooled liquid reflux, it will be left as homework assignment 6-C14.

6.9. NONCONSTANT MOLAL OVERFLOW: LATENT HEAT UNITS

Throughout this chapter we have been assuming that constant molal overflow is valid. What if this assumption is invalid? Two approaches can be used. The first is to solve the mass and energy balances simultaneously on each stage. This is the Sorel technique and can be done graphically by the Ponchon-Savarit method. The second method is to modify the McCabe-Thiele procedure by redefining the basis on which the energy balances are done. This latent heat unit method (McCabe and Thiele, 1925) will be developed here.

For constant molal overflow, the essential assumptions are

1. Well-insulated column

2. Constant heat capacities and sensible heat changes small compared to latent heat changes

3. Same latent heat of vaporization per mole for the two components $(\lambda_A = \lambda_B)$

The first assumption can always be made approximately true, and the second assumption, while never exactly true, is usually close. If the

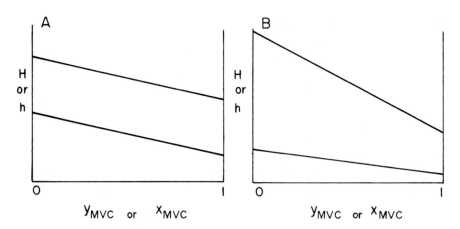

Figure 6-16.   Enthalpy-composition diagrams.   (A) Constant molal
overflow; (B) nonconstant molal overflow.

third assumption is true, then the enthalpy-composition diagram will
look like Figure 6-16A. When the third assumption is not valid but the
first two are, the enthalpy composition diagram will look like Figure 6-
16B. What we need is a transformation of the specific mass units that
will transform Figure 6-16B into 6-16A.

This may sound difficult, but we have already essentially done this
once. Figure 2-4 shows the enthalpy-composition diagram for ethanol
and water in weight units. This looks essentially like Figure 6-16B.
Changing to molal units where $\lambda_{EtOH} \sim \lambda_w$ transforms this to a figure
similar to Figure 6-16A. For systems where $\lambda_A \neq \lambda_B$, we want to find a
different unit that will transform Figure 6-16B into Figure 6-16A.[†]

This new unit is called a "latent heat unit." It is defined so that

$$\frac{\text{Heat vaporization of pure A}}{\text{Latent heat units A}} = \frac{\text{heat vaporization of pure B}}{\text{latent heat units of B}}$$

or

$$\lambda_A^* = \lambda_B^* \tag{6-24}$$

To find this latent heat unit we will arbitrarily pick component B as the
reference component (for the effects of this choice, see Problem 6-C19)

[†] Note that in some cases such as for many hydrocarbons, the appropriate unit is mass
(λ = joules/kg and weight fractions are used). Then constant mass overflow is valid.

and arbitrarily set

$$\text{Value } (MW_B^*) = \text{value } (MW_B) \tag{6-25}$$

where MW_B^* is the latent heat molecular weight in units such as grams per latent heat unit. This forces the equality

$$\text{Value } (\lambda_B^*) = \text{value } (\lambda_B) \tag{6-26}$$

where the units on the right and left sides of Eqs. (6-25) and (6-26) differ. Now if we set

$$MW_A^* = MW_A \frac{\lambda_B^*}{\lambda_A} \tag{6-27}$$

we have

$$\lambda_A^* \frac{\text{joules}}{\text{latent heat unit}} = \lambda_A \frac{\text{joules}}{\text{mole A}} \left(\frac{MW_A^* \text{ g/latent heat unit A}}{MW_A \text{ g/mole A}} \right)$$

or, substituting in Eq. (6-27),

$$\lambda_A^* = \lambda_A \left(\frac{1}{MW_A} \right)(MW_A \frac{\lambda_B^*}{\lambda_A}) = \lambda_B^* \tag{6-28}$$

Thus when the transformations in Eqs. (6-25) and (6-27) are used, constant latent heat unit overflow will be valid.

To use the latent heat units, we transform all mole fractions and all molal flow rates to constant latent heat unit fractions and constant latent heat unit flow rates. This transformation is exactly parallel to going from weight units (fractions and flow rates) to molal units.

For a binary system, pick a basis of 1 mole total. Then if the mole fraction of A is x_A, we have x_A moles of A. Then in latent heat units,

$$x_A \text{ (mole A)} \frac{MW_A \text{ g/mole}}{MW_A^* \text{ g/latent heat unit}} = x_A \left(\frac{MW_A}{MW_A^*} \right) \tag{6-29a}$$

and

$$x_B \frac{MW_B}{MW_B^*} = x_B \tag{6-29b}$$

The total is the sum of Eqs. (6-29a) and (6-29b). Thus the latent heat unit fractions are

$$x_A^* = \frac{x_A \dfrac{MW_A}{MW_A^*}}{x_A \dfrac{MW_A}{MW_A^*} + x_B} = \frac{x_A \dfrac{\lambda_A}{\lambda_B}}{x_A \dfrac{\lambda_A}{\lambda_B} + x_B} \qquad (6\text{-}30a)$$

and

$$x_B^* = \frac{x_B}{x_A \dfrac{MW_A}{MW_A^*} + x_B} = \frac{x_B}{x_A \dfrac{\lambda_A}{\lambda_B} + x_B} \qquad (6\text{-}30b)$$

Note that $x_A^* + x_B^* = 1.0$, and that $x_B^* \neq x_B$ even though the moles of B equals the number of latent heat units of B. The ratio $\dfrac{MW_A}{MW_A^*} = \dfrac{\lambda_A}{\lambda_B}$ is a constant multiplier in the transformation.

To convert flow rates, we first find the moles of A and moles of B and then convert to latent heat units of each. Thus,

$$L x_A \left(\frac{MW_A}{MW_A^*}\right) \text{ is the latent heat units of A/hr}$$

$$L x_B \frac{MW_B}{MW_B^*} = L\, x_B \text{ is latent heat units of B/hr}$$

Adding these we have

$$L^* = L\left(x_A \frac{MW_A}{MW_A^*} + x_B\right) = L\, \frac{x_B}{x_B^*} \qquad (6\text{-}31)$$

where we have substituted in Eq. (6-30b).

To solve a problem in latent heat units we must first convert all specified flow rates and compositions into latent heat units. Then the equilibrium data must be converted to latent heat unit fractions. This is easily done by setting up a table. Now the operating lines and feed lines will have exactly the same forms as before. The major difference is that we previously assumed that liquid and vapor flow rates were constant but now we have forced them to be constant. Thus in latent heat units the operating lines really are straight. The equilibrium data will be shifted.

It is best to illustrate with an example.

Example 6-3. Latent Heat Units for Methanol-Water Separation

In Example 6-1 we solved an open steam distillation example assuming constant molal overflow. We will now go back and recheck that assumption.

Solution

The column and the specified variables are shown in Figure 6-1. Since $\lambda_M \neq \lambda_W$ we will use latent heat units. Pick water as the reference component. Then from Eq. (6-25),

$$MW_W^* = MW_W = 18.016 \ , \quad \lambda_W^* = \lambda_W = 9.72$$

From Eq. (6-27),

$$MW_M^* = MW_M \left(\frac{\lambda_W^*}{\lambda_M}\right) = 32.04 \left(\frac{9.72}{8.43}\right) = 36.94$$

and the multiplier is

$$\frac{MW_M}{MW_M^*} = \frac{32.04}{36.94} = 0.867 = \frac{\lambda_M}{\lambda_W}$$

Next change all compositions to latent heat units. For instance, the feed is

$$z_M^* = \frac{z_M \dfrac{MW_M}{MW_M^*}}{z_M\left(\dfrac{MW_M}{MW_M^*}\right) + z_W} = \frac{(0.6)(0.867)}{(0.6)(0.867) + 0.4} = 0.565$$

and by similar calculations, $x_{D_M}^* = 0.943$ and $x_{B_M}^* = 0.070$.

The feed rate in latent units is

$$F^* = F\left(z_M \frac{MW_M}{MW_M^*} + z_W\right) = 322.07$$

The operating lines will be

$$y^* = \frac{L^*}{V^*} + \left(1 - \frac{L^*}{V^*}\right)x_D^*$$

and

$$y^* = \frac{\overline{L}^*}{\overline{V}^*} x^* - \frac{\overline{L}^*}{\overline{V}^*} x_B^*$$

while the feed line is $y^* = \dfrac{q^*}{q^* - 1} x^* + \dfrac{z^*}{(1 - q^*)}$

All starred quantities are in latent heat units. The top operating line slope is

$$\frac{L^*}{V^*} = \frac{L^*}{L^* + D^*} = \frac{L^*/D^*}{1 + L^*/D^*} = \frac{L/D}{1 + L/D} = \frac{3}{4}$$

where we have used $L^*/D^* = L/D$, since streams L and D are of the same composition. The value of q^* is

$$q^* = \frac{\overline{L}^* - L^*}{F^*} = \frac{\overline{V}^* - V^* + F^*}{F^*} = 1 - \frac{V_F^*}{F^*}$$

V_F^*/F^* will be very slightly different from $V_F/F = 0.3$, because V_F and F are of different composition. Since $F = 350$, we have $V_F = 105$. The vapor composition is $y_{V_F} = 0.79$ from Figure 6-2 (remember, the feed line is a solution of the flash equation). Then

$$V_F^* = 105\,[(0.79)(0.867) + 0.21] = 93.97$$

and

$$\frac{V_F^*}{F^*} = 0.29 \quad \text{so} \quad q^* = 0.71 \quad \text{and} \quad \frac{q^*}{q^* - 1} = -2.448$$

Table 6-1.   Conversion of Equilibrium Data to Latent Heat Unit Fractions. x_M and y_M Are the Mole Fraction Equilibrium Data from Table 3-3. x_M^* and y_M^* Are Calculated Using Eq. (6-30a).

x_M	x_M^*	y_M	y_M^*	T ° C
0	0	0	0	100.0
0.02	0.017	0.134	0.118	96.4
0.04	0.035	0.230	0.206	93.5
0.06	0.052	0.304	0.275	91.2
0.08	0.070	0.365	0.333	89.3
0.1	0.088	0.418	0.384	87.7
0.15	0.133	0.517	0.481	84.4
0.20	0.178	0.579	0.544	81.7
0.3	0.271	0.665	0.632	78.0
0.4	0.366	0.729	0.700	75.3
0.5	0.464	0.779	0.753	73.1
0.6	0.565	0.825	0.803	71.2
0.7	0.669	0.870	0.853	69.3
0.8	0.776	0.915	0.903	67.5
0.9	0.886	0.958	0.952	66.0
1.0	1.0	1.0	1.0	64.5

In constant molal overflow units, q was 0.7. This small difference between q and q^* can probably be safely ignored. Now we can plot the operating and feed lines.

Before stepping off the stages, we need to transform the equilibrium data into latent heat unit fractions. This is easily done as shown in Table 6-1. Note that in Table 6-1 the largest differences between mole fractions and latent heat unit fractions are near a value of 1/2. Thus the largest deviation will be near the feed stage. The equilibrium data, operating lines, and the feed line are shown in Figure 6-17. Comparison of Figure 6-17 with Figure 6-2 shows that all values are shifted slightly. Stage 3 is clearly the optimum feed stage, and 4.9 equilibrium stages are required. This is very close to the result obtained with the CMO assumption. Thus for Example 6-1, CMO is a reasonable assumption even though the latent heats per mole differ by 15.3%.

The reason CMO works so well even though the latent heats differ is that the reflux ratio is much greater than the minimum reflux ratio. The minimum internal reflux ratio from Figure 6-17 (dashed line) is $(L^*/V^*)_{min} = 0.40$ and

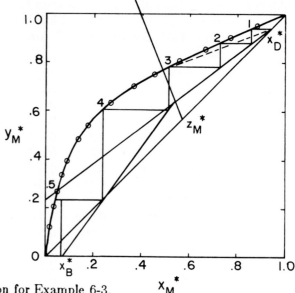

Figure 6-17. Solution for Example 6-3.

$$\left(\frac{L_0}{D}\right)_{min} = \left(\frac{L_0^*}{D^*}\right)_{min} = \frac{L^*/V^*}{1 - L^*/V^*} = 0.67$$

When CMO is assumed, the minimum external reflux ratio from Figure 6-2 is

$$\left(\frac{L}{V}\right)_{min\ CMO} = 0.37 \quad and \quad \left(\frac{L}{D}\right)_{min\ CMO} = 0.59$$

Thus $(L/D)_{min}$ is actually 14% higher than we would calculate with the CMO assumption. If we assumed CMO and used $(L/D) = 1.1\ (L/D)_{min}$, the column we designed on paper would be below the actual minimum reflux ratio and could not work regardless of the number of stages used. Thus the accuracy of the CMO assumption is critical when we are near minimum reflux.

The hardest part of this problem is converting the equilibrium data to latent heat units. Fortunately, once this has been done, the result, Table 6-1, can be used for any distillation of methanol and water.

Although equilibrium curves in general change as we transform from molal to latent heat units, it turns out that constant relative volatilities are invariant. That is,

$$\alpha^* = \frac{y^*(1 - x^*)}{x^*(1 - y^*)} = \frac{y(1 - x)}{x(1 - y)} = \alpha \tag{6-32}$$

This is also true for multicomponent systems where the latent heat unit is found as

$$x_i^* = \frac{x_i \, \lambda_i / \lambda_{ref}}{\displaystyle\sum_{i=1}^{i} \left(x_i \, \frac{\lambda_i}{\lambda_{ref}}\right)} \tag{6-33}$$

and

$$\alpha_i^* = \frac{y_i^* / x_i^*}{y_{ref}^* / x_{ref}^*} = \frac{y_i / x_i}{y_{ref} / x_{ref}} = \alpha_i \tag{6-34}$$

The proof of Eqs. (6-32) and (6-34) is left as Problem 6-C17. The significance of Eq. (6-32) is that if α is constant no change in equilibrium data is required.

6.10. ANALYTICAL AND COMPUTER CALCULATIONS

The Lewis method is exactly the same as the McCabe-Thiele method except that the calculations are not done graphically. When we wish to do the stage-by-stage calculations on the computer, we use the Lewis method. All of the methods developed in this chapter for graphical calculations are directly applicable to the Lewis method, including subcooled reflux calculations and latent heat unit calculations.

When doing the stage-by-stage calculations from the top down, we solve for x values from the equilibrium relationship and for y values from the operating equation. If we step off stages from the bottom up, then we calculate x values from the operating equation and y values from equilibrium. Check this statement out on a McCabe-Thiele diagram. When going from the bottom up, we want to solve all the operating equations for x. As noted previously, the optimum feed stage can be determined from the test in Eq. (5-36), and the point of intersection (Eq. 5-37) is more convenient to use than the feed line. For more complex situations, the point of intersection of a feed line and an operating equation can be found by simultaneously solving the equations. The biggest problem in utilizing the Lewis method on the computer is obtaining a good fit for the equilibrium data. The data can be fit to curves, or an interpolation routine can be used to interpolate between data points. Calculations should be set up on the computer when a large number of cases must be studied; this is often the case when economic analyses are done.

A flow sheet for a simple column was shown in Figure 5-15. For a more complex column this would be modified by:

1. Including equations for intermediate operating lines

2. Including additional tests for optimum feeds [change Eqs. (5-37a,b) or use alternative test discussed in Problem 6-C25]

3. Including side streams

The principles again follow McCabe-Thiele calculations step by step.

With the advent of powerful, inexpensive, and readily available computers, is the McCabe-Thiele method obsolete? We do not think so. The McCabe-Thiele method provides a powerful visual tool for analyzing what occurs inside a distillation column. The diagram demonstrates the interaction of the mass balances (operating lines), equilibrium, and energy balance on the feed (feed line).

6.11. SUMMARY - OBJECTIVES

In this chapter we applied the McCabe-Thiele and Lewis methods to a variety of binary distillation problems. The objectives you should be able to satisfy are:

1. Solve any binary distillation problem where constant molal overflow is valid. This includes:

   a. Open steam

   b. Multiple feeds

   c. Partial condensers and total reboiler

   d. Side streams

   e. Intermediate reboilers and condensers

   f. Stripping and enriching columns

   g. Total and minimum reflux

   h. Overall and Murphree efficiencies

i.   Simulation problems

j.   Any combination of the above

2. Include the effects of subcooled reflux or superheated boilup in your McCabe-Thiele and Lewis analyses.

3. Use latent heat units when constant molal overflow is not valid.

4. Develop flow charts and computer programs for any binary distillation problem.

REFERENCES

McCabe, W.L. and E.W. Thiele, "Graphical Design of Fractionating Columns," *Ind. Eng. Chem., 17,* 605 (1925).

Murphree, E.V., "Graphical Rectifying Column Calculations," *Ind. Eng. Chem., 17,* 960 (1925).

Perry, R.H. and C.H. Chilton (Eds.), *Chemical Engineer's Handbook,* 5th ed., McGraw-Hill, New York, 1973.

HOMEWORK

A.   *Discussion Problems*

A1.   If L/V with saturated reflux is the same as L/V with subcooled reflux, is $|Q_c|$ greater, the same, or less for the saturated reflux case?

A2.   What is the effect of column pressure on distillation? To explore this consider pressure's effect on the reboiler and condenser temperatures, the volumetric flow rates inside the column, and the relative volatility (which can be estimated for hydrocarbons from the DePriester charts).

A3.   What happens if we try to step off stages from the top down and E_{MV} is given? Determine how to do this calculation.

A4.   Explain and sketch on a McCabe-Thiele diagram how to determine x_B for an existing column if the given variables are F, z, q, L_0/D, saturated liquid reflux, x_D, N_{actual}, $N_{F\ actual}$, and E_{MV}.

A5.   Figure 6-12B illustrates a tangent pinch in the rectifying section.

What shape would the equilibrium data have to have for a tangent pinch in the stripping section?

A6. Figure 6-12B illustrates a subcooled liquid feed ($q > 1$). Estimate the q value at which there are two pinch points (the tangent pinch and a feed point pinch). What happens as q is decreased even more?

A7. For the simulation problem shown in Figure 6-14 why won't external balances allow you to calculate $\overline{L}/\overline{V}$ or x_B?

A8. Under what conditions will the latent heat method be invalid?

A9. When would it be safe to ignore subcooling of the reflux liquid and treat the reflux as a saturated liquid? Do a few numerical calculations for either methanol and water or ethanol and water to illustrate.

A10. Equations (6-10) and (6-11) are mass balances on particular phases. When will these equations be valid?

A11. When might you use an intermediate condenser on a column? What are the possible advantages?

A12. When would just a stripping or just an enriching column be used?

A13. Sketch the plot of number of stages versus L/D for the case where F, Z, q, x_D, x_B, and p are fixed. For the same case, sketch vapor flow rate versus L/D. Sketch Q_c and Q_R versus L/D. What conclusions can you draw about operating and capital costs? (See Chapter 14 for more details.)

A14. What is the usefulness of calculating a fractional number of equilibrium stages?

A15. Explain with a McCabe-Thiele diagram how changing feed temperature (or equivalently, q) may help an existing column achieve the desired product specifications.

A16. Develop a key relations chart for binary McCabe-Thiele distillation. That is, on one sheet of paper summarize everything you need to know about binary distillation. You will probably want to include information about operating lines, feed lines, efficiencies, subcooled reflux, and so forth.

B. *Generation of Alternatives*

B1. Invent your own problem that is distinctly different from those discussed in this chapter. Show how to solve this problem.

B2. Several ways of adapting existing columns to new uses were listed. Generate *new* methods that might allow existing systems to meet product specifications that could not be met without modification. Note that you can postulate a complex existing column such as one with an intermediate reboiler.

C. *Derivations*

C1. For Example 6-1 (open steam), show that the x intercept is at $x = x_B$.

C2. Derive the bottom operating line for a column with a total reboiler. Show that this is the same result as is obtained with a partial reboiler.

C3. Derive Eqs. (6-8) and (6-9).

C4. For a side stream below the feed:

    a. Draw a sketch corresponding to Figure 6-8A.

    b. Derive the operating equation and $y = x$ intercept.

    c. Sketch the McCabe-Thiele diagram.

C5. Derive the operating equations for the two middle operating sections when an intermediate reboiler is used (see Figures 6-9A and B). Show that the operating line with slope of L'/V' goes through the point $y = x = x_B$.

C6. Show that the total amount of cooling needed is the same for a column with one total condenser (Q_c) as for a column with a total condenser and an intermediate total condenser $(Q_c + Q_I)$. F, z, q, x_D, x_B, and Q_R are constant for the two cases. Sketch a system with an intermediate condenser. Derive the operating equations for the two middle operating lines, and sketch the McCabe-Thiele diagram.

C7. For the stripping column shown in Figures 6-10A and C, show formally that the intersection of the bottom operating line and the feed line is at y_D. In other words, solve for the intersection of these two lines.

C8. Develop the McCabe-Thiele procedure for the enriching column shown in Figure 6-10B.

C9. Derive the equations for the points of intersection of the two operating lines (y_I, x_I) for an open steam system similar to Figure 6-1.

C10. Develop a computer flow sheet for an open steam system similar to Figure 6-1.

C11. Sketch a McCabe-Thiele diagram where the tangent pinch point for minimum reflux would be in the stripping section.

C12. Show how to determine $(\overline{V}/B)_{min}$. Derive an equation for calculation of $(\overline{V}/B)_{min}$ from $(\overline{L}/\overline{V})_{max}$.

C13. Sketch the McCabe-Thiele diagram if the Murphree *liquid* efficiency is constant and $E_{ML} = 0.75$.

C14. Derive the equations to calculate $\overline{L}/\overline{V}$ when a superheated boilup is used.

C15. Derive the equations to calculate $\overline{L}/\overline{V}$ when direct superheated steam is used.

C16. If a distillation problem is given in terms of weight units (weight fractions and kg/hr), show how to convert directly to latent heat molecular weight units.

C17. Show that Eqs. (6-32) and (6-34) are valid.

C18. Show that a constant relative volatility is invariant in weight fractions.

C19. Suppose that component A was chosen as the reference component for the latent heat units. Then Eq. (6-25) becomes

$$\text{Value } (MW_A^*) = \text{value } (MW_A)$$

Show that

a. Fractions and hence equilibrium data are not changed.

b. Flow rates *do* change.

c. Final results (operating lines and number of stages) are the same.

C22. Show how f_c in Eqs. (6-21) and (6-22) is related to q.

C23. Derive the operating equation for section 2 of Figure 6-3. Show that the equations are identical whether the mass balance envelope is drawn around the top of the column or the bottom of the column.

C24. Develop a computer flow sheet for a two-feed column similar to the one in Figure 6-4. Derive all required equations.

C25. Equations (5-37) are one way to test for the optimum feed location. An alternative method is to determine at each equilibrium value which operating line is lower.

   a. Develop this test for a two-feed column. Do it both going down and going up the column.

   b. Does this test work for a column with a side stream? (See Figure 6-8.)

   c. Develop a computer flow sheet for the side-stream system illustrated in Figure 6-8.

C26. Develop a computer flow sheet for a trial-and-error simulation problem.

D.   *Problems*

D1. Find the number of stages and the optimum feed location for a column separating methanol and water. The column has a total condenser, saturated liquid reflux, and open steam heating. The steam is a saturated vapor. Constant molal overflow can be assumed. Feed composition = 0.4, $x_D = 0.96$, $x_B = 0.04$. Pressure is 1 atm.

   a. $L/D = 1.0$ and $q = 0.8$.

   b. $L/D = 1.0$ and $q = 2$.

   c. What is the minimum L/D for parts a and b?

Equilibrium data are given in Table 3-3. Do all three parts on one McCabe-Thiele diagram.

D2. For Problem 5-D7c for separation of acetone from ethanol, determine:

   a. How many stages are required at total reflux?

   b. What is $(L/V)_{min}$? What is $(L/D)_{min}$?

   c. The L/D used is how much larger than $(L/D)_{min}$?

   d. If $E_{mv} = 0.75$, how many real stages are required for $L/V = 0.8$?

D3. When water is the more volatile component we do not need a condenser but can use direct cooling with boiling water. This was shown in Problem 4-D3. We set $y_D = 0.92$, $x_B = 0.04$, $z = 0.4$ (all mole fractions water), feed is a saturated vapor, feed rate is 1000 kg moles/hr, p = 1 atm, constant molal overflow is valid, the entering cooling water (W) is a saturated liquid and is pure water, and $W/D = 3/4$. Derive and plot the top operating line. Note that external balances (that is, balances around the entire column) are not required.

D4. When water is the more volatile component we do not need a condenser but can use direct cooling. This was illustrated in Problem 4-D3. We set $y_D = 0.999$, $x_B = 0.04$, $z = 0.4$ (all mole fractions water), feed is a saturated liquid, feed rate is 1000 kg moles/hr, $p = 1$ atm, constant molal overflow is valid, the entering cooling water (W) is pure water and $W/D = 3/4$. The entering cooling water is at $100\,°F$ while its boiling temperature is $212\,°F$.

$$C_{PL,w} = 18.0\ \frac{Btu}{lb\ mole\,°F}, \quad MW_w = 18, \quad \lambda_w = 17{,}465.4\ \frac{Btu}{lb\ mole}$$

Find the slope of the top operating line, L/V.
Note: Equilibrium data are not needed.

D5. A distillation column is operating with a subcooled reflux. The vapor streams have an enthalpy of $H_1 = H_2 = 17{,}500$ Btu/lb-mole, while the saturated liquid $h_1 = 3100$ Btu/lb-mole. Enthalpy of the reflux stream is $h_0 = 1500$ Btu/lb-mole. The external reflux ratio is set at $L_0/D = 1.1$. Calculate the internal reflux ratio inside the column, L_1/V_2.

D6. We are separating methanol and water in a distillation column using open steam heating and a partial condenser. Feed is a superheated vapor where 2 moles of feed vaporize 1 mole of liquid. Feed rate is 250 moles/hr, and feed composition is 0.8. Distillate composition is $y_D = 0.96$, and bottoms composition is $x_B = 0.10$. Use a reflux ratio $L/D = 2(L/D)_{min}$, and find optimum feed plate location and total number of equilibrium stages needed. Data are in Table 3-3.

D7. We are separating methanol and water in a distillation column with a total condenser and a partial reboiler. Reflux is a saturated liquid, and constant molal overflow can be assumed. Feed is a saturated liquid of composition 0.6 and flow rate of 500 moles/hr. Distillate composition is $x_D = 0.92$, and bottoms composition is $x_B = 0.04$. The external reflux ratio is $L/D = 1.0$. The partial reboiler acts as an equilibrium contact. The stages in the column have Murphree vapor efficiencies of 0.75. Find the number of real stages needed and the optimum feed plate location in real stages. Data are in Table 3-3.

D8.   A distillation column with a total condenser and a partial reboiler is separating an ethanol-water mixture. The feed is a saturated liquid that is 20 mole % ethanol, and the feed rate is 1000 moles/hr. A bottoms composition of 2 mole % ethanol is desired. The reflux is returned as a saturated liquid, and constant molal overflow can be assumed. Find the minimum reflux ratio and the number of stages required at total reflux (minimum number of stages) if

    a. Distillate is 65 mole % ethanol.

    b. Distillate is 85 mole % ethanol.

Equilibrium data are given in Table 2-1 and Figure 2-2.

D9.   A distillation column with a total condenser and a partial reboiler is separating acetone and ethanol. The feed is 70 mole % acetone, flows at 1000 moles/hr, and is a superheated vapor for which 1 mole of liquid must vaporize on the feed plate to cool 4 moles of feed. We desire a bottoms composition of 5 mole % acetone and a distillate composition of 90 mole % acetone. Reflux is a saturated liquid. Pressure is 1 atm. Data are in Problem 5-D7.

    a. What is the minimum number of equilibrium stages required for this separation (i.e., total reflux)?

    b. What is the minimum external reflux ratio?

    c. If the actual $L/D = 2(L/D)_{min}$, find the optimum plate location and the number of stages required for separation using a Murphree vapor efficiency of 75%.

D10. A distillation column with a total condenser and partial reboiler is separating methanol and water. We desire a bottoms composition of 2% methanol and a distillate that is 95% methanol. Reflux is returned to the column as a saturated liquid, and constant molal overflow is valid.

    a. Find the number of equilibrium stages required at total reflux.

    b. If the feed is 32% methanol, flows at 1502 moles/hr, and is a subcooled liquid where 2 moles of vapor must condense on the feed plate to heat 7 moles of feed to its boiling temperature, find the minimum external reflux ratio.

    c. If $L/D = 1.6(L/D)_{min}$ is used for the problem in part b, find the number of real stages needed and the actual feed location if the Murphree vapor efficiency is 75%. Remember that the partial reboiler has an efficiency of 100%.

Data are in Table 3-3.

D11. A distillation column with a total condenser and a *total* reboiler is separating 5000 moles/hr of a mixture that is 40 mole % methanol and 60 mole % water. The feed enters as a two phase mixture that flashes as 30% vapor on the feed stage. Constant molal overflow is valid, the reflux is returned as a saturated liquid, and the boilup is returned as a saturated vapor. We desire a distillate that is 97 mole % methanol and a bottoms that is 3 mole % methanol, and wish to use a boilup ratio of $\overline{V}/B = 1.0$. If the fifth stage above the total reboiler is used as the feed stage, find the total number of stages needed for this separation. Data are in Table 3-3.

D12. A distillation column is separating 1000 moles/hr of a 32 mole % ethanol, 68 mole % water mixture. The feed enters as a sub-cooled liquid that will condense 1 mole of vapor on the feed plate for every 4 moles of feed. The column has a partial condenser and uses open steam heating. We desire a distillate product $y_D = 0.75$ and a bottoms product $x_R = 0.10$. Constant molal overflow is valid. The steam used is pure saturated water vapor. Data are in Table 2-1 and Figure 2-2.

    a. Find the minimum external reflux ratio.

    b. Use $L/D = 2.0(L/D)_{min}$, and find the number of real stages and the real optimum feed location if the Murphree vapor efficiency is 2/3 for all stages.

    c. Find the steam flow rate used.

D13. We are separating 250 kg moles/hr of a methanol-water mixture in a distillation column that has a total condenser and a partial reboiler. The feed composition is 0.3 mole fractional methanol, and the feed is a subcooled liquid. Three moles of feed will condense 1 mole of vapor in the column. Distillate is a saturated liquid with a concentration of 0.90. Bottoms concentration is 0.02. Constant molal overflow can be assumed.

    a. Find the minimum external reflux ratio for this separation.

    b. Find the minimum number of stages needed for separation (this is total reflux).

    c. If we use $L/D = 2(L/D)_{min}$, find number of equilibrium stages and the optimum feed plate location.

Equilibrium data are in Table 3-3.

D14. We have a column with a total condenser, a total reboiler, and four stages that is separating a 45 mole % methanol, 55 mole % water mixture. The feed, which is 30% liquid and 70% vapor, is fed to the third stage below the condenser at a rate of 50 kg moles/hr. Reflux is a saturated liquid and material leaving the total reboiler is a saturated vapor. Constant molal overflow can be assumed. The external reflux ratio is 0.7. Distillate composition is 0.7. Find the resulting bottoms concentration. Data are in Table 3-3.

D15. A distillation column with a total reboiler and a partial condenser is separating two mixtures of acetone and ethanol. The first mixture flows at 1000 moles/hr and is a saturated liquid; the second flows at 500 moles/hr and is a saturated vapor. Both feeds are 45 mole % acetone. We desire a distillate composition of $y_D = 0.85$ and a bottoms composition of $x_B = 0.02$. Boilup is returned to the column as a saturated vapor and CMO is valid. An external reflux ratio of $L_0/D = 2$ is used. Find the optimum feed location for each feed and the total number of stages required. Data are in Problem 5-D7.

D16. A distillation column with a partial reboiler, a total condenser, four equilibrium stages (in addition to the partial reboiler), and feed on the third stage above the partial reboiler is separating a mixture that is 62 mole % methanol and 38 mole % water. Feed is a saturated liquid that flows at 1225 gallons/hr. We desire a bottoms product that is 4 mole % methanol and a distillate that is 92 mole % methanol. What external reflux ratio must we operate at? Data are in Table 3-3.

D17. A distillation column with a total condenser and a partial reboiler is separating ethanol from water. Feed is a saturated liquid that is 25 mole % ethanol. Feed flow rate is 150 moles/hr. Reflux is a saturated liquid and constant molal overflow is valid. The column has three equilibrium stages (i.e., four equilibrium contacts), and the feed stage is second from the condenser. We desire a bottoms composition that is 5 mole % ethanol and a distillate composition that is 63 mole % ethanol. Find the required external reflux ratio. Data are in Table 2-1 and Figure 2-2.

D18. A distillation column with two equilibrium stages and a partial reboiler (3 equilibrium contacts) is separating methanol and water. The column has a total condenser. Feed, a 45 mole % methanol mixture, enters the column on the second stage below the condenser. Feed rate is 150 moles/hr. The feed is a subcooled liquid. To heat 2 moles of feed to the saturated liquid temperature, 1 mole of vapor must condense at the feed stage. A distillate concentration of 80 mole % methanol is desired. Reflux is a saturated liquid, and constant molal overflow can be assumed. An external reflux ratio of $L/D = 2.0$ is used. Find the resulting bottoms concentration x_B. Data are in Table 3-3.

D19. A distillation column is separating methanol from water at 1 atm pressure. The column has a total condenser and a partial reboiler. In addition, a saturated vapor stream of pure steam is input on the second stage above the partial reboiler (see figure).

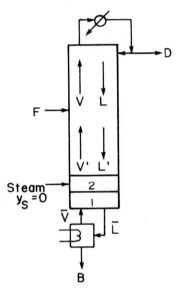

The feed flow rate is 2000 kg moles/day. Feed is 48 mole % methanol and 52 mole % water and is a subcooled liquid. For every 4 moles of feed, 1 mole of vapor must condense inside the column. Distillate composition is 92 mole % methanol. Reflux is a saturated liquid, and $L_0/D = 1.0$. Bottoms composition is 8 mole % methanol. Boilup ratio $\overline{V}/B = 0.5$. Equilibrium data are given in Table 3-3. Assume that constant molal overflow is valid. Find the optimum feed plate location and the total number of equilibrium stages required.

D20. A distillation column is separating methanol from water. The column has a total condenser that subcools the reflux so that 1 mole of vapor is condensed in the column for each 3 moles of reflux. $L_0/D = 3$. A liquid side stream is withdrawn from the second stage below the condenser. This side stream is vaporized to a saturated vapor and then mixed with the feed and input on

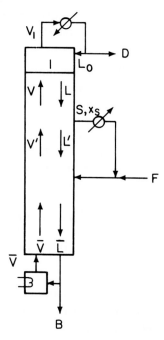

stage 4. The side withdrawal rate is S = 500 kg moles/hr. The feed is a saturated vapor that is 48 mole % methanol. Feed rate is F = 1000 kg moles/hr. A total reboiler is used, which produces a saturated vapor boilup. We desire a distillate 92 mole % methanol and a bottoms 4 mole % methanol. Assume constant molal overflow. Equilibrium data are given in Table 3-3. Find:

    a. The total number of equilibrium stages required.

    b. The value of $\overline{V}/B$.

D21. A distillation column with a total condenser and a total reboiler is separating ethanol from water. Reflux is returned as a saturated liquid, and boilup is returned as a saturated vapor. Constant molal overflow can be assumed. Assume that the stages are equilibrium stages. Column pressure is 1 atm. A saturated liquid feed that is 32 mole % ethanol is fed to the column at 1000 kg moles/hr. The feed is to be input on the optimum feed stage. We desire a distillate composition of 80 mole % ethanol and a bottoms composition that is 2 mole % ethanol. A liquid side stream is removed on the eighth stage from the top of the column at a flow rate of S = 457.3 kg moles/hr. This liquid is sent to an intermediate reboiler and vaporized to a saturated vapor, which is returned to the column at its optimum feed location. The external reflux ratio is $L_0/D = 1.86$. Find the optimum feed locations of the feed and of the vapor from the intermediate reboiler. Find the total number of equilibrium stages required. Be very neat! Data are in Table 2-1.

D22. We are using a distillation column with a partial reboiler and direct water cooling to separate water from compound A. The feed is a saturated liquid that is 80 mole % water, and the feed rate is 150 lb moles/hr. The bottoms has a concentration of $x_B = 0.3$, and the distillate $y_D = 0.98$ (both mole fractions water). In the top of the column, $L/V = 0.762$. A saturated liquid side stream is drawn off on the third stage from the top of the column. The side-stream rate is adjusted so that $L'/V' = 0.564$. The direct cooling water is a saturated liquid. Constant molal overflow can be assumed. Find the optimum feed location, the total number of stages, and the side-stream concentration. Equilibrium data are in the table. Operating pressure is 101.3 kPa. In a column with direct cooling no condenser is used and the product is taken off as a vapor (see Problem 4D-3). In the solution of this problem include the derivation and equations for the operating lines. Do the derivations; don't assume you know what the operating lines look like.

Water - Compound A Equilibrium Data

x_w	7.0	13.6	20.5	28.4	37.4	47.4	57.5	69.8	83.3	90.8
y_w	10.0	20.0	30.0	40.0	50.0	60.0	70.0	80.0	90.0	95.0

Numbers are mole %. Note: Systems such as ethylene glycol-water and acetic acid-water have water as the more volatile component.

D23. A methanol-water mixture is to be separated in a distillation column at 1 atm. Feed is 48 mole % methanol and is a saturated vapor. Distillate is 98 mole % methanol, and bottoms is 4 mole % methanol. Do *not* assume constant molal overflow. Data are given in Example 6-3 and Table 6-1.

    a. What is the minimum reflux ratio?

    b. What is the minimum number of stages?

D24. A distillation column is separating acetone from ethanol. Feed is a saturated liquid that is 40 mole % acetone. Feed rate is 50 kg moles/hr. Operation is at 1 atm and CMO can be assumed. The column has a total condenser and a partial reboiler. There are eight equilibrium stages in the column, and the feed is on the third stage above the reboiler. Three months ago the distillate flow was shut off (D = 0), but the column kept running. The boilup ratio was set at the value of $\overline{V}/B = 1.0$. Equilibrium data are given in Problem 5-D7.

    a. What is x_B?

    b. If a drop of distillate were collected, what would x_D be?

D25. A distillation column is separating acetone and ethanol. The column effectively has six equilibrium stages plus a partial reboiler. Feed is a two-phase feed that is 40% liquid and 75 mole % acetone. Feed rate is 1000 kg moles/hr, and the feed stage is fourth from the top. The column is now operating at steady state with the bottoms flow valve shut off. However, a distillate product is drawn off, and the vapor is boiled up in the reboiler. $L_0/D = 2$. Reflux is a saturated liquid. Constant molal overflow can be assumed. p = 1 atm. Equilibrium data are in Problem 5-D7. Find the distillate composition. If one drop of liquid in the reboiler is withdrawn and analyzed, predict x_B.

E. *More Complex Problems*

E1. A distillation column is separating methanol and water. The column has open (direct) steam heating and a total of five stages. A liquid side stream is withdrawn from the second plate above the bottom of the column. The feed plate is the fourth plate above the bottom of the column. The feed is 30 mole % methanol and is a subcooled liquid. One mole of vapor is condensed to heat 2 moles of feed to the saturated liquid temperature on the feed plate. Feed rate is 1000 moles/hr. A bottoms concentration of 1.5 mole % is desired. The steam used is pure saturated water vapor and the steam flow rate is adjusted so that

Steam flow rate/Bottoms flow rate = 0.833

The side stream is removed as a saturated liquid. The sidestream flow rate is adjusted so that

Side–stream flow rate/Bottoms flow rate = 0.4

A total condenser is used. Reflux is a saturated liquid, and CMO can be assumed. Find the side-stream concentration and the distillate concentration. Data are given in Table 3-3.

E2. Solve Problem 6-E1 with a computer or calculator program.

E3. We wish to distill a feed of water and acetic acid that is 70 mole % water and is a saturated liquid. Feed rate is 1000 kg moles/hr. We desire a bottoms concentration that is 28.5 mole % water and a distillate that is 90 mole % water. The column has a total condenser and a partial reboiler. Reflux is returned as a saturated liquid. Operation is at 760 mmHg. Operate at $L/D = 1.2 (L/D)_{min}$. Find the optimum feed stage and the total number of equilibrium stages required. Remember to use latent heat units.

Equilibrium Data for Water-Acetic Acid at 760 mmHg (Mole frac. water is given.)

x_w	.0530	.1250	.206	.297	.394	.510	.649	.803	.9594
y_w	.1333	.240	.338	.437	.533	.630	.751	.866	.9725

Latent heats of vaporization are: acetic acid at 391.4 K, $\lambda = 5.83$ kcal/g-mole; water at 373.16 K, $\lambda = 9.717$ kcal/g-mole.

E4. Solve Problem 6-E3 with a computer or calculator program.

E5. Write a computer program for the following problem: A distillation column with a total condenser, a partial reboiler, 13 equilibrium stages, and the feed on the seventh equilibrium stage below the condenser is separating an ethyl chloride-ethyl bromide mixture. Ethyl chloride is the more volatile. The relative volatility varies from 2.79 to 3.23. Approximate α as a linear function of the liquid composition varying from $\alpha = 2.79$ at $x = 0$ to $\alpha = 3.23$ at $x = 1.0$. The feed rate is 156 kg moles/hr, and the feed mole fraction is 0.65. The feed is a superheated vapor with $q = -1/2$. We desire $x_D = 0.98$ and $x_B = 0.01$. Reflux is a saturated liquid, and constant molal overflow can be assumed. Find the external reflux ratio that allows this separation to be made.

F. *Problems Requiring Other Resources*

F1. A distillation column separating methanol from water is using open (direct) steam heating. (Water is the less volatile component.) The entering steam is pure water at 1 atm and is superheated to 450 ° F. The bottoms leaves as a saturated liquid and is 0.0001 mole fraction methanol. Assume that CMO is valid. The column operates at 1 atm. If Steam/B = 1.2, find the slope of the bottom operating line = $\overline{L}/\overline{V}$. Use data from the Steam Tables at 1 atm.

F2. The paper by McCabe and Thiele (1925) is a classic paper in chemical engineering. Read it.

   a. Write a one-page critique of the paper.

   b. McCabe and Thiele (1925) show a method for finding the feed lines and middle operating lines for a column with two feeds that is not illustrated here. Generalize this approach when q ≠ 1.0.

chapter 7
INTRODUCTION TO MULTICOMPONENT DISTILLATION

Binary distillation problems can be solved in a straightforward manner by a stage-by-stage calculation that can be done either on a computer or graphically using a McCabe-Thiele diagram. When additional components are added, the resulting multicomponent problem becomes significantly more difficult and the solution may not be straightforward. In this chapter we will first consider why multicomponent distillation is more complex than binary distillation, and then look at the profile shapes typical of multicomponent distillation. In Chapter 8, stage-by-stage and matrix calculation methods will be applied to multicomponent distillation, and approximate methods will be developed in Chapter 9.

7.1. CALCULATIONAL DIFFICULTIES

Consider the conventional schematic diagram of a plate distillation column with a total condenser and a partial reboiler shown in Figure 7-1. Assume constant molal overflow, constant pressure, and no heat leak.

With the constant pressure and zero heat leak assumptions, a degree-of-freedom analysis around the column yields $C + 6$ degrees of freedom, where C is the number of components. For binary distillation this is 8 degrees of freedom. In a design problem we would usually specify these variables as follows (see Tables 4-1 and 4-2): F, z, feed quality q, distillate composition x_D, distillate temperature (saturated liquid), bottoms composition x_B, external reflux ratio L_0/D, and the optimum feed stage. With these variables chosen, the operating lines are defined, and we can step off stages from either end of the column using the McCabe-Thiele method.

Now if we add a third component we increase the degrees of freedom to 9. Nine variables that would most likely be specified for design of a ternary distillation column are listed in Table 7-1. Comparing this table with Tables 4-1 and 4-2, we see that the extra degree of freedom is used

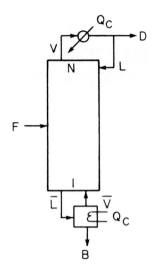

Figure 7-1. Distillation column.

to completely specify the feed composition. If there are four components, there will be 10 degrees of freedom. The additional degree of freedom must again be used to completely specify the feed composition.

Note that in multicomponent distillation neither the distillate nor the bottoms composition is completely specified because there are not enough variables to allow complete specification. This inability to completely specify the distillate and bottoms compositions has major effects on the calculation procedure. The components that do have their distillate and bottoms fractional recoveries specified (such as component 1 in the distillate and component 2 in the bottoms in Table 7-1) are called

Table 7-1. Specified Design Variables for Ternary Distillation

Number of variables	Variable
1	Feed rate, F
2	Feed composition, z_1, z_2
1	Feed quality, q (or h_F or T_F)
1	Distillate, $x_{1,dist}$ (or $x_{3,dist}$ or D or one fractional recovery)
1	Bottoms, $x_{2,bot}$ (or $x_{3,bot}$ or one fractional recovery)
1	L_0/D or $\overline{V}/B$ or Q_R
1	Saturated liquid reflux or T_{reflux}
1	Optimum feed plate location
—	
9	

Column pressure and $Q_{col} = 0$ are already specified.

key components. The most volatile of the keys is called the *light key* (LK), and the least volatile the *heavy key* (HK). The other components are *non-keys* (NK). If a non-key is more volatile (lighter) than the light key, it is a *light non-key* (LNK); if it is less volatile (heavier) than the heavy key, it is a *heavy non-key* (HNK).

The external balance equations for the column shown in Figure 7-1 are easily developed. These are the overall balance equation,

$$F = B + D \tag{7-1}$$

the component balance equations,

$$Fz_i = Bx_{i,bot} + Dx_{i,dist} \qquad i = 1, 2, \cdots C \tag{7-2}$$

and the overall energy balance,

$$Fh_F + Q_c + Q_R = Bh_B + Dh_D \tag{7-3}$$

Since we are using mole fractions, the mole fractions must sum to 1.

$$\sum_{i=1}^{C} x_{i,dist} = 1.0 \tag{7-4a}$$

$$\sum_{i=1}^{C} x_{i,bot} = 1.0 \tag{7-4b}$$

For a ternary system, Eq. (7-2) can be written three times, but these equations must add to give Eq. (7-1). Thus only two of Eqs. (7-2) plus Eq. (7-1) are independent.

Now, how do we solve the external mass balances? The unknowns are B, D, $x_{2,dist}$, $x_{3,dist}$, $x_{2,bot}$, and $x_{3,bot}$. There are 6 unknowns and 5 independent equations. Can we find an additional equation? Unfortunately, the additional equations (energy balances and equilibrium expressions) always add additional variables (see Problem 7-A1), so we cannot start out by solving the external mass and energy balances. This is the first major difference between binary and multicomponent distillation.

Can we do the internal stage-by-stage calculations first and then solve the external balances? To begin the stage-by-stage calculation procedure in a distillation column, we need to know all the compositions at one end of the column. For ternary systems with the variables specified as in Table 7-1, these compositions are unknown. To begin the analysis we would have to assume that one of them is known. Thus internal calculations for multicomponent distillation problems are trial-and-error. This is a second major difference between binary and multicomponent problems.

For ternary distillation, if one additional composition is *assumed*, then both the external and internal calculations are easily done. The results can then be compared and the assumed composition adjusted. This trial-and-error procedure is examined in more detail in Chapter 8.

Fortunately, in many cases it is easy to make an excellent first guess. If a sharp separation of the keys is required, then almost all of the heavy non-keys will appear only in the bottoms, and almost all of the light non-keys will appear only in the distillate. The obvious assumption is that all light non-keys appear only in the distillate and all heavy non-keys appear only in the bottom. Thus,

$$x_{LNK,bot} = 0 \tag{7-5a}$$

$$x_{HNK,dist} = 0 \tag{7-5b}$$

These assumptions allow us to complete the external mass balances.

Example 7-1. External Mass Balances Using Fractional Recoveries

We wish to distill 2000 kg moles/hr of a saturated liquid feed. The feed is 0.056 mole fraction propane, 0.321 n-butane, 0.482 n-pentane, and the remainder n-hexane. The column operates at 101.3 kPa. The column has a total condenser and a partial reboiler. Reflux ratio is $L_0/D = 3.5$, and reflux is a saturated liquid. The optimum feed stage is to be used. A fractional recovery of 99.4% n-butane is desired in the distillate and 99.7% of the n-pentane in the bottoms. Estimate distillate and bottoms compositions and flow rates.

Solution

A. Define. A sketch of the column is shown.

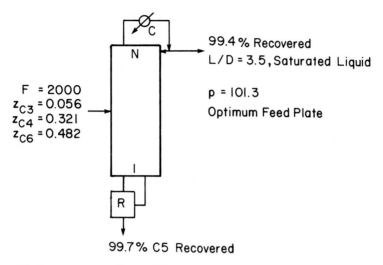

99.4 % Recovered
L/D = 3.5, Saturated Liquid

F = 2000
z_{C3} = 0.056
z_{C4} = 0.321
z_{C6} = 0.482

p = 101.3

Optimum Feed Plate

99.7% C5 Recovered

Find $x_{i,dist}$, $x_{i,bot}$, D, and B.

B. Explore. This appears to be a straightforward application of external mass balances, *except* there are two variables too many.

Thus we will have to assume the recoveries or concentrations of two of the components. A look at the DePriester charts (Figures 2-11 and 2-12) shows that the order of volatilities is propane > n-butane > n-pentane > n-hexane. Thus n-butane is the light key, and n-pentane is the heavy key. This automatically makes propane the light non-key (LNK) and n-hexane the heavy non-key (HNK). Since the recoveries of the keys are quite high, it is reasonable to assume that all of the LNK collects in the distillate and all of the HNK collects in the bottoms. We will estimate distillate and bottoms based on these assumptions.

C. Plan. Our assumptions of the non-key splits can be written either as

$$Dx_{C3,dist} = Fz_{C3} \quad \text{and} \quad Bx_{C6,bot} = Fz_{C6} \qquad (7\text{-}5a,b)$$

or

$$Bx_{C3,bot} = 0 \quad \text{and} \quad Dx_{C6,dist} = 0 \qquad (7\text{-}6a,b)$$

(see Problem 7-A2).

The fractional recovery of n-butane in the distillate can be used to write

$$Dx_{C4,dist} = (\text{frac. recovery } C_4 \text{ in distillate})(Fz_{C4}) \qquad (7\text{-}7)$$

Note that this also implies

$$Bx_{C4,bot} = (1 - \text{frac. rec. } C_4 \text{ in dist.}) (Fz_{C4}) \qquad (7\text{-}8)$$

For n-pentane the equations are

$$Bx_{C5,bot} = (\text{frac. rec. } C_5 \text{ in bot.}) Fz_{C5} \qquad (7\text{-}9)$$

$$Dx_{C5,dist} = (1 - \text{frac. rec. } C_5 \text{ in bot.}) Fz_{C5} \qquad (7\text{-}10)$$

Equations (7-5) to (7-10) represent 8 equations with 10 unknowns (4 compositions in both distillate and bottoms plus D and B). Equations (7-4) give two additional equations, which we will write as

$$\sum_{i=1}^{C} (Dx_{i,dist}) = D \qquad (7\text{-}11a)$$

$$\sum_{i=1}^{C} (Bx_{i,bot}) = B \qquad (7\text{-}11b)$$

These 10 equations can easily be solved, since distillate and bottoms can be done separately.

D. Do It. Start with the distillate.

$$Dx_{C3,dist} = Fz_{C3} = (2000)(0.056) = 112$$

$$Dx_{C6,dist} = 0$$

$$Dx_{C4,dist} = (0.9940)(2000)(0.321) = 638.5$$

$$Dx_{C5,dist} = (0.003)(2000)(0.482) = 2.89$$

Then
$$D = \sum_{i=1}^{4} Dx_{i,dist} = 753$$

Now the individual distillate mole fractions are

$$x_{i,dist} = \frac{Dx_{i,dist}}{D} \qquad (7\text{-}12)$$

Thus
$$x_{C3,dist} = \frac{112}{753.04} = 0.1487, \text{ and}$$

$$x_{C4,d} = \frac{638.15}{753.04} = 0.8474, \quad x_{C5,d} = \frac{2.89}{753.04} = 0.0038, \quad x_{C6,d} = 0$$

Check: $\sum_{i=1}^{4} x_{i,dist} = 0.9999,$ which is OK.

Bottoms can be found from Eqs. (7-56b), (7-6b), (7-8), (7-9), and (7-11b). The results are $x_{C3,bot} = 0$, $x_{C4,bot} = 0.0031$, $x_{C5,bot} = 0.7708$, $x_{C6,bot} = 0.2260$, and $B = 1247$. Remember that these are *estimates* based on our assumptions for the splits of the non-keys.

E. Check. Two checks are appropriate. The results based on our assumptions can be checked by seeing whether the results satisfy the external mass balance equations (7-1) and (7-2). These equations are satisfied. The second check is to check the assumptions, which requires internal stage-by-stage analysis and is much more difficult. In this case the assumptions are quite good.

F. Generalize. This type of procedure can be applied to many multicomponent distillation problems. It is more common to specify fractional recoveries rather than concentrations because it is more convenient.

Surprisingly, the ability to do reasonably accurate external mass balances on the basis of a first guess does not guarantee that the internal stage-by-stage calculations will be accurate. The problem given in Example 7-1 would be very difficult for stage-by-stage calculations. Let us explore why.

At the feed stage all components must be present at finite concentrations. If we wish to step off stages from the bottom up, we cannot use $x_{C3,bot} = 0$ because we would not get a nonzero concentration of propane at the feed stage. Thus $x_{C3,bot}$ must be a small but nonzero value. Unfortunately, we don't know if the correct value should be

10^{-5}, 10^{-6}, 10^{-7}, or 10^{-20}. Thus the percentage error in $x_{C3,bot}$ will be large, and it will be difficult to obtain convergence of the trial-and-error problem. If we try to step off stages from the top down, $x_{C3,dist}$ is known accurately, but $x_{C8,dist}$ is not. Thus when both heavy and light non-keys are present, stage-by-stage calculation methods are difficult. Other design procedures should be used.

If there are *only* light non-keys or *only* heavy non-keys, then an accurate first guess of compositions can be made. Suppose we specified 99.4% recovery of propane in the distillate and 99.7% recovery of n-butane in the bottoms. This makes propane the light key, n-butane the heavy key, and n-pentane and n-hexane the heavy non-keys. The assumption that all the heavy non-keys appear in the bottoms is an excellent first guess. Then we can calculate the distillate and bottoms compositions from the external mass balances (see Problem 7-D1). The composition calculated in the bottoms is quite accurate. Thus in this case we can step off stages from the bottom upward and be quite confident that the results are accurate. If only light non-keys are present, the stage-by-stage calculation should proceed from the top downward. These stage-by-stage methods are detailed in Chapter 8.

7.2. PROFILES

What do the flow, temperature, and composition profiles look like? Our intuition would tell us that these profiles will be similar to the ones for binary distillation. As we will see, this is true for the total flow rates and temperature, but not for the composition profiles.

If constant molal overflow is valid, the total vapor and liquid flow rates will be constant in each section of the column. The total flow rates can change at each feed stage or side-stream withdrawal stage. This behavior is illustrated for a computer simulation for a saturated liquid feed in Figure 7-2 and is the same behavior we would expect for a binary system. For nonconstant molal overflow, the total flow rates will vary from section to section. This is also shown in Figure 7-2. Although both liquid and vapor flow rates may vary significantly, the ratio L/V will be much more constant.

The temperature profile decreases monotonically from the reboiler to the condenser. This is illustrated in Figure 7-3 for the same computer simulation. This is again similar to the behavior of binary systems. Note that plateaus start to form where there is little temperature change between stages. When there are a large number of stages, these plateaus can be quite pronounced. They represent pinch points in the column.

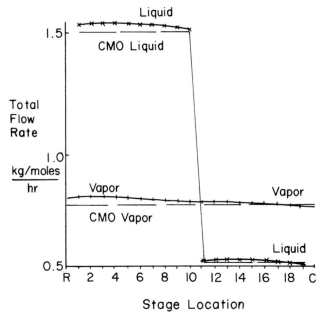

Figure 7-2.   Total liquid and vapor flow rates. Simulation for distillation of benzene-toluene-cumene. Desire 99% recovery of
benzene. Feed is 0.233 mole frac benzene, 0.333 mole frac
toluene, and 0.434 mole frac cumene and is a saturated
liquid. F = 1.0 kg mole/hr. Feed stage is number 10
above the partial reboiler, and there are 19 equilibrium
stages plus a partial reboiler. A total condenser is used. p
= 101.3 kPa. Relative volatilities: α_{ben} = 2.25, α_{tol} =
1.0, α_{cum} = 0.21.

The compositions in the column are much more complex. To study
these, we will first look at two computer simulations for the distillation
of benzene, toluene, and cumene in a column with 20 equilibrium contacts. The total flow and temperature profiles for this simulation are
given in Figures 7-2 and 7-3, respectively. With a specified 99%
recovery of benzene in the distillate, the liquid mole fractions are shown
in Figure 7-4.

At first Figure 7-4 is a bit confusing, but it will make sense after we
go through it step by step. Since benzene recovery in the distillate was
specified as 99%, benzene is the light key. Typically, the next less volatile component, toluene, will be the heavy key. Thus cumene is the
heavy non-key, and there is no light non-key. Following the benzene
curve, we see that benzene mole fraction is very low in the reboiler and

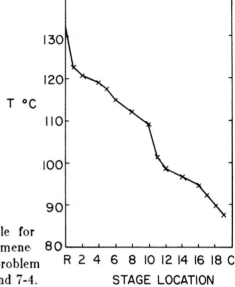

Figure 7-3. Temperature profile for benzene-toluene-cumene distillation. Same problem as in Figures 7-2 and 7-4.

Figure 7-4. Liquid-phase composition profiles for distillation of benzene-toluene-cumene. Same conditions as Figures 7-2 and 7-3 for nonconstant molal overflow. Benzene is the light key, and toluene is the heavy key. Stage 10 is the feed stage.

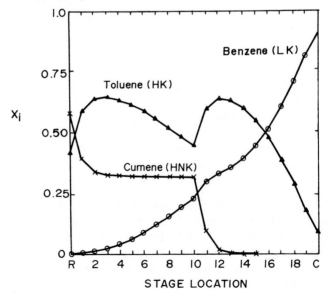

increases monotonically to a high value in the total condenser. This is essentially the same behavior as that of the more volatile component in binary distillation (for example, see Figure 5-14). In this problem benzene is always most volatile, so its behavior is simple.

Since cumene is the heavy non-key, we would typically assume that all of the cumene leaves the column in the bottoms. Figure 7-4 shows that this is essentially true (cumene distillate mole fraction was calculated as 2.45×10^{-8}). Starting at the reboiler, the mole fraction of cumene rapidly decreases and then levels off to a plateau value until the feed stage. Below the feed stage the cumene mole fraction decreases rapidly. This behavior is fairly easy to understand. Cumene's mole fraction decreases above the reboiler because it is the least volatile component. Since there is a large amount of cumene in the feed, there must be a finite concentration at the feed stage. Thus, after the initial decrease there is a plateau to the feed stage. Note that the concentration of cumene on the feed stage is *not* the same as in the feed. Below the feed stage, cumene concentration decreases rapidly because cumene is the least volatile component.

The concentration profile for the heavy key, toluene, is most complex in this example. The behavior of the heavy key can be explained by noting which binary pairs of components are distilling in each part of the column. In the reboiler and stages 1 and 2 there is very little benzene (LK) and the distillation is between the heavy key and the heavy non-key. In these stages the toluene (HK) concentration increases as we go up the column, because toluene is the more volatile of the two components distilling. In stages 3 to 10, the cumene (HNK) concentration plateaus. Thus the distillation is between the light key and the heavy key. Now toluene is the less volatile component, and its concentration decreases as we go up the column. This causes the primary maximum in HK concentration, which peaks at stage 3. Above the feed stage, in stages 11, 12, and 13, the HNK concentration plummets. The major distillation is again between the heavy key and the heavy non-key. Since the heavy key is temporarily the more volatile of these components, its concentration increases as we go up the column and peaks at stage 12. After stage 12, there is very little heavy non-key present, and the major distillation is between benzene and toluene. The toluene concentration then decreases as we continue up to the condenser. The secondary maximum above the feed stage is often much smaller than shown in Figure 7-4. The large amounts of cumene in this example cause a larger than normal secondary maximum.

In this example the heavy non-key (cumene) causes the two maxima in the heavy key (toluene) concentration profile. Since there was no

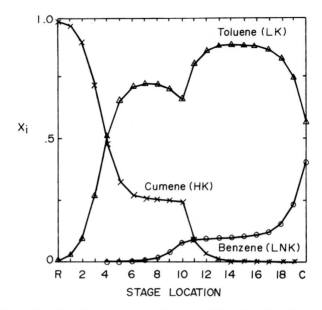

Figure 7-5. Liquid phase composition profiles for distillation of benzene (LNK), toluene (LK), and cumene (HK). Same problem as in Figures 7-2 to 7-4 except that a 99% recovery of toluene in the distillate is specified.

light non-key, the light key (benzene) has no maxima. It is informative to redo the example of Figures 7-2 to 7-4 with everything the same except for specifying 99% recovery of toluene in the distillate. Now toluene is the light key, cumene the heavy key and benzene a light non-key. The result achieved here is shown in Figure 7-5. This figure can also be explained qualitatively in terms of the distillation of binary pairs (see Problem 7-A14). Note that with no heavy non-keys, the heavy key concentration does not have any maxima.

What happens for a four-component distillation if there are light and heavy keys and light and heavy non-keys present? Since there is a light non-key, we would expect the light key curve to show maxima; and since there is a heavy non-key, we would expect maxima in the heavy key concentration profile. This is the case shown in Figure 7-6 for the distillation of a benzene-toluene-xylene-cumene mixture. Note that in this figure the secondary maxima near the feed stage are drastically repressed, but the primary maxima are readily evident.

The differences in the composition profiles for multicomponent and binary distillation can be summarized as follows:

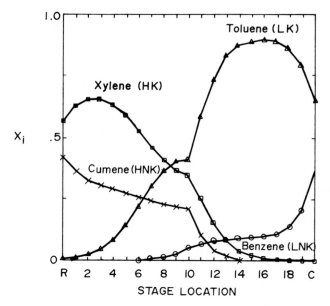

Figure 7-6. Liquid composition profiles for distillation of benzene (LNK), toluene (LK), xylene (HK), and cumene (HNK). Feed is 0.125 benzene, 0.225 toluene, 0.375 xylene, and 0.275 cumene. 99% recovery toluene in distillate. Relative volatilities: $\alpha_{ben} = 2.25$, $\alpha_{tol} = 1.0$, $\alpha_{xy} = 0.33$, $\alpha_{cum} = 0.21$.

1. In multicomponent distillation the key component concentrations can have maxima.

2. The non-keys usually do *not* distribute. That is, heavy non-keys usually appear only in the bottoms, and light non-keys only in the distillate.

3. The non-keys go through a plateau region of nearly constant composition.

4. All components must be present at the feed stage, but at that stage the primary distillation changes. Thus discontinuities occur at the feed stage.

Understanding the differences between binary and multicomponent distillation will be helpful when you are doing calculations for multicomponent distillation.

7.3. SUMMARY - OBJECTIVES

At the end of this chapter you should be able to:

1. Explain why multicomponent distillation is trial-and-error.

2. Make appropriate assumptions and solve the external mass balances.

3. Explain the flow, temperature and composition profiles for multicomponent distillation.

HOMEWORK

A. *Discussion Problems*

A1. Explain why the external mass balances cannot be solved for a ternary distillation system without an additional assumption. Why aren't the equations for the following useful?

   a. External energy balance

   b. Energy balance around the condenser

   c. Equilibrium expression in the reboiler

A2. In Example 7-1 part C we noted two ways our assumption could be written (Eqs. 7-5 or 7-6). Explain why these are equivalent.

A3. Define the following:

   a. Heavy key

   b. Heavy non-key

   c. Sandwich component (see Problem 7-A6)

   d. Optimum feed stage

   e. Minimum reflux ratio

A4. You have a four-component distillation problem with components A (most volatile), B, C, and D (least volatile),

   a. For what choice(s) of keys is a stage-by-stage calculation procedure a bad choice?

   b. For what choice(s) of keys is a stage-by-stage procedure starting from the top and calculating down the column a good procedure?

c. For what choice(s) of keys is a stage-by-stage procedure starting from the bottom and calculating up the column a good procedure?

A5. For a five-component distillation with two light non-keys, a light key, a heavy key and a heavy non-key, where in the column is the temperature the lowest?

A6. A distillation column is separating methane, ethane, propane, and butane. We pick methane and propane as the keys. This means that ethane is a *sandwich component* (a non-key component with a volatility between the two key components). *Think* when doing this problem. It is *not* a duplicate of the book.

    a. Show the approximate composition profiles for each of the four components. Label each curve.

    b. Explain in detail the reasoning used to obtain the profile for ethane.

A7. We have a feed that is 5 mole % n-propane, 25 mole % n-butane, 20 mole % n-pentane, and 50 mole % n-hexane. State whether the following choices of keys would allow one to easily do distillation calculations with a stage-by-stage method. (Answer yes if stage-by-stage is a good approach, and no if it isn't.)

    a. Light key, n-propane; heavy key, n-butane.

    b. Light key, n-propane; heavy key, n-hexane.

    c. Light key, n-butane; heavy key, n-pentane.

    d. Light key, n-butane; heavy key, n-hexane.

    e. Light key, n-pentane; heavy key, n-hexane.

A8. Draw a schematic diagram of the expected vapor composition profiles for all four components for Problem 7-A7e.

A9. Show for the problem illustrated in Figure 7-2 that L/V is more constant than either L or V when CMO is not valid. Explain why this is so.

A10. For Problem 7-A7a we have a two-phase feed that is 40% vapor. Draw a schematic diagram of the expected flow profiles (liquid and vapor). Assume constant molal overflow.

A11. Sketch the total flow profiles for a five-component distillation where feed is a superheated vapor. Label L and V.

A12. Develop a key relations chart for this chapter. You will probably want to include some sketches.

A13. In Figure 7-4, a 99% recovery of benzene does not give a high benzene purity. Why not? What would you change to also achieve a high benzene purity in the distillate?

A14. Explain Figure 7-5 in terms of the distillation of binary pairs.

A15. In Figure 7-4, the HNK and HK concentrations cross near the bottom of the columns, and in Figure 7-5 the LK and LNK concentrations do not cross near the top of the column. Explain when the concentrations of heavy key and non-key, and light key and non-key pairs will and will not cross.

A16. Figure 7-6 shows the distillation of a four-component mixture. What would you expect the profiles to look like if benzene were the light key and toluene the heavy key? What if xylene were the light key and cumene the heavy key?

D. *Problems*

D1. Repeat Example 7-1, but with a 99.4% recovery of propane in the distillate and a 99.7% recovery of n-butane in the bottoms instead of the recoveries given in Example 7-1.

D2. a. We have a feed mixture with 20 mole % propane, 35 mole % n-butane and 45 mole % n-pentane (propane is most volatile). Feed is a saturated vapor. $F = 100$ kg moles/hr, $L/D = 2.0$. We desire 98.8% recovery of n-butane in the distillate and 99.3% recovery of n-pentane in the bottoms. Find distillate and bottoms flow rates. Explicitly state any assumptions you make.

b. Draw the calculated total flow rate profiles for part a, labeling your curves. Assume CMO.

c. For part a, show the approximate concentration profiles for propane, n-butane, and n-pentane. Label your curves.

D3. We have a feed mixture of 22 mole % methanol, 47 mole % ethanol, 18 mole % n-propanol, and 13 mole % n-butanol. Feed is a saturated liquid, and $F = 10,000$ kg moles/day. We desire a 99.8% recovery of methanol in the distillate and a methanol mole fraction in the distillate of 0.99.

a. Find D and B.

b. Find compositions of distillate and bottoms.

D4. We are separating a mixture that is 40 mole% isopentane, 30 mole% n-hexane, and 30 mole% n-heptane. We desire a 98% recovery of n-hexane in the bottoms and a 99% recovery of isopentane in the distillate. F = 1000 kg moles/hr. Feed is a two-phase mixture that is 40% vapor. L/D = 2.5.

a. Find D and B. List any required assumptions.

b. Calculate L, V, $\overline{L}$, and $\overline{V}$, assuming CMO.

c. Show schematically the expected composition profiles for isopentane, n-hexane, and n-heptane. Label curves. Be neat!

F. *Problems Requiring Other Resources*

F1. Profiles for other multicomponent distillation systems are available in a variety of other books such as Brian (1972, Chapter 5), Henley and Seader (1981, Chapter 15), King (1980, Chapter 7), and Smith (1963, Chapter 10). (See references at end of Chapter 8.) Read the section on multicomponent distillation profiles in one of these books and write a one-page report comparing the profiles shown in this chapter to those in the other resource.

chapter 8
EXACT CALCULATION PROCEDURES FOR MULTICOMPONENT DISTILLATION

Since multicomponent calculations are trial-and-error, it is convenient to do them on a computer. In this chapter we will first consider how to carry out multicomponent stage-by-stage calculations when relative volatilities are constant and constant molal overflow (CMO) is valid. Then the method of utilizing a bubble-point calculation on each stage will be developed. Stage-by-stage calculations are restricted to problems where a good first guess of compositions can be made at some point in the column. In the final section of the chapter, matrix methods for multicomponent distillation will be introduced. These methods are not restricted to cases where a good guess of compositions can be made.

8.1. STAGE-BY-STAGE CALCULATIONS FOR CONSTANT MOLAL OVERFLOW

Consider the design problem for the three-component distillation shown in Figure 8-1. Component A is a light non-key (LNK), B is the light key (LK), and C is the heavy key (HK). The feed flow rate, composition, and temperature are specified, as are L_0/D, saturated liquid reflux, pressure, use of the optimum feed stage, and recoveries of the light and heavy keys in distillate and bottoms, respectively. We wish to predict the number of stages required and the separation obtained.

To start the calculation we need to assume the split for component A, the light non-key. The obvious first assumption is that all the light non-key exits in the distillate so that $x_{A,bot} = 0$ and $Dx_{A,dist} = Fz_A$. Now we can do external mass balances to find all distillate and bottoms compositions and flow rates. This was illustrated in Chapter 7. Once this is done, we can find L and V in the rectifying section. Since constant molal overflow is valid,

$$L = (\frac{L_0}{D})D, \quad V = L + D$$

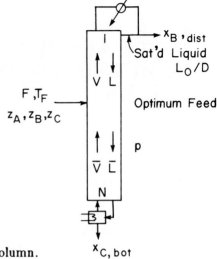

Figure 8-1.   Ternary distillation column.

At the feed stage, q can be estimated from enthalpies as

$$q = \frac{H - h_{feed}}{H - h}$$

or $q = L_F/F$ can be found from a flash calculation on the feed stream. Then $\bar{L}$ and $\bar{V}$ are determined from balances at the feed stage,

$$\bar{L} = L + qF, \quad \bar{V} = V + (1 - q)F$$

This completes the preliminary calculations for *this* assumption of how the light non-key splits in the column.

In this case, the assumed compositions are very accurate at the top of the column but not at the bottom. Thus with the light non-key present, we want to step off stages from the top down. The general procedure when CMO is valid is:

1. For total condenser, $y_{i,1} = x_{i,dist}$, where i is component A, B, or C and the second subscript is the stage location.

2. Use equilibrium to calculate $x_{i,j}$ values from known $y_{i,j}$ values (j is the stage number).

3. Use mass balances (operating equations) to calculate $y_{i,j+1}$ values from known $x_{i,j}$ values.

4. Repeat steps 2 and 3 until the feed stage is reached. Then change to the stripping section operating equations and continue.

5. The calculation is finished when $x_{HK,N+1} \geq x_{HK,bot}$ and $x_{LK,N+1} \leq x_{LK,bot}$.

If constant *mass* overflow is valid, the equations are exactly the same, but you use mass units.

We will now make this general outline specific for constant relative volatility systems.

8.2. CONSTANT RELATIVE VOLATILITY SYSTEMS

If the relative volatilities are constant, the equilibrium calculations become very simple. Let us arbitrarily choose component B as the reference. Then the definitions of the relative volatilities are

$$\alpha_{AB} = \frac{K_A}{K_B} = \frac{y_A/x_A}{y_B/x_B} , \quad \alpha_{BB} = \frac{K_B}{K_B} = 1.0 ,$$

$$\alpha_{CB} = \frac{K_C}{K_B} = \frac{y_C/x_C}{y_B/x_B} \tag{8-1}$$

As we step down the column, the y values leaving a stage will be known and the x values can be calculated from equilibrium. Thus for stage j,

$$x_{Aj} = \frac{y_{Aj}}{(y_{Aj}/x_{Aj})\big/(y_{Bj}/x_{Bj})} \frac{1}{(y_{Bj}/x_{Bj})} = \frac{y_{Aj}}{\alpha_{AB,j}} \frac{1}{K_{Bj}}$$

Note that the first equation reduces to $x_{Aj} = x_{Aj}$.

For any component i on stage j,

$$x_{ij} = \frac{y_{ij}}{\alpha_{iB,j}} \frac{1}{K_{Bj}} \tag{8-2}$$

In general, both α_{iBj} and K_{Bj} depend upon temperature and thus vary from stage to stage. When the relative volatilities are constant, only K_{Bj} varies. Since the liquid mole fractions must sum to 1.0, we can remove the dependencies on K_{Bj}. For this ternary problem,

$$x_{A,j} + x_{B,j} + x_{C,j} = \frac{y_{Bj}}{\alpha_{AB} K_{Bj}} + \frac{y_{Bj}}{\alpha_{BB} K_{Bj}} + \frac{y_{Cj}}{\alpha_{CB} K_{Bj}} = 1.0$$

Solving for K_{Bj},

$$K_{Bj} = \frac{y_{Aj}}{\alpha_{AB}} + \frac{y_{Bj}}{\alpha_{BB}} + \frac{y_{Cj}}{\alpha_{CB}} = \sum_{i=1}^{3} \left(\frac{y_{ij}}{\alpha_{iB}}\right) \tag{8-3}$$

or in general

$$K_{ref\,j} = \sum_{i=1}^{C} \left(\frac{y_{i,j}}{\alpha_{i-ref}}\right) \tag{8-4}$$

where C is the number of components. If desired, the stage temperature can be determined from the calculated $K_{ref\,j}$ value. This is not necessary, since Eq. (8-3) or (8-4) can be substituted into Eq. (8-2). Then the equilibrium expression for component A is

$$x_{Aj} = \frac{y_{Aj}/\alpha_{AB}}{\sum\limits_{i=1}^{3} \left(\dfrac{y_{ij}}{\alpha_{iB}}\right)} \tag{8-5}$$

or, in general,

$$x_{ij} = \frac{y_{ij}/\alpha_{i-ref}}{\sum\limits_{i=1}^{C} (y_{ij}/\alpha_{i-ref})} \tag{8-6}$$

Equation (8-5) or (8-6) can be used to calculate the liquid mole fractions at equilibrium. Note that the choice of the reference component is arbitrary. If Eq. (8-6) is expanded, K_{ref} divides out; thus, any component can be used as the reference as long as we are consistent.

The operating equations are essentially the same as for binary systems. These are

$$y_{i,j+1} = \frac{L}{V} x_{i,j} + \left(1 - \frac{L}{V}\right)x_{i,dist} \tag{8-7}$$

in the enriching section and

$$y_{i,k+1} = \frac{\bar{L}}{\bar{V}} x_{i,k} - (\frac{\bar{L}}{\bar{V}} - 1)x_{i,bot} \qquad (8\text{-}8)$$

in the stripping section.

The procedure is now simple. In the enriching section we use equilibrium equation (8-6) to calculate $x_{i,1}$ for each component i with values $y_{i,1} = x_{i,dist}$. Then we determine $y_{i,2}$ for each component from operating equation (8-7). Equilibrium equation (8-6) is used to find $x_{i,2}$ from values $y_{i,2}$, and so on. In the stripping section we alternate between Eqs. (8-6) and (8-8). We stop the calculation when

$$x_{HK,N+1} \geq x_{HK,bot} \quad \text{and} \quad x_{LK,N+1} \leq x_{LK,bot} \qquad (8\text{-}9)$$

This leaves us with two unanswered questions: How do we determine the optimum feed plate, and how do we correct our initial assumption?

The optimum feed plate is defined as the feed plate that gives the fewest total number of stages. To be absolutely sure you have the optimum feed plate location, use this definition. That is, pick a feed plate location and calculate N. Then repeat until you find the minimum total number of stages. The graph of your results will look like Figure 8-2A or B. Note that often several stages must be stepped off before the feed can be input. The first legal feed stage may be the optimum. This procedure sounds laborious, but it is very easy to implement on a computer.

Figure 8-2.   Optimum feed plate location. A. Typical optimum; B. optimum is first plate that works.

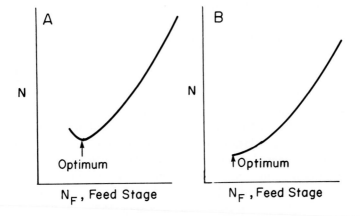

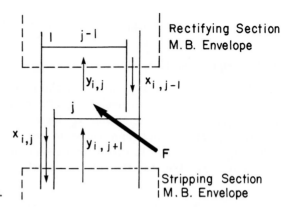

Figure 8-3.   Feed stage.

An alternative is to use an approximate test based on the proposition that we want the ratio of light key to heavy key to decrease as fast as possible as we go down the column. If, as we step down the column,

$$\left(\frac{y_{LK,j+1}}{y_{HK,j+1}}\right)_{\text{top op eq}} < \left(\frac{y_{LK,j+1}}{y_{HK,j+1}}\right)_{\text{bot op eq}} \qquad (8\text{-}10a)$$

stage j is not the feed, and we continue to use the top operating equation. When

$$\left(\frac{y_{LK,j+1}}{y_{HK,j+1}}\right)_{\text{top op eq}} > \left(\frac{y_{LK,j+1}}{y_{HK,j+1}}\right)_{\text{bot op eq}} \qquad (8\text{-}10b)$$

stage j is the feed, and we switch to the stripping section. Figure 8-3 illustrates the feed stage. The vapor compositions leaving the feed stage, $y_{i,j}$, are determined from the top operating equation. The liquid compositions leaving the feed stage, x_{ij}, are found from equilibrium; and the vapor compositions from the stage below the feed stage, $y_{i,j+1}$, are found from the bottom operating equations. Thus, use the vapor compositions leaving stage $j+1$ to check whether stage j is the feed stage.

Remember, this is an *approximate* procedure. If the optimum feed stage looks like Figure 8-2B, this procedure may predict an optimum feed stage at an illegal location. If you try to switch stages too early, the stage-by-stage calculation will eventually give negative mole fractions. This is illustrated on a binary McCabe-Thiele plot in Figure 8-4. With three or more components, switching too soon can result in a negative mole fraction but with all $x_i < 1$. With a computer program, you can guard against this mistake by checking that all mole fractions (x_i and y_i) are between zero and 1 for every stage.

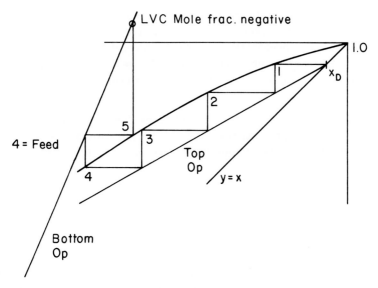

Figure 8-4.   McCabe-Thiele diagram illustrating the results of switching operating equations too soon.

The second unanswered question is, how do we check and correct our initial guess for the splits of the non-key components? If the initial guess is correct, Eqs. (8-9) should be satisfied simultaneously. Now use the calculated value of the light non-key mole fraction, $x_{LNK,N+1}$, the specified fractional recoveries of the keys, and the external mass balances to determine $x_{LNK,dist\ calc}$. This is illustrated later in Example 8-1.

If

$$\frac{\left| x_{LNK,dist\ assumed} - x_{LNK,dist\ calc} \right|}{x_{LNK,dist\ calc}} > \epsilon \qquad (8-11)$$

then a new trial is required. An appropriate ϵ value would depend upon the problem, but 0.01 is reasonable for computer trials. For the next trial we can use direct substitution and set

$$x_{LNK,dist\ next} = x_{LNK,dist\ calc} \qquad (8-12)$$

or use a damped direct substitution. The first or second trial should give convergence if the stage-by-stage approach is a good method (that is, no HNKs or no LNKs). If there are both LNKs and HNKs, convergence may be very difficult.

We could also calculate fractional numbers of stages for the light key and the heavy key; however, these values differ unless the initial guess of the light non-key concentration is perfect. Also, if the bottoms is almost pure heavy key, the error in calculating the heavy-key fraction will be quite large. Thus, in practice, it is most convenient to use an integer number of stages.

This entire discussion was for calculations down the column. If only heavy non-keys are present, then the calculation should proceed up the column. In this case the vapor compositions are determined from the equilibrium equation,

$$y_{i,j} = \frac{x_{i,j}\alpha_{i-ref}}{\sum\limits_{i=1}^{C}(x_{i,j}\alpha_{i-ref})} \tag{8-13}$$

The liquid mole fractions are found from the operating equations, which are inverted to find $x_{i,j}$ [see Eq. (8-14)]. The feed stage can be estimated from the ratio x_{LK}/x_{HK} [see Eqs. (8-15)]. Convergence now requires checking $x_{HNK,bot}$ assumed and calculated values.

Other methods of closure have been devised. For problems where both light and heavy non-keys are present, Lewis and Matheson (1932) and Thiele and Geddes (1933) calculated from both ends of the column and matched compositions at the feed stage (see Smith, 1963, Chapter 10 for details). These stage-by-stage approaches do work, but closure can be difficult. When there are both light and heavy non-keys, other calculation methods such as the matrix method discussed later are preferable.

8.3. BUBBLE-POINT OR DEW-POINT CALCULATIONS ON EACH STAGE

The assumption that relative volatilities are constant is very convenient mathematically but often unrealistic. When relative volatility is not constant, we need to do a complete equilibrium calculation on each stage. If we are stepping off stages from the bottom up, we do a bubble-point calculation on each stage. Going down the column, we use dew-point calculations. The difference between these approaches is that going up the column the $x_{i,j}$ are known and the $y_{i,j}$ are calculated from equilibrium, while going down the column the $y_{i,j}$ are known and the x_{ij} are calculated from equilibrium. (You may wish to review the bubble-point and dew-point calculations in Chapter 2.)

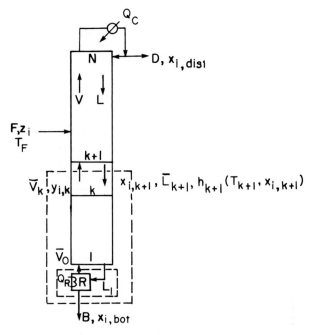

Figure 8-5.   Distillation column for stepping off stages from the bottom up.

Consider the distillation column shown in Figure 8-5, where all the non-keys are heavy non-keys. Note that the column is now numbered from the bottom up since that is the direction in which we will step off stages. Now a good first guess of concentrations can be made at the bottom of the column. With $x_{1,bot}$ known, we can start the stage-by-stage calculations with an equilibrium calculation for the partial reboiler. This bubble-point calculation is a trial-and-error calculation, since the reboiler temperature is not known. Equation (2-22),

$$K_{ref,new} = \frac{K_{ref,old}}{\sum\limits_{i=1}^{C}(K_i \, x_i)_{calc}}$$

allows us to conveniently estimate the K value and from that the reboiler temperature for the next trial. Once the bubble-point calculation has converged, the $y_{i,0}$ values are found from

$$y_{i,0} = K_i(T_{reboiler})x_{i,bot}$$

Next the operating equations are solved for $x_{i,1}$. This requires inverting the bottom operating equation. With constant molal overflow, this is

$$x_{i,k+1} = \frac{\overline{V}}{\overline{L}} \, y_{i,k} + \left(1 - \frac{\overline{V}}{\overline{L}}\right)x_{i,bot} \qquad (8\text{-}14)$$

Once the liquid mole fractions have been calculated, we determine the next stage temperature and the vapor mole fractions. Then we use Eq. (8-14). This is repeated until we reach the feed stage. Again the optimum feed stage should really be found by trial and error; however, the approximate test is, if

$$\left[\frac{x_{LK,k}}{x_{HK,k}}\right]_{top \; op \; eq} < \left[\frac{x_{LK,k}}{x_{HK,k}}\right]_{bot \; op \; eq} \qquad (8\text{-}15a)$$

and

$$\left[\frac{x_{LK,k+1}}{x_{HK,k+1}}\right]_{top \; op \; eq} > \left[\frac{x_{LK,k+1}}{x_{HK,k+1}}\right]_{bot \; op \; eq} \qquad (8\text{-}15b)$$

then stage k is probably the feed stage. Above the feed stage, $x_{i,f+1}$ is calculated from the inverted top operating line

$$x_{i,j+1} = \frac{V}{L} \, y_j - \left(\frac{V}{L} - 1\right)x_{i,dist} \qquad (8\text{-}16)$$

The alternation between equilibrium and operating lines continues until

$$y_{LK,N} \geq x_{LK,dist} \quad \text{and} \quad y_{HK,N} \leq x_{HK,dist} \qquad (8\text{-}17)$$

We can now check for convergence in the same way we did for the constant relative volatility calculations. If convergence has not been obtained, we adjust the bottoms concentrations and repeat the calculations. The second time through, we can use the previously calculated temperatures as our first guess for temperature.

If only light non-keys are present, we should step off stages going down the column. Then the liquid mole fractions are determined from dew point calculations and the vapor mole fractions are found from the operating equations. Otherwise, the procedure is very similar to going down the column with constant relative volatilities.

Remember that convergence is easy only when the non-keys are all heavy or are all light. If this is not true, other calculational methods that are not stage-by-stage methods should be used. An example is the matrix method discussed later.

Example 8-1. Analytical Stage-by-Stage Calculation

A distillation column with a partial reboiler and a total condenser is separating nC_4, nC_5, and nC_8. Operating conditions are: Feed: 10,000 kg moles/hr, saturated liquid feed. $z_{C4} = 0.15$, $z_{C5} = 0.25$, $z_{C8} = 0.60$ (mole fractions). Wanted: 99% recovery of nC_5 in the distillate and 98% recovery of nC_8 in the bottoms. Reflux is a saturated liquid and external reflux ratio $L_0/D = 1.0$. Pressure = 200 kPa. Find the optimum feed location and the total number of stages. Use constant molal overflow even though it is not strictly valid. Use the DePriester charts to calculate K values. Assume that all the nC_4 is in the distillate. Thus $x_{C4,bot} = 0$. If $x_{C4,bot\ calc} < 5 \times 10^{-4}$, do not repeat the calculation.

Solution

A. Define. Problem is sketched in the figure.

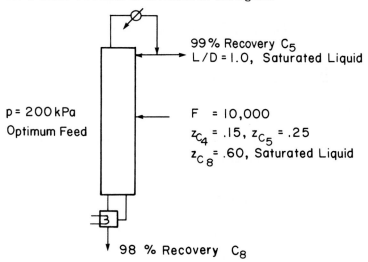

Find outlet concentrations, feed stage, and total number of stages.

B. Explore. Since this is a multicomponent system, the calculation must be by trial and error. We will assume that all of the LNK (C_4) is in the distillate. Then do external balances. Step off stages from top *down*. Thus, use dew-point calculation,

$$\sum_{i=1}^{3} x_i = \sum_{i=1}^{3} \frac{y_i}{K_i} = 1.0$$

and

$$K_{ref,new} = K_{ref,old} \left(\sum_{i=1}^{3} \frac{y_i}{K_i} \right)$$

to calculate temperatures on each stage. For feed stage, use approximate test that if

$$\left(\frac{y_{LK,j+1}}{y_{HK,j+1}} \right)_{top\ op\ eq} > \left(\frac{y_{LK,j+1}}{y_{HK,j+1}} \right)_{bot\ op\ eq}$$

then stage j is the optimum feed stage.

Done when

$$x_{C8,k+1} \geq x_{C8,bot} \quad \text{and} \quad x_{C5,k+1} \leq x_{C5,bot}$$

Then check for optimum feed stage. Finally, check that initial guess was correct.

C. Plan. (Preliminary calculations will be done here.)

External Balances:

Assume $x_{C5,bot} = 0$, which gives $Fz_{C4} = Dx_{C4,dist} = 1500$. From the fractional recoveries:

$$Dx_{C5,dist} = 0.99(Fz_{C5}) = 2475$$

$$Dx_{C8,dist} = (1 - 0.98)(Fz_{C5}) = 120$$

$$D = \sum_{i=1}^{3}(Dx_i) = 4095 \text{ kg moles/hr}$$

Then mole fractions are $x_{i,dist} = Dx_{i,dist}/D$

$$x_{C4,dist} = 0.366, \quad x_{C5,dist} = 0.605, \quad x_{C8,dist} = 0.029$$

Fractional recoveries for the bottoms are

$$Bx_{C5,bot} = (1 - 0.99)(Fz_{C5}) = 25$$

$$Bx_{C8,bot} = (0.98)(Fz_{C8}) = 5880$$

$$B = \Sigma(Bx_{i,bot}) = 5905 \text{ kg moles/hr}$$

Mole fractions: $x_{C4,bot} = 0, \quad x_{C5,bot} = 0.004, \quad x_{C8,bot} = 0.996$

At top of column, $y_{i,1} = x_{i,dist} = x_{i,0}$

Equilibrium Calculation:

Pick T_1 for which $K_{C5} = 1.0$. Calc $K_{C5,new} = (K_{C5})(\sum \dfrac{y_{i,1}}{K_{i,1}})$. Find T_1 from $K_{C5 \text{ New}}$. Repeat to convergence. With T_1 determined, $x_{i,1} = K_i(T_1)y_{i,1}$.

Top Operating Equation:

$$y_{i,j+1} = \frac{L}{V} x_{i,j} + (1 - \frac{L}{V})x_{i,dist}$$

where $\dfrac{L}{V} = \dfrac{L/D}{1 + L/D} = \dfrac{1}{2}$ since reflux is a saturated liquid.

Calculate all $y_{i,2}$. Then calculate T_2 and $x_{i,2}$ from equilibrium. Calculate $y_{i,3}$ with top operating line. Also use bottom operating line,

$$y_{i,j+1} = \frac{\overline{L}}{\overline{V}} x_{i,j} - (\frac{\overline{L}}{\overline{V}} - 1)x_{i,bot}$$

where $\overline{L} = L + F = (\dfrac{L}{D})D + F = 14{,}095$

since feed is a saturated liquid, and

$$\bar{V} = V = L + D = (\frac{L}{D} + 1)D = 8190$$

Thus $\quad \bar{L}/\bar{V} = 1.721$.

Check whether $(\frac{y_{C5,3}}{y_{C8,3}})_{\text{top op}} > (\frac{y_{C5,3}}{y_{C8,3}})_{\text{bot op}}$. If not, use rectifying operating equation values and do equilibrium calculation for stage 3. If test is satisfied, use bottom operating equation values (stage 2 is feed), and go to equilibrium for stage 3.

Repeat for each stage. Once in bottom of column, use bottom operating equation only.

Stop when $x_{C8,j+1} \geq 0.996$ and $x_{C5,j+1} \leq 0.004$. Finally, check initial assumption.

D. Do It. *Equilibrium Stage 1.* Pick $K_{C5} = 1.0$. Then $T_1 = 58.2\,^\circ C$ from DePriester chart. $K_{C5} = 2.87$, $K_{C8} = 0.065$, and

$$K_{C5,\text{new}} = 1.0(\frac{0.366}{2.87} + \frac{0.605}{1.0} + \frac{0.029}{0.065}) = 1.18$$

which gives $T_{1,\text{new}} = 65\,^\circ C$ and $K_{C4} = 3.25$, $K_{C8} = 0.082$. Continuing,

$$K_{C5,\text{new}} = 1.18(\frac{0.366}{3.25} + \frac{0.605}{1.18} + \frac{0.029}{0.082}) = 1.155$$

or $T_{1,\text{new}} = 64.3$ and $K_{C4} = 3.22$, $K_{C8} = 0.081$.

Check: $\quad \sum(\frac{y_i}{K_1}) = 0.114 + 0.524 + 0.358 = 0.995$

Top operating equation is

$$y_{i,2} = \frac{L}{V} x_{i,1} + (1 - \frac{L}{V})x_{i,\text{dist}}$$

Thus, for C_4, $y_{C4,2} = (1/2)(0.114) + (1/2)(0.366) = 0.24$ and $y_{C5,2} = 0.564$, $y_{C8,2} = 0.1935$.

Note that $\sum_{i=1}^{3} y_{i,2} = 0.998$ which is close enough to 1.0. A check on the bottom operating equation is not required, since $y_{C8,\text{bot op}} < 0$.

Stage 2. $T_2 > 64.3\,^\circ C$ since we are going down the column. Dew-point calculation is similar to first stage calculation. The result is $T = 99\,^\circ C$, and the following x values are obtained:

$$x_{C4,2} = 0.039, \quad x_{CB,2} = 0.225, \quad x_{C8,2} = 0.744, \quad \sum x_i = 1.008$$

Now we can use both operating equations to check for feed stage.

Top op eq: $\qquad y_{i,3} = \dfrac{L}{V}\, x_{1,2} + \left(1 - \dfrac{L}{V}\right) x_{i,\text{dist}}$

$$y_{C4,3} = 0.2025, \quad y_{C5,3} = 0.415, \quad y_{C8,3} = 0.3865, \quad \sum y_i = 1.004$$

Bottom op eq: $\qquad y_{i,3} = \dfrac{\overline{L}}{\overline{V}}\, x_{i,2} - \left(\dfrac{\overline{L}}{\overline{V}} - 1\right)_{\text{bot}}$

$$y_{C4,3} = 0.067, \quad y_{C5,3} = 0.384, \quad y_{C8,3} = 0.562 \quad \sum y_{i,3} = 1.01$$

Then: $\qquad \left(\dfrac{y_{C5,3}}{y_{C8,3}}\right)_{\text{top op}} = 1.07 > \left(\dfrac{y_{C5,3}}{y_{C8,3}}\right)_{\text{bot op}} = 0.683$

Use stage 2 as the feed. Then for the equilibrium calculation for stage 3, use $y_{i,3}$ values calculated with the bottom operating line.

Stage 3. Equilibrium Calculation. As first guess, try $134\,^\circ C$ (pick this by approximately balancing C_5 and C_8 to get $\sum x_i = 1.0$). Final result is $T_3 = 132\,^\circ C$ and

$$x_{C4,3} = 0.0066, \quad x_{C5,3} = 0.0853, \quad x_{C8,3} = 0.906, \quad \sum x_i = 0.998$$

Now use the bottom operating equation because we are in the bottom of the column.

$$y_{i,4} = \dfrac{\overline{L}}{\overline{V}}\, x_{i,3} - \left(\dfrac{\overline{L}}{\overline{V}} - 1\right) x_{i,\text{bot}}$$

The results are: $y_{C4,4} = 0.0114$, $\quad y_{C5,4} = 0.1439$, $\quad y_{C8,4} = 0.8410$, $\sum y_i = 0.9963$

Stage 4. Equilibrium result is $T_4 = 146\,^\circ C$ and

$x_{C4} = 0.00088$, $x_{C5} = 0.0257$, $x_{C8} = 0.967$, $\sum x_i = .993$

The bottom operating equation gives

$y_{C4,5} = 0.0015$, $y_{C5,5} = 0.0413$, $y_{C8,5} = 0.9461$

Stage 5. Equilibrium result is $T_5 = 149°C$ and

$x_{C4,5} = 0.00115$, $x_{C5,5} = 0.007$, $x_{C8,5} = 0.9959$, $\sum x_i = 1.003$

This is close to the specified condition.

The bottom operating equation gives

$y_{C4,6} = 0.000198$, $y_{C5,6} = 0.00916$, $y_{C8,6} = 0.9958$

Stage 6. Equilibrium calculation gives $T = 153°$ and

$x_{C4} = 0.0000147$, $x_{C5} = 0.0015$, $x_{C8} = 0.9997$, $\sum x_i = 1.001$

which is more separation than was required. Because of the accuracy of the DePriester charts this last number, x_{C8}, is not extremely accurate. Thus, for very tight separations you need *good* equilibrium data.

However, 6 stages (including partial reboiler) are sufficient. You could recheck the feed stage, but this is probably okay. The ratios are quite different, and stage 1 is not a possible feed stage ($y_{HK,bot\ op} < 0$). The initial assumption looks reasonable, since $x_{C4,6}$ is quite small.

E.   Check.   As a check we can do the overall mass balances for the column using the observed $x_{LNK,bot}$ value. With 6 equilibrium contacts, we have $x_{C4,bot} = 0.0000147$. We also want 99% recovery of C5 in the distillate and 98% recovery of C8 in the bottoms. From the fractional recoveries,

$$Dx_{C5,dist} = (0.99)\ Fz_{C5} = 2475$$

$$Dx_{C8,dist} = (1 - 0.98)Fz_{C8} = 120$$

$$Bx_{C5,bot} = (1 - 0.99)Fz_{C5} = 25$$

$$Bx_{C8,bot} = (0.98)\ Fz_{C8} = 5880$$

In the bottoms, $\sum x_{i,bot} = 1.0$, which is

$$25/B + 5880/B + 0.0000147 = 1.0.$$

Solving for B, we find $B = 5905/0.9999853 = 5905.868$. From the external mass balance,

$$D = 10,000 - 5905.868 = 4094.132 \text{ (not much change)}$$

Then,

$$x_{C5,dist} = \frac{Dx_{C5,dist}}{D} = \frac{2475}{4094.13} = 0.6045$$

$$x_{C8,dist} = \frac{120}{4094.132} = 0.02931$$

$$x_{C4,dist} = 1 - x_{C5,dist} - x_{C5,dist} = 0.3662$$

Since $x_{C4,dist\ calc}$ is very close to $x_{C4,dist\ assumed}$ (difference is 0.0002), another trial is not warranted.

F. Generalize. When a high fractional recovery of light key is required, the guess that all light non-key is in the distillate will be a good guess. Likewise, if a heavy non-key is present, guessing that all the heavy non-key is in the bottoms is a good guess if a high fractional recovery of the heavy key is specified.

This was a relatively easy separation because the volatilities are quite different. If we pick pentane as the reference, we can calculate the relative volatilities of α_{C8-C5} for several stages, since the temperature and K values have been calculated. Thus we have

$$(\alpha_{C8-C5})_{stage\ 1} = (\frac{K_{C8}}{K_{C5}}) = \frac{0.081}{1.155} = 0.070$$

$$(\alpha_{C8-C5})_{stage\ 2} = \frac{0.25}{2.51} = 0.10$$

$$(\alpha_{C8-C5})_{stage\ 6} = \frac{0.996}{6.2} = 0.16$$

Note that this relative volatility varies by more than a factor of 2. Thus constant relative volatility is not valid for this problem.

8.4. NONCONSTANT MOLAL OVERFLOW

Several methods can be used if the molal overflow is not constant. First, we could use latent heat units for the operating equations. Then when the equilibrium calculations are done, the known mole fractions $(x_{i,k}$ for bubble-point calculations) are determined from the calculated latent heat fractions $x_{i,k}^*$. The bubble-point calculation is done in mole fraction units, and the $y_{i,j}$ are determined. We convert this value to the latent heat unit fractions y_{ik}^* and then use the operating equations to determine $y_{i,k+1}^*$. We repeat this procedure up the column. Although two additional calculations are required on each stage, only the bubble-point calculations on each stage are trial-and-error. Programming this method for a computer is straightforward.

If the relative volatilities are constant, the calculation is even simpler, since relative volatilities are the same in molal and latent heat units $(\alpha_{i-ref}^* = \alpha_{i-ref})$. Thus if we change all mole fractions and flow rates to latent heat units, the entire analysis can be done in latent heat units using $\alpha_{i-ref}^* = \alpha_{i-ref}$.

Essentially, the latent heat unit approach includes an abbreviated energy balance that contains only the latent heat of vaporization. A second approach is to include the complete energy balance on every stage as we do the stage-by-stage analysis. At first this sounds appealing, but upon closer examination it is less so. Consider the case where we are stepping off stages from the bottom up. To use an energy balance around, say, stage k, we need the temperature of stage k plus the temperature of the stage above, stage $k+1$. We would also need the liquid mole fractions, $x_{i,k+1}$, of stage $k+1$. This is easily seen from the stripping section energy balance using the balance envelope shown in Figure 8-5.

$$\overline{V}H_k + Bh_B = \overline{L}_{k+1} h_{k+1} + Q_R$$

where liquid enthalpy h_{k+1} depends on T_{k+1} and $x_{i,k+1}$. To know $x_{i,k+1}$ and T_{k+1} we must simultaneously solve the equilibrium equations and enthalpy functions on stage $k+1$ with the mass and energy balances around stage k. This can be done, but I'd prefer not to. And I'm sure you'd prefer not to also. This procedure greatly complicates the trial-and-error calculation on each stage. One of the alternative calculation methods is normally preferred.

8.5. MATRIX SOLUTION OF MULTICOMPONENT DISTILLATION

Since distillation is a very important separation technique, considerable effort has been spent in devising better calculation procedures. Details of these procedures are available in a variety of textbooks (Henley and Seader, 1981; Holland, 1963, 1975, 1981; King, 1980; and Smith, 1963). Software for personal computers is also commercially available (e.g. see Lipowicz, 1987). A relatively simple matrix approach using the θ method of convergence will be outlined in this section.

The general behavior of multicomponent distillation columns (see Chapter 7) and the basic mass and energy balances and equilibrium relationships do not change when different calculation procedures are used. (The physical operation is unchanged; thus, the basic laws and the results are invariant.) What the different calculation procedures do is rearrange the equations to enhance convergence, particularly when it is difficult to make a good first guess. The most common approach is to group and solve the equations by *type*, not stage by stage. That is, all mass balances for component i are grouped and solved simultaneously, all energy balances are grouped and solved simultaneously, and so forth. Most of the equations can conveniently be written in matrix form. Computer routines for simultaneous solution of these equations are easily written. The advantage of this approach is that even very difficult problems can be made to converge.

The most convenient set of variables to specify are F, z_i, T_F, N, N_F, p, T_{reflux}, L/D, and D. Multiple feeds can be specified. This is then a simulation problem with distillate flow rate specified. For design problems, a good first guess of N and N_F must be made (see Chapter 9), and then a series of simulation problems are solved to find the best design.

Distillation problems converge best if a narrow-boiling feed calculation is done instead of a wide-boiling feed calculation (this is similar to the flash distillation procedures discussed in Chapter 3). The narrow-boiling or *bubble-point* procedure is shown in Figure 8-6. This procedure uses the equilibrium (bubble-point) calculations to determine new temperatures. The energy balance is used to calculate new flow rates. Temperatures are calculated and converged on first, and then new flow rates are determined. This procedure makes sense, since an accurate first guess of liquid and vapor flow rates can be made by assuming constant molal overflow. Thus temperatures are calculated using reasonable flow rate values. The energy balances are used last because they require values for $x_{i,j}$, $y_{i,j}$, and T_j.

Figure 8-6 is constructed for an ideal system where the K_i depend

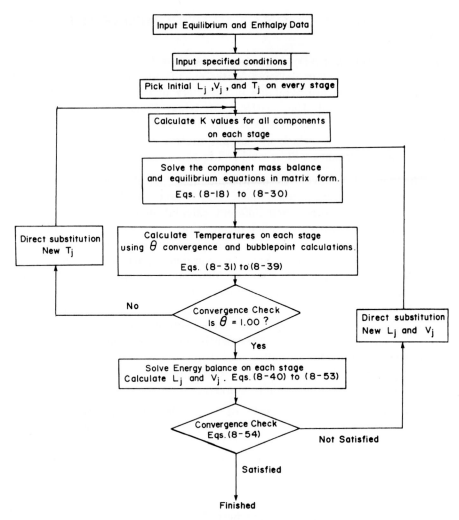

Figure 8-6.　Flow chart for multicomponent distillation.

only on temperature and pressure. If the K_i depend on compositions, then compositions must be guessed and corrected before doing the temperature calculation. We will discuss only systems where $K_i = K_i(T, p)$.

8.5.1. Component Mass Balances

To conveniently put the mass balances in matrix form, renumber the column as shown in Figure 8-7. Stage 1 is the partial reboiler, stage

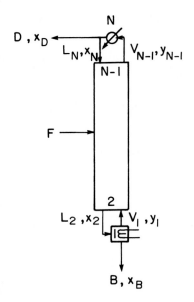

Figure 8-7. Distillation column for matrix analysis.

$N - 1$ is the top stage in the column, and the total condenser is listed as N. For a general stage j within the column (Figure 8-8), the mass balance for any component is

$$V_j y_j + L_j x_j - V_{j-1} y_{j-1} - L_{j+1} x_{j+1} = F_j z_j \qquad (8\text{-}18)$$

The unknown vapor compositions, y_j and y_{j-1}, can be replaced using the equilibrium expressions

$$y_j = K_j x_j \quad \text{and} \quad y_{j-1} = K_{j-1} x_{j-1} \qquad (8\text{-}19)$$

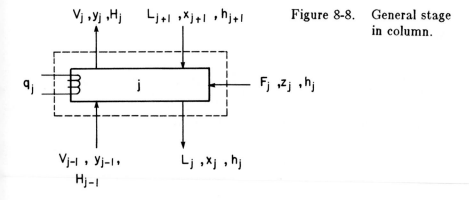

Figure 8-8. General stage in column.

where the K values depend on T and p. If we also replace x_j and x_{j+1} with

$$x_j = \ell_j/L_j, \quad x_{j+1} = \ell_{j+1}/L_{j+1} \tag{8-20}$$

where ℓ_j and ℓ_{j+1} are the liquid component flow rates, we obtain

$$\left[\frac{-V_{j-1}K_{j-1}}{L_{j-1}}\right]\ell_{j-1} + \left[1 + \frac{V_jK_j}{L_j}\right]\ell_j + (-1)\ell_{j+1} = F_jz_j \tag{8-21}$$

This equation can be written in the general form

$$A_j\ell_{j-1} + B_j\ell_j + C_j\ell_{j+1} = D_j \tag{8-22}$$

The constants A_j, B_j, C_j, and D_j are easily determined by comparing Eqs. (8-22) and (8-21).

$$A_j = -\frac{K_{j-1}V_{j-1}}{L_{j-1}}, \quad B_j = 1 + \frac{V_jK_j}{L_j}, \quad C_j = -1, D_j = F_jz_j \tag{8-23}$$

Equations (8-21) and (8-22) are valid for all stages in the column, $2 \leq j \leq N-1$, and are repeated for each of the C components. If a stage has no feed, then $F_j = D_j = 0$.

For the reboiler, the mass balance is

$$-L_2x_2 + V_1y_1 + Bx_{bot} = F_1z_1 \tag{8-24}$$

Substituting $y_1 = K_1x_1$ and $\ell_1 = L_1x_1 = Bx_{bot}$, we get

$$B_1\ell_1 + C_1\ell_2 = D_1 \tag{8-25}$$

where

$$B_1 = 1 + \frac{V_1K_1}{L_1} = 1 + \frac{V_1K_1}{B}, \quad C_1 = -1, \quad D_1 = F_1z_1 \tag{8-26}$$

For the total condenser, the mass balance is

$$L_Nx_N + Dx_D - V_{N-1}y_{N-1} = F_Nz_N \tag{8-27}$$

Since $x_N = x_D$, $y_{N-1} = K_{N-1}x_{N-1}$ and $x_N = \ell_N/L_N$, this equation becomes

$$A_N\ell_{N-1} + B_N\ell_N = D_N \tag{8-28}$$

where

$$A_N = -(K_{N-1}V_{N-1}/L_{N-1}), \quad B_N = 1 + D/L_N, \quad D_N = F_N z_N \tag{8-29}$$

Note that only B_N does not follow the general formulas of Eq. (8-23). This occurs because the total condenser is *not* an equilibrium contact.

In matrix notation, the component mass balance and equilibrium relationship for each component is

$$
\begin{bmatrix}
B_1 & C_1 & 0 & 0 & 0 & \cdot & 0 & 0 & 0 \\
A_2 & B_2 & C_2 & 0 & 0 & \cdot & 0 & 0 & 0 \\
0 & A_3 & B_3 & C_3 & 0 & \cdot & 0 & 0 & 0 \\
\cdot & \cdot & \cdot & \cdot & \cdot & \cdot & \cdot & \cdot & \cdot \\
0 & 0 & 0 & 0 & 0 & \cdot & A_{N-1} & B_{N-1} & C_{N-1} \\
0 & 0 & 0 & 0 & 0 & \cdot & 0 & A_N & B_N
\end{bmatrix}
\begin{bmatrix}
\ell_1 \\ \ell_2 \\ \ell_3 \\ \cdot \\ \ell_{N-1} \\ \ell_N
\end{bmatrix}
=
\begin{bmatrix}
D_1 \\ D_2 \\ D_3 \\ \cdot \\ D_{N-1} \\ D_N
\end{bmatrix}
\tag{8-30}
$$

This set of simultaneous linear algebraic equations can be solved by inverting the ABC matrix. This can be done using any standard matrix inversion routine on a programmable calculator or computer. The particular matrix form shown in Eq. (8-30) is a tridiagonal matrix, which is particularly easy to invert using the Thomas algorithm (see Table 8-1) (Lapidus, 1962; King, 1980). Inversion of the ABC matrix allows direct determination of the component liquid flow rate, ℓ_j, leaving each contact. You must construct the ABC matrix and invert it for each of the components.

How do we start? The A, B, and C terms in Eq. (8-30) must be calculated, but they depend on liquid and vapor flow rates and temperature (in the K values) on each stage, which we don't know. To start, guess L_j, V_j, and T_j for *every* stage j! For ideal systems the K values can be calculated for each component on every stage. Then the A, B, and C terms can be calculated for each component on every stage. Inversion of the matrices for each component gives the ℓ_j. These liquid-component flow rates are correct *for the assumed* L_j, V_j, and T_j.

Table 8-1. Thomas Algorithm for Inverting Tridiagonal Matrices

---

Consider solution of a matrix in the form of Eq. (8-30) where all A_j, B_j, C_j, and D_j are known.

1. Calculate three intermediate variables for each row of the matrix starting with $j = 1$. For $1 \leq j \leq N$,

$$(V1)_j = B_j - A_j(V3)_{j-1}$$

$$(V2)_j = [D_j - A_j(V2)_{j-1}]/(V1)_j$$

$$(V3)_j = C_j/(V1)_j$$

since $A_1 \equiv 0$, $(V1)_1 = B_1$, and $(V2)_1 = D_1/(V1)_1$.

2. Initialize $(V3)_0 = 0$ and $(V2)_0 = 0$, so you can use the general formulas.

3. Calculate all unknowns U_j [ℓ_j in Eq. (8-30), V_j in Eq. (8-48), or ΔT_j in Eq. (15-58)]. Start with $j = N$ and calculate

$$U_N = (V2)_N$$

Then going from $j = N - 1$ to $j = 1$, calculate U_{N-1}, U_{N-2}, $\cdot$ U_1 from

$$U_j = (V2)_j - (V3)_j U_{j+1}, \quad 1 \leq j \leq N-1$$

A reasonable first guess for L_j and V_j is to assume constant molal overflow [CMO was *not* assumed in Eqs. (8-18) to (8-30)]. With the CMO assumption, we can use overall mass balances to calculate all L_j and V_j. These are essentially the same preliminary calculations we used for the stage-by-stage approach. Then we can estimate the temperature from bubble-point calculations. Often it is sufficient to do a bubble-point calculation for the feed and then use this temperature on every stage. A better first guess can be obtained by estimating the distillate and bottoms compositions and doing bubble-point calculations for both. Or you can assume that temperature varies linearly from stage to stage.

8.5.2. θ METHOD

After the first guess and the solution of the matrix equations, the temperature must be corrected. This can be done with bubble-point calculations on each stage. Unfortunately, the temperature profile tends to not converge. Convergence can be obtained with a method called the θ method. The θ method first adjusts the component flow rates so that the specified distillate flow rate is satisfied. Then mole fractions are determined with these adjusted component flow rates, and bubble-point calculations on each stage are done to calculate new temperatures on each stage.

The θ method (Brian, 1972; Holland, 1963, 1975, 1981; King, 1980; Henley and Seader, 1981; Smith, 1963) defines a quantity θ that forces the following equation to be satisfied.

$$D_{\text{specified}} = \sum_{i=1}^{C} \left[\frac{Fz_i}{1 + \theta(Bx_{i,\text{bot}}/Dx_{i,\text{dist}})_{\text{calc}}} \right] \tag{8-31}$$

In other words, θ is the root of Eq. (8-31). This equation has one real, positive root which is easily found using a Newtonian convergence procedure. If we rewrite Eq. (8-31) as

$$f(\theta) = \sum_{i=1}^{C} \left[\frac{Fz_i}{1 + \theta(Bx_{i,\text{bot}}/Dx_{i,\text{dist}})_{\text{calc}}} \right] - D_{\text{spec}} = 0 \tag{8-32}$$

then the next trial value of θ is

$$\theta_{N+1} = \theta_N + \frac{f(\theta_N)}{\displaystyle\sum_{i=1}^{C} \left[\frac{Fz_i(Bx_{i,\text{bot}}/Dx_{i,\text{dist}})_{\text{calc}}}{[1 + \theta_N(Bx_{i,\text{bot}}/Dx_{i,\text{dist}})_{\text{calc}}]^2} \right]} \tag{8-33}$$

In these equations, $(Bx_{i,\text{bot}})_{\text{calc}}$ and $(Dx_{i,\text{dist}})_{\text{calc}}$ are the values calculated from the solution of Eq. (8-30).

$$(Bx_{i,\text{bot}})_{\text{calc}} = \ell_{1,i} \tag{8-34a}$$

$$(Dx_{i,\text{dist}})_{\text{calc}} = \ell_{N,i}/(L/D) \tag{8-34b}$$

Once θ has been determined, the component flow rates can be corrected. The corrected flow rates for each component are

$$(Dx_{i,dist})_{cor} = \frac{Fz_i}{1 + \theta(Bx_{i,bot}/Dx_{i,dist})_{calc}} \qquad (8\text{-}35a)$$

$$(Bx_{i,bot})_{cor} = (Dx_{i,dist})_{cor} \; \theta\left(\frac{Bx_{i,bot}}{Dx_{i,dist}}\right)_{calc} \qquad (8\text{-}35b)$$

Note that Eqs. (8-31), (8-35a) and (8-35b) give

$$\sum_{i=1}^{C} (Dx_{i,dist})_{cor} = D_{spec} \qquad (8\text{-}36a)$$

$$\sum_{i=1}^{C} (Bx_{i,bot})_{cor} = F - D_{spec} = B \qquad (8\text{-}36b)$$

Thus the θ method forces the calculation to satisfy the specified distillate flow rate.

In the rectifying section, the component flow rates are corrected:

$$(\ell_{i,j})_{cor} = \frac{(Dx_{i,dist})_{cor}}{(Dx_{i,dist})_{calc}} \; (\ell_{i,j})_{uncor} \qquad (8\text{-}37)$$

In the stripping section, use

$$(\ell_{i,j})_{cor} = \frac{(Bx_{i,bot})_{cor}}{(Bx_{i,bot})_{calc}} \; (\ell_{i,j})_{uncor} \qquad (8\text{-}38)$$

Now the corrected component flow rates are used to determine liquid mole fractions.

$$x_{i,j} = \frac{(\ell_{i,j})_{cor}}{\sum_{i=1}^{C} (\ell_{i,j})_{cor}} \qquad (8\text{-}39)$$

This procedure forces the corrected mole fractions on each stage to sum to 1.0. Once the mole fractions have been determined, the new temperatures on each stage are calculated with bubble-point calculations.

The new temperatures are used to calculate new K values (see Figure 8-6) and then new A, B, and C coefficients for Eq. (8-30). The component mass balance matrices are inverted for all components, and new $\ell_{i,j}$ are determined. The θ convergence method is used again. This procedure is continued until the temperature loop has converged, which you know has occurred when $\theta = 1.0 \pm 10^{-5}$.

8.5.3. Energy Balances

After convergence of the temperature loop, the liquid and vapor flow rates, L_j and V_j, can be corrected using energy balances (see Figure 8-6). For the general stage shown in Figure 8-8, the energy balance is

$$L_j h_j + V_j H_j = V_{j-1} H_{j-1} + L_{j+1} h_{j+1} + F_j h_{F_j} + Q_j \tag{8-40}$$

This equation is for $2 \leq j \leq N - 1$. The liquid flow rates can be substituted in from a mass balance around the bottom of the column:

$$L_j = V_{j-1} + B - \sum_{k=1}^{j-1} F_k \tag{8-41a}$$

$$L_{j+1} = V_j + B - \sum_{k=1}^{j} F_k \tag{8-41b}$$

where $2 \leq j \leq N - 1$.

Substituting Eqs. (8-41a,b) into Eq. (8-40) and rearranging, we obtain

$$(h_j - H_{j-1})V_{j-1} + (H_j - h_{j+1})V_j$$

$$= F_j h_{F_j} + Q_j + B(h_{j+1} - h_j) + (\sum_{k=1}^{j-1} F_k)h_j - (\sum_{k=1}^{j} F_k) h_{j+1} \tag{8-42}$$

For the partial reboiler $(j=1)$, the energy balance is

$$F_1 h_{F_1} + Q_1 + L_2 h_2 = B h_B + V_1 H_1 \tag{8-43}$$

while

$$L_2 = B + V_1 - F_1 \tag{8-44}$$

Substituting Eq. (8-44) into (8-43) and rearranging, we have

$$(H_1 - h_2)V_1 = F_1 h_{F_1} + Q_1 + B(h_2 - h_1) - F_1 h_2 \tag{8-45}$$

For the total condenser, the energy balance is

$$Q_N + V_{N-1}H_{N-1} + F_N h_{F_N} = V_{N-1}h_N \qquad (8\text{-}46)$$

which upon rearrangement is

$$(h_N - H_{N-1})V_{N-1} = Fh_{F_N} + Q_N \qquad (8\text{-}47)$$

If Eqs. (8-42), (8-45), and (8-47) are put in matrix form, we have

$$\begin{bmatrix} B_{E1} & 0 & 0 & 0 & \cdot & & & \\ A_{E2} & B_{E2} & 0 & 0 & \cdot & & 0 & \\ 0 & A_{E3} & B_{E3} & 0 & \cdot & & & \\ \cdot & \cdot & \cdot & \cdot & \cdot & \cdot & \cdot & \cdot \\ & & & \cdot & 0 & A_{EN-1} & B_{EN-1} & 0 \\ & 0 & & \cdot & 0 & 0 & A_{EN} & B_{EN} \end{bmatrix} \begin{bmatrix} V_1 \\ V_2 \\ V_3 \\ \cdot \\ V_{N-1} \\ V_N \end{bmatrix} = \begin{bmatrix} D_{E1} \\ D_{E2} \\ D_{E3} \\ \cdot \\ D_{EN-1} \\ D_{EN} \end{bmatrix} \qquad (8\text{-}48)$$

This matrix again has a tridiagonal form (with $C_j = 0$) and can easily be inverted to obtain the vapor flow rates. The A_E and B_E coefficients in Eq. (8-48) are easily obtained by comparing Eqs. (8-42), (8-45), and (8-47) to Eq. (8-48). For $j = 1$ these values are

$$A_{E1} = 0, \quad B_{E1} = H_1 - h_2,$$

$$D_{E1} = Fh_{F_1} + Q_1 + B(h_2 - h_1) - F_1 h_2 \qquad (8\text{-}49a)$$

For $2 \leq j \leq N-1$

$$A_{Ej} = h_j - H_{j-1}, \quad B_{Ej} = H_j - H_{j+1},$$

$$D_{Ej} = F_j h_{F_j} + Q_j + B(h_{j+1} - h_j) + \left(\sum_{i=1}^{j-1} F_i\right)h_j - \left(\sum_{i=1}^{j} F_i\right)h_{j+1} \qquad (8\text{-}49b)$$

and for $j = N$

$$A_{EN} = h_N - H_{N-1}, \quad B_{EN} = 0, \quad D_{EN} = F_N h_{F_N} + Q_N \qquad (8\text{-}49c)$$

The coefficients in Eqs. (8-49) require knowledge of the enthalpies leaving each stage and the Q values. The enthalpy values can be calculated since all x's, y's, and temperatures are known from the component mass balances and the converged temperature loop. For ideal mixtures, the enthalpies are

$$h_j = \sum_{i=1}^{C} x_{i,j} \, \tilde{h}_i(T_j) \qquad (8\text{-}50a)$$

and

$$H_j = \sum_{i=1}^{C} y_{i,j} \, \tilde{H}_i(T_j) \qquad (8\text{-}50b)$$

where $\tilde{h}_i(T_j)$ and $\tilde{H}_i(T_j)$ are the pure component enthalpies. They can be determined from data (for example, see Maxwell, 1950, or Smith, 1963) or from heat capacities and latent heats of vaporization.

Usually the column is adiabatic. Thus,

$$Q_j = 0 \quad \text{for} \quad 2 \le j \le N - 1; \quad Q_1 = Q_R, \quad Q_N = Q_C \qquad (8\text{-}51)$$

The condenser requirement can be determined from balances around the total condenser:

$$Q_N = V_{N-1}(h_D - H_{N-1}) = D(1 + \frac{L}{D})(h_D - H_{N-1}) \qquad (8\text{-}52)$$

Since D and L/D are specified and h_D and H_{N-1} can be calculated from Eqs. (8-50), Q_N is easily calculated. The reboiler heat load can be calculated from an overall energy balance:

$$Q_1 = Dh_D + Bh_B - \sum_{j=1}^{N}(F_j \, h_{F_j}) - Q_N \qquad (8\text{-}53)$$

Inversion of Eq. (8-48) gives new guesses for all the vapor flow rates. The liquid flow rates can then be determined from mass balances such as Eqs. (8-44) and (8-41). These new liquid and vapor flow rates are compared to the values used for the previous convergence of the mass balances and temperature loop. The check on convergence is, if

$$\left| \frac{L_{j,calc} - L_{j,old}}{L_{j,calc}} \right| < \epsilon \quad \text{and} \quad \left| \frac{V_{j,calc} - V_{j,old}}{V_{j,calc}} \right| < \epsilon \qquad (8\text{-}54)$$

for all stages, then the calculation has converged. For computer calculations, an ϵ of 10^{-4} or 10^{-5} is appropriate.

If the problem has not converged, the new values for L_j and V_j must be used in the mass balance and temperature loop (see Figure 8-6). Direct substitution is the easiest approach. That is, use the L_j and V_j values just calculated for the next trial.

When Eqs. (8-54) are satisfied, the calculation is finished. This is true because the mass balances, equilibrium relationships, and energy balance have all been satisfied. The solution gives the liquid and vapor mole fractions and flow rates and the temperature on each stage and in the products.

The matrix approach is easily adapted to partial condensers and to columns with side streams (see Problems 8-C4 and 8-C5). The approach will converge for normal distillation problems. Extension to more complex problems such as azeotropic and extractive distillation or very wide boiling feeds is beyond the scope of this book.

In some ways, the most difficult part of writing a multicomponent distillation program has not been discussed. This is the development of a physical properties package that will accurately predict equilibrium and enthalpy relationships. Fortunately, a considerable amount of research has been done on this (for example, see Fredenslund et al., 1977; Henley and Seader, 1981; Reid et al., 1977; and Walas, 1985). Very detailed physical properties packages can be purchased commercially.

Most companies using distillation have available canned computer programs using one of the advanced calculation procedures. Several software and design companies sell these programs. The typical engineer will use these routines and not go to the large amount of effort required to write his or her own routine. However, an understanding of the expected profiles and the basic mass and energy balances in the column can be very useful in interpreting the computer output and in determining when that output is garbage. Thus it is important to understand the principles of distillation even though the details of the computer program may not be understood.

8.6. SUMMARY - OBJECTIVES

In this chapter we have developed methods for multicomponent distillation. The objectives for this chapter are:

1. Explain and outline the general procedure for stage-by-stage analysis of multicomponent distillation.

2. Derive the appropriate equations, and solve the multicomponent distillation problems when the relative volatility is constant. Determine the optimum feed stage and closure of the trial-and-error problem.

3. Use the stage-by-stage analysis with bubble-point or dew-point calculations on every stage.

4. Develop computer flow charts (such as Figure 5-15) and programs for ternary distillation problems.

5. Use a matrix approach and the θ method of convergence to simulate multicomponent distillation.

REFERENCES

Brian, P.L.T., *Staged Cascades in Chemical Processing,* Prentice-Hall, Englewood Cliffs, NJ, 1972.

Fredenslund, A., J. Gmehling, and P. Rasmussen, *Vapor-Liquid Equilibria Using UNIFAC, A Group-Contribution Method,* Elsevier, Amsterdam, 1977.

Henley, E.J. and J.D. Seader, *Equilibrium Stage Separation Calculation in Chemical Engineering,* Wiley, New York, 1981.

Holland, C.D., *Multi-Component Distillation,* Prentice-Hall, Englewood Cliffs, NJ, 1963.

Holland, C.D., *Fundamentals and Modeling of Separation Processes: Absorption, Distillation, Evaporation, and Extraction,* Prentice-Hall, Englewood Cliffs, NJ, 1975.

Holland, C.D., *Fundamentals of Multicomponent Distillation,* McGraw-Hill, New York, 1981.

King, C.J., *Separation Processes,* 2nd ed., McGraw-Hill, New York, 1980.

Lapidus, L., *Digital Computation for Chemical Engineers,* McGraw-Hill, New York, 1962.

Lewis, W.K. and G.L. Matheson, "Studies in Distillation. Design of

Rectifying Columns for Natural and Refinery Gasoline," *Ind. Eng. Chem., 24,* 494 (1932).

Lipowicz, M., "Distillation Column Design Software," *Chem. Eng., 94* (2), 167 (Feb. 16, 1987).

Maxwell, J.B., *Data Book on Hydrocarbons,* Van Nostrand, Princeton, NJ, 1950.

Reid, R.C., J.M. Prausnitz, and T.K. Sherwood, *The Properties of Gases and Liquids,* 3rd ed., McGraw-Hill, New York, 1977.

Smith, B.D., *Design of Equilibrium Stage Processes,* McGraw-Hill, New York, 1963.

Thiele, E.W. and R.L. Geddes, "Computation of Distillation Apparatus for Hydrocarbon Mixtures," *Ind. Eng. Chem., 25,* 289 (1933).

Walas, S.M., *Phase Equilibria in Chemical Engineering,* Butterworths, Boston, 1985.

HOMEWORK

A. *Discussion Problems*

A1. If constant relative volatility is valid, what are the consequences of choosing component A or C as the reference component instead of component B?

A2. Once $K_{ref,j}$ has been determined from Eq. (8-4), how do you determine T_j?

A3. Does the test in Eqs. (8-10) work for binary systems? Check your answer on a McCabe-Thiele diagram.

A4. Explain how to use damping for the next guess for the light nonkey instead of using direct substitution, Eq. (8-12).

A5. For binary systems, illustrate the situations shown in Figures 8-2A and B on McCabe-Thiele diagrams.

A6. Figure 8-4 illustrates the results of switching operating equations too soon when stepping off stages from the top down. Illustrate on a McCabe-Thiele diagram what happens if you switch operating equations too soon when stepping off stages from the bottom up.

A7. Determine how you would introduce Murphree stage efficiencies into the stage-by-stage calculation routines.

A8. Draw a flowchart for a computer program using the stage-by-stage calculation with constant molal overflow and:

    a. Constant relative volatilities

    b. Dew-point calculation on every stage

    c. Bubble-point calculation on every stage

A9. Draw a flowchart for a computer program using a stage-by-stage calculation with constant latent heat units and a bubble-point calculation on every stage.

A10. Plot the composition and temperature profiles for Example 8-1. Explain the shapes of these curves.

A11. It has often been suggested that key components should be withdrawn as side streams at the location where their concentration maximum occurs. Explain qualitatively what this will do. Can pure light non-key be obtained?

A12. In the matrix approach, we assumed $K = K(T,p)$. How would you change the flow chart, Figure 8-6, if $K = K(T, p, x_i)$?

A13. If constant molal overflow is valid, how will Figure 8-6 be simplified?

A14. Rigorous multicomponent calculations can be done by either stage-by-stage or matrix methods. Contrast and compare these two methods for the following items:

    a. How are equations grouped?

    b. Is it a design or simulation method?

    c. List typical specified variables.

    d. What assumptions are required to start calculation?

    e. How is stage temperature determined?

A15. Develop your key relations chart for this chapter.

B. *Generation of Alternatives*

B1. All of the column configurations used for binary distillation and illustrated in Chapter 6 can be used for multicomponent distillation. Pick a particular method for a ternary distillation. Then:

a. List the specifications.

b. What initial guess would be made?

c. Determine how you would step off stages.

d. Determine the method for determining closure.

B2. Repeat Problem 8-B1 for another process.

C. *Derivations*

C1. Derive Eq. (8-13) when stages are being stepped off from the bottom up.

C2. Derive Eqs. (8-14) and (8-16) from the mass balances.

C3. Using the Newtonian convergence procedure, derive Eq. (8-33).

C4. Suppose there is a liquid side stream of composition x_j and flow rate S_j removed from stage j in Figure 8-8.

a. Derive the mass balance equations (8-21) to (8-23) for this modified column.

b. Develop new equations for the θ convergence method. Show how Eqs. (8-31), (8-32), and (8-35a) will change.

c. Develop new energy balance equations. Derive new coefficients for Eq. (8-49b).

C5. Derive the mass balance expression for the matrix approach if there is a partial condenser instead of a total condenser. Replace Eqs. (8-27) to (8-29).

C6. Derive the mass balance expression for the matrix approach if there is a total reboiler. Hint: Call the bottom stage in the column contact 1.

D. *Problems*

D1. We have a mixture of benzene, toluene, and cumene distilling in a column with a partial reboiler and a total condenser. Constant molal overflow can be assumed. The bottoms product is sampled, and the following compositions are measured: $x_B = 0.1$, $x_T = 0.3$, $x_C = 0.6$. The boilup ratio is $\overline{V}/B = 1.0$. The relative volatilities are $\alpha_{BT} = 2.5$, $\alpha_{TT} = 1.0$, $\alpha_{CT} = 0.21$. What is the composition of the vapor leaving the stage above the partial reboiler? (This is several stages below the feed stage).

D2. We have a mixture of benzene, toluene, and cumene distilling in a column that has a partial reboiler and a total condenser. Constant molal overflow can be assumed. The vapor composition leaving the feed stage is $y_B = 0.35$, $y_T = 0.20$, and $y_C = 0.45$. The external reflux ratio is $L/D = 1.7$, and reflux is returned as a saturated liquid. The relative volatilities are $\alpha_{BT} = 2.5$, $\alpha_{TT} = 1.0$, $\alpha_{CT} = 0.21$. Liquid compositions leaving the stage above the feed stage are $x_B = 0.24$, $x_T = 0.18$, $x_C = 0.58$. What is the composition of the vapor leaving the stage two stages above the feed stage?

D3. We wish to distill a mixture of benzene, toluene, and cumene in a distillation column at 101.3 kPa. The column has a total condenser and a partial reboiler. The bottoms composition is $x_B = 0.01$, $x_T = 0.37$, $x_C = 0.62$. Boilup ratio $\overline{V}/B = 2.0$. Equilibrium data can be approximated as constant relative volatilities $\alpha_{BT} = 2.25$, $\alpha_{TT} = 1.0$, $\alpha_{CT} = 0.21$. Constant molal overflow is valid. Find the composition of the liquid leaving the second stage above the partial reboiler. (The feed stage is higher than that.)

D4. We wish to distill a mixture of benzene, toluene, and cumene in a distillation column at 101.3 kPa. The column has a total condenser and a partial reboiler. Feed is a saturated liquid that is 25 mole % benzene, 35 mole % toluene, and 40 mole % cumene and is fed in at 100 kg moles/hr. Use a reflux ratio $L/D = 2.0$. We desire 95% recovery of toluene in the distillate and 96% recovery of cumene in the bottoms. The reflux is a saturated liquid, and constant molal overflow is valid. Equilibrium data can be approximated as constant relative volatilities: $\alpha_{BT} = 2.25$, $\alpha_{TT} = 1.0$, $\alpha_{CT} = 0.21$.

a. Use the optimum feed stage, and determine the total number of equilibrium stages required.

b. Use stage 5 as the feed stage, and determine the total number of equilibrium stages required.

D5. A distillation column with a partial reboiler and a total condenser is being used to separate a mixture of benzene, toluene, and cumene. The feed is input as a saturated liquid at 100 moles/hr and is 15 mole % benzene, 40 mole % toluene, and 45 mole %

cumene. We desire 95% recovery of the toluene in the distillate and 96% recovery of the cumene in the bottoms. The reflux is returned as a saturated liquid, and constant molal overflow can be assumed. Use an external reflux ratio of $L/D = 2.0$. Equilibrium can be represented as constant relative volatilities. Choosing toluene as the reference component, $\alpha_{benz} = 2.25$ and $\alpha_{cum} = 0.21$.

a. Find the number of equilibrium stages and the optimum feed location. List any assumptions used. (Include fractional number.)

b. If the feed stage is the fifth below the condenser, find the total number of equilibrium stages (including fractional number).

D6. A distillation column with a partial reboiler and a total condenser is being used to separate a mixture of benzene, toluene, and cumene. The feed is 40 mole % benzene, 30 mole % toluene, and 30 mole % cumene and is a saturated vapor. Feed rate is 1000 kg moles/hr. Reflux is returned as a saturated liquid, and $L/D = 2.0$. We desire 95% recovery of cumene in the bottoms and 95% recovery of toluene in the distillate. Pressure is 1 atm. Assume constant relative volatilities: $\alpha_{BT} = 2.25$, $\alpha_{TT} = 1.0$, $\alpha_{CT} = 0.21$. Find the optimum feed stages and the total number of equilibrium contacts required.

D7. A distillation column with nine stages plus a partial reboiler and a total condenser is separating n-butane, n-pentane, and n-hexane. Column operates at 101.3 kPa. The feed stage is the sixth from the top. $L/D = 2.0$. The distillate product is 62 mole % n-butane, 35 mole % n-pentane, and 3 mole % n-hexane. Assume that constant molal overflow is valid and use the DePriester charts or Eq. (2-12) for equilibrium. Find the temperature of the first stage in the column and the vapor composition leaving the second stage.

E. *More Complex Problems*

E1. Write and test the computer programs for Problems 8-A8 or 8-A9.

E2. Write a computer program and solve Problems 8-D4 and 8-D5 on the computer.

E3. Write a computer program and solve the following problem. A distillation column with a partial reboiler and a total condenser is being used to separate a mixture of benzene, toluene, and cumene. The feed is a two-phase mixture that is 62% liquid and is 12 mole % benzene, 55 mole % toluene, and 33 mole % cumene. Flow rate is 200 moles/hr. We desire 99.6% recovery of toluene in the distillate and 99.9% recovery of cumene in the bottoms. The reflux is returned as a saturated liquid, and constant molal overflow can be assumed. Use an $L/D = 1.0$ and $L/D = 0.50$. Choosing toluene as the reference component, equilibrium can be represented by constant relative volatilities as $\alpha_B = 2.25$, $\alpha_T = 1.00$, $\alpha_C = 0.21$. Calculate the optimum feed plate location and the number of equilibrium stages required (include fractional number of stages). Note that several stages have to be checked as optimum feed stage.

E4. We wish to distill a mixture of ethane, propane, and n-butane. The column has a partial reboiler and a partial condenser and operates at 400 kPa. The feed flow rate is 200 kg moles/hr. The feed is a saturated liquid and is 22 mole % ethane, 47 mole % propane, and 31 mole % n-butane. We wish to recover 97% of the ethane in the distillate and 99% of the propane in the bottoms. The reflux is a saturated liquid, and the external reflux ratio $L_0/D = 3.0$. Find the optimum feed stage and the total number of equilibrium contacts required. Assume constant molal overflow, and use the DePriester charts or Eq. (2-12) for K values.

E5. We wish to distill a mixture of n-butane, n-pentane, and n-hexane in a distillation column with a total condenser and a partial reboiler. The column operates at 101.3 kPa. The feed is a saturated liquid that is 0.22 wt frac n-butane, 0.35 wt frac n-pentane, and 0.43 wt frac n-hexane, and the flow rate is 1000 kg/hr. We wish to recover 98% of the n-pentane in the distillate and 97% of the n-hexane in the bottoms. The external reflux ratio is $L/D = 4.0$. Reflux is returned as a saturated liquid. Find the optimum feed stage location and the total number of equilibrium contacts required. Use the DePriester charts or Eq. (2-12) for K values. Assume that all the LNK is in the distillate and do only one iteration. The data listed below show that constant mass overflow is a better assumption than constant molal overflow. Use constant mass overflow.

	λ, kcal/g-mol	MW, g/g-mole	λ, kcal/g
nC_4	5.331	58.12	0.0917
nC_5	6.160	72.15	0.0854
nC_6	6.896	86.17	0.0800

Note: Remember to convert from weight to molal units when going from mass balances to equilibrium and vice versa. Show your work. Report temperatures ($^\circ$C), x_i and y_i (wt frac units) on each stage. Give total number of stages required and the estimated optimum feed plate location.

E6. A distillation column with a partial reboiler and a total condenser is separating nC_4, nC_5, and nC_8. The column has two equilibrium stages (a total of three equilibrium contacts), and feed is a saturated liquid fed into the bottom stage of the column. The column operates at 2 atm. Feed rate is 1000 kg moles/hr. $z_{C4} = 0.20$, $z_{C5} = 0.35$, $z_{C8} = 0.45$ (mole fractions). The reflux is a saturated liquid, and $L/D = 1.5$. The distillate rate is $D = 550$ kg moles/hr. Assume constant molal overflow. Use the DePriester chart or Eq. (2-12) for K values. For the first guess, assume that the temperatures on all stages and in the reboiler are equal to the feed bubble-point temperature. Use a matrix to solve the mass balances. Use the θ convergence method with the bubble-point arrangement to accomplish *one* iteration toward a solution for stage compositions *and* to predict new temperatures that could be used for a second iteration. Report the compositions on each stage and in the reboiler, and the temperatures of each stage and the reboiler.

E7. The Chomical Chemical Company would like to distill a mixture of dichloromethane, trichloromethane, and carbon tetrachloride. The vice president of design and development has charged the Separations Group to come up with several column designs at various operating conditions. An engineer has decided that a Cyber 205 program will be required to solve the thousands of coupled mass, momentum, and heat transport equations. Your supervisor would rather see a rough design in a few days than a stack of computer paper in a few years. Your column will have a total condenser and a partial reboiler. The desired operating pressure is 200 mmHg. The feed flow rate is 200 lb moles/hr. The feed is a saturated liquid and is 30 mole % CH_2Cl_2, 50 mole % $CHCl_3$, and 20 mole % CCl_4. In order for the process to be economically worthwhile, at least 90% of the dichloromethane and 90% of the trichloromethane must be recovered in the distillate and bottoms, respectively. The reflux is a saturated liquid, and the external reflux ratio is, $L_0/D = 50.0$.

Find the optimum feed stage and the total number of equilibrium contacts required. In addition, plot the temperature and liquid composition profiles. Assume constant molal overflow and ideal solution behavior ($K_i = VP_i/p_{tot}$).

Vapor Pressure Data

Pressure, mmHg	100	200	400	760
Component	Temperature, $^\circ$ C			
CH_2Cl_2	−6.3	8.0	24.1	40.7
$CHCl_3$	10.4	25.9	42.7	61.3
CCl_4	23.0	38.3	57.8	76.7

Source: Perry and Chilton (1973), pp. 3-57.

Note: Do only one trial through entire column based on your first guess that all HNK is in the bottoms.

E8. A distillation column with a partial reboiler and a total condenser is separating a feed that is 25 mole % nC_4, 42 mole % nC_7 and 33 mole % nC_8. Feed is a saturated liquid with a flow rate of 2500 kg moles/hr. Reflux is returned as a saturated liquid and $L_0/D = 1.3$. Pressure is 2 atm. We desire 98.2% recovery of nC_4 in the distillate and 99.1% recovery of nC_7 in the bottoms. Assume constant molal overflow. Find the total number of stages and the optimum feed plate location. Do only one trial for an initial guess of all nC_8 in the bottoms.

F. *Problems Requiring Other Resources*

F1. Do Problem 8-D6 without assuming constant relative volatility.

F2. Symbols, styles of presentation, and the general approach to doing a calculation have changed particularly since computers have become common. Despite this, the modified calculation methods often retain the names of the original investigators. Stage-by-stage methods for multicomponent distillation are often called Lewis-Matheson or Thiele-Geddes methods. Read and critique the article by Lewis and Matheson (1932). In your one-page report, also compare their calculation procedure with the stage-by-stage method in this chapter.

F3. If you have available a packaged computer program, use it to solve Problem 8-E4. Depending on the type of program, this may involve picking N and N_F until the desired separation is achieved. Then use the program to determine the effect of changing variables.

a. Try decreasing L/D.

b. Try increasing the fractional recovery of ethane.

c. Change the feed location.

    d. Solve the problem for 98.5% recovery of propane in the distillate and 99.2% recovery of n-butane in the bottoms.

    e. Try withdrawing a side stream at the concentration maximum for propane in part d.

    f. Try a feed that also contains n-pentane and repeat part d.

    g. Sketch the concentration profiles or have the computer draw the profiles and note how they differ in parts a, d, and f.

    h. Sketch the temperature profiles, and compare the different results.

    i. Sketch the total flow profiles. Does this program assume constant molal overflow? If not, how much do the flow rates differ from CMO?

F4.   (Difficult) A distillation column with two stages plus a partial reboiler and a partial condenser is separating benzene, toluene, and xylene. Feed rate is 100 kg moles/hr, and feed is a saturated vapor. Feed compositions (mole fractions) are $z_B = 0.35$, $z_T = 0.40$, $z_X = 0.25$. Reflux is a saturated liquid and p = 16 psia. A distillate flow rate of D = 30 kg moles/hr is desired. Assume $K_1 = VP_i/p$. Do *not* assume constant relative volatility, but do assume constant molal overflow. Use the matrix approach to solve mass balances and the θ method for temperature convergence. For the first guess for temperature, assume that all stages are at the dew-point temperature of the feed. Do only one iteration. See Problem 8-C5 for handling the partial condenser.

F5.   Repeat Problem 8-F4, but solve it on a computer.

chapter 9
APPROXIMATE SHORTCUT METHODS FOR MULTICOMPONENT DISTILLATION

The previous chapters served as an introduction to multicomponent distillation. The stage-by-stage method is a rigorous calculation technique, but the trial-and-error procedure can be extremely time-consuming. Matrix methods are efficient, but they still require a fair amount of time even on a fast computer. In addition, they are simulation methods and require a known number of stages and a specified feed plate location. Fairly rapid approximate methods are required for preliminary economic estimates, for recycle calculations where the distillation is only a small portion of the entire system, for calculations for control systems, and as a first estimate for more detailed simulation calculations.

In this chapter we will first develop the Fenske equation, which allows calculation of multicomponent separation at total reflux. Then we will switch to the Underwood equations, which allow us to calculate the minimum reflux ratio. To predict the approximate number of equilibrium stages we then use an empirical correlation that relates the actual number of stages to the number of stages at total reflux, the minimum reflux ratio, and the actual reflux ratio. The feed location can also be approximated from the empirical correlation.

9.1. TOTAL REFLUX: FENSKE EQUATION

Fenske (1932) derived a rigorous solution for binary and multicomponent distillation at total reflux. The derivation assumes that the stages are equilibrium stages.

Consider the multicomponent distillation column operating at total reflux shown in Figure 9-1, which has a total condenser and a partial reboiler. For an equilibrium partial reboiler for any two components A and B,

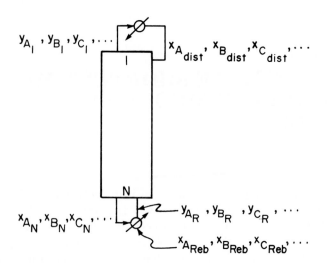

Figure 9-1.   Total reflux column.

$$\left[\frac{y_A}{y_B}\right]_R = \alpha_R \left[\frac{x_A}{x_B}\right]_R \qquad (9\text{-}1)$$

Equation (9-1) is just the definition of the relative volatility applied to the reboiler. Material balances for these components around the reboiler are

$$V_R y_{A,R} = L_N x_{A,N} - B x_{A,R} \qquad (9\text{-}2a)$$

and

$$V_R y_{B,R} = L_N x_{B,N} - B x_{B,R} \qquad (9\text{-}2b)$$

However, at total reflux, $B = 0$ and $L_N = V_R$. Thus the mass balances become

$$y_{A,R} = x_{A,N}, \quad y_{B,R} = x_{B,N} \quad \text{(at total reflux)} \qquad (9\text{-}3)$$

For a binary system this naturally means that the operating line is the y = x line. Combining Eqs. (9-1) and (9-3),

$$\left[\frac{x_A}{x_B}\right]_N = \alpha_R \left[\frac{x_A}{x_B}\right]_R \qquad (9\text{-}4)$$

If we now move up the column to stage N, the equilibrium equation is

$$\left[\frac{y_A}{y_B}\right]_N = \alpha_N \left[\frac{x_A}{x_B}\right]_N$$

The mass balances around stage N simplify to

$$y_{A,N} = x_{A,N-1} \quad \text{and} \quad y_{B,N} = x_{B,N-1}$$

Combining these equations, we have

$$\left[\frac{x_A}{x_B}\right]_{N-1} = \alpha_N \left[\frac{x_A}{x_B}\right]_N \tag{9-5}$$

Then equations (9-4) and (9-5) can be combined to give

$$\left[\frac{x_A}{x_B}\right]_{N-1} = \alpha_N \alpha_R \left[\frac{x_A}{x_B}\right]_R \tag{9-6}$$

which relates the ratio of liquid mole fractions leaving stage N−1 to the ratio in the reboiler.

Repeating this procedure for stage N-1, we obtain

$$\left[\frac{x_A}{x_B}\right]_{N-2} = \alpha_{N-1}\alpha_N \alpha_R \left[\frac{x_A}{x_B}\right]_R \tag{9-7}$$

We can alternate between the operating and equilibrium equations until we reach the top stage. The result is

$$\left[\frac{x_A}{x_B}\right]_{dist} = \alpha_1 \alpha_2 \alpha_3 \cdots \alpha_{N-1}\alpha_N \alpha_R \left[\frac{x_A}{x_B}\right]_R \tag{9-8}$$

If we define α_{AB} as the geometric average relative volatility,

$$\alpha_{AB} = [\alpha_1 \alpha_2 \alpha_3 \cdots \alpha_{N-1}\alpha_N \alpha_R]^{1/N_{min}} \tag{9-9}$$

Eq. (9-8) becomes

$$\left[\frac{x_A}{x_B}\right]_{dist} = \alpha_{AB}^{N_{min}} \left[\frac{x_A}{x_B}\right]_R \tag{9-10}$$

Solving Eq. (9-10) for N_{min}, we obtain

$$N_{min} = \frac{\ln[(\frac{x_A}{x_B})_{dist}/ (\frac{x_A}{x_B})_R]}{\ln \alpha_{AB}} \tag{9-11}$$

which is one form of the Fenske equation. N_{min} is the number of equilibrium contacts including the partial reboiler required at total reflux. If the relative volatility is constant, Eq. (9-11) is exact.

An alternative form of the Fenske equation that is very convenient for multicomponent calculations is easily derived. Equation (9-11) can also be written as

$$N_{min} = \frac{\ln[\frac{(Dx_A/Dx_B)_{dist}}{(Bx_A/Bx_B)_R}]}{\ln \alpha_{AB}} \tag{9-12}$$

$(Dx_A)_{dist}$ is equal to the fractional recovery of A in the distillate times the amount of A in the feed.

$$(Dx_A)_{dist} = (FR_A)_{dist} Fz_A \tag{9-13}$$

where $(FR_A)_{dist}$ is the fractional recovery of A in the distillate. From the definition of fractional recovery,

$$(Bx_A)_R = [1 - (FR_A)_{dist}] Fz_A \tag{9-14}$$

Substituting Eqs. (9-13) and (9-14) and the corresponding equations for component B into Eq. (9-12) gives

$$N_{min} = \frac{\ln\left\{\frac{(FR_A)_{dist} (FR_B)_{bot}}{[1 - (FR_A)_{dist}][1 - (FR_B)_{bot}]}\right\}}{\ln \alpha_{AB}} \tag{9-15}$$

Note that in this form of the Fenske equation, $(FR_A)_{dist}$ is the fractional recovery of A in the distillate, while $(FR_B)_{bot}$ is the fractional recovery of B in the bottoms. Equation (9-15) is in a convenient form for multicomponent systems.

The derivation up to this point has been for any number of components. If we now restrict ourselves to a binary system where $x_B = 1 - x_A$, Eq. (9-11) becomes

$$N_{min} = \frac{\ln\left[\dfrac{[x/(1-x)]_{dist}}{[x/(1-x)]_{bot}}\right]}{\ln \alpha_{AB}} \tag{9-16}$$

where $x = x_A$ is the mole fraction of the more volatile component. The use of the Fenske equation for binary systems is quite straightforward. With distillate and bottoms mole fractions of the more volatile component specified, N_{min} is easily calculated if α_{AB} is known. If the relative volatility is not constant, α_{AB} can be estimated from a geometric average as shown in Eq. (9-9). This can be estimated for a first trial as

$$\alpha_{avg} = (\alpha_1 \alpha_R)^{1/2}$$

where α_R is determined from the bottoms composition and α_1 from the distillate composition.

For multicomponent systems calculation with the Fenske equation is straightforward if fractional recoveries of the two keys, A and B, are specified. Equation (9-15) can now be used directly to find N_{min}. The relative volatility can be approximated by a geometric average. Once N_{min} is known, the fractional recoveries of the non-keys can be found by writing Eq. (9-15) for a non-key component, C, and either key component. Then solve for $(FR_C)_{dist}$ or $(FR_C)_{bot}$. When this is done, Eq. (9-15) becomes

$$(FR_C)_{dist} = \frac{\alpha_{CB}^{N_{min}}}{\dfrac{(FR_B)_{bot}}{1 - (FR_B)_{bot}} + \alpha_{CB}^{N_{min}}} \tag{9-17}$$

If two mole fractions are specified, say $x_{LK,bot}$ and $x_{HK,dist}$, the multicomponent calculation is more difficult. We can't use the Fenske equation directly, but several alternatives are possible. If we can assume that all non-keys are nondistributing, we have

$$Dx_{LNK,dist} = Fz_{LNK}, \quad x_{LNK,bot} = 0$$

$$Bx_{HNK,bot} = F \, z_{HNK}, \quad x_{HNK,dist} = 0 \qquad (9\text{-}18)$$

Equations (9-18) can be solved along with the light and heavy key mass balances and the equations

$$\sum_{i=1}^{c} (D \, x_{i,dist}) = D \quad \text{and} \quad \sum_{i=1}^{c} (B \, x_{i,bot}) = B \qquad (9\text{-}19)$$

Once all distillate and bottoms compositions or values for $Dx_{i,dist}$ and $Bx_{i,bot}$ have been found, Eqs. (9-11) or (9-12) can be used to find N_{min}. Use the key components for this calculation. The assumption of nondistribution of the non-keys can be checked with Eq. (9-10) or (9-17). If the original assumption is invalid, the calculated values for non-key compositions can be used to calculate the light and heavy key compositions in distillate and bottoms. Then Eq. (9-11) or (9-12) is used again.

If non-keys do distribute, a reasonable first guess for the distribution is required. This guess can be obtained by assuming that the distribution of non-keys is the same at total reflux as it is at minimum reflux. The distribution at minimum reflux can be obtained from the Underwood equation and is covered later.

Accurate use of the Fenske equation obviously requires an accurate value for the relative volatility. Smith (1963) covers in detail a method of calculating α by estimating temperatures and calculating the geometric average relative volatility. Winn (1958) developed a modification of the Fenske equation that allows the relative volatility to vary. Wankat and Hubert (1979) modified both the Fenske and Winn equations for nonequilibrium stages by including a vaporization efficiency.

Example 9-1. Fenske Equation

A distillation column with a partial reboiler and a total condenser is being used to separate a mixture of benzene, toluene, and cumene. The feed is 40 mole % benzene, 30 mole % toluene, and 30 mole % cumene and is input as a saturated vapor. We desire 95% recovery of the toluene in the distillate and 95% recovery of the cumene in the bottoms. The reflux is returned as a saturated liquid, and constant molal overflow can be assumed. Pressure is 1 atm.

Equilibrium can be represented as constant relative volatilities. Choosing toluene as the reference component, $\alpha_{BT} = 2.25$

and $\alpha_{CT} = 0.21$. Find the number of equilibrium stages required at total reflux and the recovery fraction of benzene in the distillate.

Solution

A. Define. The problem is sketched below. For A = toluene (LK), B = cumene (HK), C = benzene (LNK), we have $\alpha_{CA} = 2.25$, $\alpha_{AA} = 1.0$, $\alpha_{BA} = 0.21$, $z_A = 0.3$, $z_B = 0.3$, $z_C = 0.4$, $FR_{A,dist} = 0.95$, and $FR_{B,bot} = 0.95$.

a. Find N at total reflux.

b. Find $FR_{C,dist}$ at total reflux.

B. Explore. Since operation is at total reflux and relative volatilities are constant, we can use the Fenske equation.

C. Plan. Calculate N_{min} from Eq. (9-15), and then calculate $FR_{C,dist}$ from Eq. (9-17).

D. Do It. Equation (9-15) gives

$$N_{min} = \frac{\ln[(\frac{FR_{A,dist}}{1 - FR_{A,Dist}})/(\frac{1 - FR_{B,bot}}{FR_{B,bot}})]}{\ln \alpha_{AB}}$$

$$= \frac{\ln[\frac{0.95}{0.05}/\frac{0.05}{0.95}]}{\ln (1/0.21)} = 3.77$$

Note that $\alpha_{AB} = 1/\alpha_{BA}$. Equation (9-17) gives

$$FR_{C,dist} = \frac{(\alpha_{CB})^{N_{min}}}{\dfrac{FR_{B,bot}}{1-FR_{B,bot}} + (\alpha_{CB})^{N_{min}}}$$

$$= \frac{(2.25/0.21)^{3.77}}{\dfrac{0.95}{0.05} + (\dfrac{2.25}{0.21})^{3.77}} = 0.998$$

which is the desired benzene recovery in the distillate. Note that

$$\alpha_{CB} = \frac{K_C}{K_B} = \frac{K_C/K_A}{K_B/K_A} = \frac{\alpha_{CA}}{\alpha_{BA}}$$

E.   Check.   The results can be checked by calculating $FR_{C,dist}$ using component A instead of B. The same answer is obtained.

F.   Generalize.   We could continue this problem by calculating $Dx_{i,dist}$ and $Bx_{i,bot}$ for each component from Eqs. (9-13) and (9-14). Then distillate and bottoms flow rates can be found from Eqs. (9-19), and the distillate and bottoms compositions can be calculated.

9.2. MINIMUM REFLUX: UNDERWOOD EQUATIONS

For binary systems, the pinch point usually occurs at the feed plate. When this occurs, an analytical solution for the limiting flows can be derived (see King, 1980) that is also valid for multicomponent systems as long as the pinch point occurs at the feed stage. Unfortunately, for multicomponent systems there will be separate pinch points in both the stripping and enriching sections if there are nondistributing components. In this case an alternative analysis procedure developed by Underwood (1948) is used to find the minimum reflux ratio.

The development of the Underwood equations is quite complex and is presented in detail by Underwood (1948), Smith (1963), Holland (1975), and King (1980). Since for most practicing engineers the details of the development are not as important as the use of the Underwood equations, we will follow the approximate derivation of Thompson (1980). Thus we will outline the important points but wave our hands about the mathematical details of the derivation.

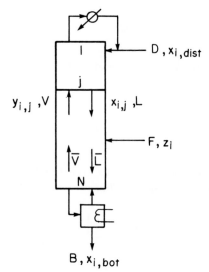

Figure 9-2.   Distillation column.

If there are nondistributing heavy non-keys present, a "pinch point" of constant composition will occur at minimum reflux in the enriching section above where the heavy non-keys are fractionated out. With non-distributing light non-keys present, a pinch point will occur in the stripping section. For the enriching section in Figure 9-2, the mass balance for component i is

$$V_{min} \, y_{i,j+1} = L_{min}x_{i,j} + Dx_{i,dist} \tag{9-20}$$

At the pinch point, where compositions are constant,

$$x_{i,j-1} = x_{i,j} = x_{i,j+1} \ , \text{and} \ \ y_{i,j-1} = y_{i,j} = y_{i,j+1} \tag{9-21}$$

The equilibrium expression can be written in terms of K values as

$$y_{i,j+1} = K_i \, x_{i,j+1} \tag{9-22}$$

Combining Eqs. (9-20) to (9-22) we obtain a simplified balance valid in the region of constant compositions.

$$V_{min}y_{i,j+1} = \frac{L_{min}}{K_i} \, y_{i,j+1} + Dx_{i,dist} \tag{9-23}$$

Defining the relative volatility $\alpha_i = K_i/K_{HK}$ and combining terms in Eq. (9-23),

$$V_{min}\, y_{i,j+1} \left(1 - \frac{L_{min}}{V_{min}\alpha_i K_{HK}}\right) = Dx_{i,dist} \qquad (9\text{-}24)$$

Solving for the component vapor flow rate, $V_{Min}\, y_{i,j+1}$, and rearranging

$$V_{min} y_{i,j+1} = \frac{\alpha_i\, Dx_{i,dist}}{\alpha_i - \dfrac{L_{min}}{V_{min} K_{HK}}} \qquad (9\text{-}25)$$

Equation (9-25) can be summed over all components to give the total vapor flow rate in the enriching section at minimum reflux.

$$V_{min} = \sum_{i=1}^{C} (V_{min}\, y_{i,j+1}) = \sum_{i=1}^{C} \left[\frac{\alpha_i Dx_{i,dist}}{\alpha_i - \dfrac{L_{min}}{V_{min} K_{HK}}}\right] \qquad (9\text{-}26)$$

In the stripping section a similar analysis can be used to derive,

$$-\overline{V}_{min} = \sum_{i=1}^{C} \left[\frac{\overline{\alpha}_i Bx_{i,bot}}{\overline{\alpha}_i - \dfrac{\overline{L}_{min}}{\overline{V}_{min}\overline{K}_{HK}}}\right] \qquad (9\text{-}27)$$

Since the conditions in the stripping section are different than in the rectifying section, in general $\alpha_i \neq \overline{\alpha}_i$ and $K_{HK} \neq \overline{K}_{HK}$.

Underwood (1948) derived generalized forms of Eqs. (9-26) and (9-27) which are equivalent to defining

$$\phi = \frac{L_{min}}{V_{min} K_{HK}} \quad \text{and} \quad \overline{\phi} = \frac{\overline{L}_{min}}{\overline{V}_{min}\, \overline{K}_{HK}} \qquad (9\text{-}28)$$

Equations (9-26) and (9-27) then become polynomials in ϕ and $\overline{\phi}$ and have C roots. The equations are now

$$V_{min} = \sum_{i=1}^{C} \frac{\alpha_i (Dx_{i,dist})}{\alpha_i - \phi} \qquad (9\text{-}29)$$

and

$$-\overline{V}_{min} = \sum_{i=1}^{C} \frac{\overline{\alpha}_i(Bx_{i,bot})}{\overline{\alpha}_i - \overline{\phi}} \tag{9-30}$$

If we assume constant molal overflow and constant relative volatilities $\alpha_i = \overline{\alpha}_i$, Underwood showed there are common values of ϕ and $\overline{\phi}$ which satisfy both equations. Equations (9-29) and (9-30) can now be added. Thus at minimum reflux

$$V_{min} - \overline{V}_{min} = \sum_{i=1}^{C} \left[\frac{\alpha_i Dx_{i,dist}}{\alpha_i - \phi} + \frac{\alpha_i Bx_{i,bot}}{\alpha_i - \phi} \right] \tag{9-31}$$

where α_i is now an average relative volatility. Equation (9-31) is easily simplified with the overall column mass balance

$$Fz_i = Dx_{i,dist} + Bx_{i,bot} \tag{9-32}$$

to

$$\Delta V_{feed} = V_{min} - \overline{V}_{min} = \sum_{i=1}^{C} \frac{\alpha_i Fz_i}{\alpha_i - \phi} \tag{9-33}$$

ΔV_{feed} is the change in vapor flow rate at the feed stage. If q is known

$$\Delta V_{feed} = F(1-q) \tag{9-34}$$

If the feed temperature is specified a flash calculation on the feed can be used to determine ΔV_{feed}.

Equation (9-33) is known as the first Underwood equation. It can be used to calculate appropriate values of ϕ. Equation (9-29) is known as the second Underwood equation and is used to calculate V_{min}. Once V_{min} is known, L_{min} is calculated from the mass balance

$$L_{min} = V_{min} - D \tag{9-35}$$

The exact method for using the Underwood equations depends on what can be assumed. Three cases will be considered.

Case A. Assume all non-keys do *not* distribute. In this case the amounts of non-keys in the distillate are:

$$Dx_{HNK,dist} = 0 \quad \text{and} \quad Dx_{LNK,dist} = Fz_{LNK}$$

while the amounts of the keys are:

$$Dx_{LK,dist} = (FR_{LK})_{dist} \, Fz_{LK} \tag{9-36}$$

$$Dx_{HK,dist} = [1-(FR_{HK})_{bot}] \, Fz_{HK} \tag{9-37}$$

Equation (9-33) can now be solved for the one value of ϕ between the relative volatilities of the two keys, $\alpha_{HK} < \phi < \alpha_{LK}$. This value of ϕ can be substituted into Eq. (9-29) to immediately calculate V_{min}. Then

$$D = \sum_{i=1}^{C} (Dx_{i,dist}) \tag{9-38}$$

And L_{min} is found from mass balance Eq. (9-35).

This assumption of nondistributing non-keys will probably not be valid for sloppy separations or when a sandwich component is present. In addition, with a sandwich component there are two ϕ values between α_{HK} and α_{LK}. Thus use Case C (discussed later) for sandwich components. The method of Shiras *et al.* (1950) can be used to check for distribution of non-keys.

Case B. Assume that the distributions of non-keys determined from the Fenske equation at total reflux are also valid at minimum reflux. In this case the $Dx_{NK,dist}$ values are obtained from the Fenske equation as described earlier. Again solve Eq. (9-33) for the ϕ value between the relative volatilities of the two keys. This ϕ, the Fenske values of $Dx_{NK,dist}$, and the $Dx_{LK,dist}$ and $Dx_{HK,dist}$ values obtained from Eqs. (9-36) and (9-37) are used in Eq. (9-29) to find V_{min}. Then Eqs. (9-38) and (9-35) are used to calculate D and L_{min}. This procedure is illustrated in Example 9-2.

Case C. Exact solution without further assumptions. Equation (9-33) is a polynomial with C roots. Solve this equation for all values of ϕ lying between the relative volatilities of all components,

$$\alpha_{LNK,1} < \phi_1 < \alpha_{LNK,2} < \phi_2 < \alpha_{LK} < \phi_3 < \alpha_{HK} < \phi_4 < \alpha_{HNK,1}$$

This gives C-1 valid roots. Now write Eq. (9-29) C-1 times; once for each value of ϕ. We now have C-1 equations and C-1 unknowns

(V_{min}, and $Dx_{i,dist}$ for all LNK and HNK). Solve these simultaneous equations and then obtain D from Eq. (9-38) and L_{min} from Eq. (9-35). A sandwich component problem which must use this approach is given in Problem 9-D15.

The hardest part of using the Underwood equations is solving Eq. (9-33) for one or more values of ϕ. In general, Eq. (9-33) will be of order C in ϕ where C is the number of components. Saturated liquid and saturated vapor feeds are special cases and after simplification are of order C-1 (see Problems 9-C10 and 9-C11). If the resulting equation is quadratic, the quadratic formula can be used to find the roots. Otherwise, a root-finding method should be employed. If only one root, $\alpha_{LK} > \phi > \alpha_{HK}$, is desired, a good first guess is to assume $\phi = (\alpha_{LK} + \alpha_{HK})/2$.

The results of the Underwood equations will only be accurate if the basic assumptions of constant relative volatility and constant molal overflow are valid. For small variations in α a geometric average calculated as

$$\alpha_i = \left(\alpha_{bot}\,\alpha_{dist}\right)^{1/2} \text{ or } \alpha_i = \left(\alpha_{bot}\,\alpha_{feed}\,\alpha_{dist}\right)^{1/3} \tag{9-39}$$

can be used as an approximation.

If constant molal overflow is not valid, the mole fractions in Eqs. (9-20) to (9-38) can be replaced by the constant latent heat units described in Chapter 6. Since the constant relative volatilities are unchanged when units are changed, the Underwood equations are also valid in latent heat units. Application of the Underwood equations to systems with multiple feeds was studied by Barnes *et al.*, (1972).

Example 9-2. Underwood Equations

For the distillation problem given in Example 9-1 find the minimum reflux ratio. Use a basis of 100 kg moles/hour of feed.

Solution

A. Define. Problem was sketched in Example 9-1. We now wish to find $(L/D)_{min}$.

B. Explore. Since the relative volatilities are approximately constant and since Example 9-1 showed that benzene does not distribute, the Underwood equations can easily be used to estimate the minimum reflux ratio.

C. Plan. This problem fits into Case A or Case B. We can then calculate $Dx_{i,dist}$ values as described in Cases A or B, Eqs. (9-36) and (9-37), and solve Eq. (9-33) for ϕ where ϕ lies between the relative volatilities of the two keys $0.21 < \phi < 1.00$. Then V_{min} can be found from Eq. (9-29), D from Eq. (9-38) and L_{min} from Eq. (9-35).

D. Do It. Follow Case B analysis. Since the feed is a saturated vapor, q = 0 and $\Delta V_{feed} = F (1 - q) = F = 100$ and Eq. (9-33) becomes

$$100 = \frac{(2.25)(40)}{2.25 - \phi} + \frac{(1.0)(30)}{1.0 - \phi} + \frac{(0.21)(30)}{0.21 - \phi}$$

Solving by trial-and-error for ϕ between 0.21 and 1.00, we obtain $\phi = 0.5454$. Equation (9-29) is

$$V_{min} = \sum_{i=1}^{C} \left(\frac{\alpha_i (Dx_{i,dist})}{\alpha_i - \phi} \right)$$

where

$$Dx_{i,dist} = F\, z_i (FR)_{i,dist}$$

For benzene this is

$$Dx_{ben,dist} = 100(0.4)(0.998) = 39.92$$

where the fractional recovery of benzene is the value calculated in Example 9-1 at total reflux. The other distillate values are

$$Dx_{tol,dist} = 100(0.3)(0.95) = 28.5$$

$$Dx_{cum,dist} = 100(0.3)(0.05) = 1.5$$

Summing the three distillate flows, D = 69.92. Equation (9-29) becomes

$$V_{min} = \frac{(2.25)(39.92)}{2.25 - 0.5454} + \frac{(1.0)(28.5)}{1.0 - 0.5454} + \frac{(0.21)(1.5)}{0.21 - 0.5454} = 114.4$$

From a mass balance, $L_{min} = V_{min} - D = 44.48$, and $(L/D)_{min} = 0.636$.

E. Check. The Case A calculation gives essentially the same result.

F. Generalize. The addition of more components does not make the calculation more difficult as long as the fractional recoveries can be accurately estimated. The value of ϕ must be accurately determined since it can have a major affect on the calculation. Since the separation is easy, $(L/D)_{min}$ is quite small in this case. $(L/D)_{min}$ will not be as dependent on the exact values of ϕ as it is when $(L/D)_{min}$ is large.

9.3. CORRELATION FOR NUMBER OF STAGES AT FINITE REFLUX RATIO: GILLILAND CORRELATION

A general shortcut method for determining the number of stages required for a multicomponent distillation at finite reflux ratios would be extremely useful. Unfortunately, such a method has not been developed. However, Gilliland (1940) noted that he could empirically relate the number of stages N at finite reflux ratio L/D to the minimum number of stages N_{min} and the minimum reflux ratio $(L/D)_{min}$. Gilliland did a series of accurate stage-by-stage calculations and found that he could correlate the function $(N-N_{min})/(N + 1)$ with the function $[L/D - (L/D)_{min}]/(L/D + 1)$. This correlation as modified by Liddle (1968) is shown in Figure 9-3. The data points are the results of Gilliland's stage-by-stage calculation and show the scatter inherent in this correlation.

To use the Gilliland correlation we proceed as follows:

1. Calculate N_{min} from the Fenske equation.

2. Calculate $(L/D)_{min}$ from Underwood equations.

3. Choose actual (L/D). This is usually done as some multiplier (1.05 to 1.5) times $(L/D)_{min}$.

4. Calculate the abscissa.

5. Determine the ordinate value.

6. Calculate the actual number of stages, N.

The Gilliland correlation should only be used for rough estimates. The

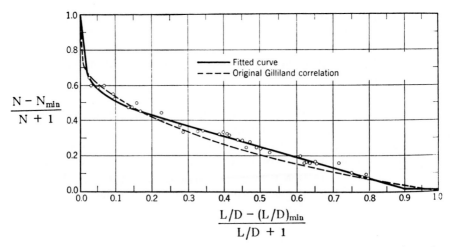

Figure 9-3. Gilliland correlation as modified by Liddle (1968). Reprinted with permission from *Chemical Engineering*, *75*(23), 137 (1968). Copyright 1968, McGraw-Hill.

calculated number of stages can be off by $\pm 30\%$ although they are usually within $\pm 7\%$.

The optimum feed plate location can also be estimated. First, use the Fenske equation to estimate where the feed stage would be at total reflux. This can be done by determining the number of stages required to go from the feed concentrations to the distillate concentrations *for the keys*.

$$N_{F,min} = \frac{\ln\left[\left[\dfrac{x_{LK}}{x_{HK}}\right]_{dist}\Big/\left[\dfrac{z_{LK}}{z_{HK}}\right]\right]}{\ln \alpha_{LK\text{--}HK}} \tag{9-40}$$

Now assume that the relative feed location is constant as we change the reflux ratio from total reflux to a finite value. Thus

$$\frac{N_{F,min}}{N_{min}} = \frac{N_F}{N} \tag{9-41}$$

The actual feed stage can now be estimated from Eq. (9-41).

The Gilliland correlation can also be fit to equations. Liddle (1968) fit the Gilliland correlation to three equations. Let $x = [L/D - (L/D)_{min}]/(L/D + 1)$. Then

$$\frac{N - N_{min}}{N + 1} = 1.0 - 18.5715x \quad \text{for } 0 \le x \le 0.01 \qquad (9\text{-}42a)$$

while for $0.01 < x < 0.90$

$$\frac{N - N_{min}}{N + 1} = 0.545827 - 0.591422x + \frac{0.002743}{x} \qquad (9\text{-}42b)$$

and for $0.90 \le x \le 1.0$

$$\frac{N - N_{min}}{N + 1} = 0.16595 - 0.16595x \qquad (9\text{-}42c)$$

For most situations Eq. (9-42b) is appropriate. The fit to the data is shown in Figure 9-3. Naturally, the equations are useful for computer calculations.

As a rough rule of thumb we can estimate $N = 2.5 \, N_{min}$. This estimate then requires *only* a calculation of N_{min} and will be useful for very preliminary estimates. Erbar and Maddox (1961) (see King, 1980 or Hines and Maddox, 1985) developed a somewhat more accurate correlation which uses more than one curve.

Example 9-3. Gilliland Correlation

Estimate the total number of equilibrium stages and the optimum feed plate location required for the distillation problem presented in Examples 9-1 and 9-2 if the actual reflux ratio is set at $L/D = 2$.

Solution

A. Define. The problem was sketched in Example 9-1. $F = 100$, $L/D = 2$, and we wish to estimate N and N_F.

B. Explore. An estimate can be obtained from the Gilliland correlation, while a more exact calculation could be done stage by stage. We will use the Gilliland correlation.

C. Plan. Calculate the abscissa $= \dfrac{L/D - (L/D)_{min}}{L/D + 1}$, determine the ordinate $= \dfrac{N - N_{min}}{N + 1}$ from the Gilliland correlation, and then find N. $(L/D)_{min}$ was found in Example 9-2, and N_{min} in Example 9-1. The feed plate location is estimated from Eqs. (9-41) and (9-40).

D. Do It.

$$\text{abscissa} = \frac{L/D - (L/D)_{min}}{L/D + 1} = \frac{2 - 0.636}{2 + 1} = 0.455$$

The corresponding ordinate $= \dfrac{N - N_{min}}{N + 1} = 0.27$ using Liddle's curve. Since $N_{min} = 3.77$, $N = 5.53$. From Eq. (9-40), $N_{F,min}$ is calculated as

$$N_{F,min} = \frac{\ln\left[\left[\dfrac{x_{LK}}{x_{HK}}\right]_{dist} \Big/ \left[\dfrac{z_{LK}}{z_{HK}}\right]\right]}{\ln \alpha_{LK-HK}} = \frac{\ln\left[\left[\dfrac{0.408}{0.021}\right] \Big/ \left[\dfrac{0.3}{0.3}\right]\right]}{\ln (1/0.21)} = 1.90$$

Where $x_{LK,dist}$ was found from Example 9-2 as

$$x_{LK,dist} = x_{tol,dist} = \frac{Dx_{tol,dist}}{D} = \frac{28.5}{69.92} = 0.408$$

and

$$x_{HK,dist} = x_{cum,dist} = 0.021$$

Then, from Eq. (9-41),

$$N_F = N \frac{N_{F,min}}{N_{min}} = 5.53\left(\frac{1.90}{3.77}\right) = 2.79 \text{ or stage 3.}$$

E. Check. A check of the Gilliland correlation can be obtained from Eq. (9-42b). With $x = 0.455$ this is

$$\frac{N - N_{min}}{N + 1} = 0.545827 - (0.591422)(0.455) + \frac{0.002743}{0.455} = 0.283$$

or $(1 - 0.283) N = N_{min} + 0.283$, which gives $N = 5.65$. The 2% difference between these two results gives an idea of the accuracy of Eq. (9-42).

A complete check would require solution by the stage-by-stage method and would be considerably more work than this calculation. This check can be made by comparing this solution with your solution to Problems 8-D6 and 8-F1.

F. Generalize. The Gilliland correlation is a rapid method for estimating the number of equilibrium stages in a distillation column. It should not be used for final designs because of its inherent inaccuracy.

9.4. SUMMARY - OBJECTIVES

In this chapter we have developed approximate shortcut methods for binary and multicomponent distillation. You should be able to satisfy the following objectives:

1. Derive the Fenske equation and use it to determine the number of stages required at total reflux and the splits of non-key components.

2. Use the Underwood equations to determine the minimum reflux ratio for multicomponent distillation.

3. Use the Gilliland correlation to estimate the actual number of stages in a column and the optimum feed stage location.

4. Develop computer flow charts and programs for the Fenske-Underwood-Gilliland design method.

REFERENCES

Barnes, F.J., D.N. Hansen, and C.J. King, "Calculation of Minimum Reflux for Distillation Columns with Multiple Feeds," *Ind. Eng. Chem. Process Des. Develop., 11,* 136 (1972).

Erbar, J.H. and R.N. Maddox, *Petrol. Refin., 40* (5), 183 (1961).

Fenske, M.R., "Fractionation of Straight-Run Pennsylvania Gasoline," *Ind. Eng. Chem., 24,* 482 (1932).

Gilliland, E.R., "Multicomponent Rectification," *Ind. Eng. Chem., 32,* 1220 (1940).

Hines, A.L. and R.N. Maddox, *Mass Transfer. Fundamentals and Applications*, Prentice-Hall, Englewood Cliffs, NJ, 1985.

Holland, C.D., *Fundamentals and Modeling of Separation Process: Absorption, Distillation, Evaporation, and Extraction*, Prentice-Hall, Englewood Cliffs, NJ, 1975.

King, C.J., *Separation Processes*, 2nd. ed., McGraw-Hill, New York, 1980.

Liddle, C.J., "Improved Shortcut Method for Distillation Calculations," *Chem. Eng.*, 75 (23), 137 (Oct.21, 1968).

Shiras, R.N., D.N. Hansen and C.H. Gibson, "Calculation of Minimum Reflux in Distillation Columns," *Ind. Eng. Chem.*, 42, 871 (1950).

Smith, B.D., *Design of Equilibrium Stage Processes*, McGraw-Hill, New York, 1963.

Thompson, R.E., "Shortcut Design Method-Minimum Reflux," *AIChE Modular Instructions*, Series B, Vol. 2, 5 (1981).

Underwood, A.J.V., "Fractional Distillation of Multicomponent Mixtures," *Chem. Eng. Prog.*, 44, 603 (1948).

Wankat, P.C. and J. Hubert, "Use of the Vaporization Efficiency in Closed Form Solutions for Separation Columns," *Ind. Eng. Chem. Process Des. Develop.*, 18, 394 (1979).

Winn, F.W., *Pet. Refiner*, 37, 216 (1958).

Yaws, C.L., C.-S. Fang, and P.M. Patel, "Estimating Recoveries in Multicomponent Distillation," *Chem. Eng.*, 86 (3), 101 (Jan. 29,1979).

HOMEWORK

A. *Discussion Problems*

A1. What assumptions were made to derive Fenske Eqs. (9-12) and (9-15)?

A2. If relative volatility is not constant, how would you use the Fenske equations [e.g. Eqs. (9-9) and (9-11)] for a design problem? List the trial-and-error procedure to use.

A3. If you want to use an average relative volatility how do you calculate it for the Fenske equation? For the Underwood equation?

A4. What variables does the Gilliland correlation not include? How might these be included? Check the Erbar-Maddox (1961) method (or see King, 1980 or Hines and Maddox, 1985) to see how they included these.

A5. Develop your key relations chart for this chapter.

C. *Derivations*

C1. Derive Eq. (9-15).

C2. Derive Eq. (9-17). Derive an equation for $(FR_C)_{bot}$ in terms of $(FR_A)_{dist}$.

C3. Derive Eq. (9-27).

C4. Derive Eq. (9-34).

C5. Check the accuracy of Eq. (9-42).

C6. If the pinch point occurs at the feed point, mass balances can be used to find the minimum flows. Derive these equations.

C7. Develop a computer flow chart for:

    a. Total reflux calculation by the Fenske equation

    b. Calculation of minimum reflux ratio by the Underwood equation

    c. Determination of actual number of stages with the Gilliland correlation

    d. Complete approximate design procedure (add together parts a, b, and c).

C8. At total reflux, the analytical stage-by-stage analysis with constant relative volatility (developed in Chapter 8) should agree with the Fenske equation. Show that it does. Hint: Write Eq. (8-13) for components A and B and use operating equations $y_i = x_i$.

C9. The choice of developing the Underwood equations in terms of V_{min} instead of solving for L_{min} is arbitrary. Rederive the Underwood equations solving for L_{min} and L_{min}. Develop the equations analogous to Eqs. (9-29) and (9-33).

C10. For binary systems, Eq. (9-33) simplifies to a linear equation for both saturated liquid and saturated vapor feeds. Prove this.

C11. For ternary systems, Eq. (9-33) simplifies to a quadratic equation for both saturated liquid and saturated vapor feeds. Prove this.

D. *Problems*

D1. We have 10 kg moles/hr of a saturated liquid feed that is 40 mole % benzene and 60 mole % toluene. We desire a distillate composition that is 0.992 mole fraction benzene and a bottoms that is 0.986 mole fraction toluene (note units). CMO is valid. Assume constant relative volatility with $\alpha_{BT} = 2.4$. Reflux is returned as a saturated liquid. The column has a partial reboiler and a total condenser.

a. Use the Fenske equation to determine N_{min}.

b. Use the Underwood equations to find $(L/D)_{min}$.

c. For $L/D = 1.1(L/D)_{min}$, use the previous results and the Gilliland correlation to estimate the total number of stages and the optimum feed stage location.

D2. We have an existing column that acts as 30 equilibrium contacts. We are separating a multicomponent mixture where $\alpha_{LK} = 1.1$ and $\alpha_{HK} = 1.0$. A recovery of 98% is required for the light key in the distillate. Find the recovery fraction of the heavy key in the bottoms. Feed is a saturated liquid that contains $z_{LNK} = 0.1$, $z_{LK} = 0.4$, $z_{HK} = 0.3$, and $z_{HNK} = 0.2$. $\alpha_{LNK} = 1.25$ and $\alpha_{HNK} = 0.75$. Operation is at total reflux.

D3. We have designed a special column that acts as exactly three equilibrium stages. Operating at total reflux, we measure vapor composition leaving the top stage and the liquid composition leaving the bottom stage. The column is separating phenol from o-cresol. We measure a phenol liquid mole fraction leaving the bottom stage of 0.36 and a phenol vapor mole fraction leaving the top stage of 0.545. What is the relative volatility of phenol with respect to o-cresol?

D4. A distillation column with a partial reboiler and a total condenser is separating chloroform (more volatile) from carbon tetrachloride at total reflux. We wish to use a column that operates as if it had 15 equilibrium contacts (including the reboiler) to determine the average relative volatility of the system. The distillate composition is measured as 0.982 mole fraction chloroform, and a bottoms composition of 0.037 is measured. What is the average relative volatility for the system chloroform-carbon tetrachloride?

D5. A column with 29 equilibrium stages and a partial reboiler is being operated at total reflux to separate a mixture of ethylene dibromide and propylene dibromide. Ethylene dibromide is more volatile, and the relative volatility is constant at a value of 1.30. We are measuring a distillate concentration that is 98.4 mole % ethylene dibromide. The column has a total condenser and saturated liquid reflux, and constant molal overflow can be assumed. Use the Fenske equation to predict the bottoms composition.

D6. We are separating 1000 moles/hr of a 40% benzene, 60% toluene feed in a distillation column with a total condenser and a partial reboiler. Feed is a saturated liquid. Constant molal overflow is valid. A distillate that is 99.3% benzene and a bottoms that is 1% benzene are desired. Use the Fenske equation to find the number of stages required at total reflux, a McCabe-Thiele diagram to find $(L/D)_{min}$, and the Gilliland correlation to estimate the number of stages required if $L/D = 1.15(L/D)_{min}$. Estimate that the relative volatility is constant at $\alpha_{BT} = 2.4$. Check your results with a McCabe-Thiele diagram.

D7. A feed mixture of benzene, toluene, xylene, and cumene is to be separated in a distillation column. Feed rate is 200 moles/hr. Feed concentrations are $z_B = 0.2$, $z_T = 0.3$, $z_X = 0.1$, $z_C = .4$. The feed enters as a two-phase mixture that is 30% vapor. The column has a partial reboiler and a total condenser. Feed is returned to the column from the condenser as a saturated liquid. The specifications require that 99.8% of the cumene be recovered in the bottoms and 99.5% of the toluene be recovered in the distillate. The equilibrium data can be represented by constant relative volatilities. We have decided to operate at total reflux. What is the recovery of xylenes in the distillate? Choosing toluene as the reference, the relative volatilities are benzene, 2.25; toluene, 1.00; xylene, 0.330; cumene, 0.210.

D8. We wish to separate a mixture of 40 mole % benzene and 60 mole % ethylene dichloride in a distillation column with a partial reboiler and a total condenser. The feed rate is 750 moles/hr, and feed is a saturated vapor. We desire a distillate product of 99.2 mole % benzene and a bottoms product that is 0.5 mole % benzene. Reflux is a saturated liquid, and constant molal overflow can be used. Equilibrium data can be approximated with an average relative volatility of 1.11 (benzene is more volatile).

a. Find the minimum external reflux ratio.

b. Use the Fenske equation to find the number of stages required at total reflux.

c. Estimate the total number of stages required for this separation, using the Gilliland correlation for $L/D = 1.2(L/D)_{min}$.

D9. a. A distillation column specially designed to measure the average relative volatilities of close-boiling binary mixtures is being run at total reflux. This column is designed so that each stage acts exactly as one equilibrium stage. The column plus reboiler act as 15 equilibrium contacts. With many equilibrium stages the precision of the measurement is increased considerably. Laboratory measurements were made on the system ethanol-isopropanol at total reflux. The ethanol concentrations obtained were: mole fraction ethanol in bottoms = 0.05, and mole fraction ethanol in distillate = 0.391. What is the average relative volatility of ethanol compared to isopropanol?

b. A distillation column is to be designed to separate ethanol from isopropanol. The column will have a partial reboiler and a total condenser. Feed is 1000 moles/hr of a saturated liquid that is 40 mole % ethanol. Reflux is returned to the column as a saturated liquid, and constant molal overflow is valid. We desire a distillate that is 99.9% ethanol and a bottoms that is 0.3% ethanol. Equilibrium stages can be assumed, and relative volatility was determined in part a. At total reflux, how many stages are needed to obtain the desired separation? What is the minimum external reflux ratio? If $L/D = 1.15(L/D)_{min}$ is used, how many stages are required? Estimate the optimum feed stage. Use the Fenske equation and Gilliland correlation for this problem.

c. Check your results to part b with a McCabe-Thiele diagram.

D10. We wish to separate three components A, B, and C. If B is picked as the reference component, $\alpha_A = 1.25$ and $\alpha_C = 0.5$. The feed rate is 100 moles/hr, and feed is saturated liquid. Feed compositions are $z_A = 0.3$, $z_B = 0.5$, $z_C = 0.2$. We desire 99.6% recovery of B in the distillate and a 98.3% recovery of C in the bottoms. The minimum L/D has been estimated as 1.04. The column has constant molal overflow, saturated liquid reflux, and equilibrium stages. If we wish to operate at $(L/D) = 1.2(L/D)_{min}$, estimate the total number of stages needed and the optimum feed plate location.

D11. A distillation column has a feed of 100 kg moles/hr. Feed is 10 mole % LNK, 55 mole % LK, and 35 mole % HK and is a saturated liquid. Reflux ratio is $L/D = 1.2(L/D)_{min}$. We desire a 99.5% recovery of the light key in the distillate. Mole fraction of the light key in the distillate should be 0.75. Use the Fenske-Underwood-Gilliland approach to design the column. Equilibrium data (the light key is chosen as reference):

$$\alpha_{LNK} = 4.0, \; \alpha_{LK} = 1.0, \; \text{and} \; \alpha_{HK} = 0.75$$

D12. a. A distillation column with a partial reboiler and a total condenser is being used to separate a mixture of benzene, toluene, and cumene. The feed is 40 mole % benzene, 30 mole % toluene and 30 mole % cumene. The feed is input as a saturated vapor. We desire 99% recovery of the toluene in the bottoms and 98% recovery of the benzene in the distillate. The reflux is returned as a saturated liquid, and constant molal overflow can be assumed. Equilibrium can be represented as constant relative volatilities. Choosing toluene as the reference component, $\alpha_B = 2.25$ and $\alpha_C = 0.210$. Use the Fenske equation to find the number of equilibrium stages required at total reflux and the recovery fraction of cumene in the bottoms.

b. For the distillation problem given in part a, find the minimum reflux ratio by use of the Underwood equations. Use a basis of 100 moles of feed/hour. Clearly state your assumptions.

c. For $L/D = 1.25(L/D)_{min}$, find the total number of equilibrium stages required for the distillation problem presented in parts a and b. Use the Gilliland correlation. Estimate the optimum feed plate location.

D13. We have a column separating benzene, toluene, and cumene. The column has a total condenser and a total reboiler and has 9 equilibrium stages. The feed is 25 mole % benzene, 30 mole % toluene, and 45 mole % cumene. Feed rate is 100 moles/hr and feed is a saturated liquid. The equilibrium data can be represented as constant relative volatilities: $\alpha_{BT} = 2.5$, $\alpha_{TT} = 1.0$, and $\alpha_{CT} = 0.21$. We desire 99% recovery of toluene in the distillate and 98% recovery of cumene in the bottoms. Determine the external reflux ratio required to achieve this separation. If $\alpha_{BT} = 2.25$ instead of 2.5, how much will L/D change?

D14. Use the Fenske-Underwood-Gilliland approach to repeat Example 8-1.

D15. A distillation column is separating benzene ($\alpha = 2.25$), toluene ($\alpha = 1.00$), and cumene ($\alpha = 0.21$). The column is operating at 101.3 kPa. The column is to have a total condenser and a partial reboiler, and the optimum feed stage is to be used. Reflux is returned as a saturated liquid, and $L_0/D = 1.2$. Feed rate is 1000 kg moles/hr. Feed is 39.7 mole % benzene, 16.7 mole % toluene, and 43.6 mole % cumene and is a saturated liquid. We desire to recover 99.92% of the benzene in the distillate and 99.99% of the cumene in the bottoms. For a first guess to this design problem, use the Fenske-Underwood-Gilliland approach to estimate the optimum feed stage and the total number of equilibrium stages. Note: The Underwood equations must be treated as a case C problem.

D16. We are separating a mixture of ethanol and n-propanol. Ethanol is more volatile and the relative volatility is approximately constant at 2.10. The feed flow rate is 1000 kg moles/hr. Feed is 60 mole % ethanol and is a saturated vapor. We desire $x_D = 0.99$ mole fraction ethanol and $x_B = 0.008$ mole fraction ethanol. Reflux is a saturated liquid. There are 30 stages in the column. Use the Fenske-Underwood-Gilliland approach to determine

a. Number of stages at total reflux

b. $(L/D)_{min}$

c. $(L/D)_{actual}$

D17. A depropanizer has the following feed and constant relative volatilities:

Methane:	$z_{C1} = 0.229$	$\alpha_{C1} = 9.92$
Propane:	$z_{C3} = 0.368$	$\alpha_{C3} = 1.0$
n-Butane:	$z_{C4} = 0.322$	$\alpha_{C4} = 0.49$
n-Hexane:	$z_{C6} = 0.081$	$\alpha_{C6} = 0.10$

$L/D = 1.5$. Reflux is a saturated liquid. The feed is a saturated liquid fed in at 1.0 kg moles/time. Assume constant molal overflow is valid. We desire to recover 98.54% of the propane in the distillate and 87.91% of the n-butane in the bottoms. Use the Fenske-Underwood-Gilliland approach to estimate the optimum feed stage and total number of equilibrium stages.

E. *More Complex Problems*

E1. Use the Fenske-Underwood-Gilliland approach to do Problem 8-E4.

E2. Write a computer or programmable calculator program for the Fenske-Underwood-Gilliland approximate design approach. Test your program on:

a. Problem 9-D9

b. Problem 9-D11

c. Problems 9-D12 and 9-D13

d. Problem 9-D15

e. Problem 9-E1

chapter 10
INTRODUCTION TO COMPLEX DISTILLATION METHODS

We have looked at binary and multicomponent mixtures in both simple and fairly complex columns. However, the chemicals separated have usually had fairly simple equilibrium behavior. In this chapter you will be introduced to a variety of more complex distillation systems used for the separation of less ideal mixtures.

Simple distillation columns are not able to completely separate mixtures when azeotropes occur, and the columns are very expensive when the relative volatility is close to 1. Distillation columns can be coupled with other separation methods to break the azeotrope. This is discussed in the first section. Extractive distillation, azeotropic distillation, and two-pressure distillation are methods for modifying the equilibrium to separate these complex mixtures. These three methods will be briefly described in the next three sections of this chapter. In the last section we will discuss the use of a distillation column as a chemical reactor, to simultaneously react and separate a mixture.

This chapter covers a number of complex topics. Some are covered at only an introductory level and not in the same depth as topics in the previous chapters.

10.1. BREAKING AZEOTROPES WITH OTHER SEPARATORS

Azeotropic systems normally limit the separation that can be achieved. For an azeotropic system such as ethanol and water (shown in Figures 2-2 and 5-13), it isn't possible to get past the azeotropic concentration of 0.8943 mole fraction ethanol with ordinary distillation. Some other separation method is required to break the azeotrope. The other method could employ adsorption, membranes, extraction, and so forth. It could also involve adding a third component to the distillation to give the azeotropic and extractive distillation systems discussed later in this chapter.

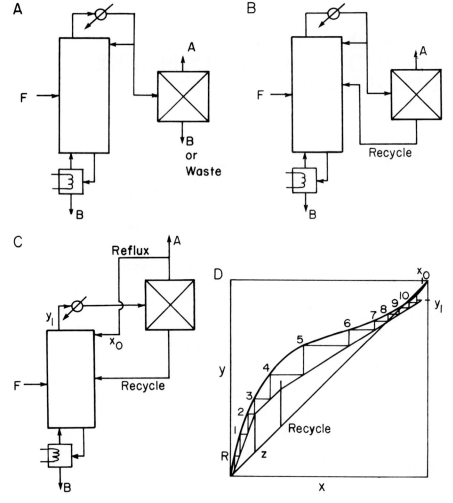

Figure 10-1. Breaking azeotropes. (A) Separator uncoupled with dis-
tillation; (B) Recycle from separator to distillation; (C)
Recycle and reflux from separator to distillation; (D)
McCabe-Thiele diagram for part C.

Three ways of using an additional separation method to break the
azeotrope are shown in Figure 10-1. The simplest, but least likely to be
used, is the completely uncoupled system shown in Figure 10-1A. The
distillate, which is near the azeotropic concentration, is sent to another
separation device, which produces both the desired products. If the
other separator can completely separate the products, why use distillation at all? If the separation is not complete, what would be done with
the waste stream?

A more likely configuration is that of Figure 10-1B. The incompletely separated stream is recycled to the distillation column, which now operates as a two-feed column, so the design procedures used for two-feed columns (Example 6-2) can be used. The arrangement shown in Figure 10-1B is commonly used industrially. The separator may actually be several separators.

An intriguing alternative is the coupled system shown in Figure 10-1C. Now some of the A product is used as reflux to the distillation column. Thus, $x_0 > y_1$. The McCabe-Thiele diagram for this is illustrated in Figure 10-1D. Note that using a fairly pure reflux stream allows the column to produce a vapor, y_1, that is greater than the azeotropic concentration. This arrangement is explored further in Problem 10-D1.

10.2. EXTRACTIVE DISTILLATION

Extractive distillation is used for the separation of azeotropes and close-boiling mixtures. In extractive distillation, a solvent is added to the distillation column. This solvent is selected so that one of the components, B, is selectively attracted to it. Since the solvent is usually chosen to have a significantly higher boiling point than the components being separated, the attracted component, B, has its volatility reduced. Thus the other component, A, becomes relatively more volatile and is easy to remove in the distillate. A separate column is required to separate the solvent and component B.

There are a variety of commercial applications for extractive distillation. For example, butane and butene are separated using furfural or furfural-water mixtures as the solvent (Shinskey, 1984). Hydrochloric acid and nitric acid are both separated from water using sulfuric acid as the solvent. Toluene is separated from paraffins using phenol as solvent (see Figure 10-3 for an example). A salt or solid sodium hydroxide can also be used as the "solvent" (Drew, 1979). The solvent does not have to be exotic; water is used as the solvent for separation of acetone and methanol.

A typical flow sheet for separation of a binary mixture is shown in Figure 10-2. In column 1 the solvent is added several stages above the feed stage and a few stages below the top of the column. In the top section, the relatively nonvolatile solvent is removed and pure A is produced as the distillate product. In the middle section, large quantities of solvent are present and components A and B are separated from each other. It is common to use 1, 5, 10, 20, or even 30 times as much solvent as feed; thus, the solvent concentration in the middle section is

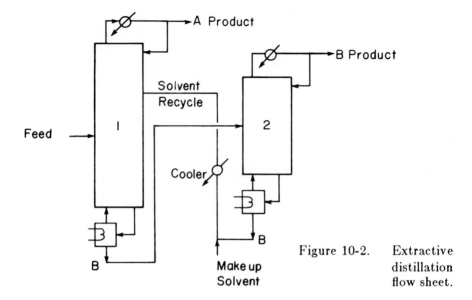

Figure 10-2. Extractive distillation flow sheet.

often quite high. Note that the A-B separation must be complete in the middle section, because any B that gets into the top section will not be separated (there is very little solvent present) and will exit in the distillate. The bottom section strips the A from the mixture so that only solvent and B exit from the bottom of the column.

The mixture of solvent and B are sent to column 2, where they are separated. If the solvent is selected correctly, the second column can be quite short, since component B is significantly more volatile than the solvent. The recovered solvent can be cooled and stored for reuse in the extractive distillation column. Note that the solvent must be cooled before entering column 1, since its boiling point is significantly higher than the operating temperature of column 1.

Column 2 is a simple distillation that can be designed by the methods discussed in Chapters 5 and 6. Column 1 is considerably more complex, but the bubble-point matrix method discussed in Chapter 8 can often be adapted. Since the system is nonideal and K values depend on the solvent concentration, a concentration loop is required in the flow chart shown in Figure 8-6. Fortunately, a good first guess of solvent concentrations can be made. Solvent concentration will be almost constant in the middle section and also in the bottom section except for the reboiler. In the top section of the column, the solvent concentration will very rapidly decrease to zero. These solvent concentrations will be relatively unaffected by the temperatures and flow rates. Thus the concentration loop should be the innermost loop. The K values can be calculated from Eq. (2-19) with the activity coefficients determined from the

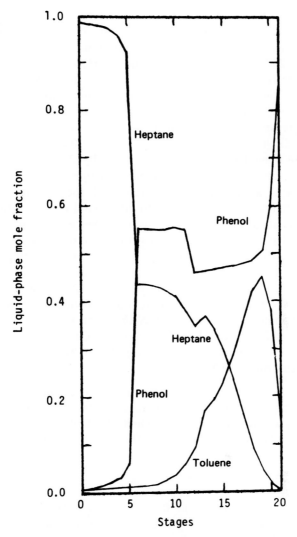

Figure 10-3.   Calculated composition profiles for extractive distillation of toluene and n-heptane. From Seader (1984). Reprinted with permission from *Perry's Chemical Engineer's Handbook,* 6th ed, Copyright 1984, McGraw-Hill.

Wilson equation, the UNIQUAC method, or other correlations. Smith (1963) discusses equilibrium calculations in considerable detail. Holland (1981) adapts the θ method to extractive distillation.

Concentration profiles for the extractive distillation of n-heptane and toluene using phenol as the solvent are shown in Figure 10-3 (Seader,

1984). The profiles were rigorously calculated using a simultaneous correction method, and activity coefficients were calculated with the Wilson equation. The feed was a mixture of 200 lb moles/hr of n-heptane and 200 lb moles/hr of toluene input as a liquid at 200°F on stage 13. The recycled solvent is input on stage 6 at a total rate of 1200 lb moles/hr. There are a total of 21 equilibrium contacts including the partial reboiler. The high-boiling phenol is attractive to the toluene, since both are aromatics. The heptane is then made more volatile and exits in the distillate (component A in Figure 10-2). Note from Figure 10-3 that the phenol concentration very rapidly decreases above stage 6 and the n-heptane concentration increases. From stages 6 to 12, phenol concentration is approximately constant and the toluene is separated from the heptane. From stages 13 to 20, phenol concentration is again constant but at a lower concentration. This change in the solvent (phenol) concentration occurs because the feed is input as a liquid. A constant solvent concentration can be obtained by vaporizing the feed or by adding some recycled solvent to it. Heptane is stripped from the mixture in stages 13 to 20. In the reboiler, the solvent is nonvolatile compared to the toluene. Thus the boilup is much more concentrated in toluene than in phenol. The result is the large increase in phenol concentration seen for stage 21 in Figure 10-3. Concentration profiles for other extractive distillation systems are shown by Robinson and Gilliland (1950) and Smith (1963).

The reason for the large increase in solvent concentration in the reboiler is easily seen if we look at an extreme case where none of the solvent vaporizes in the reboiler. Then the boilup is essentially pure component B. The liquid flow rate in the column can be split up as

$$\overline{L} = S + B_{liq} + \overline{V} \tag{10-1}$$

where S is the constant solvent flow rate; B_{liq} is the flow rate of component B, which stays in the liquid in the reboiler; and $\overline{V}$ is the flow rate of the vapor, which is essentially pure B. The bottoms flow rate consists of the streams that remain liquid,

$$Bot = S + B_{liq} \tag{10-2}$$

The mole fraction of B in the liquid in the column can be estimated as

$$x_{col} = \frac{B_{liq} + \overline{V}}{S + B_{liq} + \overline{V}} \tag{10-3}$$

and the mole fraction B in the bottoms is

$$x_{bot} = \frac{B_{liq}}{S + B_{liq}} \qquad (10\text{-}4)$$

For the usual flow rates this gives $x_{col} >> x_{bot}$. Even when the solvent is fairly volatile, as in Figure 10-3, the toluene (component B) concentration drops in the reboiler.

Selection of the solvent is extremely important. The process is similar to that of selecting a solvent for liquid-liquid extraction, which is discussed in Chapter 16. By definition, the solvent should not form an azeotrope with any of the components. If the solvent does form an azeotrope, the process becomes azeotropic distillation, which is discussed in the next section. Usually, a solvent is selected that is more similar to the heavy key. Then the volatility of the heavy key will be reduced. Exceptions to this rule exist; for example, in the n-butane - 1-butene system, furfural decreases the volatility of the 1-butene, which is more volatile (Shinskey, 1984). Lists of extractants (Van Winkle, 1967) for extractive distillation are helpful in finding a general structure that will effectively increase the volatility of the keys.

Solvent selection can be aided by considering the polarities of the compounds to be separated. A short list of classes of compounds arranged in order of increasing polarity is given in Table 10-1. If two compounds of different polarity are to be separated, a solvent can be selected to attract either the least polar or the most polar of the two. For example, suppose we wish to separate acetone (a ketone boiling at $56.5\,^{\circ}C$) from methanol (an alcohol boiling at $64.7\,^{\circ}C$). This system forms an azeotrope. We could add a hydrocarbon to attract the acetone, but if enough hydrocarbon were added, the methanol would become more volatile. A simpler alternative is to add water, which

Table 10-1. Increasing Polarities of Classes of Compounds

Hydrocarbons
Ethers
Aldehydes
Ketones
Esters
Alcohols
Glycols
Water

attracts the methanol and makes acetone more volatile. The methanol and water are then separated in column 2. In this example, we could also add a higher molecular weight alcohol such as butanol to attract the methanol.

When two hydrocarbons are to be separated, the larger the difference in the number of double bonds the better a polar solvent will work to change the volatility. For example, furfural will decrease the volatility of butenes compared to butanes. Furfural (a cyclic alcohol) is used instead of water because the hydrocarbons are miscible with furfural.

A more detailed analysis of solvent selection shows that hydrogen bonding is more important than polarity (Berg, 1969; Smith, 1963). Thus, more detailed analyses of solvents should be based on hydrogen bonding.

Once a general structure has been found, homologs of increasing molecular weight can be checked to find which has a high enough boiling point to be easily recovered in column 2. However, too high a boiling point is undesirable, because the solvent recovery column would have to operate at too high a temperature. The solvent should be completely miscible with both components over the entire composition range of the distillation.

It is desirable to use a solvent that is nontoxic, nonflammable, non-corrosive, and nonreactive. In addition, it should be readily available and inexpensive since solvent makeup and inventory costs can be relatively high. As usual, the designer must make tradeoffs in selecting a solvent. One common compromise is to use a solvent that is used elsewhere in the plant or is a by-product of a reaction even if it may not be the optimum solvent otherwise. This can be particularly useful if the same solvent can be used for both an extraction step and extractive distillation.

For isomer separations, extractive distillation usually fails, since the solvent has the same effect on both isomers. For example, Berg (1969) reported that the best entrainer for separating m- and p-xylene increased the relative volatility from 1.02 to 1.029. An alternative to normal extractive distillation is to use a solvent that preferentially and reversibly reacts with one of the isomers (Terrill *et al.*, 1985). The process scheme will be similar to Figure 10-2, with the light isomer being product A and the heavy isomer product B. The forward reaction occurs in the first column, and the reaction product is fed to the second column. The reverse reaction occurs in column 2, and the reactive solvent is recycled to column 1. This procedure is quite similar to the combined reaction-distillation discussed in Section 10.5.

10.3. AZEOTROPIC DISTILLATION PROCESSES

The presence of an azeotrope can be used to separate an azeotropic system. This is most convenient if the azeotrope is heterogeneous; that is, the vapor from the azeotrope will condense to form two liquid phases that are immiscible. Azeotropic distillation is often performed by adding a solvent or entrainer that forms an azeotrope with one or both of the components. Before discussing these more complex azeotropic distillation systems, let us consider the simpler binary systems that form a heterogeneous azeotrope.

10.3.1. Binary Heterogeneous Azeotropes

Figure 2-8 was the y-x diagram for n-butanol and water, which form a heterogeneous azeotrope. For this type of heterogeneous azeotrope the two column systems shown in Figure 10-4A can provide a complete separation. Column 1 is a stripping column that receives liquid of composition x_α from the liquid-liquid settler. It operates on the left-hand side of the equilibrium diagram shown in Figure 10-4B. In this region, species B is more volatile and the bottoms from column 1 is almost pure A $(x_{bot1} \sim 0)$. The overhead vapor from column 1, y_1^1, is condensed and then goes to the liquid-liquid settler where it separates into two liquid phases. Liquid of composition x_α is refluxed to column 1, while liquid of composition x_β is refluxed to column 2.

The second column operates on the right-hand side of Figure 10-4B, where species B is the less volatile component. Thus the bottoms from this column is almost pure B $(x_{bot2} \sim 1.0)$. The overhead vapor, which is richer in species A, is condensed and sent to the liquid-liquid separator. The liquid-liquid separator takes the condensed liquid, $x_\alpha < x < x_\beta$, and separates it into the two liquid phases in equilibrium. These liquids are used as reflux to columns 1 and 2. The liquid-liquid separator allows one to get past the azeotrope and is therefore a necessary part of the equipment.

The overall external mass balance for the two column system shown in Figure 10-4A is

$$F = B_1 + B_2 \tag{10-5}$$

while the external mass balance on component B is

$$Fz = B_1 x_{bot1} + B_2 x_{bot2} \tag{10-6}$$

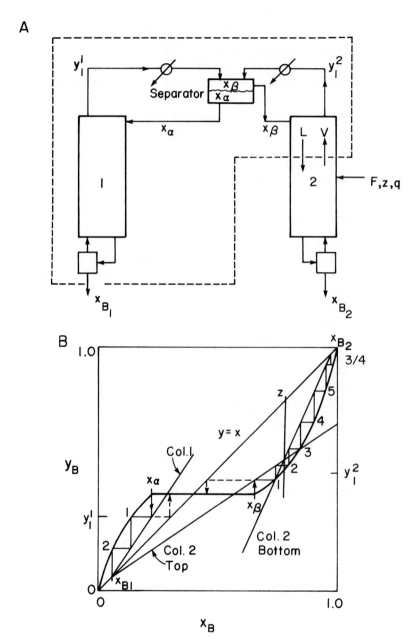

Figure 10-4.   Binary heterogeneous azeotrope. (A) Two column distillation system; (B) McCabe-Thiele diagram.

Solving these equations simultaneously for the unknown bottoms flow rates, we obtain

$$B_1 = \frac{F(z - x_{bot2})}{(x_{bot1} - x_{bot2})} \qquad (10\text{-}7)$$

$$B_2 = \frac{F(z - x_{bot1})}{(x_{bot2} - x_{bot1})} \qquad (10\text{-}8)$$

Note that this result does not depend on the details of the distillation system.

Analysis of stripping column 1 is also straightforward. The bottom operating equation is

$$y = (\frac{\overline{L}}{\overline{V}})_1 x - [(\frac{\overline{L}}{\overline{V}})_1 - 1]x_{bot1} \qquad (10\text{-}9)$$

The feed to this column is the saturated liquid reflux of composition x_α. This is a vertical feed line. Then the overhead vapor y_1^1 is found on the operating line at x_α.

The bottom operating equation for column 2 is

$$y = (\frac{\overline{L}}{\overline{V}})_2 x - [(\frac{\overline{L}}{\overline{V}})_2 - 1]x_{bot2} \qquad (10\text{-}10)$$

The top operating line is a bit different. The easiest mass balance to write uses the mass balance envelope shown in Figure 10-4A. Then the top operating equation is

$$y = (\frac{L}{V})_2 x + (1 - \frac{L}{V})_2 x_{bot1} \qquad (10\text{-}11)$$

This is somewhat unusual, because it includes a bottoms concentration leaving the first column. The reflux for this top operating line is liquid of composition x_β.

The McCabe-Thiele diagram for this system is shown in Figure 10-4B, where column 2 appears upside down because we have plotted y_B versus x_B, and B is the *less volatile* component in column 2. (If this is not clear, return to Problem 5-A5.) The two reflux streams are x_α and x_β, and reflux is *not* at the usual value of $y = x = x_D$.

The two dashed lines in Figure 10-4B show the route of the overhead vapor streams as they are condensed to saturated liquids (made into x values) and then sent to the liquid-liquid separator. The lever-arm rule [see Eqs. (2-8) and (2-9) and Figures 2-9 and 2-10] can be applied to the liquid-liquid separator.

Several modifications of the basic arrangement shown in Figure 10-4A can be used. If the feed composition is less than x_α, then column 1 would be a complete column and column 2 would be just a stripping column (see Problem 10-C1). The liquids may be subcooled so that the liquid-liquid separator operates below the boiling temperature. This can be advantageous, since the partial miscibility of the system depends on temperature. When the liquids are subcooled, the separator calculation must be done at the temperature of the settler. Then the reflux concentrations can be plotted on the McCabe-Thiele diagram. More details for heterogeneous azeotropes are explored by Hoffman (1964) and Shinskey (1984). Single-column configurations for binary heterogeneous azeotropes are discussed by Luyben (1973). These configurations are illustrated in Example 10-1.

10.3.2. Drying Organic Compounds That Are Partially Miscible with Water

For partially miscible systems, a single phase is formed only when the water concentration is low or very high. For example, a small amount of water can dissolve in gasoline. If more water is present, two phases will form. In the case of gasoline, the water phase is detrimental to the engine, and in cold climates it can freeze in gas lines, immobilizing the car. Since the solubility of water in gasoline decreases as the temperature is reduced, it is important to have dry gasoline.

Fortunately, the water can easily be removed by distillation or adsorption. During distillation the water acts as a very volatile component, so a mixture of water and organics is taken as the distillate. After condensation, two liquid phases form, and the organic phase can be refluxed. The system is a type of heterogeneous azeotropic system similiar to those discussed in the previous section. Drying differs from the previous systems since a pure water phase is usually not desired, the relative solubilities are often quite low, and the water phase is sent to waste. Thus the system will look like Figure 10-5 with a single column and a phase separator. With very high relative volatilities, one equilibrium stage may be sufficient and a flash system can be used.

Simplified equilibrium theories are useful for partially immiscible

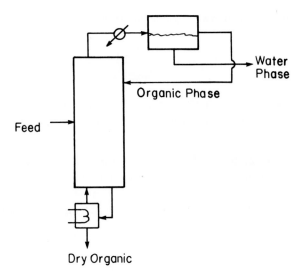

Figure 10-5.   Distillation to dry organic that is partically miscible with water.

liquids. There is always a range of concentrations where the species are miscible even though the concentrations may be quite small. For the water phase, it is reasonable as a first approximation to assume that the water follows Raoult's law and the organic components follow Henry's law (Robinson and Gilliland, 1950). Thus,

$$p_w = (VP_w)x_{w\,in\,w}, \quad p_{org} = H_{org}\,x_{org\,in\,w} \tag{10-12}$$

where H_{org} is the Henry's law constant for the organic component in the aqueous phase, VP_w is the vapor pressure of water, and $x_{w\,in\,w}$ and $x_{org\,in\,w}$ are the mole fractions of water and organic in the water phase, respectively. In the organic phase it is reasonable to use Raoult's law for the organic compounds and Henry's law for the water.

$$p_w = H_w x_{w\,in\,org}, \quad p_{org} = (VP_{org})x_{org\,in\,org} \tag{10-13}$$

where H_w is the Henry's law constant for water in the organic phase. At equilibrium, the partial pressure of water in the two phases must be equal. Thus, equating p_w in Eqs. (10-12) and (10-13) and solving for H_w we obtain

$$H_w = \frac{(VP_w)x_{w\,in\,w}}{x_{w\,in\,org}} \tag{10-14}$$

Similar manipulations for the organic phase give

$$H_{org} = \frac{(VP_{org})x_{org\ in\ org}}{x_{org\ in\ w}} \qquad (10\text{-}15)$$

Using Eqs. (10-14) and (10-15), we can calculate the Henry's law constants from the known solubilities (which give the mole fractions) and the vapor pressures.

Equations (10-12) to (10-15) are valid for both drying organic compounds and steam distillation. The ease of removing small amounts of water from an organic compound that is immiscible with water can be seen by estimating the relative volatility of water in the organic phase.

$$\alpha_{w\text{-}org\ in\ org} = \frac{y_w/x_{w\ in\ org}}{y_{org}/x_{org\ in\ org}}$$

$$= \frac{(p_w/p_{tot})/x_{w\ in\ org}}{(p_{org}/p_{tot})/x_{org\ in\ org}} = \frac{p_w/x_{w\ in\ org}}{p_{org}/x_{org\ in\ org}} \qquad (10\text{-}16)$$

In the organic phase Eqs.(10-13) and (10-14) can be substituted into Eq. (10-16) to give

$$\alpha_{w\text{-}org\ in\ org} = \frac{H_w}{VP_{org}} = \frac{(VP_w)x_{w\ in\ w}}{(x_{w\ in\ org})(VP_{org})} \qquad (10\text{-}17)$$

This calculation is illustrated in Example 10-1.

If data for the heterogeneous azeotrope (y and $x_{w\ in\ org}$) is available, $\alpha_{w\text{-}org\ in\ org}$ is easily estimated from the definition of relative volatility [the first equals sign in Eq. (10-16)]. This will be more accurate than assuming Raoult's law.

Organics can be dried either by continuous distillation or by batch distillation. In both cases the vapor will condense into two phases. The water phase can be withdrawn and the organic phase refluxed to the distillation system. For continuous systems, the McCabe-Thiele design procedure can be used. The McCabe-Thiele graph can be plotted from Eq. (10-17), and the analysis is the same as in the previous section. This is illustrated in Example 10-1.

Example 10-1. Drying Benzene by Distillation

A benzene stream contains 0.01 mole fraction water. Flow rate is

1000 kg moles/hr, and feed is a saturated liquid. Column has saturated liquid reflux of the organic phase from the liquid-liquid separator (see Figure 10-5) and uses $L/D = 2 \, (L/D)_{min}$. We want the outlet benzene to have $x_{w \, in \, benz,bot} = 0.001$. Design the column.

Solution

A. Define. The column is the same as Figure 10-5. Find the total number of stages and optimum feed stage.

B. Explore. Need equilibrium data. From Robinson and Gilliland (1950), $x_{benz \, in \, w} = 0.00039$, $x_{w \, in \, benz} = 0.015$. Use Eq. (10-17) for equilibrium. At the boiling point of benzene (80.1 °C), $VP_{benz} = 760$ mmHg and $VP_w = 356.6$ mmHg (Perry and Green, 1984). Operation will be at a different temperature, but the ratio of vapor pressures will be approximately constant.

C. Plan. Calculate equilibrium from Eq. (10-17):

$$\alpha_{w-benz} = \frac{(VP_w) \, x_{w \, in \, w}}{x_{w \, in \, benz} \, (VP_{benz})} = \frac{(356.6)(1 - 0.00039)}{(0.015)(760)} = 31.3$$

This is valid for $x_{w \, in \, benz} < 0.015$. After that, we have a heterogeneous azeotrope. Plot this on a McCabe-Thiele diagram. (Two diagrams will be used for accuracy.) Solve with the McCabe-Thiele method as a heterogeneous azeotrope problem.

D. Do It. Plot equilibrium: $y_w = \dfrac{\alpha x_w}{1 + (\alpha - 1)x_w} = \dfrac{31.3 x_w}{1 + 30.3 x_w}$
where y and x are mole fractions of water in the benzene phase. This is valid for $x_w \leq 0.015$. See Figure 10-6A. Since Figure 10-6A is obviously not accurate, we use Figure 10-6B. Calculate

$$(L/V)_{min} = \frac{0.9996 - 0.24}{0.9996 - 0.01} = 0.7676$$

$$\left[L/D\right]_{min} = \left[\frac{L/V}{1 - L/V}\right]_{min} = 3.303$$

$$\left[L/D\right]_{act} = 2 \, (3.303) = 6.606$$

$$\left[L/V\right]_{act} = \frac{L/D}{1 + L/D} = 0.868$$

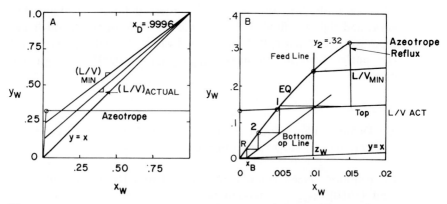

Figure 10-6.   Solution for Example 10-1. (A) McCabe-Thiele diagram for entire range; (B) McCabe-Thiele diagram for low concentrations.

Top Operating Line:

$$y = \frac{L}{V} x + \left(1 - \frac{L}{V}\right) x_D = 0.868x + (0.132)(.9996)$$

$$y \text{ intercept } (x = 0) = 0.132$$

Plot top operating line.

Feed Line: Saturated liquid.

$$\text{Slope} = \frac{q}{q - 1} = \infty \text{ at } y = x = z_w = 0.01$$

Bottom Operating Line: $y = \dfrac{\overline{L}}{\overline{V}} x - \left[\dfrac{\overline{L}}{\overline{V}} - 1\right] x_B$

goes through $y = x = x_B = 0.001$ and intersection of top operating line and feed line.

Reflux is the benzene phase from the liquid-liquid separator; thus, $x_{reflux} = x_{w \text{ in benz}} = 0.015$. Use this to start stepping off stages. Optimum feed stage is top stage of column. We need 2 stages plus a partial reboiler.

E. Check. All of the internal consistency checks work. The value of $\alpha_{\text{w-benz}}$ agrees with the calculation of Robinson and Gilliland (1950). The best check on $\alpha_{\text{w-benz}}$ would be comparison with data.

F. Generalize. Since the solubility of organics in water is often very low, this type of heterogeneous azeotrope system requires only one distillation column.

Even though water has a higher boiling point than benzene, the relative volatility of water dissolved in benzene is extremely high. This happens because water dissolved in an organic cannot hydrogen bond as it does in an aqueous phase, and thus it acts as a very small molecule that is quite volatile. The practical consequence of this is that small amounts of water can easily be removed from organics if the liquids are partially immiscible. There are alternative methods for drying organics such as adsorption that may be cheaper than distillation in many cases.

10.3.3. Azeotropic Distillation with Added Solvent

When a homogeneous azeotrope is formed or the mixture is very close boiling, the procedures shown in the previous section cannot be used. However, the engineer can add a solvent (or entrainer) that forms a binary or ternary azeotrope and use this to separate the mixture. The trick is to pick a solvent that forms an azeotrope that is either heterogeneous (then the procedures of the previous section are useful) or easy to separate by other means such as extraction with a water wash. Since there are now three components, it is possible to have one or more binary azeotropes or a ternary azeotrope. The flow sheet depends upon the equilibrium behavior of the system. A few typical examples will be illustrated here.

Figure 10-7 shows a simplified flow sheet (extensive heat exchange is not shown) for the separation of butadiene from butylenes using liquid ammonia as the entrainer (Poffenberger *et al.,* 1946). Note the use of the intermediate reboiler in the azeotropic distillation column to minimize polymerization. At $40\,^{\circ}$C the azeotrope is homogeneous. The ammonia can be recovered by cooling, since at temperatures below $20\,^{\circ}$C two liquid phases are formed. The colder the operation of the settler the purer the two liquid phases. At the $-40\,^{\circ}$C used in commercial plants during World War II, the ammonia phase contained about 7 wt % butylene. This ammonia is recycled to the azeotropic column either as reflux or on stage 30. The top phase is fed to the stripping column and contains about 5 wt % ammonia. The azeotrope produced in the stripping column is recycled to the separator. This example illustrates the following general points: (1) The azeotrope formed is often

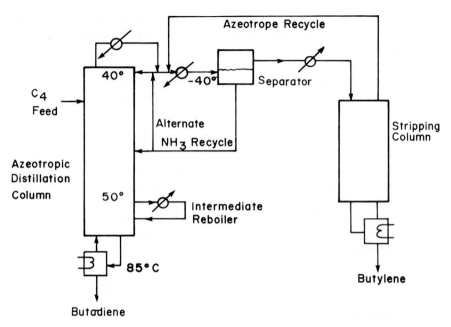

Figure 10-7.   Separation of butadiene from butylenes using ammonia
as entrainer. (Poffenberger *et al.*, 1946).

cooled to obtain two phases and/or to optimize the operation of the
liquid-liquid settler. (2) Streams obtained from a settler are seldom
pure and have to be further purified. This is illustrated by the stripping
column in Figure 10-7. (3) Product (butylene) can often be recovered
from solvent (NH_3) in a stripping column instead of a complete distillation column because the azeotrope is recycled.

Another system with a single binary azeotrope is shown in Figure
10-8 (Smith, 1963). In the azeotropic column, component A and the
entrainer form a minimum boiling azeotrope, which is recovered as the
distillate. The other component, B, is recovered as a pure bottoms product. In this case the azeotrope formed is homogeneous, and a water
wash (extraction using water) is used to recover the solvent from the
desired component with which it forms an azeotrope. Pure A is the product from the water wash column. A simple distillation column is
required to recover the solvent from the water. Chemical systems using
flow diagrams similar to this include the separation of cyclohexane (A)
and benzene (B), using acetone as the solvent, and the removal of
impurities from benzene with methanol as the solvent.

A third example that is quite common is the separation of the
ethanol water azeotrope using a hydrocarbon as the entrainer. Benzene

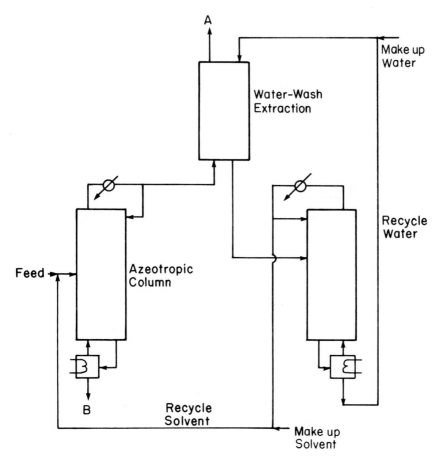

Figure 10-8.   Azeotropic distillation with one minimum boiling binary
azeotrope. Use of water wash for solvent recovery.
From Smith (1963).

used to be the most common entrainer, but because of its toxicity it has
been replaced by diethyl ether, n-pentane, and n-hexane. A heterogeneous ternary azeotrope is removed as the distillate product from the azeotropic distillation column. A typical flow sheet for this system is shown
in Figure 10-9 (Black, 1980; Robinson and Gilliland, 1950; Seader, 1984;
Shinskey, 1984; Smith, 1963). The feed to the azeotropic distillation
column is the distillate product from a binary ethanol-water column and
is close to the azeotropic composition. The composition of the ternary
azeotrope will vary slightly depending upon the entrainer chosen. For
example, when n-hexane is the entrainer the azeotrope contains 85 wt%
hexane, 12 wt % ethanol, and 3 wt % water (Shinskey, 1984). The
water/ethanol ratio in the ternary azeotrope must be greater than the

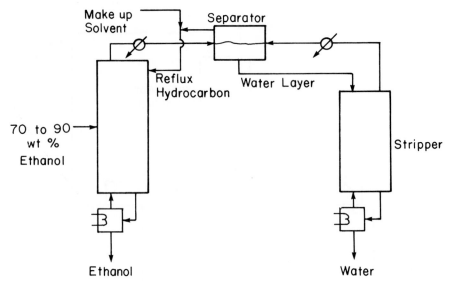

Figure 10-9.   Ternary   azeotropic   distillation   for   separation   of
ethanol-water with hydrocarbon entrainer.

water/ethanol ratio in the feed so that all the water can be removed
with the azeotrope and excess ethanol can be removed as a pure bottoms
product. The upper layer in the separator is 96.6 wt % hexane, 2.9 wt
% ethanol and 0.5 wt % water, while the bottom layer is 6.2 wt % hexane, 73.7 wt % ethanol and 20.1 wt % water. The upper layer from the
separator is refluxed to the azeotropic distillation column, while the bottom layer is sent to a stripping column to remove water.

Calculations for any of the azeotropic distillation systems are considerably more complex than for simple distillation or even for extractive
distillation.   The complexity arises from the obviously very nonideal
equilibrium behavior and from the possible formation of three phases
(two liquids and a vapor) inside the column. Calculation procedures for
azeotropic distillation are reviewed by Prokopakis and Seider (1983).
Applications of the θ method are presented by Holland (1981).

Results of simulations have been presented by Black (1980), Hoffman
(1964), Holland (1981), Prokopakis and Seider (1983), Robinson and Gilliland (1950), Seader (1984), and Smith (1963). Seader's (1984) results
for the dehydration of ethanol using n-pentane as the solvent are plotted
in Figure 10-10. The system used is similiar to the flow sheet shown in
Figure 10-9. The feed to the column contained 0.8094 mole fraction
ethanol. The column operated at a pressure of 331.5 kPa to allow con-

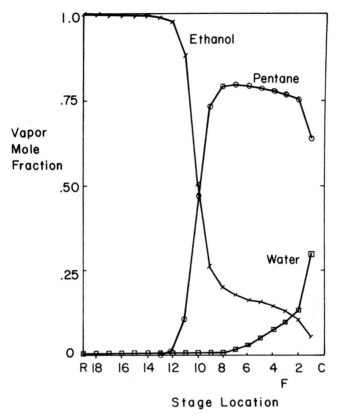

Figure 10-10.   Composition profiles for azeotropic distillation column separating water and ethanol with n-pentane entrainer. (Seader, 1984).

densation of the distillate with cooling water and had 18 stages plus a partial reboiler and a total condenser. The third stage below the condenser was the feed stage. Note that the composition profiles are different from those shown in Chapter 7. The pentane appears superficially to be a light key except that none of it appears in the bottoms. Instead, a small amount of the water exits in the bottoms with the ethanol.

Selecting a solvent for azeotropic distillation is often more difficult than for extractive distillation. There are usually fewer solvents that will form azeotropes that boil at a low enough temperature to be easy to remove in the distillate or boil at a high enough temperature to be easy to remove in the bottoms. In addition, the binary or ternary azeotrope formed must be easy to separate. In practice, this requirement is met by heterogeneous azeotropes and by azeotropes that are easy to separate with a water wash. The chosen entrainer must also satisfy the usual requirements of being nontoxic, noncorrosive, chemically stable, readily available, and inexpensive. Because of the difficulty in finding suitable solvents, azeotropic distillation systems with unique solvents are patentable.

10.3.4. Steam Distillation

In steam distillation, water (as steam) is intentionally added to the distilling organic mixture to reduce the required temperature and to keep suspended any solids that may be present. Steam distillation may be operated with one or two liquid phases in the column. In both cases the overhead vapor will condense into two phases. Thus the system can be considered a type of azeotropic distillation where the added solvent is water and the *separation is between volatiles and nonvolatiles*. Steam distillation is commonly used for purification of essential oils in the perfume industry, for distillation of organics obtained from coal, for hydrocarbon distillations, and for removing solvents from solids in waste disposal (Ellerbe, 1979; Woodland, 1978).

For steam distillation with a liquid water phase present, both the water and organic layers exert their own vapor pressures. At 1 atm pressure the temperature must be less than $100°C$ even though the organic material by itself might boil at several hundred degrees. Thus one advantage of steam distillation is lower operating temperatures. With two liquid phases present and in equilibrium, their compositions will be fixed by their mutual solubilities. Since each phase exerts its own vapor pressure, the vapor composition will be constant regardless of the average liquid concentration. A heterogeneous azeotrope is formed. As the amount of water or organic is increased, the phase concentrations do not change; only the amount of each liquid phase will change. Since an azeotrope has been reached, no additional separation is obtained by adding more stages. Thus only a reboiler is required. This type of steam distillation is often done as a batch operation (see Chapter 11).

Equilibrium calculations are similiar to those for drying organics except that now two liquid phases are present. Since each phase exerts its own partial pressure, the total pressure is the sum of the partial pressures. With one volatile organic present, this is

$$p_{org} + p_w = p_{tot} \qquad (10\text{-}18)$$

Substituting in Eqs. (10-12) and (10-13), we obtain

$$(VP_{org})x_{\text{volatile in org}} + (VP_w)x_{w\text{ in }w} = P_{tot} \qquad (10\text{-}19)$$

The compositions of the liquid phases are set by equilibrium. If total pressure is fixed, then Eq. (10-19) enables us to calculate the temperature. Once the temperature is known the vapor composition is easily calculated as

$$y_i = \frac{p_i}{P_{tot}} \qquad (10\text{-}20)$$

The number of moles of water carried over in the vapor is easily estimated, since the ratio of moles of water to moles organic is equal to the ratio of vapor mole fractions.

$$\frac{n_{org}}{n_w} = \frac{y_{\text{volatile}}}{y_w} \qquad (10\text{-}21)$$

Substituting in Eq. (10-20), this is

$$\frac{n_{org}}{n_w} = \frac{p_{org}}{p_w} = \frac{p_{org}}{P_{tot} - p_{org}} = \frac{(VP)_{org}\, x_{\text{volatile}}}{P_{tot} - (VP)_{org}\, x_{\text{volatile}}} \qquad (10\text{-}22)$$

If several organics are present, y_{org} and p_{org} are the sums of the respective values for all the organics. The total moles of steam required is n_w plus the amount condensed to heat and vaporize the organic.

Example 10-2. Steam Distillation

A cutting oil that has approximately the properties of n-decane ($C_{10}H_{22}$) is to be recovered from nonvolatile oils and solids in a steady-state single-stage steam distillation. Operation will be with liquid water present. The feed is 50 mole % n-decane. A bottoms that is 15 mole % n-decane in the organic phase is desired. Feed rate is 10 kg moles/hr. Feed enters at the temperature of the boiler. Pressure is atmospheric pressure, which in your plant is approximately 745 mmHg. Find:

a. The temperature of the still
b. The moles of water carried over in the vapor
c. The moles of water in the bottoms

Solution

A. Define. The still is sketched in the figure. Note that there is no reflux.

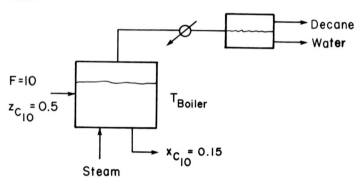

B. Explore. Equilibrium is given by Eq. (10-19). Assuming that the organic and water phases are completely immiscible, we have $x_{C10 \text{ in org}} = 0.15$ and $x_{w \text{ in } w} = 1.0$. Vapor pressure data as a function of temperature are available in Perry and Green (1984). Then Eq. (10-19) can be solved by trial and error to find T_{boiler}. Equation (10-22) and a mass balance can be used to determine the moles of water and decane vaporized. The moles of water condensed to vaporize the decane can be determined from an energy balance. Latent heat data are available in Perry and Green (1984).

C. Plan. On a water-free basis the mass balances around the boiler are

$$F = n_{C10,vapor} + B$$

$$z_{C10} F = n_{C10,vapor} + x_{bot} B$$

where B is the bottoms flow rate of the organic phase. Since $F = 10$, $x_{bot} = 0.15$, and $z_{C10} = 0.5$, we can solve for $n_{C10,vapor}$ and B. Equation (10-22) gives n_w once T_{boiler} is known. Since the feed, bottoms, and vapor are all at T_{boiler}, the energy balance simplifies to

$$n_{C10}\lambda_{C10} = \lambda_w(\text{moles water condensed})$$

D. Do It. a. Perry and Green (1984) give the following n-decane vapor pressure data (vapor pressure in mm Hg and T in °C):

VP	5	10	20	40	60	100	200	400	760
T	42.3	55.7	69.8	85.5	95.5	108.6	128.4	150.6	174.1

A very complete table of water vapor pressures is given in that source (see Problem 10-D6). As a first guess, try 95.5 °C, where $(VP)_w = 645.7$ mmHg. Then Eq. (10-19),

$$(VP)_{C10}\, x_{C10} + (VP)_w\, x_w = p_{tot}$$

becomes

$$(60)(0.15) + (645.7)(1.0) = 654.7 < 745 = p_{tot}$$

where we have assumed completely immiscible phases so that $x_w = 1.0$. This temperature is too low. At 100 °C, a plot of $(VP)_{C10}$ vs T gives $(VP)_{C10} = 70$. Then Eq. (10-19) is

$$70\,(0.15) + (760)(1.0) = 770.5 > 745$$

Lowering the temperature to $T_{boiler} = 99$ °C, we estimate $(VP)_{C10} = 68$, $(VP)_w = 733.2$, and

$$68(0.15) + 733.2(1.0) = 743.4 \sim 745$$

which is close enough to p_{tot}. Thus $T_{boiler} = 99$ °C.
b. Solving the mass balances, the kg moles/hr of vapor are

$$n_{C10,vapor} = \frac{F(z_{C10} - x_{bot})}{1 - x_{bot}} = \frac{10(0.5 - 0.15)}{1 - 0.15} = 4.12$$

which is 586.2 kg/hr. Equation (10-22) becomes

$$n_w = \frac{n_{C10,vapor}}{(VP)_{C10}\, x_{bot}} \left[p_{tot} - (VP)_{C10}\, x_{bot} \right]$$

or

$$n_w = \frac{4.12}{(68)(0.15)} \, [745 - (68)(0.15)] = 296.8 \text{ kg moles/hr}$$

which is 5347.1 kg/hr.

c. The moles of water to vaporize the decane is

$$\text{Moles water condensed} = \frac{n_{C10} \, \lambda_{C10}}{\lambda_w}$$

$$= \frac{(4.12 \text{ kg moles})\left(319\dfrac{kJ}{kg}\right)\left(142.28\dfrac{kg}{kg \text{ mole}}\right)}{\left(2260\,\dfrac{kJ}{kg}\right)\left(18.016\dfrac{kg}{kg \text{ mole}}\right)} = 4.59 \text{ kg moles}$$

where λ_{C10} at 99°C is interpolated from Table 3-231 in Perry and Green (1984), and λ_w at 99°C is interpolated from Table 3-302 in Perry and Green (1984).

E. Check. A check for complete immiscibility is advisable since all the calculations are based on this assumption.

F. Generalize. Obviously, the decane is boiled over at a temperature well below its boiling point, but a large amount of water is required. Most of this water is carried over in the vapor. On a weight basis, the kilograms of total water required per kilogram decane vaporized is 9.26. Less water will be used if the boiler is at a higher temperature and there is no liquid water in the still. Less water is also used for higher values of $x_{org,bot}$ (see Problems 10-D6 and 10-F1).

Additional separation can be obtained by operating without a liquid water phase in the column. Reducing the number of phases increases the degrees of freedom by one. Operation must be at a temperature higher than that predicted by Eq. (10-19), or a liquid water layer will form in the column. Thus the column must be heated with a conventional reboiler and/or the sensible heat available in superheated steam. The latent heat available in the steam cannot be used, because it would produce a layer of liquid water. Operation without liquid water in the column reduces the energy requirements but makes the system more complex.

10.4. PROCESSES USING CHANGES IN PRESSURE

Pressure affects vapor-liquid equilibrium, and in systems that form azeotropes it will affect the composition of the azeotrope. For example, Table 2-1 shows that the ethanol-water system has an azeotrope at 0.8943 mole fraction ethanol at 1 atm pressure. If the pressure is reduced, the azeotropic concentration increases (Seader, 1984). At pressures below 70 mmHg, the azeotrope disappears entirely, and the distillation can be done in a simple column. Unfortunately, use of this disappearance of the azeotrope for the separation of ethanol and water is not economical because the column requires a large number of stages and has a large diameter (Black, 1980). However, the principle of finding a pressure where the azeotrope disappears may be useful in other distillations. The effect of pressure on the azeotropic composition and temperature can be estimated (Barduhn, 1985).

Even though the azeotrope may not disappear, in general, pressure affects the azeotropic composition. If the shift in composition is large enough, a two-column process using two different pressures can be used to completely separate the binary mixture. A schematic of the flow chart for this two-pressure distillation process is shown in Figure 10-11 (Drew, 1979; Shinskey, 1984; Van Winkle, 1967). Column 1 usually operates at atmospheric pressure, while column 2 is usually at a higher pressure but can be at a lower pressure.

Figure 10-11. Two-pressure distillation for azeotropic separation.

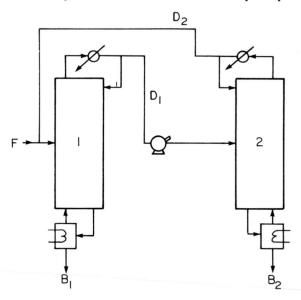

To understand the operation of this process, consider the separation of methyl ethyl ketone (MEK) and water (Drew, 1979). At 1 atm the azeotrope contains 35% water, while at 100 psia the azeotrope is 50% water. If a feed containing more than 35% water is fed to the first column, the bottoms will be pure water. The distillate from this atmospheric column will be the 35% azeotrope. When this azeotrope is sent to the high-pressure column, an azeotrope containing 50% water comes off as the distillate; this distillate is recycled to column 1. Since the feed (the 35% azeotrope) contains less water than this distillate, the bottoms from column 2 is pure MEK. Note that the water is less volatile in column 1 and the MEK is less volatile in column 2. The McCabe-Thiele diagram for one of the columns will have the equilibrium curve and the operating lines below the $y = x$ line.

Mass balances for the system shown in Figure 10-11 are of interest. The external mass balances are identical to Eqs. (10-5) and (10-6). Thus, the bottoms flow rates are given by Eqs. (10-7) and (10-8). Although the processes shown in Figures 10-4A and 10-11 are very different, they look the same to the external mass balances. Differences in the processes become evident when balances are written for individual columns. For instance, for column 2 the mass balances are

$$D_1 = D_2 + B_2 \qquad (10\text{-}23)$$

and

$$D_1 x_{dist1} = D_2 x_{dist2} + B_2 x_{bot2} \qquad (10\text{-}24)$$

Solving these equations simultaneously and then inserting the values in Eq. (10-8), we obtain

$$D_2 = \frac{B_2(x_{bot2} - x_{dist1})}{x_{dist1} - x_{dist2}} = F\left(\frac{z - x_{b1}}{x_{bot2} - x_{bot1}}\right)\left(\frac{x_{bot2} - x_{dist1}}{x_{dist1} - x_{dist2}}\right) \qquad (10\text{-}25)$$

This is of interest since D_2 is the recycle flow rate. As the two azeotrope concentrations at the two different pressures approach each other, $x_{dist1} - x_{dist2}$ will become small. According to Eq. (10-25), the recycle flow rate D_2 becomes large. This increases both operating and capital costs and makes this process too expensive if the shift in the azeotrope concentration is small.

The two-pressure system is also used for the separation of tetrahydrofuran-water, methanol-MEK, and methanol-acetone. In the latter application the second column is at 200 torr. Realize that these applications are rare. For most azeotropic systems the shift in the azeotrope with pressure is small, and use of the system shown in Figure 10-9 will involve a very large recycle stream. This causes the first column to be rather large, and costs become excessive.

10.5. DISTILLATION WITH CHEMICAL REACTION

Distillation columns are occasionally used as chemical reactors. The advantage of this approach is that distillation and reaction can take place simultaneously in the same vessel, and the products can be removed to drive the reversible reaction to completion. The most common industrial application is for the formation of esters from a carboxylic acid and an alcohol. This method was first patented by Backhaus in 1921 and has been the subject of several patents since then (see Neumann and Sasson, 1984 or Terrill *et al.*, 1985 for references). Reaction in a distillation column may also be undesirable when one of the desired products decomposes.

Distillation with reaction is useful for reversible reactions. Examples would be reactions such as

$$A = C$$

$$A = C + D$$

$$A + B = C + D$$

The purposes of the distillation are to separate the product(s) from the reactant(s) to drive the reactions to the right, and to recover purified product(s).

Depending on the equilibrium properties of the system, different distillation configurations can be used as shown in Figure 10-12. Figure 10-12A shows the case where the reactant is less volatile than the product (Belck, 1955). If several products are formed, no attempt is made to separate them in this system. The bleed is used to prevent the buildup of nonvolatile impurities or products of secondary reactions. If the feed is more volatile than the desired product, the arrangement shown in Figure 10-12B can be used (Belck, 1955). This column is essentially at total reflux except for a small bleed, which may be needed to remove volatiles or gases.

Figures 10-12C, D, and E all show systems where two products are formed and the products are separated from each other and from the

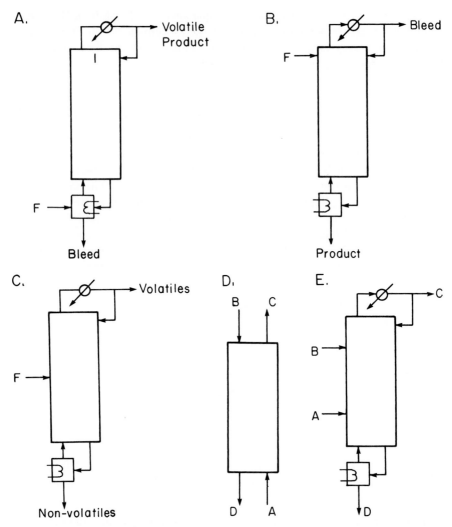

Figure 10-12.   Schemes for distillation plus reaction. (A) Volatile product, reaction is A = C; (B) nonvolatile product, reac-
tion is A = D; (C) Two products, reactions are A = C
+ D or A + B = C + D; (D,E) Reaction A + B = C
+ D with B and D nonvolatile.

reactants in the distillation column. In Figure 10-12C the reactant(s)
are of intermediate volatility between the two products. Then the reactants will stay in the middle of the column until they are consumed,
while the products are continuously removed, driving the reaction to the
right. If the reactants are not of intermediate volatility, some of the

reactants will appear in each product stream (Suzuki *et al.*, 1971; Kinoshita *et al.*, 1983; Neumann and Sasson, 1984). The alternative schemes shown in Figures 10-12D (Nelson, 1971) and 10-12E (Suzuki *et al.*, 1971; Neumann and Sasson, 1984) will often be advantageous for the reaction

$$A + B = C + D$$

In these two figures, species A and C are relatively volatile while species B and D and relatively nonvolatile. Since reactants are fed in at opposite ends of the column, there is a much larger region where both reactants are present. Thus, the residence time for the reaction will be larger in Figures 10-12D and E than in Figure 10-12C, and higher yields can be expected. The systems shown in Figures 10-12C and D have been used for esterification reactions such as

Acetic acid + ethanol → ethyl acetate + water

(Suzuki *et al.*, 1971) and

Acetic acid + methanol → methyl acetate + water

(Neumann and Sasson, 1984).

When a reaction occurs in the column, the mass and energy balance equations must be modified to include the reaction terms. The general mass balance equation for stage j (Eq. 8-18) can be modified to

$$V_j y_j + L_j x_j - V_{j-1} y_{j-1} - L_{j+1} x_{j+1} = F_j z_j + r_j \qquad (10\text{-}26)$$

where the reaction term r_j is positive if the component is a product of the reaction. To use Eq. (10-26), the appropriate rate equation for the reaction must be used for r_j. In general, the reaction rate will depend on both the temperature and the liquid compositions.

Modern computer solutions for distillation with reaction have written the mass balances in matrix form (Holland, 1981; Kinoshita *et al.*, 1983; Nelson, 1971; Neumann and Sasson, 1984; Suzuki *et al.*, 1971; and Terrill *et al.*, 1985). The reaction term can conveniently be included with the feed in the D term in Eqs. (8-23) to (8-30). This retains the tridiagonal form of the mass balance, but the D term now depends upon liquid concentration and stage temperature. The convergence procedures to solve the resulting set of equations must be modified, because the procedures outlined in Chapter 8 may not be able to converge. The

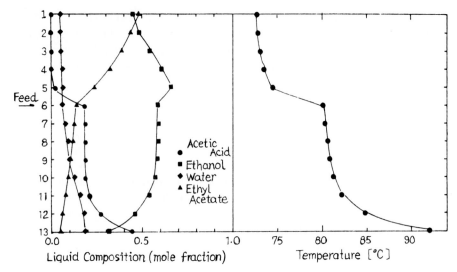

Figure 10-13. Composition and temperature profiles for the reaction acetic acid + ethanol = ethyl acetate + water. From Suzuki *et al.* (1971). Copyright 1971. Reprinted with permission from *Journal of Chemical Engineering of Japan.*

equations have become highly nonlinear because of the reaction rate term. Newton-Rapheson iteration procedures (Kinoshita *et al.*, 1983; Nelson, 1971; Neumann and Sasson, 1984; and Suzuki *et al.*, 1971) and a modified θ convergence procedure (Holland, 1981) have been successful.

A sample of composition and temperature profiles for the esterification of acetic acid and ethanol is shown in Figure 10-13 (Suzuki *et al.*, 1971) for the distillation system of Figure 10-12C. The distillation column is numbered with 13 stages including the total condenser (No.1) and the partial reboiler (No.13). Reaction can occur on every stage of the column and in both the condenser and the reboiler. Feed is introduced to stage 6 as a saturated liquid. The feed is mainly acetic acid and ethanol with a small amount of water. A reflux ratio of 10 is used. The top product contains most of the ethyl acetate produced in the reaction plus ethanol and a small amount of water. All of the non-reacted acetic acid appears in the bottoms along with most of the water and a significant fraction of the ethanol. Reaction is obviously not complete.

Neumann and Sasson (1984) looked at the esterification of acetic acid and methanol. They found that the configuration shown in Figure 10-12D was superior to that of Figure 10-12C. By controlling the reflux ratio and the molar feed ratio, they were able to obtain conversions as high as 98%. This is significantly higher than can be achieved in a reactor without separation.

A somewhat different type of distillation with reaction is "catalytic distillation" (Lander *et al.*, 1983). In this process bales of catalyst are stacked in the column. The bales serve both as the catalyst and as the column packing (see Chapter 13). This process is used commercially for production of methyl tert-butyl ether (MTBE) from the liquid-phase reaction of isobutylene and methanol. The heat generated by the exothermic reaction is used to supply much of the heat required for the distillation.

Although many reaction systems do not have the right reaction equilibrium or vapor-liquid equilibrium characteristics for distillation with reaction, for those that do this technique is a very valuable industrial tool.

10.6. SUMMARY-OBJECTIVES

In this chapter we have looked at azeotropic and extractive distillation systems plus distillation with simultaneous chemical reaction. At the end of this chapter you should be able to satisfy the following objectives:

1. Analyze binary distillation systems using other separation schemes to break the azeotrope.

2. Explain the purpose of extractive distillation, select a suitable solvent, and explain the expected concentration profiles.

3. Solve binary heterogeneous azeotrope problems, including the drying of organic solvents, using McCabe-Thiele diagrams.

4. Explain azeotropic distillation with an added solvent including steam distillation and develop flow sheets for the separation of various chemicals.

5. Use McCabe-Thiele diagrams to solve problems where two pressures are used to separate azeotropes.

6. Explain qualitatively the purpose of doing a reaction in a distillation column, and discuss the advantages and disadvantages of the different column configurations.

334

REFERENCES

Barduhn, A.J., "The Effect of Pressure on Azeotropes," *Chem. Eng. Commun., 38,* 9 (1985).

Belck, L.H., "Continuous Reactions in Distillation Equipment," *AIChE J., 1,* 467 (1955).

Berg, L., "Selecting the Agent for Distillation Processes," *Chem. Eng. Prog., 65* (9), 52 (Sept. 1969).

Black, C., "Distillation Modeling of Ethanol Recovery and Dehydration Processes for Ethanol and Gasohol," *Chem. Eng. Prog., 76* (9), 78 (Sept. 1980).

Drew, J.W., "Solvent Recovery," in P.A. Schweitzer (Ed.), *Handbook of Separation Techniques for Chemical Engineers,* McGraw-Hill, New York, 1975, Section 1.6.

Ellerbe, R.W., "Steam Distillation/Stripping," in P.A. Schweitzer (Ed.), *Handbook of Separation Techniques for Chemical Engineers,* McGraw-Hill, New York, 1979, pp. 1-169 to 1-178.

Hoffman, E.J., *Azeotropic and Extractive Distillation,* Interscience, New York, 1964.

Holland, C.D., *Fundamentals of Multicomponent Distillation,* McGraw-Hill, New York, 1981.

Kinoshita, M., I. Hashimoto, and T. Takamatsu, "A New Simulation Procedure for Multicomponent Distillation Column Processing Nonideal Solutions or Reactive Solutions," *J. Chem. Eng. Japan, 16,* 370 (1983).

Lander, E.P., J.N. Hubbard, and L.A. Smith, "Revving-up Refining Profits with Catalytic Distillation," *Chem. Eng., 90* (8), 36 (April 18, 1983).

Luyben, W.L., "Azeotropic Tower Design by Graph," *Hydrocarbon Processing, 52*(1), 109 (Jan. 1973).

Nelson, P.A., "Countercurrent Equilibrium Stage Separation with Reaction," *AIChE J., 17,* 1043 (1971).

Neumann, R. and Y. Sasson, "Recovery of Dilute Acetic Acid by

Esterification in a Packed Chemorectification Column," *Ind. Eng. Chem. Process Design Develop., 23,* 654 (1984).

Poffenberger, N., L.H. Horsley, H.S. Nutting, and E.C. Britton, "Separation of Butadiene by Azeotropic Distillation with Ammonia," *Trans. Amer. Inst. Chem. Eng., 42,* 815 (1946).

Prokopakis, G.J. and W.D. Seider, "Dynamic Simulation of Azeotropic Distillation Towers," *AIChE J., 29,* 1017 (1983).

Robinson, C.S. and E.R. Gilliland, *Elements of Fractional Distillation,* 4th ed., McGraw-Hill, New York, 1950, Chapt. 10.

Seader, J.D., "Distillation," in R.H. Perry and D.W. Green (Eds.), *Perry's Chemical Engineers' Handbook,* 6th ed., McGraw-Hill, New York, 1984, Section 13.

Shinskey, F.G., *Distillation Control, For Productivity and Energy Conservation,* 2nd ed., McGraw-Hill, New York, 1984, Chapts. 9 and 10.

Smith, B.D., *Design of Equilibrium Stage Processes,* McGraw-Hill, New York, 1963, Chapter 11.

Suzuki, I., H. Yagi, H. Komatsu, and M. Hirata, "Calculation of Multicomponent Distillation Accompanied by a Chemical Reaction," *J. Chem. Eng. Japan, 4,* 26 (1971).

Terrill, D.L., Sylvestre, L.F., and M.F. Doherty, "Separation of Closely Boiling Mixtures by Reactive Distillation. 1. Theory," *Ind. Eng. Chem. Process Des. Develop., 24,* 1062 (1985).

Van Winkle, M., *Distillation,* McGraw-Hill, New York, 1967.

Woodland, L.R., "Steam Distillation," in D.J. DeRenzo (Ed.), *Unit Operations for Treatment of Hazardous Industrial Wastes,* Noyes Data Corp., Park Ridge, NJ, 1978, pp. 849 to 868.

HOMEWORK

A.   *Discussion Problems*

A1.   Compare the systems shown in Figures 10-1A, B, and C. What are the advantages and disadvantages of each system?

A2. Explain the differences between extractive and azeotropic distillation. What are the advantages and disadvantages of each procedure?

A3. Why is a cooler required in Figure 10-2? Can this energy be reused in the process?

A4. Explain the purpose of the liquid-liquid settler in Figures 10-4A, 10-5, and 10-8.

A5. Explain why the external mass balances are the same for Figures 10-4A and 10-10.

A6. Why are makeup solvent additions shown in Figures 10-7, 10-8, and 10-9?

A7. Explain why the pentane composition profile shows a maximum in Figure 10-10.

A8. Explain in your own words the advantages of doing reaction and distillation simultaneously.

A9. When doing distillation with reaction, the column should be designed both as a reactor and as a distillation column. In what ways might these columns differ from normal distillation columns?

A10. Reactions are usually not desirable in distillation columns. If there is a reaction occurring, what can be done to minimize it?

A11. Develop your key relations chart for this chapter.

B. *Generation of Alternatives*

B1. Sketch how you would separate a mixture of ethyl acetate and ethanol (this mixture forms a homogeneous azeotrope).

C. *Derivations*

C1. Derive Eq. (10-11) for the two-column, binary, heterogeneous azeotrope system.

C2. For a binary heterogeneous azeotrope, draw the column arrangement if the feed composition is less than the azeotrope concentration $(z < x_\alpha)$. Show the McCabe-Thiele diagram for this system.

C3. For a binary heterogeneous azeotrope separation, the feed can be introduced into the liquid-liquid separator. In this case two stripping columns are used.

a. Sketch the column arrangement.

b. Draw the McCabe-Thiele diagram for this system.

c. Compare this system to the system in Figure 10-4A and Problem 10-C2.

C4. Show how to apply the lever-arm rule on Figure 10-4B for the liquid-liquid separator.

C5. Develop a computer flow sheet for a binary heterogeneous azeotrope system.

C6. If no makeup solvent is required, show that the external mass balances for Figure 10-9 are the same as for Figures 10-4A and 10-11.

C7. Sketch the McCabe-Thiele diagram for a two-pressure system similar to that of Figure 10-11.

C8. An equation for $\alpha_{\text{org-w in w}}$ similar to Eq. (10-17) is easy to derive; do it. Compare the predicted equilibrium in water with the butanol-water equilibrium data given in Problem 10-D2. Comment on the fit. Vapor pressure data are in Perry and Green (1984). Use the data in Problem 10-D2 for solubility data.

C9. Derive Eq. (10-25).

D. *Problems*

D1. A distillation column is separating isopropanol and water at 101.3 kPa. We need $x_p = 0.96$ mole fraction isopropanol, which the distillation column cannot produce. However, if the column gives $y_1 = 0.80$, a membrane separator can produce a product with $x_p = 0.96$. Some of this is returned as a saturated liquid reflux ($x_0 = 0.96$). The initial feed to the column is $F = 1000$ kg moles/hr and is 20 mole % isopropanol. This feed is a saturated vapor. The recycle stream is a saturated liquid and has a mole fraction $= 0.4$. We desire $x_B = 0.01$. The column uses open steam, which is a saturated vapor and is pure water. We set the internal reflux ratio in the top section, $L_0/V_1 = 5/9$. Find the two optimum feed plate locations and the total number of stages. You may assume CMO. The column is similar to Figure 10-1C except that open steam heating is used. Isopropanol water equilibrium data are in Perry and Green (1984, p. 13-13). Isopropanol values are listed below.

x	.0045	.0069	.0127	.0357	.0678	.1330	.1651	.3204
y	.0815	.1405	.2185	.3692	.4647	.5036	.5153	.5456
x	.3752	.4720	.5197	.5945	.7880	.8020	.9303	.9660
y	.5615	.5860	.6033	.6330	.7546	.7680	.9010	.9525

D2. Vapor-liquid equilibrium data for water-n-butanol are given in Table 10-2. We wish to distill 5000 kg moles/hr of a mixture that is 28 mole % water and 30% vapor in a two-column azeotropic distillation system. A butanol phase that contains 0.04 mole fraction water and a water phase that is 0.995 mole fraction water are desired. Pressure is 101.3 kPa. Reflux is a saturated liquid. Use $L/V = 1.23(L/V)_{min}$ in the column producing almost pure butanol. Both columns have partial reboilers. $(\overline{V}/B)_2 = 0.132$ in the column producing water.

Table 10-2. Vapor-Liquid Equilibrium Data for Water and n-Butanol at 1 atm

Liquid	Vapor	T °C	Liquid	Vapor	T °C
3.9	26.7	111.5	57.3	75.0	92.8
4.7	29.9	110.6	97.5	75.2	92.7
5.5	32.3	109.6	98.0	75.6	93.0
7.0	35.2	108.8	98.2	75.8	92.8
25.7	62.9	97.9	98.5	77.5	93.4
27.5	64.1	97.2	98.6	78.4	93.4
29.2	65.5	96.7	98.8	80.8	93.7
30.5	66.2	96.3	99.2	84.3	95.4
49.6	73.6	93.5	99.4	88.4	96.8
50.6	74.0	93.4	99.7	92.9	98.3
55.2	75.0	92.9	99.8	95.1	98.4
56.4	75.2	92.9	99.9	98.1	99.4
57.1	74.8	92.9	100	100	100

(*Source:* Chu *et al.* (1950). Mole % water)

a. Find flow rates of the products.

b. Find the optimum feed location and number of stages in the columns.

Note: Draw two McCabe-Thiele diagrams.

D3. A saturated vapor feed that is 20 mole % water is input to an enriching section. The column has no reboiler, but it has a total condenser and a liquid-liquid separator. The water phase is withdrawn as distillate product, while the butanol phase is refluxed to the column. We desire $x_{bot} = 0.04$ mole fraction water. Pressure is 1 atm. Reflux is a saturated liquid. How many stages are required? What external reflux ratio L/D must be used? Data are in Table 10-2.

D4. The distillation arrangement shown in Figure 10-5 is to be used to dry 100 kg moles/hr of benzene containing 0.006 mole fraction water. The feed is a saturated liquid. A bottoms that is 0.0005 mole fraction water is required. Boilup ratio $\overline{V}/B$ is 0.0444. Reflux is a saturated liquid. Use $\alpha_{wb} = 31.3$. Solubility data are in Example 10-1. Find the optimum feed stage and the total number of stages required.

D5. We have a feed of 15,000 kg/hr of diisopropyl ether $(C_6 H_{14} 0)$ that contains 0.004 wt frac water. We want a diisopropyl ether product that contains 0.0004 wt frac water. Feed is a saturated liquid. Use the system shown in Figure 10-5, operating at 101.3 kPa. Use $L/D = 1.5 \ (L/D)_{min}$. Determine $(L/D)_{min}$, L/D, optimum feed stage, and total number of stages required. Assume that CMO is valid. The following data for the diisopropyl ether - water azeotrope is given (*Trans. AIChE, 36,* 593, 1940):

y = 0.959 , Separator: Top layer, x = 0.994 ; bottom layer x = 0.012; at 101.3 kPa and 62.2 ° C. All compositions are weight fractions of diisopropyl ether.

Estimate $\alpha_{w-ether \, in \, ether}$ from these data (in mole fraction units). Assume that this relative volatility is constant.

D6. A single-stage steam distillation system is recovering n-decane from a small amount of nonvolatile organics. Pressure is 760 mmHg. If the still is operated with liquid water present and the organic layer in the still is 99 mole % n-decane, determine:

a. The still temperature
b. The moles of water vaporized per mole of n-decane vaporized

Decane vapor pressure is in Example 10-2. Water vapor pressures are (T in ° C and VP in mm Hg) (Perry and Green, 1984)

T	95.5	96.0	96.5	97.0	97.5	98.0	98.5	99.0	99.5
VP	645.67	657.62	669.75	682.07	694.57	707.27	720.15	733.24	746.52

D7. Seader (1984) gives data for the azeotrope composition of the system ethanol-benzene. At 101.3 kPa the azeotrope is at 67.9 ° C and is 0.449 mole frac ethanol. At 1333 kPa the azeotrope is at 159 ° C and is 0.75 mole frac ethanol. Ethanol is more volatile at mole fractions below the azeotrope concentration. Sketch a two-pressure system to do this separation. If the feed flow rate is 100 kg moles/hr and the feed is 0.35 mole frac ethanol, determine the product flow rates for an ethanol product that is 99% ethanol and a benzene product that is 99% benzene. Determine the recycle flow rate required.

E. *More Complex Problems*

E1. We wish to separate water from nitromethane $(CH_3 NO_2)$. The first feed (to column 1) is 24 mole % water, is a saturated vapor, and flows at 1000 kg mole/day. The second feed (to column 2) is 96 mole % water, is a saturated liquid, and flows at 500 kg moles/day. A two-column system with a liquid-liquid separator will be used. The bottoms from column 1 is 2 mole % water; use $(L/V)_1 = 0.887$ in this column. The bottoms from column 2 is 99 mole % water; use $(\overline{L}/\overline{V})_2 = 6$ in this column. Data are given in Table 10-3.

Table 10-3. Water-Nitromethane Equilibrium Data at 1 atm

Liquid	Vapor	Liquid	Vapor
7.9	34.0	81.4	50.1
16.2	45.0	88.0	50.2
23.2	49.7	91.4	50.3
31.2	50.0	92.5	51.3
39.2	50.2	93.5	52.9
45.0	50.2	95.1	55.9
55.2	50.2	95.9	57.5
63.8	50.2	97.5	65.8
73.0	50.2	97.8	68.4
77.0	50.2	98.6	79.2

Source: Chu *et al.* (1950). (Mole % water)

a. Find the number of stages and the optimum feed location in both columns.

b. Calculate L/V in column 2. Explain your result.

F. *Problems Requiring Other Resources*

F1. A single-stage steam distillation apparatus is to be used to recover n-nonane ($C_9 H_{20}$) from nonvolatile organics. Operation is at 1 atm (760 mmHg), and the still will operate with liquid water present. If the still bottoms are to contain 99 mole % nonane in the organic phase (the remainder is non-volatiles), determine:

    a. The temperatures of the distillation

    b. The moles of water vaporized per mole of nonane vaporized

    c. The moles of water condensed per mole of nonane vaporized for a liquid entering at the still temperature

    d. Repeat parts a and b if still bottoms will contain 2.0 mole % nonane

Data are available in Perry and Green (1984).

F2. Look up the vapor pressure data for diisopropyl ether and for water. Using the liquid-liquid separator compositions given in

Problem 10-D5, estimate $\alpha_{w-ether\,in\,ether}$ from Raoult's law. Compare this to the value of α obtained from the heterogeneous azeotrope data.

chapter 11
BATCH DISTILLATION

Continuous distillation is a thermodynamically efficient method of producing large amounts of material of constant composition. When small amounts of material or varying product compositions are required, batch distillation has several advantages. In batch distillation a charge of feed is loaded into the reboiler, the steam is turned on, and after a short startup period, product can be withdrawn from the top of the column. When the distillation is finished, the heat is shut off and the material left in the reboiler is removed. Then a new batch can be started. Usually the distillate is the desired product.

Batch distillation is versatile. A run may last from a few hours to several days. Batch distillation is the choice when the plant does not run continuously and the batch must be completed in one or two shifts (8 to 16 hours). It is often used when the same equipment distills several different products at different times. If distillation is required only occasionally, batch distillation would again be the choice.

Equipment can be arranged in a variety of configurations. In simple batch distillation (Figure 11-1), the vapor is withdrawn continuously from the reboiler. The system differs from flash distillation in that there is no continuous feed input and the liquid is drained only at the end of the batch.

In a multistage batch distillation, a staged or packed column is placed above the reboiler as in Figure 11-2. Reflux is returned to the column. In the usual operation, distillate is withdrawn continually (Robinson and Gilliland, 1950; Pratt, 1962; Luyben, 1971; Ellerbe, 1979) until the column is shut down and drained. In an alternative method (Treybal, 1970), no distillate is withdrawn; instead, the composition of liquid in the accumulator changes. When the distillate in the accumulator is of the desired composition in the desired amount, both the accumulator and reboiler are drained. Luyben (1971) indicated that the usual method should be superior; however, the alternative method may be simpler to operate.

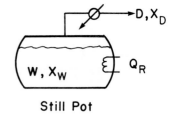

Still Pot

Figure 11-1.   Simple batch distillation.

Figure 11-2.   Multistage batch distillation. (A) Schematic. (B) Photograph of packaged batch distillation/solvent recovery system of approximately 400-gallon capacity. Courtesy of APV Equipment Inc., Tonawanda, NY.

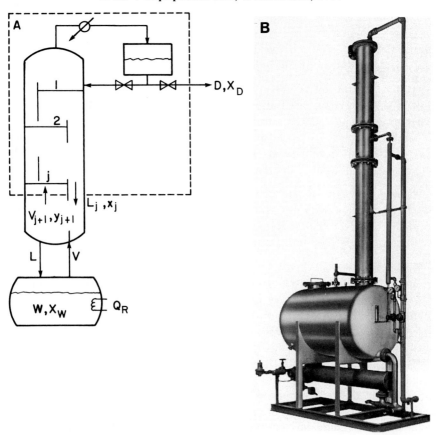

Another alternative is called inverted batch distillation (Pratt, 1967; Robinson and Gilliland, 1950) because bottoms are withdrawn continuously while distillate is withdrawn only at the end of the distillation. In this case the charge is placed in the accumulator and a reboiler with a small holdup is used. Inverted batch distillation is seldom used, but it is useful when quite pure bottoms product is required.

11.1. MASS BALANCES AND RAYLEIGH EQUATION

The mass balances for batch distillation are somewhat different from those for continuous distillation. In batch distillation we are more interested in the total amounts of bottoms and distillate collected than in the rates. For a binary batch distillation, mass balances around the entire system for the entire operation time are

$$F = W_{final} + D_{total} \tag{11-1}$$

$$Fx_F = x_{W,final}W_{final} + D_{total}x_{D,avg} \tag{11-2}$$

The feed into the column is F kg moles of mole fraction x_F of the more volatile component. The final moles in the reboiler at the end of the batch is W_{final} of mole fraction $x_{W,final}$. The symbol W is used since the material left in the reboiler is often a waste. D_{total} is the total kilogram moles of distillate of average concentration $x_{D,avg}$. Equations (11-1) and (11-2) are applicable to simple batch and normal multistage batch distillation. Some minor changes in variable definitions are required for inverted batch distillation.

Usually F, x_F, and the desired value of either $x_{W,final}$ or $x_{D,avg}$ are specified. An additional equation is required to solve for the three unknowns D_{total}, W_{final}, and $x_{W,final}$ (or $x_{D,avg}$). This additional equation, known as the Rayleigh equation (Rayleigh, 1902), is derived from a differential mass balance. Assume that the holdup in the column and in the accumulator is negligible. Then if a differential amount of material, $-dW$, of concentration x_D is removed from the system, the differential mass balance is

$$- \text{Out} = \text{accumulation in reboiler}$$

or

$$- x_D \, dW = - d(Wx_W) \tag{11-3}$$

Expanding Eq. (11-3),

$$- x_D \, dW = - W \, dx_W - x_W \, dW \qquad (11\text{-}4)$$

Then rearranging and integrating,

$$\int_{W=F}^{W_{final}} \frac{dW}{W} = \int_{x_F}^{x_{W,final}} \frac{dx_W}{x_D - x_W} \qquad (11\text{-}5)$$

which is

$$\ln\Big[\frac{W_{final}}{F}\Big] = - \int_{x_{W,final}}^{x_F} \frac{dx_W}{x_D - x_W} \qquad (11\text{-}6)$$

The minus sign comes from switching the limits of integration. Equation (11-6) is a form of the Rayleigh equation that is valid for both simple and multistage batch distillation. Of course, to use this equation we must relate x_D to x_W and do the appropriate integration. This is covered in sections 11.2 and 11.4.

Time does not appear explicitly in the derivation of Eq. (11-6), but it is implicitly present since W, x_W, and usually x_D are all time-dependent.

11.2. SIMPLE BATCH DISTILLATION

In the simple batch distillation system shown in Figure 11-1, the vapor product is in equilibrium with liquid in the still pot at any given time. Since we use a total condenser, $y = x_D$. Substituting this into Eq. (11-6), we have

$$\ln\Big(\frac{W_{final}}{F}\Big) = - \int_{x_{W,final}}^{x_F} \frac{dx}{y - x} = - \int_{x_{W,final}}^{x_F} \frac{dx}{f(x) - x} \qquad (11\text{-}7)$$

where y and x are now in equilibrium and the equilibrium expression is $y = f(x,p)$. For any given equilibrium expression, Eq. (11-7) can be integrated analytically, graphically, or numerically.

The general graphical integration procedure for Eq. (11-7) is:

1. Plot y-x equilibrium curve.

2. At a series of x values, find $y - x$.

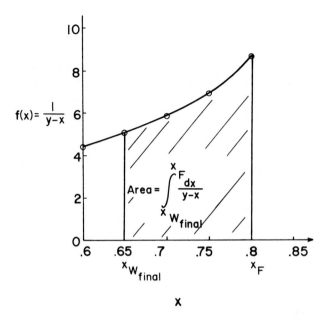

Figure 11-3. Graphical integration for simple batch distillation, Example 11-1.

3. Plot $1/(y - x)$ versus x.

4. Graphically integrate from x_F to $x_{W,final}$. This is shown in Figure 11-3.

5. From Eq. (11-7), find the final charge of material in the still pot:

$$W_{final} = F \exp \left(- \int_{x_{W,final}}^{x_F} \frac{dx}{y - x}\right) \qquad (11\text{-}8)$$

or

$$W_{final} = F e^{-Area} \qquad (11\text{-}9)$$

where the area is shown in Figure 11-3.

6. The average distillate concentration, $x_{D\,avg}$, can be found from the mass balances. Solving Eqs. (11-1) and (11-2),

$$x_{D,avg} = \frac{F x_F - W_{final}\, x_{W,final}}{F - W_{final}} \qquad (11\text{-}10)$$

$$D_{total} = F - W_{final} \qquad (11\text{-}11)$$

The Rayleigh equation can also be integrated numerically. One convenient method for doing this is to use Simpson's rule (e.g., see Mickley *et al.*, 1957, pp. 35 - 42). If the ordinate in Figure 11-3 is called $f(x)$, then one form of Simpson's rule is

$$\int_{x_{W,final}}^{x_F} f(x)dx = \frac{x_F - x_{W,final}}{6}[f(x_{W,final}) + 4f(\frac{x_{W,final}+x_F}{2}) + f(x_F)] \qquad (11\text{-}12)$$

where terms are shown in Figure 11-3. Simpson's rule is exact if $f(x)$ is cubic or lower order. For smooth curves such as in Figure 11-3, Simpson's rule will be quite accurate (see Example 11-1). For more complex shapes, Simpson's rule will be more accurate if the integration is done in two or more pieces (see Example 11-2). Other integration formulas that are more accurate can be used.

If the average distillate concentration is specified, a trial-and-error procedure is required. This involves guessing the final still pot concentration, $x_{W,final}$, and calculating the area in Figure 11-3 either graphically or with Simpson's rule. Then Eq. (11-9) gives W_{final} and Eq. (11-10) is used to check the value of $x_{D,avg}$. This trial-and-error procedure can be conveniently carried out by starting with a guess for $x_{W,final}$ that is too high. Then every time $x_{W,final}$ is decreased, the additional area is added to the area already calculated.

If the equilibrium expression is given as a constant relative volatility, α, the Rayleigh equation can be integrated analytically. In this case the equilibrium expression is

$$y = \frac{\alpha x}{1 + (\alpha - 1)x} \qquad (11\text{-}13)$$

Substituting Eq. (11-13) into Eq. (11-7) and integrating, we obtain

$$\ln\left(\frac{W_{final}}{F}\right) = \frac{1}{\alpha - 1} \ln\left(\frac{x_{W,final}(1 - x_F)}{x_F(1 - x_{W,final})}\right) + \ln\left(\frac{1 - x_F}{1 - x_{W,final}}\right) \qquad (11\text{-}14)$$

When it is applicable, Eq. (11-14) is obviously easier to apply than graphical or numerical integration.

Example 11-1. Simple Rayleigh Distillation

We wish to use a simple batch still (one equilibrium stage) to separate methanol from water. The feed charge to the still pot is 50 moles of an 80 mole % methanol mixture. We desire an average distillate concentration of 89.2 mole % methanol. Find the amount of distillate collected, the amount of material left in the still pot, and the concentration of material in the still pot. Pressure is 1 atm. Methanol-water equilibrium data at 1 atm are given in Table 3-3.

Solution

A. Define. The apparatus is shown in Figure 11-1. The conditions are: $p = 1$ atm, $F = 50$, $x_F = 0.80$, and $x_{D,avg} = 0.892$. We wish to find $x_{W,final}$, D_{tot}, and W_{final}.

B. Explore. Since the still pot acts as one equilibrium contact, the Rayleigh equation takes the form of Eqs. (11-7) to (11-9). To use these equations, either a plot of $1/(y-x)_{equil}$ versus x is required for graphical integration or Simpson's rule can be used. Both will be illustrated. Since $x_{W,final}$ is unknown, a trial-and-error procedure will be required for either integration routine.

C. Plan. First plot $1/(y-x)$ vs x from the equilibrium data. The trial-and-error procedure is as follows:

Guess $x_{W,final}$

Integrate to find Area $= \displaystyle\int_{x_{W,final}}^{x_F} \frac{dx}{y-x}$

Calculate W_{final} from Rayleigh equation and $x_{D,calc}$ from mass balance.
Check: Is $x_{D,calc} = x_{D,avg}$? If not, continue trial-and-error.

D. Do It. From the equilibrium data the following table is easily generated:

x	y	y − x	$\dfrac{1}{y-x}$
.8	.915	.115	8.69
.75	.895	.145	6.89
.70	.871	.171	5.85
.65	.845	.195	5.13
.60	.825	.225	4.44
.50	.780	.280	3.57

This is plotted in Figure 11-3. For the numerical solution a large graph on millimeter graph paper was constructed.

First guess: $x_{W,final} = 0.70$.

From Figure 11-3, Area $= \int_{x_{W,final}}^{x_F} \dfrac{dx}{y-x}$, which is $= 0.7044$.

Then $W_{final} = F \exp(-Area) = 50\, e^{-0.7044} = 24.72$

$$D_{calc} = F - W_{final} = 25.28$$

$$x_{D,calc} = \frac{Fx_F - W_{final}x_{W,final}}{D_{calc}} = 0.898$$

The alternative integration procedure using Simpson's rule gives

$$Area = \left(\frac{x_F - x_{W,final}}{6}\right)\left[\left(\frac{1}{y-x}\right)\Big|_{x_{W,final}}\right.$$

$$\left. + 4\left(\frac{1}{y-x}\right)\Big|_{(x_{W,final}+x_F)/2} + \left(\frac{1}{y-x}\right)\Big|_{x_F}\right]$$

$$Area = \left(\frac{0.1}{6}\right)[5.85 + 4(6.89) + 8.69] = 0.70166$$

Then for Simpson's rule, $W_{final} = 24.79$, $D_{calc} = 25.21$, and $x_{D,calc} = 0.898$. Thus Simpson's rule appears to be quite accurate. W_{final} is off by 0.3%, and $x_{D,calc}$ is the same as the more exact calculation. These values appear to be close to the desired value, but we don't know the sensitivity of the calculation.

Second guess: $x_{W,final} = 0.60$. Calculations similar to the first trial give

Area $= 1.2084$, $W_{final} = 14.93$, $D_{calc} = 35.07$, $x_{D,calc} = 0.885$ from Figure 11-3, and $x_{D,calc} = 0.884$ from the Simpson's rule calculation. These are also close, but they are low. The value of x_D is insensitive to $x_{W,final}$.

Third guess: $x_{W,final} = 0.65$. Calculations give

Area = 0.971, W_{final} = 18.94, D_{calc} = 31.06, $x_{D,calc}$ = 0.891 from Figure 11-3 and $x_{D,calc}$ = 0.890 from the Simpson's rule calculation of the area, which are both close to the specified value of 0.892.

Thus use $x_{W,final}$ = 0.65 as the answer.

E. Check. The overall mass balance should check. This gives: $W_{final} x_{W,final} + D_{calc} x_{D,calc}$ = 39.985 as compared to Fx_F = 40. Error is $(40-39.985)/40 \times 100$, or 0.038%, which is acceptable.

F. Generalize. The integration can also be done numerically on a computer using Simpson's rule or an alternative integration method. This is an advantage, since then the entire trial-and-error procedure can be programmed. Note that large differences in $x_{W,final}$ and hence in W_{final} cause rather small differences in $x_{D,avg}$. Thus, for this problem, exact control of the batch system may not be critical. This problem illustrates a common difficulty of simple batch distillation - a pure distillate and a pure bottoms product cannot be obtained unless the relative volatility is very large.

11.3. BATCH STEAM DISTILLATION

In batch steam distillation, steam is sparged directly into the still pot as shown in Figure 11-4. This is normally done for systems that are immiscible with water. The reasons for adding steam directly to the still pot are that it keeps the temperature below the boiling point of water, it eliminates the need for heat transfer surface area and it helps keep slurries and sludges well mixed so that they can be pumped. The major use is in treating wastes that contain valuable volatile organics.

Figure 11-4.   Batch steam distillation.

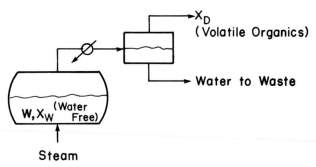

These waste streams are often slurries or sludges that would be difficult to process in an ordinary batch still. Compounds that are often steam distilled include glycerine, lube oils, fatty acids, and halogenated hydrocarbons (Woodland, 1978).

Batch steam distillation is usually operated with liquid water present in the still. Then both the liquid water and liquid organic phases exert their own partial pressure. Equilibrium is given by Eqs. (10-18) to (10-22) when there is one volatile organic and some nonvolatile organics present. As long as there is minimal entrainment, there is no advantage to having more than one stage. For low-molecular-weight organics, vaporization efficiencies

$$E = \frac{P_{volatile}}{P^*_{volatile}} = \frac{P_{volatile}}{(VP)_{volatile} \, x_{volatile}} \tag{11-15}$$

are often in the range from 0.9 to 0.95 (Carey, 1950). This efficiency is close enough to equilibrium that equilibrium calculations are adequate.

The system shown in Figure 11-4 can be analyzed with mass balances on a water-free basis. The mass balances are Eqs. (11-1) and (11-2), which can be solved for W_{final} if $x_{W,final}$ is given.

$$W_{final} = F \left(\frac{x_D - z}{x_D - x_{W,final}} \right) \tag{11-16}$$

With a single volatile organic, $x_D = 1.0$ if entrainment is negligible. Then,

$$W_{final} = F \left(\frac{1 - z}{1 - x_{W,final}} \right) \tag{11-17}$$

and the flow rate of the organic distillate product is

$$D = F - W_{final} \tag{11-18}$$

The Rayleigh equation can also be used and will give the same results.

At any moment the instantaneous moles of water dn_w carried over in the vapor can be found from Eq. (10-22). This becomes

$$dn_w = dn_{org} \frac{P_{tot} - (VP)_{volatile} x_{volatile \, in \, org}}{(VP)_{volatile} x_{volatile \, in \, org}} \tag{11-19}$$

The total moles of water carried over in the vapor can be obtained by integrating this equation:

$$n_w = \int_0^D \frac{P_{tot} - (VP)_{\text{volatile}} \, x_{\text{volatile in org}}}{(VP)_{\text{volatile}} \, x_{\text{volatile in org}}} \tag{11-20}$$

During the batch steam distillation, the mole fraction of the volatile organics in the still varies, and thus the still temperature determined by Eq. (10-20) varies. Equation (11-20) can be integrated numerically in steps. The total moles of water required is n_w plus the moles of water condensed to heat the feed and vaporize the volatile organics.

11.4. MULTISTAGE BATCH DISTILLATION

The separation achieved in a single equilibrium stage is often not large enough to both obtain the desired distillate concentration and a low enough bottoms concentration. In this case a distillation column is placed above the reboiler as shown in Figure 11-2. The calculation procedure will be detailed here for a staged column, but packed columns can easily be designed using the procedures explained in Chapter 13.

For multistage systems x_D and x_W are no longer in equilibrium. Thus the Rayleigh equation, Eq. (11-6), cannot be integrated until a relationship between x_D and x_W is found. This relationship can be obtained from stage-by-stage calculations. We will assume that there is negligible holdup on each plate, in the condenser, and in the accumulator. Then at any specific time we can write mass and energy balances around stage j and the top of the column as shown in Figure 11-2. These balances simplify to

$$\text{Input} = \text{output}$$

since accumulation was assumed to be negligible everywhere except the reboiler. Thus, at a given time t,

$$V_{j+1} = L_j + D \tag{11-21}$$

$$V_{j+1}y_{j+1} = L_jx_j + Dx_D \tag{11-22}$$

$$Q_c + V_{j+1}H_{j+1} = L_jh_j + Dh_D \tag{11-23}$$

In these equations V, L and D are now molal flow rates. These balances are essentially the same equations we obtained for the rectifying section of a continuous column except that Eqs. (11-21) to (11-23) are time-dependent. If we can assume constant molal overflow, the vapor and liquid flow rates will be constant and the energy balance is not needed. Combining Eqs. (11-21) and (11-22) and solving for y_{j+1}, we obtain the operating equation for constant molal overflow:

$$y_{j+1} = \frac{L}{V} x_j + \left(1 - \frac{L}{V}\right)x_D \qquad (11\text{-}24)$$

At any specific time Eq. (11-24) represents a straight line on a y-x diagram. The slope will be L/V, and the intercept with the $y = x$ line will be x_D. Since either x_D or L/V will have to vary during the batch distillation, the operating line will be continuously changing.

The most common operating method is to use a constant reflux ratio and allow x_D to vary. This procedure corresponds to simple batch operation where x_D also varies. The relationship between x_D and x_W can now be found from a stage-by-stage calculation using a McCabe-Thiele analysis. Operating equation (11-24) is plotted on a McCabe-Thiele diagram for a series of x_D values. Then we step off the specified number of equilibrium contacts on each operating line starting at x_D to find the x_W value corresponding to that x_D. This procedure is shown in Figure 11-5 and Example 11-2.

The McCabe-Thiele analysis gives x_W values for a series of x_D values. We can now calculate $1/(x_D - x_W)$. The integral in Eq. (11-6) can be determined by either numerical integration such as Simpson's rule given in Eq. (11-12) or by graphical integration. Now the same procedure used for simple batch distillation can be used. Thus W_{final} is found from Eq. (11-9), $x_{D\ avg}$ from Eq. (11-10), and D_{total} from Eq. (11-11). If $x_{D\ avg}$ is specified, a trial-and-error procedure will again be required.

Example 11-2. Multistage Batch Distillation

We wish to batch distill 50 kg moles of a 32 mole % ethanol, 68 mole % water feed. The system has a still pot plus two equilibrium stages and a total condenser. Reflux is returned as a saturated liquid, and we use $L/D = 2/3$. We desire a final still pot composition of 4.5 mole % ethanol. Find the average distillate composition, the final charge in the still pot, and the amount of distillate collected. Pressure is 1 atm.

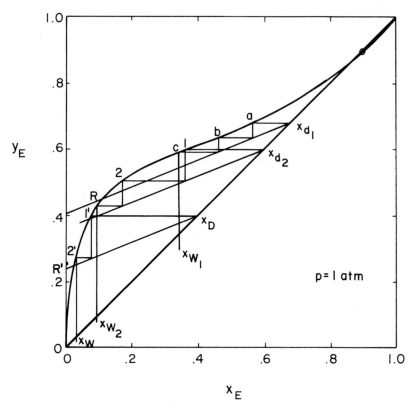

Figure 11-5.   McCabe-Thiele diagram for multistage batch distillation with constant L/D, Example 11-2.

Solution

A. Define. The system is shown in the figure.

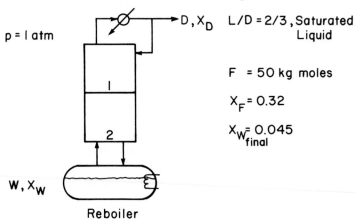

Find W_{final}, D_{total}, $x_{D,avg}$.

B. Explore. Since we can assume CMO, a McCabe-Thiele diagram (Figure 2-2) can be used. This will relate x_D to x_W at any time. Since x_F and $x_{W,final}$ are known, the Rayleigh equation (11-6) or (11-9) can be used to determine W_{final}. Then $x_{D,avg}$ and D_{avg} can be determined from Eqs. (11-10) and (11-11), respectively. A trial-and-error procedure is not needed for this problem.

C. Plan. Worked out during the Explore step.

D. Do It. The McCabe-Thiele diagram for several arbitrary values of x_D is shown in Figure 11-5. The top operating line is

$$y = \frac{L}{V} x + \left(1 - \frac{L}{V}\right) x_D$$

where

$$\frac{L}{V} = \frac{L/D}{1 + L/D} = \frac{2/3}{5/3} = \frac{2}{5}$$

The corresponding x_W and x_D values are used to calculate $x_D - x_W$ and then $1/(x_D - x_W)$ for each x_W value. These values are plotted in Figure 11-6 (some values not shown in Figure 11-5 are shown in Figure 11-6). The area under the curve (going down to an ordinate value of zero) from $x_F = 0.32$ to $x_{W,final} = 0.045$ is 0.608 by graphical integration.

Then from Eq. (11-9):

$$W_{final} = Fe^{-Area} = (50) \exp(-0.608) = 27.21$$

From Eq. (11-11): $\quad D_{total} = F - W_{final} = 22.79$

and from Eq. (11-10):

$$x_{D,avg} = \frac{Fx_F - W,final\ x_{W,final}}{F - W_{final}} = 0.648$$

The area can also be determined by Simpson's rule. However, because of the shape of the curve in Figure 11-6 it will probably be less accurate than in Example 11-1. Simpson's rule gives

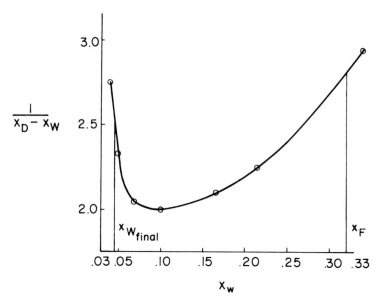

Figure 11-6.   Graphical integration, Example 11-2.

$$\text{Area} = \left(\frac{x_F - x_{W,f}}{6} \right) \left[\left(\frac{1}{x_D - x_w} \right) \Big| x_{W,f} \right.$$

$$+ \; 4\left(\frac{1}{x_D - x_w} \right) \Big| \frac{(x_{W,f} + x_F)}{2} + \left(\frac{1}{x_D - x_w} \right) \Big| x_F \Bigg]$$

where $(x_{W,\text{final}} + x_F)/2 = 0.1825$ and $1/(x_D - x_W) = 2.14$ at this midpoint.

$$\text{Area} = \left(\frac{0.275}{6} \right)[2.51 + 4(2.14) + 2.82] = 0.6366$$

This can be checked by breaking the area into two parts and using Simpson's rule for each part. Do one part from $x_{W,\text{final}} = 0.045$ to $x_W = 0.10$ and the other part from 0.1 to $x_F = 0.32$. Each of the two parts should be relatively easy to fit with a cubic. Then,

$$\text{Area part 1} = \left(\frac{0.10 - 0.045}{6} \right)[2.51 + 4(2.03) + 2.00] = 0.1158$$

$$\text{Area part 2} = (\frac{0.32 - 0.10}{6})[2.00 + 4(2.23) + 2.82] = 0.5038$$

Total area = 0.6196

Note that Figure 11-6 is very useful for finding the values of $1/(x_D - x_W)$ at the intermediate points $x_W = 0.0725$ and $x_W = 0.21$. The total area calculated is closer to the answer obtained graphically (1.9% difference compared to 4.7% difference for the first estimate).

Then, doing the same calculations as previously [Eqs. (11-9), (11-11), and (11-10)] with Area = 0.6196,

$$W_{final} = 26.91 , \quad D_{total} = 23.09 , \quad x_{D,avg} = 0.640$$

E. Check. The mass balances for an entire cycle, Eqs. (11-1) and (11-2), should be and are satisfied. Since the graphical integration and Simpson's rule give similar results, this is another reassurance.

F. Generalize. Note that we did not need to find the exact value of x_D for x_F or $x_{W,final}$. We just made sure that our calculated values went beyond these values. This is true for both integration methods. Our axes in Figure 11-6 were selected to give maximum accuracy; thus we did not need to graph parts of the diagram that we didn't use. For more accuracy, Figure 11-5 should be expanded. Note that the graph in Figure 11-6 is very useful for interpolation to find values for Simpson's rule. If Simpson's rule is to be used for very sharply changing curves, accuracy will be better if the curve is split into two or more parts. Comparison of the results obtained with graphical integration to those obtained with the two-part integration with Simpson's rule shows a difference in $x_{D,avg}$ of 0.008. This is within the accuracy of the equilibrium data.

The batch distillation column can also be operated with variable reflux ratio to keep x_D constant. The operating equation (11-24) is still valid. Now the slope will vary, but the intersection with the y=x line will be constant at x_D. The McCabe-Thiele diagram for this case is shown in Figure 11-7. This diagram relates x_W to x_D, and the Rayleigh equation can be integrated as in Figure 11-6. Since x_D is kept constant, the calculation procedure is somewhat different.

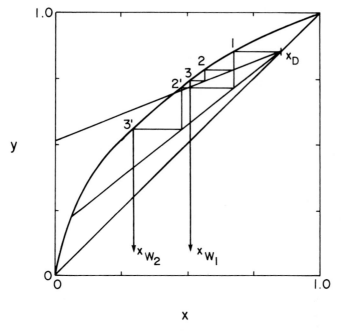

Figure 11-7.   McCabe-Thiele diagram for multistage batch distillation with constant x_D.

With x_D and the number of stages specified, the initial value of L/V is found by trial and error to give the feed concentration x_F. The final value $x_{W,final}$ occurs when L/V equals $(L/V)_{max}$ or $L/V = 1.0$, which is total reflux. $(L/V)_{max}$ can be determined from the maximum Q_R or maximum $|Q_c|$ if D is constant; or from the minimum D, which is acceptable if Q_R and $|Q_c|$ are constant; or from the maximum acceptable operating time. Once $x_{W,final}$ is determined, W_{final} is found from Eqs. (11-8) or (11-9) by graphical integration.

If the assumption of negligible holdup is not valid, then the holdup on each stage and in the accumulator acts like a flywheel and retards changes. A different calculational procedure is required for this case and for multicomponent systems (Holland and Liapis, 1983; Robinson, 1969; Stewart *et al.*, 1973; Sadotomo and Miyahara, 1983). If constant molal overflow is not valid, the entire analysis can easily be done in latent heat units.

11.5. OPERATING TIME

The operating time and batch size may be controlled by economics or other factors. For instance, it is not uncommon for the entire batch including startup and shutdown to be done in one 8-hour shift. If the same apparatus is used for several different chemicals, the batch sizes may vary. Also the time to change over from one chemical to another may be quite long, since a rigorous cleaning procedure may be required.

The total batch time, t_{batch}, is

$$t_{batch} = t_{down} + t_{op} \tag{11-25}$$

The down time, t_{down}, includes dumping the bottoms, cleanup, loading the next batch, and heating the next batch until reflux starts to appear. This time can be estimated from experience. The operating time, t_{op}, is the actual period during which distillation occurs, so it must be equal to the total amount of distillate collected divided by the distillate flow rate.

$$t_{op} = \frac{D_{total}}{D} \tag{11-26}$$

D_{total} is calculated from the Rayleigh equation calculation procedure, with F set either by the size of the still pot or by the charge size. For an existing apparatus the distillate flow rate cannot be set arbitrarily. The column was designed for a given maximum vapor velocity, u_{flood}, which corresponds to a maximum molal flow rate, V_{max}. Then, from the mass balance around the condenser,

$$D_{max} = \frac{V_{max}}{1 + \dfrac{L}{D}} \tag{11-27}$$

We usually operate at some fraction of this flow rate such as $D = 0.75$ D_{max}.

V_{max} can be calculated using the design procedures used for diameter calculations (see Chapter 12). Then Eqs. (11-26) and (11-27) can be used to estimate t_{op}. If the resulting t_{batch} is not convenient, adjustments must be made.

Batch distillation has somewhat different design and process control requirements than continuous distillation. In addition, startup and troubleshooting are somewhat different. These aspects are discussed by Ellerbe (1979).

11.6. SUMMARY - OBJECTIVES

In this chapter we have explored binary batch distillation calculations. At this time you should be able to satisfy the following objectives:

1. Explain the operation of simple and multistage batch distillation systems.

2. Discuss the differences between batch and continuous operation.

3. Derive and use the Rayleigh equation for simple batch distillation.

4. Solve problems in batch steam distillation.

5. Use the McCabe-Thiele method to analyze multistage batch distillation for:

   a. Batch distillation with constant reflux ratio

   b. Batch distillation with constant distillate composition

   c. Inverted batch distillation

6. Determine the operating time for a batch distillation.

7. Develop computer flow charts and programs for simple and multistage batch distillation.

REFERENCES

Carey, J.S., "Distillation" in J.H. Perry (Ed.), *Chemical Engineer's Handbook,* 3rd ed., McGraw-Hill, New York, 1950, pp. 582-585.

Ellerbe, R.W., "Batch Distillation," in P.A. Schweitzer (Ed.), *Handbook of Separation Techniques for Chemical Engineers,* McGraw-Hill, New York, 1979, p. 1.147.

Holland, C.D. and Liapis, A.I., *Computer Methods for Solving Dynamic Separation Problems,* McGraw-Hill, New York, 1983, Chapt. 5.

Luyben, W.L., "Some Practical Aspects of Optimal Batch Distillation," *Ind. Eng. Chem. Process Des. Develop., 10,* 54 (1971).

Mickley, H.S., Sherwood, T.K. and Reed, C.E., *Applied Mathematics in Chemical Engineering,* McGraw-Hill, New York, 1957, pp. 35-42.

Pratt, H.R.C., *Countercurrent Separation Processes*, Elsevier, New York, 1967.

Rayleigh, Lord, *Phil. Mag.* [vi], *4* (23), 521 (1902).

Robinson, C.S. and E.R. Gilliland, *Elements of Fractional Distillation*, 4th ed., McGraw-Hill, New York, 1950, Chapts. 6 and 16.

Robinson, E.R., "Review of Batch Distillation," *Chem. and Process Eng.*, *40*, 83 (Dec. 1969).

Sadotomo, H. and K. Miyahara, "Calculation Procedure for Multicomponent Batch Distillation," *Int. Chem. Eng.*, *23*, 56 (Jan. 1983).

Stewart, R.R., E. Weisman, B.M. Goodwin, and C.E. Speight, "Effect of Design Parameters in Multicomponent Batch Distillation," *Ind. Eng. Chem. Process Des. Develop.*, *12*, 130 (1973).

Treybal, R.E., "A Simple Method for Batch Distillation," *Chem. Eng.*, *77* (21), 95 (Oct. 5, 1970).

Woodland, L.R., "Steam Distillation," in D.J. DeRenzo (Ed.), *Unit Operations for Treatment of Hazardous Industrial Wastes*, Noyes Data Corp., Park Ridge, NJ, 1978, pp. 849-868.

HOMEWORK

A. *Discussion Problems*

A1. Batch distillation tends to be labor-intensive. Discuss how you could automate each step of the batch process.

A2. In the derivation of the Rayleigh equation:

   a. In Eq. (11-3), why do we have $-x_D \, dW$ instead of $-x_D \, dD$?

   b. In Eq. (11-3), why is the left-hand side $-x_D \, dW$ instead of $-d(x_D W)$?

A3. Explain how the graphical integration shown in Figures 11-3 and 11-5 could be done numerically on the computer.

A4. Discuss the advantages and disadvantages of batch distillation compared to continuous distillation. Would you expect to see batch or continuous distillation in the following industries? Why?

a. Large basic chemical plant

b. Plant producing fine chemicals

c. Petroleum refinery

d. Petrochemical plant

e. Pharmaceutical plant

f. Cryogenic plant for recovering oxygen from air

g. Still for solvent recovery in a painting operation

h. Ethanol production facilities for a farmer

Note that continuous distillation is often used in *campaigns*. That is, the plant produces a product continuously, but for a relatively short period of time. How does this change your answers?

A5. Assuming that either batch or continuous distillation could be used, which would use less energy? Explain why.

A6. If there is holdup on the trays, how would Eqs. (11-21) to (11-23) be changed? Qualitatively, what will this do to the batch distillation?

A7. Develop a key relations chart for this chapter.

B. *Generation of Alternatives*

B1. List all the different ways a binary batch or inverted batch problem can be specified. Which of these will be trial-and-error?

B2. What can be done if an existing batch system cannot produce the desired values of x_D and x_W even at total reflux? Generate ideas for both operating and equipment changes.

C. *Derivations*

C1. Derive Eq. (11-14).

C2. Assume that holdup in the column and in the total reboiler is negligible in an inverted batch distillation. See sketch in Problem 11-E2.

a. Derive the appropriate form of the Rayleigh equation.

b. Derive the necessary operating equations for constant molal overflow. Sketch the McCabe-Thiele diagrams.

C3. Equation (11-3) could be written as

$$d(x_D \, dW) = D(W x_W) \qquad (11\text{-}3a)$$

a. Explain this equation with reference to the basic mass balance.

b. Expand Eq. (11-3a) and determine what assumptions are necessary to obtain Eqs. (11-3) and (11-4).

D. *Problems*

D1. A mixture of benzene and toluene is to be batch distilled in a simple batch still. $F = 2$ kg mole, $x_F = 0.55$ mole frac benzene. Relative volatility is $\alpha_{BT} = 2.5$.

a. For $x_{W,final} = 0.2$, find $x_{D,avg}$.

b. For $x_{D,avg} = 0.7$, find $x_{W,final}$ and W_{final}.

Hint. Use Eq. (11-14) - it's easier.

D2. We wish to use a simple batch still (one equilibrium stage) to separate methanol from water. The feed charge to the still pot is 100 moles of a 75 mole % methanol mixture. We desire a final bottoms concentration of 55 mole % methanol. Find the amount of distillate collected, the amount of material left in the still pot, and the average concentration of distillate. Pressure is 1 atm. Equilibrium data are given in Table 3-3.

D3. We wish to use a distillation system with a still pot plus a column with one equilibrium stage to batch distill a mixture of methanol and water. A total condenser is used. The feed is 57 mole % methanol. We desire a final bottoms concentration of 15 mole % methanol. Pressure is 101.3 kPa. Reflux is a saturated liquid, and L_0/D is constant at 1.85. Find W_{final}, D, and $x_{D,avg}$. Methanol-water equilibrium data is given in Table 3-3. Calculate on the basis of 1 kg mole of feed.

D4. We wish to do a normal batch distillation of methanol and water. The system has a still pot that acts as an equilibrium stage and a column with two equilibrium stages (total of three equilibrium contacts). The column has a total condenser, and reflux is returned as a saturated liquid. The column is operated with a varying reflux ratio so that x_D is held constant. The initial charge is $F = 10$ kg moles and is 40 mole % methanol. We desire a final still-pot concentration of 8 mole % methanol, and the distillate concentration should be 85 mole % methanol. Pressure is 1 atm and constant molal overflow is valid. Equilibrium data is given in Table 3-3.

a. What initial external reflux ratio, L_0/D, must be used?

b. What final external reflux ratio must be used?

c. How much distillate product is withdrawn, and what is the final amount, W_{final}, left in the still pot?

D5. Write a computer or calculator program to solve Problem 11-D1 using Eq. (11-14). Try varying α_{BT} from 2.20 to 2.60 to see how sensitive this problem is to the value of relative volatility used.

D6. Write a computer or calculator program for the following problem. 100 moles of a 40 mole % mixture of benzene and toluene is to be batch distilled. Assuming that the equilibrium relation is adequately described by $\alpha_{BT} = 2.5$, calculate the amount of material to be distilled if an average distillate concentration of 85 mole % benzene is desired. Also find final bottoms concentration. Do this for a column with three stages plus the still pot with $L/V = 0.7$ and 0.9. Column pressure is 1 atm. Check your program.

D7. We wish to batch distill a mixture of 1-butanol and water. Since this system has a heterogeneous azeotrope (see Chapter 10), we will use the system shown in the figure. The liquid with 97.5 mole % water is removed as product, and the liquid layer, which is 57.3 mole % water, is returned to the still pot. Pressure is 1 atm. The feed is 20 kg moles and is 40 mole % water. Data are given in Problem 10-D2.

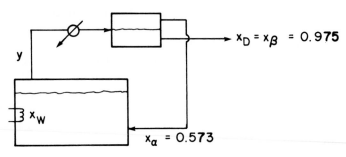

    a. If the final concentration in the still pot should be $x_{W,final} = 0.28$, what is the final amount of distillate D collected, in kg moles?

    b. If the batch distillation is continued what is the lowest value of $x_{W,final}$ that can be obtained while still producing a distillate $x_D = 0.975$?

D8.   A simple steam distillation is being done in the apparatus shown in Figure 10-4. The organic feed is 90 mole% n-decane and 10% nonvolatile organics. The system is operated with liquid water in the still. Distillation is continued until the organic layer in the still is 10 mole % n-decane. F = 10 kg moles. Pressure is 760 mmHg.

    a. At the final time, what is the temperature in the still?

    b. What is W_{final}? What is D?

    c. Estimate the moles of water passed overhead per mole of n-decane at the end of the distillation.

Data: Assume that water and n-decane are completely immiscible. Vapor pressure data for nC_{10} is in Example 10-2. Vapor pressure data for water are listed in Problem 10-D6.

E.   *More Complex Problems*

E1.   We wish to batch distill a mixture of water and acetic acid. The feed is 40 mole % water. We desire an average distillate composition of 70 mole % water and a final bottoms composition of 28.5 mole % water. There are four equilibrium stages in the column, and the still pot also acts as an equilibrium contact. Feed charge is 100 kg moles. Pressure is 1 atm. What constant internal reflux ratio, L/V, is required? Equilibrium data are given in Problem 6-E3. Remember to use latent heat units.

E2.   We wish to batch distill an acetone-water mixture at 1 atm. The batch size is 10 kg moles, and the feed is 70 mole % acetone. An inverted batch distillation system will be used (see figure). The condenser is a total condenser, and the column has three equilibrium stages. The boilup ratio is constant at $\overline{V}/B = 0.3$. We desire an average bottoms concentration of $x_{B,avg} = 0.24$ mole fraction acetone. Assume that holdup in the column and in the total reboiler is negligible. Find amount of bottoms, B, amount of distillate, D_{final}, and final distillate composition $x_{D,final}$. Use latent heat units.

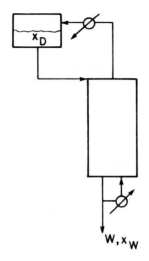

Data for Acetone-Water Vapor-Liquid Equilibrium
(Values are mole fraction acetone)

x	0.0100	0.0230	0.041	0.120	0.264	0.300	0.444
y		0.355	0.462	0.585	0.756	0.802	0.809	0.832
T, °C	87.8	83.0	76.5	66.2	61.8	61.1	60.0

x	0.506	0.538	0.609	0.661	0.793	0.850	1.0
y	0.837	0.840	0.847	0.860	0.900	0.917	1.0
T, °C	59.7	59.5	58.9	58.5	57.4	57.1	

$$\lambda_w = 9.171 \ \frac{kcal}{g \ mole} \ , \qquad MW_w = 18.016$$

$$\lambda_a = 124.4 \ \frac{cal}{g} \ , \qquad MW_a = 58.08$$

F.   *Problems Requiring Other Resources*

F1.   Read Treybal's (1970) article. Compare that method of doing
batch distillation with the more typical method. Write a one-
page report critiquing Treybal's method and recommending either
it or the standard operating procedure.

G. *Open-Ended or Synthesis Problems*

G1. About once every 2 weeks your plant has batches of a methanol-water mixture to distill. The batches vary in size from about 450 to 1025 kg. They range in concentration from 32 to 61 wt % methanol. For reuse, the methanol must be 95 wt % purity or higher. The waste will be sent to the sewer. The sewage charge is proportional to flow rate and to the average concentration of organics (methanol). The plant runs from 8:00 am to 4:30 pm, Monday through Friday. Design a batch distillation system for this problem.

chapter 12
STAGED COLUMN DESIGN

In previous chapters we saw how to determine the number of equilibrium stages and the separation in distillation columns. In this chapter we will discuss the details of design such as tray geometry, determination of column efficiency, calculation of column diameter, downcomer sizing, and tray layout. In Chapter 14 these design details will be converted into economics. We will start with a qualitative description of column internals and then proceed to a quantitative description of efficiency prediction, determination of column diameter, sieve tray design, and valve tray design. Determinations of efficiency and column diameter are required for determining the economics, and thus they are necessary for optimization studies. New engineers are expected to be able to do these calculations. The internals of distillation columns are usually designed under the supervision of experts with many years of experience.

This chapter is not a shortcut to becoming an expert. However, upon completion of this chapter you should be able to finish a preliminary design of the column internals, and you should be able to discuss distillation designs intelligently with the experts.

12.1. STAGED COLUMN EQUIPMENT DESCRIPTION

A very basic picture of staged column equipment was presented in Chapter 4. In this section, a much more detailed qualitative picture will be presented. Much of the material included here is from the series of articles by Kister (1980, 1981), the book by Ludwig (1979), and the review by Zenz (1979). These sources should be consulted for more details.

Sieve trays, which were illustrated in Figure 4-7, are easy to manufacture and are inexpensive. The holes are punched or drilled (a more expensive process) in the metal plate. Considerable design information is available, and since the designs are not proprietary anyone can build a sieve tray column. The efficiency is good at design conditions. However, turndown (the performance when operating below the designed flow rate) is relatively poor. This means that operation at significantly lower rates than the design condition will result in low efficiencies. For sieve plates, efficiency drops markedly for gas flow rates that are less than about 60% of the design value. Thus these trays are not extremely flexible. Sieve trays are very good in fouling applications or when there are solids present, because they are highly resistant to clogging, they can have large holes, and they are easy to clean. Sieve trays are a standard item in industry, but new columns are more likely to have valve trays.

Valve trays are designed to have better turndown properties than sieve trays, and thus they are more flexible when the feed rate varies. There are many different proprietary valve tray designs, of which one type is illustrated in Figure 12-1. The valve tray is similar to a sieve tray in that it has a deck with holes in it for gas flow and downcomers for liquid flow. The difference is that the holes, which are quite large, are fitted with "valves," flat covers that can move up and down as the pressures of the vapor and the liquid change. Each valve has feet or a cage that restrict its upward movement. At high vapor velocities, the valve will be fully open, which provides a maximum slot for gas flow (see Figure 12-1). When the gas velocity drops, the valve will drop. This keeps the gas velocity through the slot close to constant, which keeps efficiency close to constant and prevents weeping. An individual valve is stable only in the fully closed or fully open position. At intermediate velocities some of the valves on the tray will be open and some will be closed. Usually, the valves alternate between the open and closed positions. The Venturi valve has the lip of the hole facing upwards to produce a Venturi opening, which will minimize the pressure drop.

At the design vapor rate, valve trays have about the same efficiency as sieve trays. However, their turndown characteristics are generally better, and the efficiency remains high as the gas rate drops. They can also be designed to have a lower pressure drop than sieve trays, although the standard valve tray will have a higher pressure drop. The disadvantages of valve trays are they are about 20% more expensive than sieve trays (Glitsch, 1985) and they are more likely to foul or plug if dirty solutions are distilled.

Bubble cap trays are illustrated in Figure 12-2. In a bubble cap there is a riser or weir around each hole in the tray. A cap with slots or holes is placed over this riser, and the vapor bubbles through these holes. This design is quite flexible and will operate satisfactorily at very high and very low liquid flow rates. However, entrainment is about three times that of a sieve tray, and there is usually a significant liquid

A

B

Figure 12-1. (A) Valve assembly for Glitsch A-1 valve. (B) Small Glitsch A-1 Ballast tray. Courtesy of Glitsch, Inc., Dallas, TX.

gradient across the tray. The net result is that tray spacing must be significantly greater than for sieve trays. Average tray spacing in small columns is about 18 inches, while 24 to 36 inches is used for vacuum distillations. Efficiencies are usually the same or less than for sieve trays, and turndown characteristics are often worse. The bubble-cap has prob-

Figure 12-2.   Different bubble-cap designs made by Glitsch, Inc. Courtesy of Glitsch, Inc., Dallas, TX.

lems with coking, polymer formation, or high fouling mixtures. Bubble cap trays are approximately four times as expensive as valve trays (Glitsch, 1985). Very few new bubble-cap columns are being built. However, new engineers are likely to see older bubble-cap columns still operating. Lots of data are available for the design of bubble-cap trays. Excellent discussions on the design of bubble-cap columns are available (for example, Bolles, 1963; Ludwig, 1979). More details will not be given here.

Perforated plates without downcomers look like sieve plates but with significantly larger holes. The plate is designed so that liquid weeps through the holes at the same time that vapor is passing through the center of the hole. The advantage of this design is that the cost and space associated with downcomers are eliminated. Its major disadvantage is that it is not robust. That is, if something goes wrong the column may not work at all instead of operating at a lower efficiency. These columns are usually designed by the company selling the system. Some design details are presented by Ludwig (1979).

12.1.1. Trays, Downcomers and Weirs

In addition to choosing the type of tray, the designer must select the flow pattern on the trays and design the weirs and downcomers. This section will continue to be mainly qualitative.

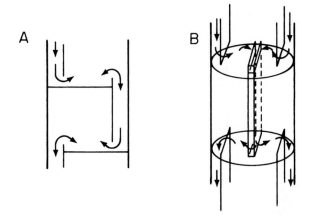

Figure 12-3.   Flow patterns on trays. (A) Cross flow; (B) double pass.

The most common flow pattern on a tray is the cross-flow pattern shown in Figure 4-7 and repeated in Figure 12-3A. This pattern works well for average flow rates and can be designed to handle suspended solids in the feed. Cross-flow trays can be designed by the user on the basis of information in the open literature (Bolles, 1963; Fair, 1963, 1984, 1985; Kister, 1980, 1981; Ludwig, 1979), from information in company design manuals (Glitsch, 1974; Koch, 1982), or from any of the manufacturers of staged distillation columns. Design details for cross-flow trays are discussed later.

Multiple-pass trays are used in large-diameter columns with high liquid flow rates. A double-pass tray is shown in Figure 12-3B. The liquid flow is divided into two sections (or passes) to reduce the liquid gradient on the tray and to reduce the downcomer loading. With even larger liquid loadings, three- or four-pass trays are used. This type of tray is also usually designed by experts, although preliminary designs can be obtained by following the design manuals published by some of the equipment manufacturers (Glitsch, 1974; Koch, 1982).

Which flow pattern is appropriate for a given problem? As the gas and liquid rates increase, the tower diameter increases. However, the ability to handle liquid flow increases with weir length, while the gas flow capacity increases with the square of the tower diameter. Thus, eventually multiple-pass trays are required. A selection guide is given in Figure 12-4 (Huang and Hodson, 1958), but it is only approximate, particularly near the lines separating different types of trays.

Downcomers and weirs are very important for the proper operation of staged columns, since they control the liquid distribution and flow. A

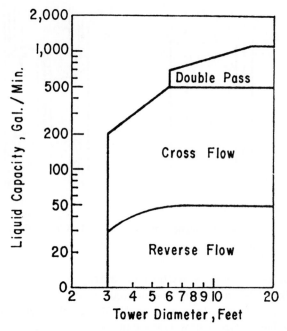

Figure 12-4.   Selection guide for sieve trays. Reprinted with permission from Huang and Hodson, *Petroleum Refiner, 37* (2), 104 (1958). Copyright 1958, Gulf Pub. Co.

Figure 12-5.   Downcomer and weir designs. (A) Circular pipe; (B) straight segmental; (C) sloped downcomers; (D) envelope.

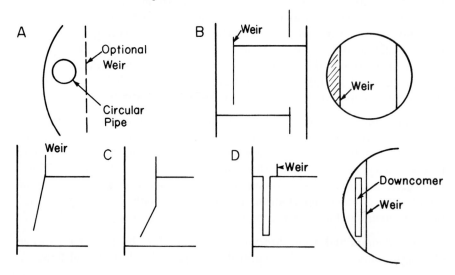

ety of designs are used, four are shown in Figure 12-5 (Kister,)e). In small columns and pilot plants the circular pipe shown in ire 12-5A is commonly used. The pipe may stick out above the tray r to serve as the weir, or a separate weir may be used. In the ershaw design commonly used in pilot plants, the pipe is in the er of the sieve plate and is surrounded by holes. The most common gn in commercial columns is the segmented vertical downcomer vn in Figure 12-5B. This type is inexpensive to build, easy to ill, almost impossible to install incorrectly, and can be designed for a variety of liquid flow rates. If liquid-vapor disengagement is cult, one of the sloped segmental designs shown in Figure 12-5C can iseful. These designs help retain the active area of the tray below. irtunately, they are more expensive and are easy to install back-ds. For very low liquid flow rates, the envelope design shown in ire 12-5D is occasionally used.

The simplest weir design is the straight horizontal weir shown in Fig-4-7 and 12-5. This type is the cheapest but does not have the best down properties. The adjustable weir shown in Figure 12-6A is a seductive design, since it appears to solve the problem of turndown. irtunately, if maladjusted, this weir can cause lots of problems such xcessive weeping or trays running dry, so it should probably be ded. When flexibility in liquid rates is desired, one of the notched s shown in Figure 12-6B will work well; they are not much more nsive than a straight weir. Notched weirs are particularly useful low liquid flow rates.

Trays, weirs, and downcomers need to be mechanically supported. is illustrated in Figure 12-7 (Zenz, 1979). The trick is to ade-ely support the weight of the tray plus the highest possible liquid ing it can have without excessively blocking either the vapor flow or the active area on the tray. As the column diameter increases, support becomes more critical. See Ludwig (1979), Kister (1980d), enz (1979) for more details.

re 12-6.   Weir designs. (A) Adjustable; (B) notched.

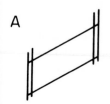

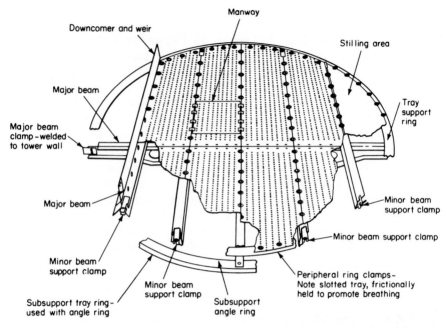

Figure 12-7.   Mechanical supports for sieve trays. From Zenz (1979).
Reprinted with permission from Schweitzer, *Handbook
of Separation Techniques for Chemical Engineers.*
Copyright 1979, McGraw-Hill, New York.

12.1.2. Inlets and Outlets

Inlet and outlet ports must be carefully designed to prevent problems
(Kister, 1980a,b; Glitsch, 1985). Inlets should be designed to avoid both
excessive weeping and entrainment when a high-velocity stream is
added. Several acceptable designs for a feed or reflux to the top tray
are shown in Figure 12-8. The baffle plate or pipe elbow prevents high-
velocity fluid from shooting across the tray. The designs shown in Figures 12-8D and E can be used if there is likely to be vapor in the feed.
These two designs will not allow excessive entrainment. Intermediate
feed introduction is somewhat similar, and several common designs are
shown in Figure 12-9. Low-velocity liquid feeds can be input through
the side of the column as shown in Figure 12-9A. Higher velocity feeds
and feed containing vapor require baffles as shown in Figures 12-9B and
C. The vapor is directed sideways or downwards to prevent excessive
entrainment. When there is a large quantity of vapor in the feed, the
feed tray should have extra space for disengagement of liquid and vapor.
For large diameter columns some type of distributor such as the one
shown in Figure 12-9D is often used.

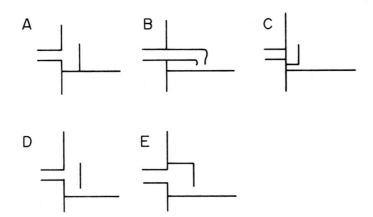

Figure 12-8.   Inlets for reflux or feed to top tray.

The vapor return at the bottom of the column should be at least 12 in above the liquid surge level. The vapor inlet should be parallel to the seal pan and parallel to the liquid surface as shown in Figure 12-10A. The purpose of the seal pan is to keep liquid in the downcomer. The vapor inlet should *not* el-down to impinge on the liquid as shown in Figure 12-10B. When a thermosiphon reboiler (a common type of total reboiler) is used, the split drawoff shown in Figure 12-10C is useful.

Figure 12-9.   Intermediate feed systems.   (A)   Side inlet;   (B, C) baffles; (D) distributor.

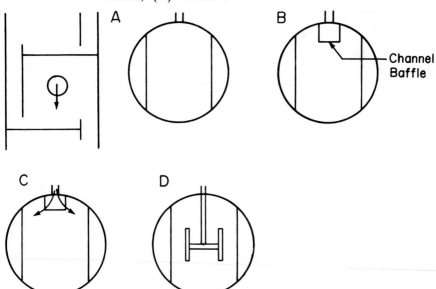

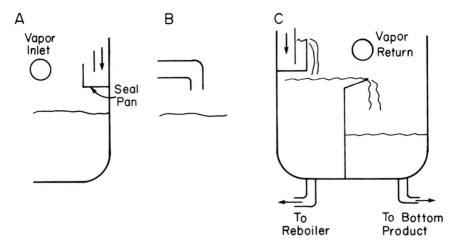

Figure 12-10.   Bottom vapor inlet and liquid drawoffs. (A) Correct. Inlet vapor parallel to seal pan. (B) Incorrect. Inlet vapor el-down onto liquid. (C) Bottom draw-off with thermosiphon reboiler.

Intermediate liquid drawoffs require some method for disengaging the liquid and vapor. The cheapest way to do this is with a downcomer tapout as shown in Figure 12-11A. A more expensive but surer method is to use a chimney tray (Figure 12-11B). The chimney tray provides enough liquid volume to fill lines and start pumps. There are no holes or valves in the deck of this tray; thus it doesn't provide for mass transfer and should not be counted as an equilibrium stage. Chimney trays must often support quite a bit of liquid; therefore, mechanical

Figure 12-11.   Intermediate liquid draw-off. (A) Downcomer sump; (B) chimney tray (downcomer to next tray not shown).

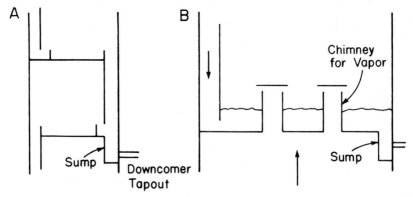

design is important. An alternative to the chimney tray is to use a downcomer tapout with an external surge drum.

The design of vapor outlets is relatively easy. The main consideration is that the line must be of large enough diameter to have a modest pressure drop. At the top of the column a demister may be used to prevent liquid entrainment. An alternative is to put a knockout drum in the line before any compressors.

12.2. TRAY EFFICIENCIES

Tray efficiencies were introduced in Chapter 6. In this section they will be discussed in more detail, and methods for estimating the value of the efficiency will be explored. The effect of mass transfer rates on the stage efficiency is discussed in Chapter 19.

The overall efficiency, E_o, is

$$E_o = \frac{N_{equil}}{N_{actual}} \tag{12-1}$$

The determination of the number of equilibrium stages required for the given separation should not include a partial reboiler or a partial condenser. The overall efficiency is extremely easy to measure and use; thus, it is the most commonly used efficiency value in the plant. However, it is difficult to relate overall efficiency to the fundamental heat and mass transfer processes occurring on the tray, so it is not generally used in fundamental studies.

The Murphree vapor and liquid efficiencies were also introduced in Chapter 6. The Murphree vapor efficiency is defined as

$$E_{MV} = \frac{y_{out} - y_{in}}{y^*_{out} - y_{in}} = \frac{\text{actual change in vapor}}{\text{change in vapor at equilibrium}} \tag{12-2a}$$

while the Murphree liquid efficiency is

$$E_{ML} = \frac{x_{out} - x_{in}}{x^*_{out} - x_{in}} = \frac{\text{actual change in liquid}}{\text{change in liquid at equilibrium}} \tag{12-2b}$$

The physical model used for both of these efficiencies is shown in Figure 12-12. The gas streams and the downcomer liquids are assumed to be perfectly mixed. Murphree also assumed that the liquid on the tray is

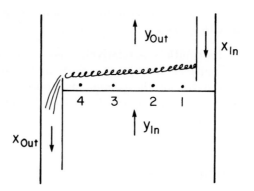

Figure 12-12.   Murphree efficiency model.

perfectly mixed, which means $x = x_{out}$. The y^*_{out} is the vapor composition that would be in equilibrium with the actual liquid composition leaving the tray, x_{out}. In the liquid efficiency, x^*_{out} is in equilibrium with the actual leaving vapor composition, y_{out}. The Murphree efficiencies are popular because they are relatively easy to measure and they are very easy to use in calculations (see Figure 6-13B). Unfortunately, there are some difficulties with their definitions. In large columns the liquid on the tray is not well mixed; instead there will be a cross-flow pattern. If the flow path is long, the more volatile component will be preferentially removed as liquid flows across the tray. Thus, in Figure 12-12,

$$x_{out} < x_4 < x_3 < x_2 < x_1$$

Note that x_{out} and hence y^*_{out} are based on the lowest concentration on the tray. Thus it is possible to have $y^*_{out} < y_{out}$, because y_{out} is an average across the tray. Then the numerator in Eq. (12-2a) will be greater than the denominator and $E_{MV} > 1$. This is often observed in large-diameter columns. Although not absolutely necessary, it is desirable to have efficiencies defined so that they range between zero and 1. A second and more serious problem with Murphree efficiencies is that the efficiencies of different components must be different for multicomponent systems. Fortunately, for binary systems the Murphree efficiencies are the same for the two components. In multicomponent systems, not only are the efficiencies different, but on some trays they may be negative. This is both disconcerting and extremely difficult to predict. Despite these problems, Murphree efficiencies are still popular.

The point efficiency is defined in a fashion very similar to the Murphree efficiency,

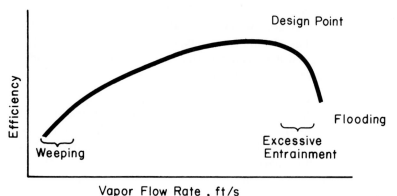

Figure 12-13.   Efficiency as a function of vapor velocity.

$$E_{pt} = \frac{y'_{out} - y'_{in}}{y'^* - y'_{in}} \qquad (12\text{-}3)$$

where the prime indicates that all the concentrations are determined at a specific point on the tray. The Murphree efficiency can be determined by integrating all of the point efficiencies (which will vary from location to location) on the tray. Typically, the point and Murphree efficiencies are not equal. The point efficiency is difficult to measure in a commercial column, but it can often be predicted from heat and mass transfer calculations. Thus it is used for prediction and for scale-up.

In general, the efficiency of a tray depends on the vapor velocity, which is illustrated schematically in Figure 12-13. The trays are designed to give a maximum efficiency at the design condition. At higher vapor velocities, entrainment increases. When entrainment becomes excessive, the efficiency plummets. At vapor velocities less than the design rate the mass transfer is less efficient. At very low velocities the tray starts to weep and efficiency again plummets. Trays with good turndown characteristics have a wide maximum, so there is little loss in efficiency when vapor velocity decreases.

The best way to determine efficiency is to have data for the chemical system in the same type of column of the same size at the same vapor velocity. If velocity varies, then the efficiency will follow Figure 12-13. The Fractionation Research Institute (FRI) has reams of efficiency data, but they are available to members only. Most large chemical and oil companies belong to FRI. The second best approach is to have efficiency data for the same chemical system but with a different type of tray. Much of the data available in the literature are for bubble-cap or sieve trays. Usually, the efficiency of valve trays is equal to or better than sieve tray efficiency, which is equal to or better than bubble cap tray efficiency. Thus, if bubble-cap efficiencies are used for a valve tray column, the design will be conservative. The third best approach is to use efficiency data for a similar chemical system.

If data are not available, a detailed calculation of the efficiency can be made on the basis of fundamental mass and heat transfer calculations. With this method, you first calculate point efficiencies from heat and mass transfer calculations and then determine Murphree and overall efficiencies from flow patterns on the tray. Unfortunately, the results are often not extremely accurate. A simple application of this method is developed in Chapter 19.

The simplest approach is to use a correlation to determine the efficiency. The most widely used is the O'Connell correlation shown in Figure 12-14 (O'Connell, 1946), which gives an estimate of the overall efficiency as a function of the relative volatility of the key components times the liquid viscosity at the feed composition. Both α and μ are determined at the average temperature and pressure of the column.

Figure 12-14.   O'Connell correlation for overall efficiency of distillation columns. From O'Connell (1946). Reprinted from *Transactions Amer. Inst. Chem. Eng., 42,* 741 (1946). Copyright 1946, American Institute of Chemical Engineers.

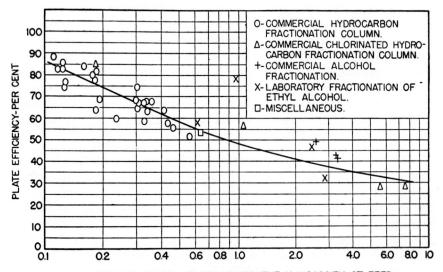

RELATIVE VOLATILITY OF KEY COMPONENT X VISCOSITY OF FEED
(AT AVERAGE COLUMN CONDITIONS)

Efficiency drops as viscosity increases, since mass transfer rates are lower. Efficiency drops as relative volatility increases, since the mass that must be transferred to obtain equilibrium increases. The scatter in the 38 data points is evident in the figure. O'Connell was probably studying bubble-cap columns; thus the results are conservative for sieve and valve trays. For computer and calculator use it is convenient to fit the data points to an equation. When this was done using a nonlinear least squares routine, the result (Kessler and Wankat, 1987) was

$$E_o = 0.52782 - 0.27511 \log_{10}(\alpha\mu) + 0.044923[\log_{10}(\alpha\mu)]^2 \quad (12\text{-}4)$$

Viscosity is in centipoise (cP) in Eq. (12-4) and in Figure 12-14. This equation is not an exact fit to O'Connell's curve, since O'Connell apparently used an eyeball fit. MacFarland *et al.* (1972) obtained a correlation for the Murphree vapor efficiency, but it is not as widely used as O'Connell's.

Efficiencies can be scaled up from laboratory data taken with an Oldershaw column (a laboratory-scale sieve-tray column) (Fair *et al.*, 1983). The overall efficiency measured in the Oldershaw column is very close to the point efficiency measured in the large commercial column. This is illustrated in Figure 12-15, where the vapor velocity has been

Figure 12-15.   Overall efficiency of 1-in diameter Oldershaw column compared to point efficiency of 4.0 ft diameter FRI column. System is cyclohexane/n-heptane. From Fair *et al.* (1983). Reprinted with permission from *Ind. Eng. Chem. Process Des. Develop.*, *22*, 53 (1983). Copyright 1983, American Chemical Society.

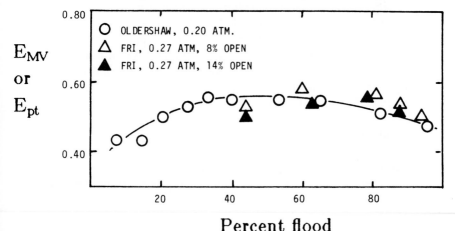

normalized with respect to the fraction of flooding (Fair *et al.*, 1983). The point efficiency can be converted to Murphree and overall efficiencies once a model for the flow pattern on the tray has been adopted. One commonly used model assumes that the vapor is completely mixed and the liquid flows in plug flow across the tray. Then the Murphree vapor efficiency can be determined (see Chapter 19) from

$$E_{MV} = \frac{L}{mV} \left[\exp\left(\frac{mV}{L} E_{pt}\right) - 1 \right] \tag{12-5}$$

where m is the local slope of the equilibrium line. The overall efficiency can be estimated as

$$E_o = \frac{\log[1 + E_{MV}(\frac{mV}{L} - 1)]}{\log(mV/L)} \tag{12-6}$$

If the flow pattern is more complex, the equations become significantly more complicated.

For very complex mixtures, the entire distillation design can be done using the Oldershaw column by changing the number of trays and the reflux rate until a combination that does the job is found. Since the commercial column will have an overall efficiency equal to or greater than that of the Oldershaw column, this combination will also work in the commercial column. This approach eliminates the need to determine vapor-liquid equilibrium data (which may be quite costly), and it also eliminates the need for complex calculations.

Example 12-1. Overall Efficiency Estimation

A sieve-plate distillation column is separating a feed that is 50 mole % n-hexane and 50 mole % n-heptane. Feed is a saturated liquid. Plate spacing is 24 in. Average column pressure is 1 atm. Distillate composition is $x_D = 0.999$ (mole fraction n-hexane) and $x_B = 0.001$. Feed rate is 1000 lb moles/hr. Internal reflux ratio $L/V = 0.8$. The column has a total reboiler and a total condenser. Estimate the overall efficiency.

Solution

To use the O'Connell correlation we need to estimate α and μ at the average temperature and pressure of the column. The column temperature can be estimated from equilibrium (the DePriester chart). The following values are easily generated from Figure 2-12.

x_{C6}	0	0.341	0.398	0.50	1.0
y_{C6}	0	0.545	0.609	0.70	1.0
T, °C	98.4	85	83.7	80	69

Relative volatility is $\alpha = (y/x)/[(1-y)/(1-x)]$. The average temperature can be estimated several ways:

Arithmetic average T = $(98.4 + 69)/2 = 83.7$, $\alpha = 2.36$

Average at x = 0.5, T = 80, $\alpha = 2.33$

Not much difference. Use $\alpha = 2.35$ corresponding to approximately 82.5 °C.

The liquid viscosity of the feed can be estimated (Reid et al., 1977, p. 462) from

$$\ln \mu_{mix} = x_1 \ln \mu_1 + x_2 \ln \mu_2$$

The pure component viscosities can be estimated from

$$\log_{10} \mu = A \left[\frac{1}{T} - \frac{1}{B} \right]$$

where μ is in centipoise and T is in kelvins (Reid et al., 1977, App. A)

nC_6: A = 362.79, B = 207.08

nC_7: A = 436.73, B = 232.53

These equations give $\mu_{C6} = 0.186$, $\mu_{C7} = 0.224$, and $\mu_{mix} = 0.204$. Then $\alpha \mu_{mix} = 0.480$. From Eq. (12-4), $E_o = 0.62$, while from Figure 12-14, $E_o = 0.59$. To be conservative, the lower value would probably be used.

Note that once T_{avg}, α_{avg}, and μ_{feed} have been estimated, calculating E_o is easy.

12.3. CALCULATING THE COLUMN DIAMETER

To design a sieve tray we need to calculate the column diameter that prevents flooding, design the tray layout, and design the downcomers. Several procedures for designing column diameters have been published in the open literature (Fair, 1963, 1984, 1985; Ludwig, 1977; McCabe *et al.*, 1985). In addition, each equipment manufacturer has its own procedure. We will follow Fair's procedure, since it is widely known. This procedure first estimates the vapor velocity that will cause flooding due to excessive entrainment, then uses a rule of thumb to determine the operating velocity, and from this calculates the column diameter. Column diameter is very important in controlling costs (see Chapt. 14) and has to be estimated even for preliminary designs. The column diameter is calculated to allow a vapor velocity that will prevent excessive entrainment. The method is applicable to sieve, valve, and bubble cap trays.

The flooding velocity is determined from

$$u_{\text{flood}} = K \sqrt{\frac{\rho_L - \rho_v}{\rho_v}}, \quad \text{ft/s} \tag{12-7}$$

This is similar to Eq. (3-50), which was used to size vertical flash drums. The factor K in Eq. (12-7) is found from

$$K = C_{sb} \left[\frac{\sigma}{20} \right]^{0.2} \tag{12-8}$$

where σ is the surface tension in dynes/cm and C_{sb} is the capacity factor. C_{sb} is a function of the flow parameter

$$F_{lv} = \frac{W_L}{W_v} \sqrt{\frac{\rho_v}{\rho_L}} \tag{12-9}$$

where W_L and W_v are the mass flow rates of liquid and vapor. The correlation for C_{sb} is shown in Figure 12-16 (Fair and Matthews, 1958). For computer use it is convenient to fit the curves in Figure 12-16 to equations. The results of a nonlinear least squares regression analysis (Kessler and Wankat, 1987) for 6-in tray spacing are

$$\log_{10} C_{sb} = -1.1977 - 0.53143 \log_{10} F_{lv} - 0.18760 (\log_{10} F_{lv})^2 \tag{12-10a}$$

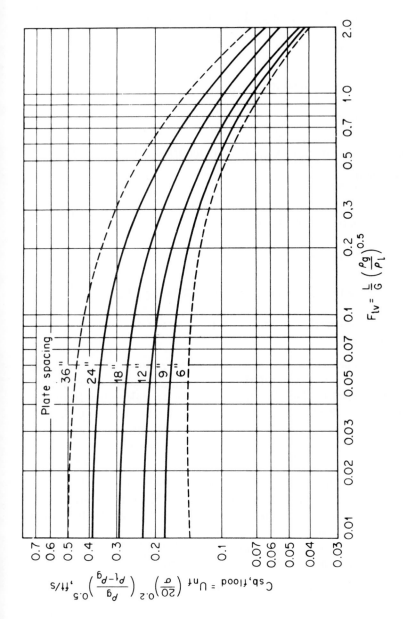

Figure 12-16. Capacity factor for flooding of sieve trays. From Fair and Matthews (1958). Reprinted with permission from *Petroleum Refiner*, *37* (4), 153 (1958). Copyright 1958, Gulf Pub. Co.

for 9-in tray spacing

$$\log_{10} C_{sb} = -1.1622 - 0.56014 \log_{10} F_{lv} - 0.18168(\log_{10} F_{lv})^2 \qquad (12\text{-}10b)$$

for 12-in tray spacing

$$\log_{10} C_{sb} = -1.0674 - 0.55780 \log_{10} F_{lv} - 0.17919(\log_{10} F_{lv})^2 \qquad (12\text{-}10c)$$

for 18-in tray spacing

$$\log_{10} C_{sb} = -1.0262 - 0.63513 \log_{10} F_{lv} - 0.20097(\log_{10} F_{lv})^2 \qquad (12\text{-}10d)$$

for 24-in tray spacing

$$\log_{10} C_{sb} = -0.94506 - 0.70234 \log_{10} F_{lv} - 0.22618(\log_{10} F_{lv})^2 \qquad (12\text{-}10e)$$

and for 36-in tray spacing

$$\log_{10} C_{sb} = -0.85984 - 0.73980 \log_{10} F_{lv} - 0.23735(\log_{10} F_{lv})^2 \qquad (12\text{-}10f)$$

The same flooding correlation but in metric units is available in graphical form (Fair, 1985).

The flooding correlation assumes that β, the ratio of the area of the holes, A_{hole}, to the active area of the tray, A_{active}, is equal to or greater than 0.1. If $\beta < 0.1$, then the flooding velocity calculated from Eq. (12-7) should be multiplied by a correction factor (Fair, 1984). If $\beta = 0.08$, the correction factor is 0.9; while if $\beta = 0.06$, the correction factor is 0.8. Note that this is a linear correction and can easily be interpolated. The resulting value for the flooding velocity will be conservative.

Tray spacing, which is required for the flooding correlation, is selected according to maintenance requirements. Sieve trays are spaced 6 to 36 in apart with 12 to 16 in a common range for smaller (less than 5 ft) towers. Tray spacing is usually greater in large-diameter columns. A minimum of 18 in, with 24 in typical, is used if it is desirable to have a worker crawl through the column for inspection.

The operating vapor velocity is determined as

$$u_{op} = (\text{fraction}) \, u_{flood}, \quad \text{ft/s} \qquad (12\text{-}11)$$

where the fraction can range from 0.65 to 0.9. Jones and Mellbom (1982) suggest using a value of 0.75 for the fraction for all cases. Higher fractions of flooding do not greatly affect the overall system cost, but they do restrict flexibility. The operating velocity u_{op} can be related to the molar vapor flow rate,

$$u_{op} = \frac{V \overline{MW_v}}{\rho_V A_{net} (3600)}, \quad ft/s \qquad (12\text{-}12)$$

where the 3600 converts from hours (in V) to seconds (in u_{op}). The net area is

$$A_{net} = \frac{\pi (Dia)^2}{4}\eta, \quad ft^2 \qquad (12\text{-}13)$$

where η is the fraction of the column cross-sectional area that is available for vapor flow above the tray. Then $1 - \eta$ is the fraction of the column area taken up by one downcomer. Typically η lies between 0.85 and 0.95; its value can be determined exactly once the tray layout is finalized. Equations (12-12) and (12-13) can be solved for the diameter of the column.

$$Dia = \sqrt{\frac{4 \, V(\overline{MW_v})}{\pi \, \eta \, \rho_v(fraction) \, u_{flood} \, (3600)}}, \quad ft \qquad (12\text{-}14)$$

If the ideal gas law holds,

$$\rho_v = \frac{p \, \overline{MW_v}}{R \, T} \qquad (12\text{-}15)$$

and Eq. (12-14) becomes

$$Dia = \sqrt{\frac{4VRT}{\pi \, \eta(3600)p(fraction)u_{flood}}}, \quad ft \qquad (12\text{-}16)$$

Note that these equations are dimensional, since C_{sb} is dimensional.

The terms in Eqs. (12-7) to (12-16) vary from stage to stage in the column. If the calculation is done at different locations, different diameters will be calculated. The largest diameter should be used and rounded off to the next highest 1/2-ft increment. (For example, a 9.18-ft column is rounded off to 9.5 ft). Ludwig (1977) suggests using a minimum column diameter of 2.5 ft; that is, if the calculated diameter is 2.0 ft, use 2.5 ft instead, since it is usually no more expensive. Columns with diameters less than 2.0 ft are usually constructed as packed columns (see Chapter 13). If diameter calculations are done at the top and bottom of the column and above and below the feed, one of these locations will be very close to the maximum diameter, and the design based on the largest calculated diameter will be satisfactory. If there is a very large change in the vapor velocity in the column, the calculated diameters can be quite different. Occasionally, columns are built in two sections of different diameter to take advantage of this situation, but this solution is economical only for large changes in diameter. If a column with a single diameter is constructed, the efficiencies in different parts of the column may vary considerably (see Figure 12-13). This variation in efficiency has to be included in the design calculations.

This design procedure sizes the column to prevent flooding caused by excessive entrainment. Flooding can also occur in the downcomers, and this case is discussed later. Excessive entrainment can also cause a large drop in stage efficiency because liquid that has not been separated is mixed with vapor. The effect of entrainment on the Murphree vapor efficiency can be estimated from

$$E_{MV,\text{entrainment}} = E_{MV} \left[\frac{1}{1 + E_{MV}\psi/(1 - \psi)} \right] \qquad (12\text{-}17)$$

where E_{MV} is the Murphree efficiency without entrainment and ψ is the fractional entrainment defined as

$$\psi = \frac{e}{L + e} = \frac{\text{absolute entrainment}}{\text{total liquid flow rate}} \qquad (12\text{-}18)$$

where e is the moles/hour of entrained liquid. The relative entrainment ψ for sieve trays can be estimated from Figure 12-17 (Fair, 1963). Once the corrected value of E_{MV} is known, the overall efficiency can be determined from Eq. (12-6). Usually, entrainment is not a problem until operation is in the range of 85 to 100% of flood (Ludwig, 1979). Thus a 75% of flood value should have a negligible correction for entrainment. This can be checked during the design procedure (see Example 12-3).

Example 12-2. Diameter Calculation

Determine the required diameter at the top of the column for the distillation column in Example 12-1.

Solution

We can use Eq. (12-16) with 75% of flooding. Since the distillate

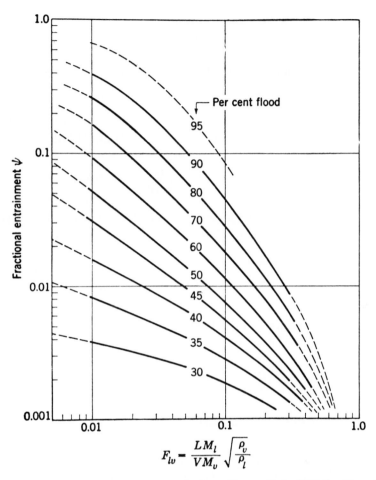

Figure 12-17.   Entrainment correlation. From Fair (1963). Reprinted with permission from Smith, B.D., *Design of Equilibrium Stage Processes*, Copyright 1963. McGraw-Hill, New York.

is almost pure n-hexane, we can approximate properties as pure n-hexane at 69 ° C. Physical properties are from Perry and Green (1984). T = 69 ° C = 342 K; liquid sp grav. = 0.659 (at 20 °); viscosity = 0.22 cP; MW = 86.17.

$$\rho_v = \frac{p(MW)}{RT} = \frac{(1 \text{ atm})(86.17\frac{lb}{lb \text{ mole}})}{(1.314 \frac{atm \text{ ft}^3}{K \text{ lbmole}})(342 \text{ K})} = 0.1917 \text{ lb/ft}^3$$

$\rho_L = (0.659)(62.4) = 41.12 \text{ lb/ft}^3$ (will vary, but not a lot)

$$\frac{W_L}{W_v} = \frac{L}{V}\frac{MW_L}{MW_v} = \frac{L}{V} = 0.8$$

Surface tension $\sigma = 13.2$ dynes/cm (Reid *et al.*, 1977, p. 610)

Flow parameter, $F_{lv} = \frac{W_L}{W_v}(\frac{\rho_v}{\rho_L})^{0.5} = 0.0546$

Ordinate from Figure 12-16 for 24 in tray spacing, $C_{sb} = 0.36$ while $C_{sb} = 0.38$ from Eq. (12-10e). From Eq. (12-8),

$$K = C_{sb}(\frac{\sigma}{20})^{0.2} = 0.36(\frac{13.2}{20})^{0.2} = 0.331$$

$K = 0.35$ if Eq. (12-10e) is used. The lower value of K is used for a conservative design. From Eq. (12-7),

$$u_{flood} = K\sqrt{\frac{\rho_L - \rho_v}{\rho_v}} = (0.331)(\frac{41.12 - 0.1917}{0.1917})^{0.5} = 4.836$$

We will estimate η as 0.90. The vapor flow rate $V = L + D$. From external mass balances, $D = 500$. Since $L = V(L/V)$,

$$V = \frac{D}{1 - L/V} = \frac{500}{0.2} = 2500 \text{ lb moles/hr}$$

The diameter equation (12-16) becomes

$$Dia = \left[\frac{4(2500)(1.314)(342)}{\pi(0.90)(3600)(1)(0.75)(4.836)}\right]^{1/2} = 11.03 \text{ ft}$$

An 11-ft diameter column would probably be used. If $\eta = 0.95$, Dia = 10.74 feet; thus the value of η is not extremely important. Note that this is a large-diameter column for this feed rate. The reflux rate is quite high, and thus V is high, which leads to a larger diameter. The effect of location on the diameter calculation can be explored by doing Problem 12-D2. The value of $\eta = 0.9$ will be checked in Example 12-3. The effect of column pressure is explored in Problem 12-F2.

12.4. SIEVE TRAY LAYOUT AND TRAY HYDRAULICS

Tray layout is an art with its own rules. This section follows the presentation of Ludwig (1979), Bolles (1963), and Fair (1963, 1984, 1985), and more details are available in those sources. The holes on a sieve plate are not scattered randomly on the plate. Instead, a detailed pattern is used to ensure even flow of vapor and liquid on the tray. The punched holes in the tray usually range in diameter from 1/8 to 1/2 in. The 1/8 in holes with the holes punched from the bottom up are often used in vacuum operation to reduce entrainment and minimize pressure drop. In normal operation, holes are punched from the bottom down since this is much safer for maintenance personnel. In fouling applications, holes are 1/2 in or larger. For clean service, 3/16 in is a reasonable first guess for hole diameter.

A common tray layout is the equilateral triangular pitch shown in Figure 12-18. The holes are spaced from $2.5d_o$ to $5d_o$, with $3.8d_o$ a reasonable average. The region containing holes should have a minimum 2 to 3 in clearance from the column shell and from the inlet downcomer. A 3 to 5 in minimum clearance is used before the downcomer weir because it is important to allow for disengagement of liquid and vapor. Since flow on the tray is very turbulent, the vapor does not go straight up from the holes. The active hole area is considered to be 2 to 3 in from the peripheral holes; thus the area up to the column shell is active. The fraction of the column that is taken up by holes depends upon the hole size, the pitch, the hole spacing, the clearances, and the size of the downcomers. Typically, 4 to 15% of the entire tower area is hole area. This corresponds to a value of β, of 6 to 25%. The average value of β is between 7 and 16%, with 10% a reasonable first guess.

The value of β is selected so that the vapor velocity through the holes, v_o, lies between the weep point and the maximum velocity. The exact design point should be selected to give maximum flexibility in operation. Thus if a reduction in feed rate is much more likely than an

Figure 12-18.   Tray geometry. (A) Equilateral triangular pitch; (B) downcomer area geometry.

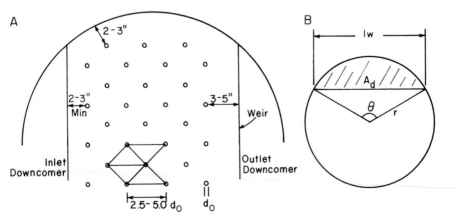

increase in feed rate, the design vapor velocity will be close to the maximum. The vapor velocity through the holes, v_o, in ft/s can be calculated from

$$v_o = \frac{V \overline{MW}_v}{3600 \; \rho_v \; A_{hole}} \tag{12-19}$$

where V is the lb moles/hr of vapor, ρ_v is the vapor density in lb/ft^3, and A_{hole} is the total hole area on the tray in ft^2. Obviously, A_{hole} can be determined from the tray layout.

$$A_{hole} = (\text{No. of holes}) \; (\pi \frac{d_o^2}{4}) \tag{12-20a}$$

or

$$A_{hole} = \beta \; A_{active} \tag{12-20b}$$

The active area can be estimated as

$$A_{active} \sim A_{total} \; (1 - 2\eta) \tag{12-20c}$$

since there are two downcomers. Obviously,

$$A_{total} = \frac{\pi (\text{Dia})^2}{4} \tag{12-20d}$$

The downcomer geometry is shown in Figure 12-18B. From this and geometric relationships, the downcomer area A_d can be determined from

$$A_d = \frac{1}{2} \; r^2 \; (\theta - \sin \theta) \tag{12-21a}$$

where θ is in radians. The downcomer area can also be calculated from

$$A_d = (1 - \eta) A_{total} \tag{12-21b}$$

Combining Eqs. (12-20d), (12-21a), and (12-21b), we have

$$2(1 - \eta)\pi = \theta - \sin \theta \tag{12-21c}$$

Table 12-1. Geometric Relationship Between
η and l_{weir}/Diameter

η	0.8	0.825	0.85	0.875	0.900	0.925	0.95	0.975
l_{weir}/Dia	0.871	0.843	0.811	0.773	0.726	0.669	0.593	0.478

Once this equation is solved for the angle θ, the length of the weir l_{weir} can be found from

$$\frac{l_{weir}}{Dia} = \sin\left(\frac{\theta}{2}\right) \tag{12-22}$$

Solving Eqs. (12-21c) and (12-22) we obtain the results given in Table 12-1. Typically the ratio of l_{weir}/Dia falls in the range of 0.6 to 0.75.

If the liquid is unable to flow down the downcomer fast enough, the liquid level will increase, and if it keeps increasing until it reaches the top of the weir of the tray above, the tower will flood. This downcomer flooding must be prevented. Downcomers are designed on the basis of pressure drop and liquid residence time, and their cost is relatively small. Thus downcomer design is done only in the final equipment sizing.

The tray and downcomer are drawn schematically in Figure 12-19, which shows the pressure heads caused by various hydrodynamic effects.

Figure 12-19.   Pressure heads on sieve trays.

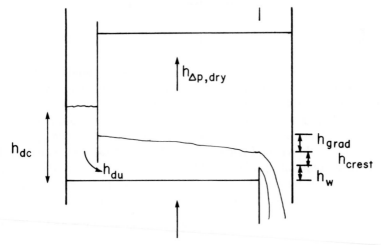

The head of clear liquid in the downcomer, h_{dc}, can be determined from the sum of heads that must be overcome.

$$h_{dc} = h_{\Delta p,dry} + h_{weir} + h_{crest} + h_{grad} + h_{du} \qquad (12\text{-}23)$$

The head of liquid required to overcome the pressure drop of gas on a dry tray, $h_{\Delta p,dry}$, can be measured experimentally or estimated (Ludwig, 1979) from

$$h_{\Delta p,dry} = 0.003 \, v_o^2 \, \rho_v \left(\frac{\rho_{water}}{\rho_L} \right)(1 - \beta^2)/C_o^2 \qquad (12\text{-}24)$$

where v_o is the vapor velocity through the holes in ft/s from Eq. (12-19). The orifice coefficient, C_o, can be determined from the correlation of Hughmark and O'Connell (1957). This correlation can be fit by the following equation (Kessler and Wankat, 1987):

$$C_o = 0.85032 - 0.04231 \frac{d_o}{t_{tray}} + 0.0017954\left(\frac{d_o}{t_{tray}}\right)^2 \qquad (12\text{-}25)$$

where t_{tray} is the tray thickness. The minimum value for d_o/t_{tray} is 1.0. Equation (12-24) gives $h_{\Delta p,dry}$ in inches.

The weir height, h_{weir}, is the actual height of the weir. The minimum weir height is 0.5 in with 1 to 3 in more common. The weir must be high enough that the opposite downcomer remains sealed and always retains liquid. The height of the liquid crest over the weir, h_{crest}, can be calculated from the Francis weir equation.

$$h_{crest} = 0.092 \, F_{weir} \left(L_g/l_{weir}\right)^{2/3} \qquad (12\text{-}26)$$

where h_{crest} is in inches. In this equation, L_g is the liquid flow rate in gal/min that is due to both L and e. The entrainment e can be determined from Figure 12-17. l_{weir} is the length of the straight weir in feet. The factor F_{weir} is a modification factor to take into account the curvature of the column wall in the downcomer (Bolles, 1946, 1963; Ludwig, 1979). This is shown in Figure 12-20 (Bolles, 1946). An equation for this figure is available (Bolles, 1963). For large columns where l_{weir} is large, F_{weir} approaches 1.0. On sieve trays, the liquid gradient, h_{grad}, across the tray is often very small and is usually ignored.

There is a frictional loss due to flow in the downcomer and under the

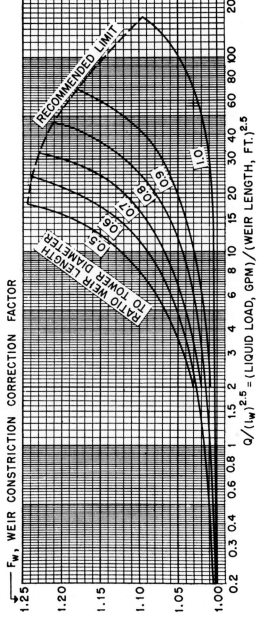

Figure 12-20.   Weir correction factor, F_{weir}, for segmental weirs. From Bolles (1946). Reprinted with permission of *Petroleum Processing*.

downcomer onto the tray. This term, h_{du}, can be estimated from the empirical equation (Ludwig, 1979; Bolles, 1963).

$$h_{du} = 0.56 \left(\frac{L_g}{449 \, A_{du}}\right)^2 \qquad (12\text{-}27)$$

where h_{du} is in inches and A_{du} is the flow area under the downcomer apron in ft^2. The downcomer apron typically has a 1-in gap above the tray.

$$A_{du} = (gap)l_{weir} \qquad (12\text{-}28)$$

The value of h_{dc} calculated from Eq. (12-23) is the head of clear liquid in inches. In an operating distillation column the liquid in the downcomer is aerated. The density of this aerated liquid will be less than that of clear liquid, and thus the height of aerated liquid in the downcomer will be greater than h_{dc}. The expected height of the aerated liquid in the downcomer, $h_{dc,aerated}$, can be estimated (Fair, 1984) from the equation

$$h_{dc,aerated} = h_{dc}/\phi_{dc} \qquad (12\text{-}29)$$

where ϕ_{dc} is the relative froth density. For normal operation, a value of $\phi_{dc} = 0.5$ is satisfactory, while 0.2 to 0.3 should be used in difficult cases (Fair, 1984). To avoid downcomer flooding, the tray spacing must be greater than $h_{dc,aerated}$. Thus in normal operation the tray spacing must be greater than $2h_{dc}$.

The downcomer is designed to give a liquid residence time of 3 to 7 s. Minimum residence times are listed in Table 12-2 (Kister, 1980e). The residence time in a straight segmental downcomer is

Table 12-2. Minimum Residence Times in Downcomers

Foaming Tendency	Examples	Residence Time(s)
Low	Alcohols, low-MW hydrocarbons	3
Medium	Medium-MW hydrocarbons	4
High	Mineral oil absorbers	5
Very high	Amines, glycols	7

Source: Kister (1980c)

$$t_{res} = \frac{A_d h_{dc}(3600)\rho_L}{(L+e)(\overline{MW}_L)(12)}, \quad sec \tag{12-30}$$

where the 3600 converts hours to seconds (from $L+e$) and the 12 converts h_{dc} in inches to feet. Density is the density of clear liquid, and h_{dc} is the height of clear liquid. Equation (12-30) is used to make sure there is enough time to disengage liquid and vapor in the downcomers.

Figure 12-13 showed that the two limits to acceptable tray operation are excessive entrainment and excessive weeping. Weep and dump points are difficult to determine exactly. An approximate analysis can be used to ensure that operation is above the weep point. Liquid will not drain through the holes as long as the sum of heads due to surface tension, h_σ, and gas flow, $h_{\Delta p,dry}$, are greater than a function depending on the liquid head. This condition for avoiding excessive weeping can be estimated (Kessler and Wankat, 1987) as

$$h_{\Delta p,dry} + h_\sigma \geq 0.10392 + 0.25119\,x - 0.021675\,x^2 \tag{12-31}$$

where $x = h_{weir} + h_{crest} + h_{grad}$ and β ranges from 0.06 to 0.14. The dry tray pressure drop is determined from Eqs. (12-24) and (12-25). The surface tension head, h_σ, can be estimated (Fair, 1963) from

$$h_\sigma = \frac{0.040\sigma}{\rho_L d_o} \tag{12-32}$$

where σ is in dynes/cm, ρ_L in lb/ft^3, d_o in inches, and h_σ in inches of liquid. Equation (12-31) is conservative.

Example 12-3. Tray Layout and Hydraulics

Determine the tray layout and pressure drops for the distillation column in Examples 12-1 and 12-2. Determine if entrainment or weeping is a problem. Determine if the downcomers will work properly. Do these calculations only at the top of the column.

Solution

This is a straightforward application of the equations in this section. We can start by determining the entrainment. In Example 12-2 we obtained $F_{lv} = 0.0546$. Then Figure 12-17 at 75% of flooding gives $\psi = 0.045$. Solving for e,

$$e = \frac{\psi L}{1 - \psi} \qquad (12\text{-}33)$$

Since L = (L/V) V = 2000 lb moles/hr,

$$e = \frac{(0.045)(2000)}{1 - 0.045} = 94.24 \text{ lb moles/hr}$$

and L + e = 2094.24 lb moles/hr. This amount of entrainment is quite reasonable.

The geometry calculations proceed as follows:

Eq. (12-20d), $A_{total} = (\pi)(11.0)^2/4 = 95.03 \text{ ft}^2$

Eq. (12-21b), $A_d = (1 - 0.9)(95.03) = 9.50 \text{ ft}^2$

Eq. (12-21c) gives $\theta = 1.627$ radians. Then Eq. (12-22) gives $l_{weir}/Dia = 0.726$ or $l_{weir} = 8.0$ ft.

Eq. (12-20c), $A_{active} = (95.03)(1 - 0.2) = 76 \text{ ft}^2$

Eq. (12-20b), $A_{hole} = (0.1)(76) = 7.6 \text{ ft}^2$

We will use 14 gauge standard tray material ($t_{tray} = 0.078$ in) with 3/16-in holes. Thus, $d_o/t_{tray} = 2.4$.

From Eq. (12-19)

$$v_o = \frac{V \overline{MW_V}}{3600 \, \rho_V \, A_{hole}} = \frac{2500(86.17)}{3600(0.192)(7.6)} = 41.1 \text{ ft/s}$$

where ρ_V is from Example 12-2. This hole velocity is reasonable.

The individual pressure drop terms can now be calculated. The orifice coefficient C_o is, from Eq. (12-25),

$$C_o = 0.85032 - 0.04231 \, (2.4) + 0.0017954 \, (2.4)^2 = 0.759$$

From Eq. (12-24),

$$h_{\Delta p, dry} = (0.003)(41.1)^2(0.192)(\frac{61.03}{41.12}) \frac{1 - 0.01}{(0.759)^2} = 2.472 \text{ in}$$

In this equation ρ_w at 69° C was found from Perry and Green (1984, p 3-75). A weir height of $h_{weir} = 2$ in will be selected.

The correction factor F_{weir} can be found from Figure 12-20. L_g in the abscissa is the liquid flow rate including entrainment in gallons per minute.

$$L_g = (2094.24)(86.17\frac{lb}{lb\ mole})(\frac{1\ ft^3}{41.12\ lb})(7.48\frac{gal}{ft^3})(\frac{1\ hr}{60\ min})$$

$$= 547.1\ gal/min$$

$$Abscissa = \frac{L_g}{l_{weir}^{2.5}} = \frac{547.1}{(8)^{2.5}} = 3.022$$

Parameter $l_{weir}/Dia = 0.727$. Then $F_{weir} = 1.025$

From Eq. (12-26), $h_{crest} = (0.092)(1.025)(\frac{547.1}{8})^{2/3} = 1.577$ in.

We will assume $h_{grad} = 0$. The area under the downcomer is determined with an 1-in gap.

From Eq. (12-28), $A_{du} = (1/12)(8) = 2/3\ ft^2$.

From Eq. (12-27), $h_{du} = 0.56\left[\frac{547.1}{449(2/3)}\right]^2 = 1.871$ in.

Total head from Eq. (12-23),

$$h_{dc} = 2.472 + 2 + 1.577 + 0 + 1.871 = 7.92$$

This is inches of clear liquid. For the aerated system,

$$h_{dc,aerated} = \frac{7.92}{0.5} = 15.84$$

Since this is much less than the 24-in tray spacing, there should be no problem. The residence time is, from Eq. (12-30),

$$t_{res} = \frac{(9.51)(7.92)(3600)(41.12)}{(2094.24)(86.17)(12)} = 5.15\ s$$

This is greater than the minimum residence time of 3 s.

Weeping can be checked. From Eq. (12-32),

$$h_\sigma = \frac{(0.040)(13.2)}{(41.12)(3/16)} = 0.068$$

Then the left-hand side of Eq. (12-31) is

$$h_{\Delta p, dry} + h_\sigma = 2.472 + 0.068 = 2.54$$

The function $x = 2 + 1.577 + 0 = 3.577$, and the right-hand side of Eq. (12.31) is 0.725. The inequality is obviously satisfied. Weeping should not be a problem.

Note that the design should be checked at other locations in the column. Problem 12-D3 shows that backup of liquid in the downcomers is a problem in the bottom of the column. This occurs because $\bar{L} = L + F = 3000$ lb moles/hr, which is significantly greater than the liquid flow in the top of the column. This problem can be handled by an increase in the gap between the downcomer and the tray.

12.5. VALVE TRAY DESIGN

Valve trays, which were illustrated in Figure 12-1, are proprietary devices, and the final design would normally be done by the supplier. However, the supplier will not do the optimization studies that the buyer would like without receiving compensation for the additional work. In addition, it is always a good idea to know as much as possible about equipment before making a major purchase. Thus the nonproprietary valve tray design procedure of Bolles (1976) is very useful for estimating performance.

Bolles's (1976) design procedure uses the sieve tray design procedure as a basis and modifies it as necessary. One major difference between valve and sieve trays is in their pressure drop characteristics. The dry tray pressure drop in a valve tray is shown in Figure 12-21 (Bolles, 1976). As the gas velocity increases, Δp first increases and then levels off at a plateau level. In the first range of increasing Δp, all valves are closed. At the closed balance point, some of the valves open. Additional valves open in the plateau region until all valves are open at the open balance point. With all valves open, Δp increases as the gas velocity increases further. The head loss in inches of liquid for both closed and open valves can be expressed in terms of the kinetic energy,

$$h_{\Delta p, valve} = K_v \frac{\rho_v}{\rho_L} \frac{v_o^2}{2g} \tag{12-34a}$$

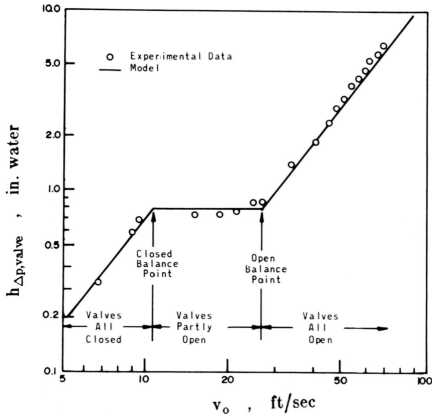

Figure 12-21.   Dry tray pressure drop for valve tray. From Bolles (1976). Reprinted from *Chemical Engineering Progress,* Sept. 1976. Copyright 1976, American Institute of Chemical Engineers.

where v_o is the velocity of vapor through the holes in the deck in ft/s, g = 32.2 ft/s^2, and K_v is different for closed and open valves. For the data shown in Figure 12-21,

$$K_{v,closed} = 33, \quad K_{v,open} = 5.5 \tag{12-34b}$$

Note that Eq. (12-34a) has the same dependence on $v_o^2 \rho_v / \rho_L$ as $h_{\Delta p,dry}$ for sieve trays in Eq. (12-24).

The closed balance point can be determined by noting that the pressure must support the weight of the valve, W_{valve}, in pounds. Pressure is then W_{valve}/A_v, where A_v is the valve area in square feet. The pressure drop in terms of feet of liquid of density ρ_L is then

$$h_{\Delta p, valve} = C_v \frac{W_{valve}}{A_v \, \rho_L} \tag{12-35}$$

where the valve coefficient C_v is introduced to include turbulence losses. For the data in Figure 12-21, $C_v = 1.25$. Setting Eqs. (12-33) and (12-35) equal allows solution of both the closed and open balance points.

$$v_{o,bal} = \sqrt{\frac{C_v \, W_{valve} \, 2g}{K_v \, A_v \, \rho_v}} \tag{12-36}$$

where $v_{o,bal}$ is the closed balance point velocity if $K_{v,closed}$ is used. The values of $K_{v,closed}$, $K_{v,open}$, and C_v depend upon the thickness of the deck and, to a small extent, on the type of valve (Bolles, 1976).

Much of the remainder of the preliminary design of valve trays is the same or slightly modified from sieve tray design. Flooding and diameter calculations are the same except that the correction factor for $A_{hole}/A_a < 0.1$ is replaced by a correction factor for $A_{slot}/A_a < 0.1$. The same values for the correction factors are used. The slot area A_{slot} is the vertical area between the tray deck and the top of the valve through which the vapor passes in a horizontal direction. In the region between the balance points the slot area is variable and can be determined from the fraction of valves that are open. The pressure drop equation (12-23) is the same, while Eq. (12-24) for $h_{\Delta p, dry}$ is replaced by Eq. (12-33) or (12-35). Equations (12-26) to (12-30) are unchanged. The gradient across the valve tray, h_{grad}, is probably larger than on a sieve tray but is usually ignored. Valve trays usually operate with higher weirs. An h_{weir} of 3 in is normal.

The efficiency of a valve tray depends upon the vapor velocity, the valve design, and the chemical system being distilled. Except at vapor flow rates near flooding, the efficiencies of valve trays are equal to or higher than sieve tray efficiencies, which are equal to or higher than bubble cap tray efficiencies. Thus the use of the efficiency correlations discussed earlier will result in a conservative design.

12.6. SUMMARY AND OBJECTIVES

In this chapter both qualitative and quantitative aspects of column design are discussed. At the end of this chapter you should be able to satisfy the following objectives.

1. Describe the equipment used for staged distillation columns.

2. Define different definitions of efficiency, predict the overall efficiency, and scale-up the efficiency from laboratory data.

3. Determine the diameter of sieve and valve trays.

4. Determine tray pressure drop terms for sieve and valve trays and design downcomers.

5. Lay out a tray that will work.

REFERENCES

Bolles, W.L., *Pet. Refiner, 25,* 613 (1946).

Bolles, W.L., "Tray Hydraulics, Bubble-Cap Trays," in B.D. Smith, (Ed.), *Design of Equilibrium Stage Processes,* McGraw-Hill, New York, 1963, Chapt. 14.

Bolles, W.L. "Estimating Valve Tray Performance," *Chem. Eng. Prog., 72* (9), 43 (Sept. 1976).

Fair, J.R., "Tray Hydraulics. Perforated Trays," in B.D. Smith, *Design of Equilibrium Stage Processes,* McGraw-Hill, New York, 1963, Chapt. 15.

Fair, J.R., "Gas-Liquid Contacting," in R.H. Perry and D. Green (Eds.), *Perry's Chemical Engineer's Handbook,* 6th ed., McGraw-Hill, New York, 1984, Section 18.

Fair, J.R., "Stagewise Mass Transfer Processes," in A. Bisio and R.L. Kabel (Eds.), *Scaleup of Chemical Processes,* Wiley, New York, 1985, Chapt. 12.

Fair, J.R., H.R. Null, and W.L. Bolles, "Scale-up of Plate Efficiency from Laboratory Oldershaw Data," *Ind. Eng. Chem. Process Des. Develop., 22,* 53 (1983).

Glitsch, Inc., *Ballast Tray Design Manual,* Glitsch, Inc., Dallas, TX, 1974.

Glitsch, Inc., "17 Critical Questions and Answers About Trays, Column Internals and Accessories," Glitsch, Inc., Dallas, TX, 1985.

Huang, C.-J. and J.R. Hodson "Perforated Trays," *Pet. Ref., 37* (2), 104 (1958).

Hughmark, G.A. and H.E. O'Connell, "Design of Perforated Plate Fractionating Towers," *Chem. Eng. Prog.,* 53, 127-M (1957).

Jones, E.A. and M.E. Mellborn, "Fractionating Column Economics," *Chem. Eng. Prog., 75*(5), 52 (1982).

Kessler, D.P. and P.C. Wankat, "Correlations for Column Parameters," *Chem. Eng.,* 71 (Sept. 26, 1988).

Kister, H.Z., "Guidelines for Designing Distillation - Column Intervals," *Chem. Eng., 87* (10), 138 (1980a).

Kister, H.Z., "Outlets and Internal Devices for Distillation Columns," *Chem. Eng., 87* (15), 79 (1980b).

Kister, H.Z., "Design and Layout for Sieve and Valve Trays," *Chem. Eng., 87*(18), 119 (1980c).

Kister, H.Z., "Mechanical Requirements for Sieve and Valve Trays," *Chem. Eng., 87* (23), 283 (1980d).

Kister, H.Z., "Downcomer Design for Distillation Tray Columns," *Chem. Eng. 87*(26), 55 (1980e).

Kister, H.Z., "Inspection Assures Troublefree Operation," *Chem. Eng., 88*(3), 107 (1981a).

Kister, H.Z., "How to Prepare and Test Columns Before Startup," *Chem. Eng., 88* (7), 97 (1981b).

Koch Engineering Co., *Design Manual Flexitray,* Koch Engineering Co., Wichita, KS, 1982.

Ludwig, E.E., *Applied Process Design for Chemical and Petrochemical Plants,* 2nd ed., Vol. 2, Gulf Pub., Houston, TX, 1979.

MacFarland, S.A., P.M. Sigmund, and M. Van Winkle, "Predict Distillation Efficiency," *Hydrocarbon Proc.,* 51 (7), 111 (1972).

McCabe, W.L., J.C. Smith, and P. Harriott, *Unit Operations of Chemical Engineering,* 4th ed., McGraw-Hill, New York, 1985.

O'Connell, H.E., "Plate Efficiency of Fractionating Columns and Absorbers," *Trans. Amer. Inst. Chem. Eng.*, *42*, 741 (1946).

Reid, R.C., J.M. Prausnitz, and T.K. Sherwood, *The Properties of Gases and Liquids,* 3rd ed., McGraw-Hill, New York, 1977.

Zenz, F.A., "Design of Gas Absorption Towers," in P.A. Schweitzer (Ed.), *Handbook of Separation Techniques for Chemical Engineers,* McGraw-Hill, New York, 1979, Section 3.2.

HOMEWORK

A. *Discussion Problems*

A1. What effect does increasing the spacing between trays have on:

   a. Column efficiency

   b. K and column diameter

   c. Column height

A2. Define η and (fraction) in Eqs. (12-14) and (12-16).

A3. Several different column areas are used in this chapter. Define and contrast: total cross-sectional area, net area, downcomer area, active area, and hole area.

A4. Relate the head of clear liquid to a pressure drop in psig.

A5. Explain why the notched weirs in Figure 12-6B have better turn-down characteristics than straight weirs.

A6. Explain the difference between flooding due to entrainment and downcomer flooding. Explain how these phenomena interact.

A7. Explain Figure 12-21. What would the pressure drop curve for a sieve tray look like [see Eq. (12-24)]?

A8. Valve trays are often constructed with two different weight valves. What would this do to Figure 12-21? What are the probable advantages of this design?

A9. a. Check Eq. (12-4) versus Figure 12-14.

   b. Check Eqs. (12-10) versus Figure 12-16.

A10. The basic design method for determining column diameter determines u_{flood} from Eq. (12-7). Is this a vapor or a liquid velocity?

How is the flow rate of the other phase (liquid or vapor) included in the design procedure?

A11. Write your key relations chart for this chapter.

A12. Intermediate feeds should not be introduced into a downcomer. Explain why not.

B. *Generation of Alternatives*

B1. Several types of valves are shown in Figure 12-1. Brainstorm alternative ways in which valves could be designed.

C. *Derivations*

C1. You need to temporarily increase the feed rate to an existing column without flooding. Since the column is now operating at about 90% of flooding, you must vary some operating parameter. The column has 18-in tray spacing, is operating at 1 atm, and has a flow parameter

$$F_{lv} = \frac{L}{G} \left(\frac{\rho_v}{\rho_L}\right)^{0.5} = 0.05$$

The column is rated for a total pressure of 10 atm. $L/D = $ constant. The relative volatility for this system does not depend on pressure. The condenser and reboiler can easily handle operation at a higher pressure. Downcomers are large enough for larger flow rates. Will increasing the column pressure increase the feed rate that can be processed? It is likely that:

$$\text{Feed rate} \propto (\text{pressure})^{\text{exponent}} \qquad (12\text{-}37)$$

Determine the value of the exponent for this situation. Use the ideal gas law.

C2. Derive Eq. (12-30).

C3. Show that the column diameter is proportional to $F^{1/2}$ and to $(1 + L/D)^{1/2}$.

D. *Problems*

D1. Repeat Example 12-1 for an average column pressure of 700 kPa.

D2.   Repeat Example 12-2, except calculate the diameter at the bottom of the column. For n-heptane: MW = 100.2, bp 98.4°C, sp grav = 0.684, viscosity (98.4°C) = 0.205 cP, σ (98.4°C) = 12.5 dynes/cm.

D3.   The calculations in Example 12-3 were done for conditions at the top of the column. Physical properties will vary throughout the column, but columns are normally constructed with identical trays, downcomers, weirs, etc., on every stage (this is simpler and cheaper). For the dimensions calculated in Example 12-3, calculate entrainment, pressure drops, downcomer residence time, and weeping at the bottom of the column. (See Problem 12-D2 for data.) Note that this operation is no longer at 75% of flooding. The results of Problem 12-D2 are required. If the column will not operate, will it work if the gap between the tray and downcomer apron is increased to 1.5 in?

D4.   We wish to repeat the distillation in Examples 12-2 and 12-3 except that valve trays will be used. The valves have a 2-in diameter head. For the top of the column, estimate the pressure drop versus hole velocity curve. Assume that K_v and C_v values are the same as in Figure 12-21. Each valve weighs approximately 0.08 lb.

E.   *More Complex Problems*

E1.   Design a sieve-plate column to separate 1000 lb moles/hr of a feed that is 45 mole % water and 55% acetic acid. The feed is a saturated liquid. We desire a distillate that is 96 mole % water and a bottoms that is 6 mole % water. The column has a partial reboiler and a total condenser. Reflux is a saturated liquid. Pressure is 1 atm. Use an external reflux ratio $L/D = 1.25$ $(L/D)_{min}$. Use the optimum feed plate. Design single pass trays. Determine:

   a.   Optimum feed plate

   b.   Total number of equilibrium stages and number of real stages

   c.   Column height

   d.   Column diameter

   e.   Check that entrainment and weeping are not excessive

   f.   Determine head of liquid in the downcomer

Data: See Problem 6-E3. Note that CMO is not valid, and use latent heat units. $MW_{acetic\ acid} = 60.05$, plate spacing is 24 in, operate at 80% of flooding, $d_o = 3/16$ in, $t_{tray} = 0.078$ in, $\eta = 0.85$, $h_{weir} = 2$ in, $\beta = 0.1$. For liquid densities, see p. 3-89 in Perry and Green (1984); for vapor densities, assume ideal gas; for viscosities, see pp. 3-251 and 3-252 in Perry and Green (1984). Surface tension σ, dynes/cm, for acetic acid in water at 30 °C (wt % HOAc):

HOAc	1.00	2.475	5.001	10.01	30.09	49.96	69.91
σ	68.00	64.40	60.10	54.60	43.60	38.40	34.30

Surface tensions, pure acetic acid:

10 °C, $\sigma = 28.8$; 20 °C, $\sigma = 27.8$; 50 °C, $\sigma = 24.8$.

F.   *Problems Requiring Other Resources*

F1.   Plot Eq. (12-25). Check this with the plot in Hughmark and O'Connell (1957) or Ludwig (1979).

F2.   Repeat Example 12-2 for an average column pressure of 700 kPa.

F3.   Repeat Example 12-3 for an average column pressure of 700 kPa. Note that column diameter is obtained in Problem 12-F2.

F4.   We are separating an ethanol-water mixture in a column operating at atmospheric pressure with a total condenser and a partial reboiler. Constant molal overflow can be assumed, and the reflux is a saturated liquid. The feed rate is 100 lb moles/hr of a 30 mole % ethanol mixture. The feed is a subcooled liquid, and 3 moles of feed will condense 1 mole of vapor at the feed plate. We desire an $x_D = 0.8$, $x_B = 0.01$ and use $L/D = 2.0$. Use a plate spacing of 18 in. What diameter is necessary if we will operate at 75% of flooding? How many real stages are required, and how tall is the column?

The downcomers can be assumed to occupy 10% of the column cross-sectional area. Surface tension data are available on page F-29 of *Handbook of Chemistry and Physics*, 48th edition, and are available in other editions. The surface tension may be extrapolated as a linear function of temperature. Liquid densities are given in *Perry's*. Vapor densities can be found from the perfect gas law. The overall efficiency can be estimated from the O'Connell correlation. Note that the diameter calculated at different locations in the column will vary. The largest diameter calculated should be used. Thus you must either calculate a diameter at several locations in the column or justify why a given location will give the largest diameter.

F5. We have a distillation column that has 30 real sieve plates. The feed plate is the 18th plate from the top. The column has a total condenser and a kettle-type partial reboiler that acts as one equilibrium contact. The column is 6 ft in diameter and has single-pass trays. The two downcomers take up 15% of the cross-sectional area of the column. The trays are 24 in apart. The plant this column was used in is now shut down. In the past the column operated satisfactorily at 75% of flooding; use this value again. We now wish to use this column for separating methanol and water. A very large amount of a methanol-water solution that is 21.2 mole % methanol and is a saturated liquid is available. We desire a bottoms product that is 0.5 mole % methanol and a distillate that is 98 mole % methanol.

    a. What overall efficiency do you expect?

    b. What reflux ratio, L/D, should be used?

    c. How much feed can be processed per hour?

Assume that pressure is 1 atm and that reflux is returned as a saturated liquid. Do NOT assume constant molal overflow. Use latent heat units. See Table 6-1 for equilibrium data.

chapter 13
PACKED COLUMN DESIGN

Instead of staged columns we often use packed columns for distillation, absorption, stripping, and occasionally extraction. Packed columns are used for smaller diameter columns since it is expensive to build a staged column that will operate properly in small diameters. Packed columns are definitely more economical for columns less then 2 ft in diameter. In larger packed columns the liquid may tend to channel, and without careful design packed towers may not operate very well; in many cases large-diameter staged columns are cheaper. Packed towers have the advantage of a smaller pressure drop and are therefore useful in vacuum fractionation.

In designing a packed tower, the choice of packing material is based on economic considerations. A wide variety of packings are available. Once the packing has been chosen it is necessary to know the column diameter and the height of packing needed. The column diameter is sized on the basis of either the approach to flooding or the acceptable pressure drop. Packing height can be found either from an equilibrium stage analysis or from mass transfer considerations. The equilibrium stage analysis using the height equivalent to a theoretical plate (HETP) procedure will be considered here; the mass transfer design method is discussed in Chapter 19.

13.1. COLUMN INTERNALS

In a packed column used for vapor-liquid contact, the liquid flows over the surface of the packing and the vapor flows in the void space inside the packing and between pieces of packing. The purpose of the packing is to provide for intimate contact between vapor and liquid with a very large surface area for mass transfer. At the same time, the packing should provide for easy liquid drainage and have a low pressure drop for gas flow. Since packings are often randomly dumped into the column, they also have to be designed so that one piece of packing will not cover up and mask the surface area of another piece.

413

Packings are available in a large variety of styles, some of which are shown in Figure 13-1. The simpler styles such as Raschig rings are usually cheaper on a volumetric basis but will often be more expensive on a performance basis, since some of the proprietary packings are much more efficient. The individual rings and saddles are dumped into the column and are distributed in a random fashion. The structured or arranged packings (Glitsch grid, Goodloe, and Koch Sulzer) are placed carefully into the column. The structured packings usually have lower pressure drops and are more efficient than dumped packings, but they are often more expensive. Packings are available in a variety of materials including plastics, metals, ceramics, and glass. One of the advantages of packed columns is they can be used in extremely corrosive service.

The packing must be properly held in the column to fully utilize its separating power. A schematic diagram of a packed distillation column

Figure 13-1.　Types of column packings. (A) Ceramic Intalox[R] saddle. (B) Plastic Super Intalox[R] saddle. (C) Pall ring. (D) Gempak[TM] cartridge. E. Glitsch EF-25AGrid[R]. Figures A, B and C courtesy of Norton Chemical Process Products, Akron, OH. Figures D and E courtesy of Glitsch, Inc., Dallas, TX.

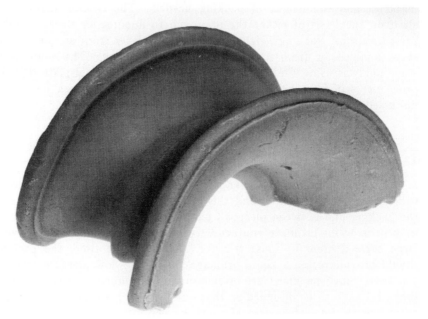

A

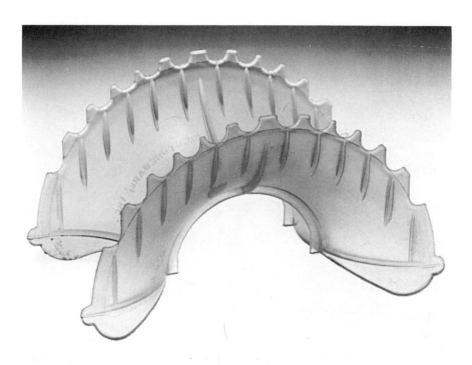

B

C

D

E

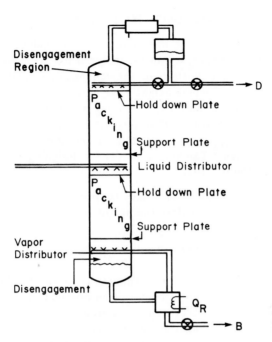

Figure 13-2.   Packed distillation column.

is shown in Figure 13-2. In addition to the packed sections where
separation occurs, sections are needed for distribution of the reflux, feed,
and boilup and for disengagement between liquid and vapor. The liquid
distributors are very important for proper operation of the column. An
absolute minimum of 20 drip points per square meter is recommended,
and 50 drip points per square meter is preferable (Fair, 1985). Scale-up
of distributors is done by keeping the number of drip points per square
meter constant. Redistribution systems may be required on large
columns. The packing is supported by a support plate, which may be a
grid or series of bars. A hold-down plate is often employed to prevent
packing movement when surges in the gas rate occur. Since liquid and
vapor are flowing countercurrently throughout the column, there are no
downcomers. The packed tower internals must be carefully designed to
obtain good operation. (See Fair, 1985; Ludwig 1979, Chapt. 9; Perry
and Chilton, 1973, Chapt. 18; or Perry and Green, 1984, Chapt. 18).
Additional details are given in manufacturers' literature (see "Packed
Column Internals," 1965 and "Tower Packings and Internals," 1975).
The internals of a packed column for absorption or stripping would be
similar to the distillation column shown in Figure 13-2 but without the
center feed, reboiler, or condenser.

13.2. HEIGHT OF PACKING: HETP APPROACH

Even though a packed tower has continuous instead of discontinuous contact of liquid and vapor, it can be analyzed like a staged tower. We assume that the packed portion of the column can be divided into a number of segments of equal height. Each segment acts as an equilibrium stage, and liquid and vapor leaving the segment are in equilibrium. It is important to note that this staged model is *not* an accurate picture of what is happening physically in the column, but the model can be used for design. The staged model for designing packed columns was first used by Peters (1922).

We calculate the number of stages from either a McCabe-Thiele or Lewis analysis and then calculate the height as

$$\text{Height} = \text{number of equilibrium stages} \times \text{HETP} \qquad (13\text{-}1)$$

The HETP, which is measured experimentally, is the height of packing needed to obtain the change in composition obtained with one theoretical equilibrium contact. HETPs can vary from 1/2 in (very low gas flow rates in self-wetting packings) to several feet (large Raschig rings). In normal industrial equipment the HETP varies between 1 and 4 feet. The smaller the HETP, the shorter the column and the more efficient the packing.

To measure the HETP, determine the top and bottom compositions at total reflux and then calculate the number of equilibrium stages. Then

$$\text{HETP} = \frac{\text{height of packing}}{\text{number of theoretical stages}} \qquad (13\text{-}2)$$

A partial reboiler is usually used but should not be included in the calculation of HETP.

The HETP determined at total reflux is then used at the actual reflux ratio. Ellis and Brooks (1971) found that there is an increase in the HETP for internal reflux ratios below 1.0, but the increase is usually quite small until L/V approaches 1/2. Thus the usual measurement procedure can be used for most design situations.

The HETP varies with the packing type and size, chemicals being separated, and gas flow rate. Some typical HETP curves are shown in Figure 13-3. The HETP values for several types of packing are listed in the manufacturers' bulletins and have been compared by Perry (1950, p.

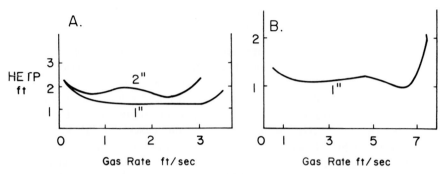

Figure 13-3.   HETP versus vapor rate for metal pall rings. (A) Iso-Octane-Toluene; (B) acetone-water. Reprinted with permission from "Pall Rings in Mass Transfer Operations," 1968. Courtesy of Norton Chemical Process Products, Akron, OH.

620) and Ellis and Brooks (1971) while mass transfer results are compared by Furter and Newstead (1973), Perry and Chilton (1973), and Perry and Green (1984). Correlations to determine HETP values were developed by Murch (1953) and Whitt (1959), but these do not include modern packings (see Ludwig, 1979). An improved mass transfer model for packings and HETP data for a variety of packings are presented by Bolles and Fair (1982) and discussed in Chapter 19. The HETPs are different for different chemical systems and are higher for larger size packing. Most packings of the same size will have approximately the same HETP. Note from Figure 13-3 that the HETP for a given system and packing size is roughly constant over a wide range of gas flow rates. Then as flooding is approached the efficiency of the contact decreases and the HETP rises. Also, at very low gas flow rates the HETP often rises. This occurs because the packing is not completely wet. For self-wetting packings, where capillary action keeps the packing wet, the HETP usually drops at very low gas flow rates.

HETP values are most accurate if determined from data. If no data are available, generalized mass transfer correlations are used (see Chapter 19). If no information is available, Ludwig (1979) suggests using an average of 1.5 to 2.0 ft for dumped packings. If the column diameter is greater than 1 ft, an HETP greater than 1 ft should be used. Another approximate approach is to set HETP equal to column diameter (Ludwig, 1979). Eckert (1979) notes that HETP values for 1, $1\frac{1}{2}$ and 2 in pall rings are 1, $1\frac{1}{2}$ and 2 ft, respectively. These HETP values are almost independent of the system distilled. If liquid distribution is not excellent, a 30 to 50% safety factor is suggested.

Although packed columns operate with a continuous change in vapor and liquid concentrations, the staged model is still a useful design method. Since the HETP is often almost constant throughout the usual design range for gas flow rates, concentrations, and reflux ratios, a single HETP value can usually be used in comparing many different designs. This greatly facilitates design. In certain cases HETP can vary significantly within the column because of changes in composition; it can then be estimated for each stage from the mass transfer coefficient (see Sherwood *et al.*, 1965, pp. 521-523 for an example). Alternatively, a mass transfer design approach can be used and is preferred (see Chapter 19).

13.3. COLUMN DIAMETER

The column diameter is sized to operate at 65 to 90% of flooding or to have a given pressure drop per foot of packing. Flooding can be more easily measured in a packed column than in a plate column and is usually signaled by a break in the curve of pressure drop versus gas flow rate.

The generalized flooding correlation developed by Sherwood *et al.* (1938) as modified by Eckert (1970, 1979) is shown in Figure 13-4. The packing factor, F, depends on the type and size of packing. The higher the value of F, the larger the pressure drop per foot of packing. F values for several types of dumped packing are given in Table 13-1 (Eckert, 1970, 1979; Ludwig, 1979), and F values for structured packings are in Table 13-2 (Fair, 1985). As the packing size increases, the F value decreases, and thus pressure drop per foot will decrease. Ceramic packings have thicker walls than plastic, and the plastic have thicker walls than metal; ceramic packings thus have the lowest free space, highest pressure drops, and highest F values. Generally, the lower the F value the smaller the column diameter. Figure 13-4 is not a perfect fit of all the data. Better results can be obtained using pressure drop curves measured for a given packing.

The flooding curve can be fit by the equation (Kessler and Wankat, 1988; see Chapter 12)

$$\log_{10}\left(\frac{G'^2 F \psi \mu^{0.2}}{\rho_G \rho_L g_c}\right) = -1.6678 - 1.085 \log(F_{lv}) - 0.29655 [\log F_v]^2 \quad (13\text{-}3a)$$

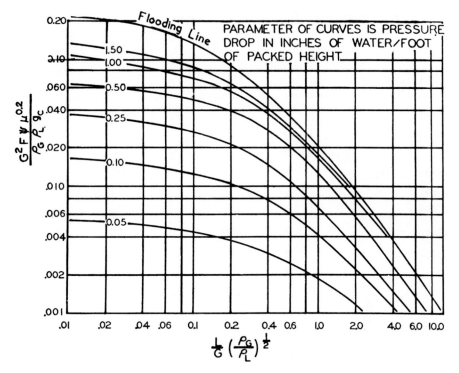

Figure 13-4. Generalized flooding and pressure drop correlation for packed columns. Reprinted with permission from Eckert, *Chem. Eng. Prog., 66* (3), 39 (1970). Copyright 1970 AIChE. Reproduced by permission of the American Institute of Chemical Engineers.

where μ is the liquid viscosity in centipoise, $\psi = \rho_{water}/\rho_L$, $g_c = 32.2$,

$$F_{lv} = \frac{L'}{G'}\left[\frac{\rho_G}{\rho_L}\right]^{1/2}$$

is the abscissa of Figure 13-4, densities are in lb/ft³ and $G' = $ lb/s-ft.²

In the region below the flooding curves, the pressure drop can be correlated with an equation of the form

$$\Delta p = \alpha(10^{\beta L'})(\frac{G'^2}{\rho_G}) \tag{13-3b}$$

where Δp is the pressure drop in inches of water per foot of packing.

422

Table 13-1. Parameters for Dumped Packings

Packing Type		1/4	3/8	1/2	Nominal 5/8
Raschig rings	F	700	390	300	170
(metal, 1/32" wall)	α				1.20
	β				0.28
Raschig rings	F	--	--	410	290
(metal, 1/16" wall)	α			1.59	1.01
	β			0.29	0.39
Raschig rings	F	1600	1000	580	380
(ceramic)	α			1.96	1.31
	β			0.56	0.39
Pall rings	F	--	--	--	97
(plastic)	α				
	β				
Pall rings	F	--	--	--	70
(metal)	α				0.43
	β				0.17
Berl saddles	F	900	--	240	--
(ceramic)	α			1.16	
	β			0.46	
Intalox saddles	F	725	330	200	--
(ceramic)	α			1.04	
	β			0.37	
Intalox saddles	F	--	--	--	--
(plastic)					
Flexirings (plastic)	F				78
Ballast ring (plastic)	F				
Cascade miniring	F				--
(plastic)					

Source: Eckert (1970); Ludwig (1979).

Packing	Size,	in			
3/4	1	1 1/4	1 1/2	2	3
155	115	--	--	--	--
220	137	110	83	57	32
0.80	0.53	--	0.29	0.23	--
0.30	0.19	--	0.20	0.14	--
255	155	125	95	65	37
0.82	0.53	--	0.31	0.23	0.18
0.38	0.22	--	0.21	0.17	0.15
--	52	--	32	25	--
	0.22			0.10	--
	0.14		--	0.12	--
--	48	--	28	20	--
	0.15		0.08	0.06	--
	0.15		0.16	0.12	--
170	110	--	65	45	--
0.56	0.53	--	0.21	0.16	--
0.25	0.18	--	0.16	0.12	--
145	98	--	52	40	22
0.52	0.52		0.13	0.14	--
0.25	0.16		0.15	0.10	--
--	33	--	--	21	16
--	45	--	28	22	--
97	--	52	32	25	--
30	25	18	--	15	--

Table 13-2. F Values for Structured Packings

	Koch Flexipac				Koch	Munters	
	1	2	3	4	Sulzer	12060	19060
F factor	108	72	52	30	66	90	49

Source: Fair (1985)

L' and G' are fluxes in lb/s-ft². Constants α and β are also given in Table 13-1 for dumped packings (Ludwig, 1979). This equation is obviously appropriate for use on calculator or computer.

The generalized correlation in Figure 13-4 or Eqs. (13-3) are used as follows. The designer first picks a point in the column and determines gas and liquid densities (ρ_G and ρ_L), viscosity (μ), value of ψ, and packing factor for the packing of interest. The ratio of liquid to vapor fluxes, L'/G', is equal to the internal reflux ratio, L/V, if the liquid and vapor are of same composition, because the area terms divide out and molecular weights cancel. If liquid and vapor mole fractions are significantly different at this point, then

$$\frac{L'}{G'} \frac{lb/s-ft^2}{lb/s-ft^2} = \frac{L}{V} \frac{\dfrac{moles}{s} \left(MW \text{ liquid } \dfrac{lb}{mole}\right)}{\dfrac{moles}{s} \left(MW \text{ vapor } \dfrac{lb}{mole}\right)} \tag{13-4}$$

In the first design method the designer chooses the pressure drop per unit length of packing. This number ranges from 0.1 to 0.4 in of water per foot for vacuum columns, from 0.2 to 0.6 in of water per foot for absorbers and strippers, and from 0.5 to 1.0 in of water per foot for atmospheric and high-pressure columns. With the value of the abscissa and the parameter known, the ordinate can be determined. G' is the only unknown in the ordinate. Once G' is known, the area is

$$\text{Area} = \frac{\left(V \dfrac{lb \text{ moles}}{s}\right)(M.W. \text{ vapor } \dfrac{lb}{lb \text{ mole}})}{G' \dfrac{lb}{s \text{ ft}^2}} \tag{13-5}$$

A programmable calculator routine for doing these calculations is available (Blackwell, 1984).

In the second design method, the flooding curve or Eq. (13-3a) is used. Then G'_{flood} is calculated from the ordinate. The actual operating vapor flux will be some percent of G'_{flood}. The usual range is 65 to 90% of flooding with 70 to 80% being most common. The area is then determined from Eq. (13-5). The flooding correlation is not perfect. To have 95% confidence a safety factor of 1.32 should be used for the calculated cross-sectional area (Bolles and Fair, 1982; Fair, 1985).

The diameter is easily calculated once the area is known. Since the liquid and vapor properties and gas and liquid flow rates all vary, the designer must calculate the diameter at several locations and use the largest value. Usually, variations in vapor flow rate dominate diameter calculations.

Example 13-1. Diameter Calculation

A distillation column is separating n-hexane from n-heptane using 1-in ceramic Intalox saddles. The allowable pressure drop in the column is 0.5 in H_2O/ft. Average column pressure is 1 atm. Separation in the column is essentially complete, so the distillate is almost pure hexane and the bottoms is almost pure heptane. Feed is a 50-50 mixture and is a saturated liquid. In the top, $L/V = 0.8$. If F = 1000 lb moles/hr and D = 500 lb moles/hr, estimate the column diameter required at the top.

Solution

A. Define. Find the diameter that gives $\Delta p = 0.5$ for top of column.

B. Explore. Need physical properties. From Perry and Green (1984):

n-Hexane: MW 86.17, bp 69°C = 342 K, sp grav = 0.659, viscosity (at 69°) = 0.22 cP

n-Heptane: MW 100.2, bp 98.4°C = 371.4 K, sp grav = 0.684, viscosity (at 98.4°) = 0.205 cP

The ideal gas law can be used to estimate vapor densities, $\rho_v = \dfrac{n(MW)}{V} = \dfrac{p(MW)}{RT}$. Water density at 69° = 0.9783 g/ml. We can use Figure 13-4 with F from Table 13-1 or Eq. (13-3) with α and β from Table 13-1. We will use both methods.

C. Plan. Figure 13-4 and Eq. (13-3) can both be used to determine the required diameter.

D. Do It. The *top* is essentially pure n-hexane. Then,

$$\rho_V = \frac{p(MW)}{RT} = \frac{(1\ atm)(86.17\frac{lb}{lb\ mole})}{(1.314\frac{atm\ ft^3}{K\ lb\ mole})(342\ K)} = 0.1917\frac{lb}{ft^3}$$

$$\frac{L'}{G'} = (\frac{L}{V})(\frac{MW\ liquid}{MW\ vapor}) = (0.8)(\frac{86.17}{86.17}) = 0.8$$

Abscissa for Figure 13-4 is

$$\frac{L'}{G'}(\frac{\rho_v}{\rho_L})^{1/2} = (0.8)\left[\frac{0.1917\ lb/ft^3}{(0.659\frac{g}{cm^3})(\frac{62.4\ lb/ft^3}{g/cm^3})}\right]^{1/2} = 0.055$$

From Figure 13-4 (at $\Delta p = 0.5$), ordinate $= \dfrac{G'^2F\ \psi\mu^{0.2}}{\rho_G\rho_L g_c} = 0.055$
(Obtaining the same value for ordinate and abscissa is an accident!) Then $G' = (\dfrac{0.055\rho_G\rho_L g_c}{F\ \psi\ \mu^{0.2}})^{1/2}$. From Table 13-1, $F = 98$. Thus

$$G' = \left[\frac{(0.055)(0.1917)(0.659)(62.4)(32.2)}{98(\frac{0.9783}{0.659})(0.22)^{.2}}\right]^{1/2} = 0.360\ \frac{lb}{s\ ft^2}$$

From Eq. (13-5), Area $= \dfrac{(V\frac{lb\ mole}{s})(MW)}{G'}$

Calculate V from $V = L + D = (\dfrac{L}{D} + 1)D$, where
$$\frac{L}{D} = \frac{L/V}{1 - L/V} = \frac{0.8}{0.2} = 4$$
$$V = (5)\ D = 5(500\frac{lb\ moles}{hr})(\frac{1\ hr}{3600\ s}) = 0.6944\frac{lb\ mole}{s}$$

This gives

$$Area = \frac{(0.6944)(86.17)}{0.360} = 166\ ft^2$$

Diameter $= (\frac{4 \text{ Area}}{\pi})^{1/2} = (\frac{4}{\pi} 166)^{1/2} = 14.54 \text{ ft}$

Alternative: Use Eq. (13-3). First we must rearrange the equation. Since $L'/G' = L/V$, have $L' = (\frac{L}{V})G'$. Then Eq. (13-3) becomes

$$\Delta p = \alpha(10^{\beta(\frac{L}{V})G'})(\frac{G'^2}{\rho_G})$$

From Table 13-1, $\alpha = 0.52$ and $\beta = 0.16$. Then the equation is

$$0.50 = (0.52)(10^{(0.16)(0.8)G'})(\frac{G'^2}{0.1917})$$

This is an equation with one unknown, G', so it can be solved for G'. Rearranging the equation we have

$$G' = (\frac{0.1843}{10^{0.128G'}})^{1/2}$$

Using our previous answer, $G' = 0.360$, as the first guess and using direct substitution, we obtain $G' = 0.404$ as the answer in two trials.

$$\text{Then, } \quad \text{Area} = \frac{V(\text{MW})}{G'} = \frac{(0.6944)(86.17)}{0.404} = 148.1 \text{ ft}^2$$

$$\text{Diameter} = (\frac{4 \text{ Area}}{\pi})^{1/2} = 13.73 \text{ ft}$$

Note that there is a 6% difference between this answer and the one we obtained graphically. To be conservative, we would use the larger value. Additional safety factors (see Fair, 1985) might be employed if the pressure drop is critical.

E. Check. Solving the problem two different ways is a good, but incomplete, check. The check is incomplete because the same values for several variables (e.g., ρ_G, V, and MW) were used in both solutions. Errors in these variables will not be evident in the comparison of the two solutions. This check did pick up an error I made in reading the ordinate of Figure 13-4 the first time I solved this problem.

F. Generalize. Either Figure 13-4 or Eq. (13-3) can be used for pressure drop calculations in packed beds. The use of both is a good check procedure the first time you calculate a diameter (or Δp). Remember, the required diameters should also be estimated at other locations in the column. It is interesting to compare this design with a design at 75% of flooding. The 75% of flooding design (Problem 13-D5) requires a diameter of 12.4 ft and has a pressure drop of approximately 1.5 in of water per foot of packing. It is also interesting to compare this example with Example 12-2, which is a sieve-tray column for the same distillation problem. At 75% of flooding the sieve tray was 11.03 ft in diameter. The packed column is a larger diameter because a small packing was used. If a larger diameter packing were used, the packed column would be smaller (see Problem 13-D7). The effect of location on the calculated column diameter is explored in Problem 13-D6. If pressure drop is absolutely critical, the column area should be multiplied by a safety factor of 2.2 (Bolles and Fair, 1982).

13.4. ECONOMIC TRADE-OFFS

In the design of a packed column the designer has many trade-offs that are ultimately reflected in the operating and capital costs. After deciding that a packed column will be used instead of a staged column, the designer must choose the packing type. There is no single packing that is most economical for all separations. For most distillation systems the more efficient packings (low HETP and low F) are most expensive per volume but may be cheaper overall. The designer must then pick the material of construction. Since commercial packings of the same size will all have an HETP in the range of 1 to 2 ft, the major difference between them is the packing factor, F. Perusal of Table 13-1 shows that there is a very large effect of packing size and a lesser but still up to fourfold effect of packing type on F value. The material of construction can also change F by a factor that can be as high as 3. From Figure 13-4,

$$G' = \left(\frac{(\text{ordinate})\rho_G\rho_L g_c}{F \, \psi \, \mu^{0.2}} \right)^{1/2}$$

(13-6)

A fourfold increase in F would cause a halving of G' and a doubling of the required area [(Eq. (13-5)]. The diameter can then be calculated,

$$\text{Diameter} \propto (F)^{1/4} \tag{13-7}$$

Thus the major advantage of more efficient packings, structured packings, and the larger size packings is that they can be used with a smaller diameter column, which is not only less expensive but will also require less packing.

At this point, the designer can pick the packing size and determine both the HETP and packing factor. Larger size packing will have a larger HETP (require a larger height), but a smaller F factor and hence a smaller diameter. Thus there is a trade-off between packing sizes. The larger size packings are cheaper per cubic foot but can't be used in very small diameter columns. As a rule of thumb,

$$\text{Column diameter/Packing diameter} > 8 \text{ to } 12 \tag{13-8}$$

depending on whose thumb you are using. The purpose of this rule is to prevent excessive channeling in the column. Structured packings are purchased for the desired diameter column and are not restrained by Eq. (13-8). In small-diameter columns (say less than 6 in), structured packings allow low F factors and hence low Δp without violating Eq. (13-8).

The designer can also select the pressure drop per foot. Operating costs in absorbers and strippers will increase as $\Delta p/ft$ increases, but the diameter decreases and hence capital costs for the column decrease. Operation should be in the range of 20 to 90% of flood and is usually in the range of 65 to 90% of flooding. Since columns are often made in standard diameters, the pressure drop per foot is usually adjusted to give a standard size column.

The reflux ratio is a critical variable for packed columns as it is for staged columns. An L/D between 1.05 $(L/D)_{min}$ and 1.25 $(L/D)_{min}$ would be an appropriate value for the reflux ratio. The exact optimum point depends upon the economics of the particular case.

13.5. SUMMARY - OBJECTIVES

In this brief chapter we have studied the design of packed columns. The objectives you should be able to satisfy are:

1. Describe the parts of a packed column and explain the purpose of each part.

430

2. Use the HETP method to design a packed column. Determine the HETP from data.

3. Calculate the required diameter of a packed column.

4. Determine an appropriate range of operating conditions for a packed column.

REFERENCES

Blackwell, W.W., *Chemical Process Design on a Programmable Calculator,* McGraw-Hill, New York, 1984, Chapt. 4.

Bolles, W.L. and J.R. Fair, "Improved Mass-Transfer Model Enhances Packed-Column Design," *Chem. Eng., 89* (14), 109 (July 12, 1982).

Eckert, J.S., "Selecting the Proper Distillation Column Packing," *Chem. Eng. Prog., 66* (3), 39 (1970).

Eckert, J.S., "Design of Packed Columns," in P.A. Schweitzer (Ed.), *Handbook of Separation Techniques for Chemical Engineers,* McGraw-Hill, New York, 1979, Section 1.7.

Ellis, S.R.M. and F. Brooks, "Performance of Packed Distillation Columns Under Finite Reflux Conditions," *Birmingham Univ. Chem. Eng., 22,* 113 (1971).

Fair, J.R., "Continuous Mass Transfer," in A.R. Bisio and R.L. Kabel (Eds.), *Scaleup of Chemical Processes,* Wiley, New York, 1985, Chapt. 13.

Furter, W.F. and W.T. Newstead, "Comparative Performance of Packings for Gas-Liquid Contacting Columns," *Can. J. Chem. Eng., 51,* 326 (1973).

Ludwig, E.E., *Applied Process Design for Chemical and Petrochemical Plants,* Vol. 2, 2nd ed., Gulf Pub. Co., Houston, TX, 1979.

Murch, D.P., "Height of Equivalent Theoretical Plate in Packed Fractionation Columns," *Ind. Eng. Chem., 45,* 2616 (1953).

"Packed Column Internals," Bulletin No. KI-4, Koch Engineering Co., Wichita, KS (no date).

"Packed Tower Internals," Design Manual TA-60, U.S. Stoneware division of Norton, Akron, Ohio, 1965.

Perry, J.H. (Ed.), *Chemical Engineer's Handbook,* 3rd ed., McGraw-Hill, New York, 1950.

Perry, R.H., C.H. Chilton and S.D. Kirkpatrick (Eds.), *Chemical Engineer's Handbook,* 4th ed., McGraw-Hill, New York, 1963.

Perry, R.H. and C.H. Chilton (eds.), *Chemical Engineer's Handbook,* 5th ed., McGraw-Hill, New York, 1973.

Perry, R.H. and D. Green (Eds.), *Perry's Chemical Engineer's Handbook,* 6th ed., McGraw-Hill, New York, 1984.

Peters, W.A., "The Efficiency and Capacity of Fractionating Columns," *Ind. Eng. Chem., 14,* 476 (1922).

Sherwood, T.K., G.H. Shipley, and F.A.L. Holloway, "Flooding Velocities in Packed Columns," *Ind. Eng. Chem., 30,* 765 (1938).

Sherwood, T.K., R.L. Pigford, and C.R. Wilke, *Mass Transfer,* McGraw-Hill, New York, 1975.

"Tower Packings and Internals," Bulletin Number 217, 3rd ed., Glitsch Inc., Dallas, TX, 1975.

Whitt, F.R., "A Correlation for Absorption Column Packing," *Brit. Chem. Eng.,* July, 1959, p. 395.

HOMEWORK

A. *Discussion Problems*

A1. What are the characteristics of a good packing? Why are marbles a poor packing material?

A2. Develop your key relations chart for this chapter.

A3. Would the addition of a demister in the disengagement region above the packing reduce the required column diameter? Explain your answer.

A4. Why do packed columns become cheaper to build than staged columns for small-diameter systems? For larger diameters, why are staged columns often cheaper? (Hint: What is cost of trays proportional to, and what is cost of packing proportional to?) Under what other conditions would packed columns be preferred?

A5. If HETP varies significantly with the gas rate, how would you design a packed column?

A6. Explain why pressure drop can be detrimental when you are operating a vacuum column.

A7. What effect will an increase in viscosity have on

    a. Pressure drop in a packed column

    b. HETP (consider mass transfer effects)?

A8. Refer to Table 13-1.

    a. Which is more desirable, a high or low packing factor, F?

    b. As packing size increases, does F increase or decrease? What is the functional form of this change (linear, quadratic, cubic, etc.)?

    c. Why do ceramic packings have higher F factors than plastic or metal packings of the same type and size? When would you choose a ceramic packing?

A9. Why can't large-size packings be used in small-diameter columns? What is the reason for the rule of thumb given in Eq. (13-8)?

B. *Generation of Alternatives*

B1. What other ways of contacting in packed columns can you think of? After brainstorming this, see Problem 13-F1.

B2. a. A farmer friend of yours is going to build his own distillation system to purify ethanol made by fermentation. He wants to make his own packing. Suggest 30 different things he could make or buy cheaply to use as packing (set up a brainstorming group to do this - make no judgments as you list ideas).

b. Look at your list in part a. Which idea is the craziest? Use this idea as a trigger to come up with 20 more ideas (some of which may be reasonable).

c. Go through your two lists from parts a and b. Which ideas are technically feasible? Which ideas are also cheap and durable? List about 10 ideas that look like the best for further exploration.

C. *Derivations*

C1. Convert the typical pressure drop per length of packing numbers [given after Eq. (13-3)] to Pa/m.

C2. Since packings are often tested at total reflux, derive an expression to determine HETP from measurements of x_D and x_B for a binary constant relative volatility system at total reflux.

C3. If the packing factor were unknown, you could measure Δp at a series of gas flow rates. How would you determine F from this data?

D. *Problems*

D1. We are testing a new type of packing. A methanol-water mixture is distilled at total reflux and a pressure of 101.3 kPa. The packed section is 1 m long. We measure a concentration of 96 mole % methanol in the liquid leaving the condenser and a composition of 4 mole % methanol in the reboiler liquid. What is the HETP of this packing at this gas flow rate? Equilibrium data are in Table 3-3.

D2. We are testing a new packing for separation of benzene and toluene. The column is packed with 3.5 m of packing and has a total condenser and a partial reboiler. Operation is at 760 mmHg, where α varies from 2.61 for pure benzene to 2.315 for pure toluene (Perry *et al.*, 1963, p. 13-3). At total reflux we measure a benzene mole fraction of 0.987 in the condenser and 0.008 in the reboiler liquid. Find HETP:

    a.   Using $\alpha = 2.315$

    b.   Using $\alpha = 2.40$

    c.   Using $\alpha = 2.50$

    d.   Using $\alpha = 2.61$

    e.   Using a geometric average α.

Use either the results of Problem 13-C2 or a McCabe-Thiele diagram. Compare the differences in parts a to e.

D3. We wish to distill an ethanol-water mixture to produce 2250 lb of distillate product per day. The distillate product is 80 mole % ethanol and 20 mole % water. An L/D of 2.0 is to be used. The column operates at 1 atm. A packed column will use 5/8-in plastic Pall rings. Calculate the diameter at the top of the column.

Physical properties: $MW_E = 46$, $MW_W = 18$, assume ideal gas, $\mu = 0.52$ cP at 176 °F, $\rho_L = 0.82$ g/ml.

    a. Operation is at 75% of flooding. What diameter is required?

    b. Operation is at a pressure drop of 0.25 in water per foot of packing. What diameter is required?

    c. Repeat part a, but for a feed that is 22,500 lb of distillate product per day. Note: it is NOT necessary to redo the entire calculation, since D and hence V and hence diameter are related to the feed rate.

D4. A distillation system is a packed column with 5.0 ft of packing. A saturated vapor feed is added to the column (which is only an enriching section). Feed is 23.5 mole % water with the remainder nitromethane. F=10 kg moles/hr. An L/V of 0.8 is required. $x_D = x_\beta = 0.914$.

    a. Find HETP and water mole fraction in bottoms. Water-nitromethane data are given in Problem 10-E1.

    b. Suppose we try to operate the same system but with 3.0 ft of packing (same HETP). What will happen?

D5. Repeat Example 13-1 for operation at 75% of flood. Calculate the column diameter and the pressure drop per foot of packing.

D6. Repeat Example 13-1, except calculate the diameter at the bottom of the column.

D7. Repeat Example 13-1, except use 3-in Intalox saddles.

E. *More Complex Problems*

E1. Repeat Problem 11-E2 (batch distillation) but use 1-in metal Pall rings instead of a staged column.

F. *Problems Requiring Other Resources*

F1. (Extension) An interesting alternative to normal packed columns is the Higee system. Read either of the short papers listed below and write a short critique (maximum one page). How does this idea fit into your understanding of packed column design? What would a curve of HETP versus mean acceleration look like? Explain the observed mass transfer effects.

Ramshaw, C., "Higee Distillation. An Example of Process Intensification," *The Chemical Engineer (London)*, p. 13-14 (Feb. 1983).

Short, H., "New Mass-Transfer Find Is a Matter of Gravity," *Chemical Engineering, 90* (4) 23 (Feb. 21, 1983).

F2. Repeat Problem 5-D2 (you can use a McCabe-Thiele diagram), except design a packed column instead of a staged column.

    a. Use 2-in ceramic Raschig rings.

    b. Use 1-in ceramic Raschig rings.

    c. Use 1/2-in ceramic Raschig rings.

F3. We wish to distill a mixture of acetone and water. The column will have a total condenser and a partial reboiler. Feed flow rate is 100 kg moles/hr. Feed is 55 mole % acetone and is at 120 °F. We desire a distillate composition of 97 mole % acetone and a bottoms composition of 0.5 mole % acetone (0.005 mole fraction). Operation is at 1 atm. Reflux is returned as a saturated liquid, and $L_0/D = 1.2(L_0/D)_{min}$. The column will be well insulated. Use the optimum feed location. See Problem 11-E2 for equilibrium data. Determine the column diameter and heights of each section for a packed column using 1-in metal pall rings.
Note: The hardest part of this assignment is estimating physical properties. I suggest you use *Perry's Chemical Engineer's Handbook* and the *CRC Handbook of Chemistry and Physics* as resources.

F4. Repeat Problem 12-F4, except design a packed column using 1-in metal pall rings. Approximate HETP for ethanol-water is 1.2 ft.

F5. Refer to Problem 11-E2 for operation at 75% of flooding and a pressure of 700 kPa. Determine the column diameter.

G. *Open-Ended and Synthesis Problems*

G1. We are separating an ethanol-water mixture. The feed rate is

100 kg moles/hr of a 30 mole % ethanol mixture. The feed is a subcooled liquid at 22 ° C but can be preheated by heat exchange. We desire an x_D of 0.8 or above. Bottoms will be sent to the sewer, so ethanol in the bottoms should be quite low. Select the desired separation method. Decide on the type of equipment to use. Design a system to do this separation at minimum cost.

chapter 14
THE ECONOMICS OF DISTILLATION AND ENERGY CONSERVATION IN DISTILLATION

Now that we have considered the design of the entire column we can explore the effect of design and operating parameters on the cost of operation. A brief review of economics will be helpful (for complete coverage, see a design or economics text such as Peters and Timmerhaus, 1980; Rudd and Watson, 1968; or Woods, 1976).

14.1. CAPITAL AND OPERATING COSTS

Total cost per year is the sum of the operating costs plus some fraction of the capital cost of the column. This latter figure is usually determined as depreciation rate times the capital cost. (The number of years to depreciate and allowable depreciation methods are set by government tax laws.) The simplest way to calculate depreciation is by the straight-line method (other methods are covered in design texts). Then

$$\text{Depreciation rate} = \frac{1}{\text{no. years to depreciate to zero value}} \qquad (14\text{-}1)$$

Then,

$$\text{Capital cost/yr} = (\text{depreciation rate})(\text{total capital cost}) \qquad (14\text{-}2)$$

and

$$\text{Total cost/yr} = \text{capital cost/yr} + \text{op cost/yr} + \text{admin. cost/yr} \qquad (14\text{-}3)$$

The total capital cost is the sum of the costs for the condenser, reboiler, tower, and trays and miscellaneous costs including controls, piping, pumps, and so forth. Operating costs include labor costs plus the cost of steam for the reboiler plus the cost of cooling water, plus any additional costs for electricity, compressed air, and so forth. For our purposes, administrative costs include the costs of administration and selling.

Capital costs can be determined by estimating delivered equipment costs and adding on installation, building, piping, engineering, contingency, and indirect costs. These latter costs are often estimated as a factor times the delivered equipment cost for major items of equipment.

$$\text{Total capital cost} = (\text{factor})(\text{delivered equipment cost}) \qquad (14\text{-}4)$$

where the factor ranges from approximately 3.1 to 4.8, with 4.0 a reasonable average (Rudd and Watson, 1968). Thus these "extra" costs greatly increase the capital cost.

Costs of major equipment are often estimated from a power law formula:

$$\text{Cost for size A} = (\text{cost size B})\left(\frac{\text{size A}}{\text{size B}}\right)^{\text{exponent}} \qquad (14\text{-}5)$$

Some equipment will not follow this power law. The appropriate size term depends on the type of equipment. For example, for shell and tube heat exchangers such as condensers, the size used is the area of heat exchanger surface and the exponent is approximately 0.48 (Rudd and Watson, 1968),

$$\text{Condenser cost, size A} = (\text{cost, size B})\left(\frac{\text{area A}}{\text{area B}}\right)^{0.48} \qquad (14\text{-}6)$$

As the area becomes larger, the cost per square foot decreases.

Exact cost determination requires very exact determination of factors, but for determining the effect of operating and design variables, average values can be used. The cost equations for distillation equipment are given in Table 14-1. These tables are for carbon steel columns. Corrosive chemicals may require special materials of construction which will cause the equipment to cost more. These values should be used only for estimating the effects of changing variables and not for accurate economic calculations.

The exponent in Eq. (14-5) is usually less than 1. This means that as size increases, the cost per unit size decreases. This is also shown for Berl saddles where the cost per cubic foot is given and the exponent is negative. This will be translated into a lower cost per kilogram of pro-

Table 14-1. Cost Equations for Distillation Equipment

Condenser cost, Eq. (14–6)*	=	$4080 \left(\dfrac{\text{area, ft}^2}{50}\right)^{0.48}$, \$
Reboiler cost*	=	$12{,}300 \left(\dfrac{\text{area, ft}^2}{400}\right)^{0.25}$, \$
Tower casing cost*	=	$8750 \left(\dfrac{\text{wt, lb}}{6000}\right)^{0.75}$, \$
Sieve tray cost*	=	$151 \left(\dfrac{\text{diameter, ft}}{3}\right)^{1.63}$, \$
Porcelain Berl saddles[†]	=	$29 \left(\dfrac{\text{size, in}}{1}\right)^{-1.12}$, \$/ft^3
Stainless steel pall ring[‡]	=	$72 \left(\dfrac{\text{size, in}}{1}\right)^{-0.63}$, \$/ft^3

* Costs are for carbon steel systems operating near atmospheric pressure. Costs updated from costs in Rudd and Watson (1968) to 1982 cost indices using *Chemical Engineering* plant cost index, which was 312.9. Limiting conditions for use of these equations are discussed by Rudd and Watson (1968).

† Determined from graph in Hall *et al.* (1982), December 1981 costs. Accuracy ±10%. *Chemical Engineering* plant cost index = 305.6.

‡ Determined from data in Peters and Timmerhaus (1980); January 1979 cost. *Chemical Engineering* plant cost index = 229.8.

duct. This "economy of scale" is the major reason that large plants have been built in the past. However, there is currently a trend toward smaller, more flexible plants that can change when the economy changes.

The cost at one size can be estimated by updating published sources or from current vendors' quotes. The method for updating costs is to use a cost index.

$$\text{Cost at time 2} = (\text{cost at time 1})\left(\frac{\text{index, time 2}}{\text{index, time 1}}\right) \qquad (14\text{-}7)$$

The Marshall and Stevens equipment cost index or the *Chemical*

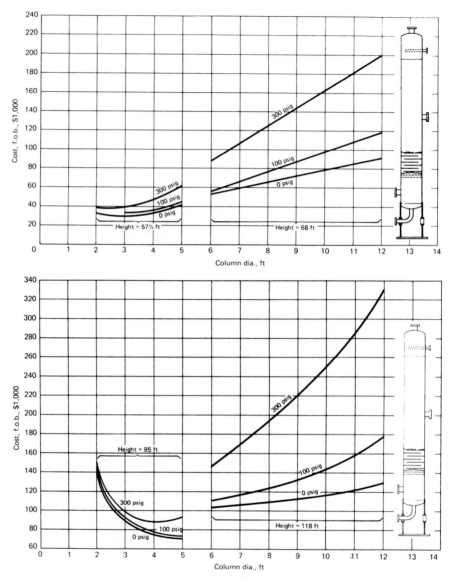

Figure 14-1. Costs of sieve-tray distillation columns. (A) 25 trays; (B) 50 trays. Excerpted by special permission from *Chemical Engineering, 89* (7), 80 (April 5, 1982). Copyright 1982, McGraw-Hill, Inc., New York, NY 10020.

Engineering plant cost index are usually used. Current values are given in each issue of *Chemical Engineering* magazine. The total uninstalled equipment cost will be the sum of condenser, reboiler, tower casing, and

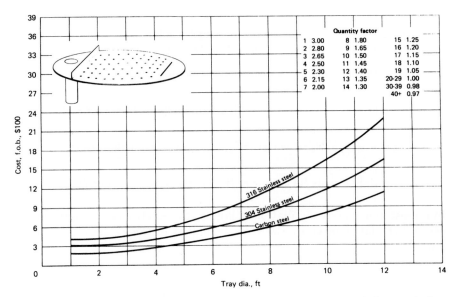

Figure 14-2.   Costs for sieve trays. Excerpted by special permission from *Chemical Engineering, 89* (7), 80 (April 5, 1982). Copyright 1982, McGraw-Hill, Inc., New York, NY 10020.

tray costs. The total capital cost is then found from Eq. (14-4) and capital cost per year from Eq. (14-2).

An alternative method of estimating costs is to use graphs. For example, Hall *et al.* (1982) present graphs for a variety of equipment items including heat exchangers, staged distillation columns and trays, a complete packaged distillation system, and packed towers and packing. These costs were current as of October, November, or December, 1981. They can be updated using Eq. (14-7). Column costs not including trays are given in Figures 14-1A and B (Hall *et al.*, 1982). Figure 14-1A is for a column with 25 trays, while Figure 14-1B is for a column with 50 trays. The jump in height when diameter increases from 5 to 6 ft occurs because the tray spacing must increase to provide for maintenance. Note that for small-diameter columns the cost can actually decrease as the diameter increases, because the tall slender columns must have heavier bases to withstand wind-load stresses. In these small diameters, Eq. (14-5) is obviously not followed. Figures 14-1 are for carbon steel and should be accurate to ±15%. Sieve tray costs are shown in Figure 14-2 (Hall *et al.*, 1982). The graph is based on 20 trays. Costs have to be adjusted using the quantity factor if a different number of trays are ordered. The cost per tray is less when a large number are ordered because the same design and jig can be used to make all of the trays. Note that Figures 14-1 and 14-2 will disagree somewhat with Table 14-1. This is not unexpected, since cost estimates can vary by up to 35%.

The total operating costs per year can be determined as

$$\text{Total operating cost/yr} = [(\text{lb steam/hr})(\text{cost of steam,\$/lb})$$

$$+ (\text{gal water/hr})(\text{cost water, \$/gal}) + (\text{kw elec./hr})(\text{cost elec./hr})$$

$$+ (\text{labor costs/hr})(\text{labor hr/hr oper.})] \times [\text{hours oper./yr}] \quad (14\text{-}8)$$

For most continuous distillation columns, the electricity costs are modest and the labor costs are the same regardless of the values of operating variables.

14.2. CALCULATION OF SIZES AND COSTS

This section will collect together the calculation methods discussed in previous chapters and will combine them with cost factors. We can use the methods developed in Chapters 5, 6, and 8 to calculate the number of equilibrium stages, N_{equil} required. Then

$$N_{actual} = \frac{N_{equil}}{E_o} \quad (14\text{-}9)$$

The height of the column is

$$H = (N_{actual})(\text{tray spacing}) + \text{disengagement heights} \quad (14\text{-}10)$$

The column diameter is found using the methods in Chapter 12. Equation (12-16) shows that for higher pressures the diameter will be somewhat reduced. Conversely, for vacuum operation the diameter will be increased. Increases in tray spacing increase K, which also increases u_{op}, and thus the diameter will decrease while the column height increases.

As $L/D \rightarrow \infty$ (total reflux), the number of stages approaches a minimum which minimizes the column height, but the diameter goes to infinity. As $L/D \rightarrow (L/D)_{min}$, the number of stages and the height become infinite while the diameter becomes a minimum. Both these limits will have infinite capital costs. Thus we expect an optimum L/D to minimize capital costs. Column height is independent of feed flow rate, while diameter is proportional to $F^{1/2}$ and $(L/D)^{1/2}$.

With number of trays and tower diameter known, the tower cost can be determined from Table 14-1. The column weight is

$$\text{Weight} = \pi(\text{Dia})(\text{H})(\text{shell thickness})(\text{steel density}) \qquad (14\text{-}11)$$

The density of carbon steel is about 490 lb/ft^3. The thickness of the shell can be estimated (Perry *et al.*, 1963) for towers above atmospheric pressure as

$$t = \frac{pR}{SE - 0.6p} \qquad (14\text{-}12)$$

where R = radius, p = internal pressure in psi, E = weld efficiency, and S = maximum allowable stress value in psi (13,750 for average carbon steel). Note that standard thicknesses of steel plate would usually be used. Once the weight has been estimated the tower cost can be determined from Table 14.1. An alternative is to use Figure 14-1.

The effect of pressure can be explored by dividing Eq. (12-16) by 2.0 and substituting the result into Eq. (14-12). The result shows t proportional to $p^{3/4}$. Thus, increasing the pressure increases the shell thickness (at modest pressure the $-0.6p$ term in the denominator is negligible). At pressures near 1 atm there will be no effect of pressure, since standard designs are for approximately 100 psig. Substituting these equations into Eq. (14-11) gives us that weight is proportional to $p^{1/4}$. Thus, increasing the pressure increases the weight and hence the cost of the tower casing. The cost of the sieve trays will decrease, because the diameter decreases as pressure increases. If the sieve trays are more expensive than the column casing, then the total cost of the column will be less if the pressure is modestly increased. This is true when there are a lot of stages.

Condenser and reboiler sizes depend on Q_c and Q_R. These values can be determined from external mass and energy balances around the column. Equations (4-14) and (4-16) allow us to calculate Q_c and Q_R for columns with a single feed. Note that Q_c and hence Q_R both increase linearly with L/D and with F. The amount of cooling water required is easily determined from an energy balance on the cooling water.

$$\text{kg cooling water/hr} = |\,Q_c\,|\,/(C_{p,w}\,\Delta T_w) \qquad (14\text{-}13)$$

where $\Delta T_w = T_{w,hot} - T_{w,cold}$, and reasonable values for ΔT_w are in the range of 30 to 40 °F. The cooling water cost per year is

$$\text{Cooling water cost, \$/yr} = (\text{kg water/hr})(\frac{\text{gal}}{\text{kg}})(\frac{\text{cost, \$}}{\text{gal}})(\frac{\text{hr}}{\text{yr}}) \qquad (14\text{-}14)$$

This will be a linear function of L/D and of F if cost per gallon is constant.

In the reboiler the steam is usually condensed from a saturated or superheated vapor to a saturated liquid. Then the steam rate is

$$\text{Steam rate, kg/hr} = \frac{Q_R}{H_{steam} - h_{liquid}} \tag{14-15a}$$

In many applications, $H_{steam} = H_{saturated\ vapor}$ and

$$\text{Steam rate, kg/hr} = \frac{Q_R}{\lambda} \tag{14-15b}$$

where λ is the latent heat of vaporization of water. The steam rate will increase linearly with L/D and with F. The value of λ can be determined from the steam tables. Then the steam cost per year is

$$\text{Steam cost, \$/yr} = (\text{kg steam/hr})(\frac{\text{cost, \$}}{\text{kg steam}})(\frac{\text{hr operation}}{\text{yr}}) \tag{14-16}$$

Note that Q_c and hence Q_R depend linearly on L/D and F; thus, increases in L/D or F linearly increase cooling water and steam rates.

The sizes of the heat exchangers can be determined from the heat transfer equation

$$|\,Q\,| = UA\,\Delta T \tag{14-17}$$

where U = overall heat transfer coefficient, A = heat transfer area, and ΔT is the temperature difference between the fluid being heated and the fluid being cooled. How to use Eq. (14-17) is explained in detail in books on transport phenomena and heat transfer (e.g., Bennett and Myers, 1982; Greenkorn and Kessler, 1972; Bird et al., 1960; Whitaker, 1976; and Kern, 1950). For condensers and reboilers, the condensing fluid is at constant temperature. Then

$$\Delta T = \Delta T_{avg} = T_{hot} - T_{cold,avg} \tag{14-18}$$

where T_{hot} = condensing temperature of fluid or of steam, and $T_{cold,avg}$ = $(1/2)(T_{cold,1} + T_{cold,2})$. For a reboiler, the cold temperature will be constant at the boiling temperature, and $\Delta T = T_{steam} - T_{bp}$. The values of the heat transfer coefficient U depend upon the fluids being

Table 14-2. Approximate Heat Transfer Coefficients

		U, Btu/hr-ft^2– °F
Reboiler:	Steam to boiling aqueous solution	300 to 800 (average ~ 600)
	Steam to boiling oil	20 to 80
Condenser:	Condensing aqueous mixture to water	150 to 800
	Condensing organic vapor to water	60 to 300

Source: Greenkorn and Kessler (1972).

heated and cooled and the condition of the heat exchangers. Tabulated values and methods of calculating U are given in the references. Approximate ranges are given in Table 14-2 (Greenkorn and Kessler, 1972).

If we use average values for U and for the water temperature in the condenser, we can estimate the condenser area. With the steam pressure known, the steam temperature can be found from the steam tables; then, with an average U, the area of the reboiler can be found. Condenser and reboiler costs are then determined from Table 14-1. Since area is directly proportional to Q, which depends linearly on L/D and F, the heat exchanger areas increase linearly with L/D or F.

If the column pressure is raised, the condensation temperature in the condenser will be higher. This is desirable, since ΔT in Eqs. (14-17) and (14-18) will be larger and the required condenser area will be less. In addition, higher pressures will often allow the designer to cool with water instead of using refrigeration. This can result in a large decrease in cooling costs because refrigeration is expensive. With increased column pressure, the boiling point in the reboiler will be raised. Since this is the cold temperature, the value of ΔT in Eqs. (14-17) and (14-18) is reduced and the reboiler area will be increased. An alternative solution is to use a higher pressure steam so that the steam temperature is increased and a larger reboiler won't be required. This approach does increase operating costs, though, since higher pressure steam is more expensive.

The total operating cost per year is given by Eq. (14-8). This value can be estimated as steam costs [(Eq. (14-16)] plus cooling water costs (Eq. (14-14)). A look at these equations shows the effects of various variables; these effects are summarized in Table 14-3.

The capital cost per year can be found from Eqs. (14-2) and (14-4)

Table 14-3. Effect of Changes in Operating Variables on Operating Costs

Cost Item	Change in Variable	Effect on Other Variables	Effect on Cost
Cooling water	L/D increases	$|Q_c|$ increases linearly	Up
	F increases	$|Q_c|$ increases linearly	Up
	Column pressure increases	ΔT_{water} may become larger since T_{bp} in condenser increases or may be unchanged	Down or no effect
		Cooling water may be used instead of refrigeration	Down significantly
	Water costs increase		Up
Steam	L/D increases	Q_R increases linearly	Up
	F increases	Q_R increases linearly	Up
	Column pressure increases	Temp. in reboiler increases; steam pressure may increase to compensate	Up if steam press increases
	Steam costs increase		Up

and Table 14-1. This cost is then

Capital cost per year = (deprec. rate) (factor)   (14-19)

$$\times \text{[condenser cost + reboiler cost}$$

$$\text{+ tower casing cost + (N)(sieve tray cost)]}$$

The individual equipment costs depends on the condenser area, reboiler area, tower weight, and tray diameter. Some of the variable effects on the capital costs are complex. These are outlined in Table 14-4. The net result of changing L/D is shown in Figure 14-3; capital cost goes through a minimum.

The total cost per year is the sum of capital and operating costs and is illustrated in Figure 14-3. Note that there is an optimum reflux ratio. As operating costs increase (increased energy costs), the optimum will shift closer to the minimum reflux ratio. As capital cost increases due to special materials or very high pressures, the total cost optimum will shift toward the capital cost optimum.

The column pressure also has complex effects on the costs. If two pressures both above 1 atm are compared and cooling water can be used for both pressures, then total costs can be either higher or lower for the higher pressure. The effect depends on whether casing costs or tray costs dominate. If refrigeration would be required for condenser cooling at the lower pressure and cooling water can be used at the higher pressure, then the operating costs and the total costs will be less at the higher pressure. See Tables 14-3 and 14-4 for more details.

The effects of other variables are somewhat simpler than the effect of L/D or pressure. For example, when the design feed rate increases, all costs go up; however, the capital cost per pound of feed drops significantly. Thus total costs per pound can be significantly cheaper in large plants than in small plants. The effects of other variables are also summarized in Tables 14-3 and 14-4.

The equations presented here can be used to estimate column costs for sieve tray columns operating near atmospheric pressure for carbon steel columns. For other circumstances, chemical engineering design and economics textbooks should be consulted.

Example 14-1. Cost Estimate for Distillation Column

Table 14-4. Effects of Changes in Design Variables on Capital Costs

Equipment Item	Change in Design Variable	Effect on Other Variables	Effect on Cost
Condenser	L/D increases	$|Q_c|$ increases linearly, area increases linearly	Up
	F increases	$|Q_c|$ increases linearly, area increases linearly	Up
	Column pressure increases	ΔT (Eqs. 14-17 and 14-18) increases, area decreases, but more expensive construction may be needed	Down or up
	U increases	Area decreases	Down
Reboiler	Increase L/D	Q_R increases linearly, area increases linearly	Up
	Increase F	Q_R increases linearly, area increases linearly	Up
	Increase feed temp.	h_F increases and Q_R drops	Down
	Column pressure increases	ΔT (Eq. (14-17 and 14-18) drops and area increases (if steam pressure const.).	Up
		If more expensive construction req'd.	Up
	U increased	area drops	Down

Tower casing	Increase L/D	Diameter increases, weight increases height drops	Down, then up
	Increase F	Diameter increases, weight increases	Up
	Column pressure up	Diameter down, thickness up, weight up	Down, then up
	Higher tray spacing	K up, diameter down, height up, weight up	Up
Sieve tray costs	Increase L/D	Number of trays down as diameter becomes excessive	Down / Up
	Increase F	Diameter up	Up
	Column pressure up	Diameter drops	Down
	Feed temperature increases	More trays at constant (L/D) (see Problem 14-A3)	Up
	Higher tray spacing	Entrainment down, better efficiency	Down
	Trays fancier than sieve trays with increased tray efficiency	Cost per tray up / Fewer trays but may cost more per tray	Calculate for each case

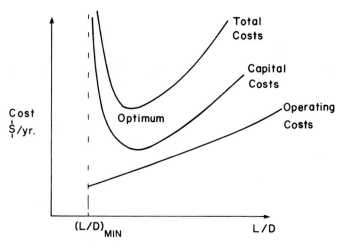

Figure 14-3.　Effect of reflux ratio on costs.

Estimate the cost of the distillation column designed in Examples 12-1 to 12-3.

Solution

The number of equilibrium stages can be calculated from a McCabe-Thiele diagram or estimated by the Fenske-Underwood-Gilliland approach. We will use the latter approach. In Example 12-1, $\alpha = 2.35$ was used. Then from the Fenske equation (9-16),

$$N_{min} = \frac{\ln\left[\left(\frac{x}{1-x}\right)_{dist} \middle/ \left(\frac{x}{1-x}\right)_{bot}\right]}{\ln \alpha}$$

$$= \frac{\ln\left[\frac{0.999}{(1-0.999)} \middle/ \frac{0.001}{(1-0.001)}\right]}{\ln 2.35} = 16.2$$

From Example 16-1, $y = 0.7$ when $x = 0.5$, which is the feed concentration. Then for a saturated liquid feed,

$$\left(\frac{L}{V}\right)_{min} = \frac{x_D - y}{x_D - z} = \frac{0.999 - 0.7}{0.999 - 0.5} = 0.5992$$

$$\left(\frac{L}{D}\right)_{\text{min}} = \frac{(L/V)_{\text{min}}}{1 - (L/V)_{\text{min}}} = 1.495$$

$$\left(\frac{L}{D}\right)_{\text{act}} = \frac{L/V}{1 - L/V} = \frac{0.8}{0.2} = 4$$

Using Eq. (9-42b) for the Gilliland correlation, we have

$$x = \frac{L/D - (L/D)_{\text{min}}}{L/D + 1} = \frac{4 - 1.495}{5} = 0.501$$

and

$$\frac{N - N_{\text{min}}}{N + 1} = 0.545827 - 0.591422(0.501) + \frac{0.002743}{(0.501)} = 0.255$$

$$N = \frac{0.255 + N_{\text{min}}}{(1 - 0.255)} = 22.09 \text{ equil. stages}$$

Subtract 1 for a partial reboiler. From Example 12-1, the overall efficiency $E_o = 0.59$. Then

$$N_{\text{act}} = \frac{N_{\text{equil}}}{0.59} = \frac{21.09}{0.59} = 35.7 \text{ or } 36 \text{ stages}$$

The column diameter was 11 ft in Example 12-2 and 12 ft in Problem 12-D2. Use the larger value.

The shell cost for a 12-ft diameter 25-tray column at 0 psig is $90,000 from Figure 14-1A. For a 50-tray column, the cost is $130,000 from Figure 14-1B. A linear interpolation for 36 stages gives $107,600. An alternative interpolation is to use Eq. (14-5), which becomes

$$\text{Cost size A} = (\text{cost 25 trays})\left(\frac{\text{No. trays size A}}{25}\right)^{\text{exponent}}$$

If A is 50 trays, the cost information from Figures 14-1A and B can be used to estimate the exponent.

$$\text{exponent} = \frac{\ln\left[\dfrac{\text{cost 50 trays}}{\text{cost 25 trays}}\right]}{\ln\left[\dfrac{50}{25}\right]} = 0.53$$

Then,   Shell cost 36 trays = $(\text{cost 25 trays})\left[\dfrac{36}{25}\right]^{0.53}$ = \$109,200

This is close to the linear interpolation. To be conservative, we use the higher value.

The tray cost can be obtained from Figure 14-2. For 12-ft diameter carbon steel trays, the cost per tray is \$1140 times the quantity factor 0.98. This gives a value of \$1120. Then total tray cost is (1120)(36) = \$40,300.

Total column cost = 109,200 + 40,300 = \$149,500 or \$150,000

This cost does not include pumps, instrumentation and controls, reboiler, condenser, or installation. Cost is as of November 1981. It can be updated to current cost with the cost index.

14.3. CHANGES IN PLANT OPERATING RATES

Plants are designed for some maximum nameplate capacity but commonly produce less. The operating cost per kilogram can be found by dividing Eq. (14-8) by the total number of kilograms produced.

$$\text{Operating cost/kg} = \left(\frac{\text{kg steam/hr}}{F}\right)(\text{cost steam})$$

$$+ \left(\frac{\text{gal water/hr}}{F}\right)(\text{water cost}) + \left(\frac{\text{kW elec/h}}{F}\right)(\$/\text{kW})$$

$$+ \left(\frac{\text{worker hr labor/hr}}{F}\right)(\text{labor cost, \$/man hour}) \qquad (14\text{-}20)$$

The kg steam/hr, gal water/hr, and kW elec/hr are all directly proportional to F. Thus, except for labor costs, the operating cost per kilogram will be constant regardless of the feed rate. Since labor costs are often a small fraction of total costs in automated continuous chemical plants, we can treat the operating cost per kilogram as constant.

Capital cost per kilogram does depend on the total amount of feed produced per year. Then, from Eq. (14-19),

$$\text{Capital cost/kg} = \frac{(\text{depreciation rate})(\text{factor})}{(F, \text{ kg/hr})(\text{No. hrs operation/yr})} \qquad (14\text{-}21)$$

Operation at half the designed feed rate doubles the capital cost per kilogram.

The total cost per kilogram is

$$\text{Total cost/kg} = \text{cap. cost/kg} + \text{op. cost/kg} + \text{admin. cost/kg} \qquad (14\text{-}22)$$

The effect of reduced feed rates depends on what percent of the total cost is due to capital cost.

If the cost/kilogram for the entire plant is less than the selling price, the plant will be losing money. However, this does not mean that it should be shut down. If the selling cost is greater than the operating cost plus administrative costs, then the plant is still helping to pay off the capital costs. Since the capital charges are present even if $F = 0$, it is usually better to keep operating. Of course, a new plant would not be built under these circumstances.

14.4. ENERGY CONSERVATION IN DISTILLATION COLUMNS

Distillation columns are often the major user of energy in a plant. Mix et al. (1978) estimated that approximately 3% of the *total* U. S. energy consumption is used by distillation! Thus, energy conservation in distillation systems is extremely important, regardless of the current energy price. Although the cost of energy oscillates, the long-term trend has been up and will probably continue to be up for many years. Several energy-conservation schemes have already been discussed in detail. Most important among these are optimization of the reflux ratio and choice of the correct operating pressure.

What can be done to reduce energy consumption in an existing, operating plant? Since the equipment already exists, there is an incentive to make rather modest, inexpensive changes. Retrofits like this are a favorite assignment to give new engineers, since they serve to familiarize the new engineer with the plant and failure will not be critical. The first thing to do is to challenge the operating conditions (Geyer and Kline, 1976). If energy can be saved by changing the operating conditions, the change may not require any capital. When the feed rate to the column changes, is the column still operating at vapor rates that are near those for optimum efficiency? If not, explore the possibility of varying the column pressure to change the vapor flow rate and thus operate closer to the optimum. This will allow the column to have the equivalent of more equilibrium contacts and allow the operator to reduce the reflux ratio. Reducing the reflux ratio saves energy in the system. Challenge the specifications for the distillate and bottoms products. When products are very pure, rather small changes in product purities can mean significant changes in the reflux ratio.

Second, look at modifications that require capital investment. Improving the controls and instrumentation can increase the efficiency of the system (Geyer and Kline, 1976; Mix *et al.*, 1978; Shinskey, 1984). Better control allows the operator to operate much closer to the required specifications, which means a lower reflux ratio. Payback on this investment can be as short as 6 months. If the column has relatively inefficient trays (e.g., bubble caps) or packing (e.g., Raschig rings), putting in new highly efficient trays (e.g., valve trays) or new high-efficiency packing (e.g., modern rings, saddles, or structured packing) will usually pay even though it is fairly expensive. Old columns and any column with damaged insulation should have the insulation removed and replaced. Current standards for insulation quality and thickness are much more stringent than they were before 1973 (the year of the first oil crisis). Heat exchange and integration of columns are probably far from optimum in existing distillation systems. An upgrade of these facilities should be considered to determine whether it is economical.

When designing new facilities, many energy conservation approaches can be used that might not be economical in retrofits. Heat exchange between streams and integration of processes should be used extensively to minimize overall energy requirements. These methods have been known for many years (e.g., see Robinson and Gilliland, 1950 or Rudd and Watson, 1968), but they were not economical when energy costs were very low. When energy costs shot up in the 1970s and early 1980s, a lot of effort was expended in saving energy in distillation columns (Geyer and Kline, 1976; Henley and Seader, 1981; King, 1981; Kline, 1974; Mix *et al.*, 1978; Null, 1976; O'Brien, 1976; Radian Corp., 1980; Rathore, 1982; and Shinskey, 1984). The sources just cited and many other references discussed by these authors will give more details.

The basic idea of heat exchange is to use hot streams that need to be cooled to heat cold streams that need to be heated. The optimum way to do this depends upon the configuration of the entire plant, since streams from outside the distillation system can be exchanged. The goal is to use exothermic reactions to supply all or at least as much as possible of the heat energy requirements in the plant. If only the distillation

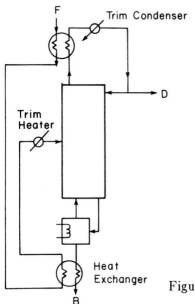

Figure 14-4.   Heat exchange for an isolated distillation column.

system is considered, there are two main heat exchange locations, as illustrated in Figure 14-4. The cold feed is preheated by heat exchange with the hot distillate; this partially or totally condenses the distillate. The trim condenser is used for any additional cooling that's needed and for improved control of the system. The feed is then further heated with the sensible heat from the bottoms product. The heat exchange is done in this order since the bottoms is hotter than the distillate. Further heating of the feed is done in a trim heater to help control the distillation. The system shown in Figure 14-4 may not be optimum, though, particularly if several columns are integrated. Nevertheless, the heat exchange ideas shown in Figure 14-4 are quite basic.

A technique similar to that of Figure 14-4 is to produce steam in the condenser. If there is a use for this low-pressure steam elsewhere in the plant, this can be a very economical use of the energy available in the overhead vapors.

Heat exchange integration of columns is an important concept for reducing energy use. The basic idea is to condense the overhead vapor from one column in the reboiler of a second column. This is illustrated in Figure 14-5. (In practice, heat exchanges like those in Figure 14-4 will also be used, but they have been left off Figure 14-5 to keep the figure simple.) Obviously, the condensation temperature of stream D_1 must be higher than the boiling temperature of stream B_2. When distillation is used for two rather different separations the system shown in Figure 14-

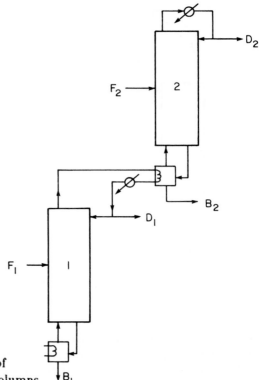

Figure 14-5.   Integration of
distillation columns.

5 can be used without modification. However, in many cases stream D_1 is the feed to the second column. The system shown in Figure 14-5 will work if the first column is at a higher pressure than the second column so that stream D_1 condenses at a higher temperature than that at which stream B_2 boils.

Many variations of the basic idea shown in Figure 14-5 have been developed. If a solvent is recovered from considerably heavier impurities, some variant of the multi-effect system shown in Figure 14-6 is useful (Henley and Seader, 1981; King, 1981; O'Brien, 1976). After preheating, the solvent is first recovered as the distillate product in the first column, which operates at low pressure. The bottoms from this column is pumped to a higher pressure, preheated, and fed to the second column. Since the second column is at a higher pressure, the overheads can be used in the reboiler of the low-pressure column. Thus the steam used in the reboiler of the higher pressure column serves to heat both columns. The steam efficiency is almost doubled. Since the separation is easy, not too many stages are required and the two distillate products are both essentially pure solvent. This system is closely related to multieffect evaporation (Mehra, 1986).

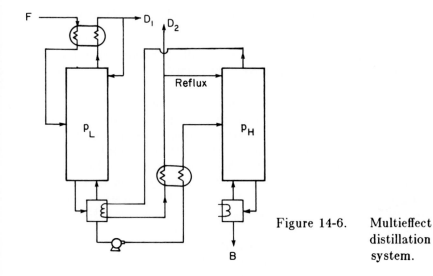

Figure 14-6. Multieffect distillation system.

The condensing vapor from the overhead can be used to heat the reboiler of the same column if vapor recompression or a heat pump is used (Henley and Seader, 1981; King, 1981; Meili and Stuecheli, 1987; Null, 1976; Robinson and Gilliland, 1950). One arrangement for this is illustrated in Figure 14-7. The overhead vapors are condensed to a pressure at which they condense at a higher temperature than that at which the bottoms boil. Vapor recompression works best for close-boiling distillations, since modest pressure increases are required. Generally, vapor recompression is more expensive than heat integration of columns. Thus vapor recompression is used when the column is an isolated installation or is operating at extremes of high or low temperatures.

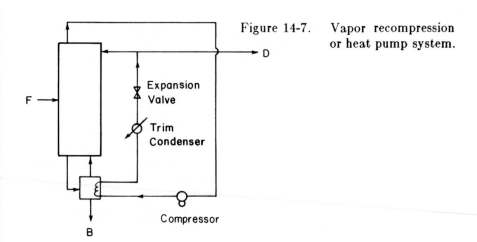

Figure 14-7. Vapor recompression or heat pump system.

14.5. COUPLING OF COLUMNS FOR MULTICOMPONENT PURIFICATIONS

A continuous distillation column is essentially a binary separator; that is, it separates a feed into two parts. For binary systems, both parts can be the desired pure products. However, for multicomponent systems, a single column is unable to separate all the components. For ternary systems, two columns are required to produce pure products; for four-component systems, three columns are required; and so forth. There are many ways in which these multiple columns can be coupled together for multicomponent separations. The choice of cascade can have a large effect on both capital and operating costs. In this section we will briefly look at the coupling of columns. More detailed presentations are available in other books (Henley and Seader, 1981; King, 1981; Rudd et al., 1973), in reviews (Nishida et al., 1981), and in a large number of research papers (Ellashi and Luyben, 1983; Freshwater and Henry, 1975; Henry, 1978; Minderman and Tedder, 1982; Malone et al., 1985; Naka et al., 1982; Rathore, 1982; Seader and Westerberg, 1977; Stupin and Lockhart, 1972).

How many ways can columns be coupled for multicomponent distillation? Lots! For example, Figure 14-8 illustrates nine ways in which columns can be coupled for a ternary system that does not form azeotropes. With more components, the number of possibilities increases geometrically. Figure 14-8A shows the "normal" sequence, where the more volatile components are removed in the distillate one at a time. This is probably the most commonly used sequence, particularly in older plants. Scheme B shows an inverted sequence, where products are removed in the bottoms one at a time. With more components, a wide variety of combinations of these two schemes are possible. The scheme in Figure 14-8C is similiar to the one in part A except that the reboiler has been removed and a return vapor stream from the second column supplies boilup to the first column. Capital costs will be reduced, but the columns are coupled, which will make control and startup more difficult. Scheme D is similiar to B, except that a return liquid stream supplies reflux.

The scheme in Figure 14-8E uses a side enricher while the one in part F uses a side stripper to purify the intermediate component B. The stream is withdrawn at the location where component B has a concentration maximum. These schemes are often used in petroleum refineries. Figure 14-8G illustrates a thermally coupled system (sometimes called Petyluk columns). The first column separates A from C, which is the easiest separation, and the second column then produces three pure products. This system will often have the lowest energy requirements, but it will be more difficult to start up and control. This system may also

459

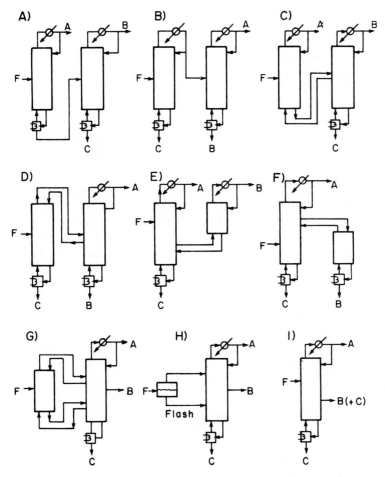

Figure 14-8.   Sequences for distillation of ternary mixtures. No azeotropes. Component A is most volatile, and C is least
volatile.

require an excessive number of stages if either the A-B or B-C separations is difficult. A variant of this scheme is shown in Figure 14-8H,
where the A-C separation is so easy that a flash drum can be used
instead of the first column.

The scheme in Figure 14-8I is quite different from the others, since a
single column with a side stream is used. The side stream cannot be
completely pure B, although it may be pure enough to meet the product
specifications. This or closely related schemes are most likely to be useful when the concentration of C in the feed is quite low. Then at the
point of peak B concentration there will not be much C present.

Methods using side streams to connect columns can also be used (Glinos *et al.*, 1986).

These sequences are only the start of what can be done. For example, the heat exchange and energy integration schemes discussed in the previous section can be interwoven with the separation scheme. Obviously, the system becomes quite complex (Henley and Seader, 1981; Henry, 1978; Naka *et al.*, 1982).

Which method is the best to use depends upon the separation problem. The ease of the various separations, the required purities, and the feed concentrations are all important in determining the optimum configuration. The optimum configuration may also depend upon how "best" is defined. The engineer in charge of operating the plant will prefer the uncoupled systems, while the engineer charged with minimizing energy consumption may prefer the coupled and integrated systems. The only way to be assured of finding the best method is to model all the systems and try them. This is difficult to do, because it involves a large number of interconnected multicomponent distillation columns. Shortcut methods are often used for the calculations to save computer time and money. Unfortunately, the result may not be optimum. Many studies have ignored some of the arrangements shown in Figure 14-8 and thus may not have come up with the optimum scheme. Conditions are always changing, and a distillation cascade may not be optimum when the plant is built because of changes in plant operating conditions such as feed rates and feed or product concentrations. Changes in economics such as energy costs or interest rates may also alter the optimality of the system. Sometimes it is best to build a nonoptimum system because it is more versatile.

An alternative approach to design is to use *heuristics*, which are *rules of thumb* used to exclude many possible systems. The heuristic approach may not result in the optimum separation scheme, but it usually produces a scheme that is close to optimum. Heuristics have been developed by doing a large number of simulations and then looking for ideas that connect the best schemes. Some of the most common heuristics are:

1. Remove the lightest component first.

2. Remove components one by one in the distillate.

3. Remove first the components that require very high or very low temperatures or pressures.

4. Do the most difficult separation last or without non-keys present.

5. Remove the component in excess first.

6. Favor 50:50 splits.

7. Do not use distillation if $\alpha_{LK-HK} < \alpha_{min}$, where $\alpha_{min} \sim 1.05$ to 1.10.

8. Do the easiest remaining separation next.

9. For sloppy separations, consider using side streams to withdraw products.

There are rational reasons for each of the heuristics. Heuristics 1 to 3 ensure that the component hardest to condense is removed first, which will minimize the use of high-pressure columns or refrigerated systems. In addition, heuristic 2 ensures that almost all components are removed as distillate products and are thus less likely to be contaminated with degradation products. The fourth heuristic ensures that flow rates will be minimized in the tall columns required for difficult separations; this will minimize the column diameter for these tall columns and thus reduce costs. Heuristic 7 seeks to avoid the use of excessively tall columns. Heuristic 5 seeks to minimize the flow rates in the cascade, while heuristic 6 will produce columns without large changes in flow rates. This means that both rectifying and stripping sections can operate at flow rates near the optimum. Heuristic 8 will also tend to minimize flow rates in the more difficult separations. The last heuristic states that when high purity is not required, single-column schemes should be considered.

Each heuristic should be preceded with the words "All other things being equal." Unfortunately, all other things usually are not equal, and the heuristics often conflict with each other. For example, the most concentrated component may not be the most volatile. When there are conflicts between the heuristics, the cascade schemes suggested by both of the conflicting heuristics should be generated and then compared with more exact calculations. The heuristics will usually not generate thermally coupled column sequences unless heuristic 8 is used to force their generation, so thermally coupled sequences should be added to the list of sequences to be compared in more detail.

Example 14-2. Use of Heuristics

A feed with 25 mole % ethanol, 15 mole % isopropanol, 35 mole % n-propanol, 10 mole % isobutanol, and 15 mole % n-butanol is

to be distilled. 98% purity of each alcohol is desired. Determin
possible optimum column configurations.

Solution

A, B, C. Define, Explore, Plan. With five components there ar
a huge number of possibilities; thus, we will use heuristics to gen
erate possible configurations. Equilibrium data can be approxi
mated as constant relative volatilities (King, 1981). Ethanol
$\alpha = 2.09$, Isopropanol, $\alpha = 1.82$, n-propanol, $\alpha = 1.0$; isobutanol
$\alpha = 0.677$; n-butanol, $\alpha = 0.428$.

D. Do It.

Case 1. Heuristics 1 and 2 give the scheme

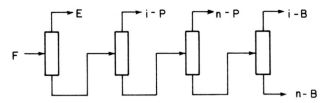

(Reboilers and condensers are not shown.) This will certainly
work, but it is not very inventive.

Case 2. Heuristic 4 is often very important. Which separation i
most difficult? This can be determined by finding the relativ
volatilities of all adjacent pairs of compounds. For example,

$$\alpha_{\text{EtOH,i-P}} = \frac{2.09}{1.82} = 1.15$$

This is the hardest separation. If we also use heuristic 6 fo
column A and heuristic 5 or 6 for column C, we obtain th
scheme shown in the figure.

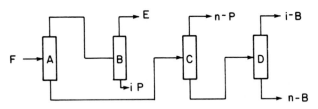

Naturally, other alternatives are possible.

Case 3. Heuristic 8 can be used to generate an entirely thermally coupled system (see Problem 14-A17). This would be difficult to operate. However, we can use heuristics 4, 6, and 8 to obtain a modification of case 2 (see figure).

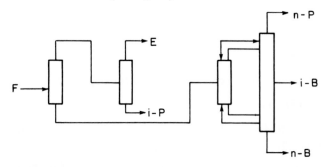

Other systems can be generated, but one of the three shown here is probably close to optimal.

E. Check. Finding the optimum configuration requires a simulation of each alternative. This can be done for cases 1 and 2 using the Fenske-Underwood-Gilliland approach. For case 3 the thermally coupled columns are more complex and probably should be simulated in detail.

F. Generalize. It is likely that one of these designs is close to optimum. Because of the low relative volatility between ethanol and iso-propanol, heuristic 4 is probably most important. Thus the case 2 or case 3 systems are probably closest to optimum. Use of the heuristics does avoid having to look at several hundred other alternatives. Note that heuristics 3, 7, and 9 were not used since they were not applicable to this problem.

14.6. SUMMARY - OBJECTIVES

In this chapter we have looked briefly at the economics of distillation and the effects of changing operating variables. At the end of this chapter you should be able to meet the following objectives:

1. Use the basic economics.

2. Estimate the capital and operating costs for a distillation column.

3. Predict the effect of the following variables on column capital and operating costs:

a. Feed rate

b. Column pressure

c. External reflux ratio

4. Estimate the effects that external factors have on capital and operating costs. External factors would include:

   a. Energy costs

   b. Government schedule allowed for depreciation

   c. The general state of the economy

5. Discuss methods for reducing energy in distillation systems. Develop flow sheets with appropriate heat exchange.

6. Use heuristics to develop alternative cascades for the distillation of multicomponent mixtures.

REFERENCES

Bennett, C.O. and J.E. Myers, *Momentum, Heat and Mass Transfer*, 3rd ed., McGraw-Hill, New York, 1982.

Bird, R.B., W.E. Stewart, and E.N. Lightfoot, *Transport Phenomena*, Wiley, New York, 1960.

Elsahi, A. and W.L. Luyben, "Alternative Distillation Configurations for Energy Conservation in Four-Component Separations," *Ind. Eng. Chem. Process Des. Develop., 22,* 80 (1983).

Freshwater, D.C. and B.D. Henry, "The Optimal Configuration of Multicomponent Distillation Trains," *The Chemical Engineer* (London), No. 301, 533 (Sept. 1975).

Geyer, G.R. and P.E. Kline, "Energy Conservation Schemes for Distillation Processes," *Chem. Eng. Prog., 72* (5), 49 (May 1976).

Glinos, K.N., J.P. Nikolaides, and M.F. Malone, "New Complex Column Arrangements for Ideal Distillation," *Ind. Eng. Chem. Process Des. Develop, 25,* 694 (1986).

Greenkorn, R.A. and D.P. Kessler, *Transfer Operations*, McGraw-Hill, New York, 1972.

Hall, R.S., J. Matley, and K.J. McNaughton, "Current Costs of Process Equipment," *Chem. Eng., 89* (7), 80-116 (April 5, 1982).

Henley, E.J. and J.D. Seader, *Equilibrium-Stage Separation Operations in Chemical Engineering,* Wiley, New York, 1981.

Henry, B.D., "Economies Possible in the Separation of Multicomponent Mixtures by Distillation," in *Alternatives to Distillation,* Institute of Chemical Engineers Symp. Ser., No. 54, 75 (1978).

Kern, D.Q., *Process Heat Transfer,* McGraw-Hill, New York, 1950.

King, C.J., *Separation Processes,* 2nd ed., McGraw-Hill, New York, 1981.

Kline, P.E., "Technical Task Force Approach to Energy Conservation," *Chem. Eng. Prog., 70* (2), 23 (Feb. 1974).

Malone, M.F., K. Glinos, F.E. Marquez, and J.M. Douglas, "Simple, Analytical Criteria for the Sequencing of Distillation Columns," *AIChE J., 31,* 683 (1985).

Mehra, D.K., "Selecting Evaporators," *Chem. Eng., 93* (3), 56 (Feb. 3, 1986).

Meili, A. and A. Stuecheli, "Distillation Columns with Direct Vapor Recompression," *Chem. Eng., 94*(2), 167 (Feb. 16, 1987).

Minderman, P.A. and Tedder, D.W., "Comparisons of Distillation Networks: Extensively State Optimized Versus Extensively Energy Integrated," *AIChE Symp. Ser., 78* (217), 69 (1982).

Mix, T.J., J.S. Dweck, M. Weinberg, and R.C. Armstrong, "Energy Conservation in Distillation," *Chem. Eng. Prog., 74* (4), 49 (April 1978).

Naka, Y., M. Terashita and T. Takamatsu, "A Thermodynamic Approach to Multicomponent Distillation System Synthesis," *AIChE J., 28,* 812 (1982).

Nishida, N., G. Stephanopoulous, and A.W. Westerberg, "A Review of Process Synthesis," *AIChE J., 27,* 321 (1981).

Null, H.R., "Heat Pumps in Distillation," *Chem. Eng. Prog., 72* (7), 58 (July 1976).

O'Brien, N.G., "Reducing Column Steam Consumption," *Chem. Eng. Prog.*, *72* (7), 65 (July 1976).

Perry, R.H., C.H. Chilton, and S.D. Kirkpatrick (Eds.), *Chemical Engineer's Handbook*, 4th ed., McGraw-Hill, New York, 1963.

Peters, M.S. and K.D. Timmerhaus, *Plant Design and Economics for Chemical Engineers*, 3rd ed., McGraw-Hill, New York, 1980.

Radian Corp., "Energy Conservation in Distillation," DOE/CS/4431-T2, Oak Ridge, TN, 1980.

Rathore, R.N.S., "Reusing Energy Lowers Fuel Needs of Distillation Towers," *Chem. Eng.*, *89* (12), 155 (June 14, 1982).

Rathore, R.N.S., "Process Resequencing for Energy Conservation," *Chem. Eng. Prog.*, *78* (12), 75 (Dec. 1982).

Robinson, C.S. and E.R. Gilliland, *Elements of Fractional Distillation*, 4th ed., McGraw-Hill, New York, 1950, Chapt. 7.

Rudd, D.F., G.J. Powers, and J.J. Sirola, *Process Synthesis*, Prentice-Hall, Englewood Cliffs, NJ, 1973.

Rudd, D.F. and C.C. Watson, *Strategy of Process Engineering*, Wiley, New York, 1968.

Seader, J.D. and A.W. Westerberg, "A Combined Heuristic and Evolutionary Strategy for Synthesis of Simple Separation Sequences," *AIChE J.*, *23*, 951 (1977).

Shinskey, F.G., *Distillation Control, For Productivity and Energy Conservation*, 2nd ed., McGraw-Hill, New York, 1984.

Stupin, W.J. and F.J. Lockhart, "Thermally Coupled Distillation - A Case History," *Chem. Eng. Prog.*, *68* (10), 71 (Oct. 1972).

Whitaker, S., *Fundamental Principles of Heat Transfer*, Pergamon, Elmsford, NY, 1976.

Woods, D.R., *Financial Decision Making in the Process Industries*, Prentice-Hall, Englewood Cliffs, NJ, 1976.

HOMEWORK

A.   *Discussion Problems*

A1.   If valve trays cost more than sieve trays, why are they often advertised as a way of decreasing tower costs?

A2.   Develop your key relations chart for this chapter.

A3.   What is the effect of increasing the feed temperature if

   a. L/D is constant

   b. $L/D = 1.15 (L/D)_{min}$.   Note that $(L/D)_{min}$ will change.

   Include effects on Q_R and number of stages. Use a McCabe-Thiele diagram.

A4.   If the government slows down depreciation schedules, what effect will this have on:

   a.   Design of new plants

   b.   Operation of existing plants

A5.   Working capital is the money required for day-to-day operation of the plant, and interest on working capital is an operating expense. If the feed rate drops, is the working capital cost per kilogram constant? What happens to working capital if customers are slow paying their bills? Some companies give a 5% discount for immediate payment of bills; explain why this might or might not be a good idea.

A6.   How does the general state of the economy affect:

   a.   Design of new plants

   b.   Operation of existing plants

A7.   Why is the dependence on size less than linear [in other words, why is the exponent in Eq. (14-5) less than 1]?

A8.   It is common to design columns at reflux ratios slightly above $(L/D)_{opt}$. Use a curve of total cost/yr versus L/D to explain why an $L/D > (L/D)_{opt}$ is used. Why isn't there a large cost penalty?

A9.   Discuss the concept of economies of scale. What happens to economies of scale if the feed rate is half the design value?

A10. Use a McCabe-Thiele diagram to explain how reducing the product concentration allows the use of a lower L/D for an existing column.

A11. Referring to Figure 14-5, if D_1 is the feed to column 2, explain what would be done to make this system work.

A12. Explain how vapor recompression works.

A13. Figure 14-8I shows an arrangement that is useful when the feed concentration of the heavy component, C, is low. Sketch an arrangement for use when the light component, A, feed concentration is low.

A14. In the desert, cooling water is very expensive. What does this do to your costs? What alternatives can be used?

A15. Explain how Figures 14-6 and 14-5 differ.

A16. Figure 14-7 shows a heat pump system in which the distillate vapor is used as the working fluid. It may be desirable to use a separate working fluid. Sketch this. What are the advantages and disadvantages?

A17. Draw the entirely thermally coupled system (an extension of Figure 14-8G) for Example 14-2.

A18. Preheating the feed will often increase the number of stages required for the separation (F, z, x_D, x_B, L/D constant). Use a McCabe-Thiele diagram to explain why this happens.

B. *Generation of Alternatives*

B1. Sketch the possible column arrangements for separation of a four-component system. Note that there are a large number of possibilities.

B2. Repeat Example 14-2 but for a 99.9% recovery of n-propanol. Purities can be lower.

B3. Multieffect distillation or column integration can be done with more than two columns. Use the basic ideas in Figures 14-5 and 14-6 to sketch as many ways of thermally connecting three columns as you can.

B4. We wish to separate a feed that is 10 mole % benzene, 55 mole % toluene, 10 mole % xylene, and 25 mole % cumene. Use heuristics to generate desirable alternatives. Average relative volatilities are $\alpha_B = 2.5$, $\alpha_T = 1.0$, $\alpha_X = 0.33$, $\alpha_C = 0.21$. 98% purity of all products is required.

B5. Repeat Problem 14-B4 for an 80% purity of the xylene product.

C. *Derivations*

C1. Show that cooling water and steam costs are directly proportional to the feed rate.

C2. Calculate the effect of L/D on the tower casing cost. Show that there is a minimum in this cost.

C3. Show that sieve tray costs go through a minimum as L/D increases. Is this minimum significant?

C4. Packing costs are usually directly proportional to the volume of packing. Show that packing costs go through a minimum as L/D increases.

D. *Problems*

D1. Problems 12-D1 and 12-F2 explored the same distillation as in Examples 12-1, 12-2, and 14-1 except that the pressure was increased to 700 kPa. If the resulting column has $E_o = 0.73$ and is 9 ft in diameter, estimate the cost of the shell and the trays. Other parameters are the same as in Examples 12-1 and 14-1. Note that the relative volatility is a function of pressure.

D2. Estimate the cost of the condenser and reboiler for the distillation of Examples 12-1, 12-2, and 14-1. Pressure is 101.3 kPa. $C_{PL,C7} = 50.8$ Btu/lb-mole-$°$F, $\lambda_{C7} = 14,908$ Btu/lb-mole. Data for hexane are given in Problem 4-D6. The saturated steam in the reboiler is at $110°$C. $\lambda_{steam} = 958.7$ Btu/lb. Cooling water enters at $70°$F and leaves at $110°$F. $C_{P,w} = 1.0$ Btu/lb-$°$F. Use heat transfer coefficients from Table 14-2 and cost data from Table 14-1. Watch your units.

D3. Determine the steam and water operating costs per hour for Problem 14-D2. Cost of steam is $8.00/1000 lb, and cost of cooling water is $1.25/1000 gal.

D4. Example 13-1 and Problem 13-D6 sized the diameter of a packed column doing the separation in Examples 12-1 and 14-1. Suppose a 15-ft diameter column is to be used. The 1-in Intalox saddles have an HETP of 1.1 ft. In January 1979 the 1-in Intalox saddle cost $14.20/ft^3. Estimate the packing costs. Pressure is 101.3 kPa.

D5. Problem 13-F5 showed that a 9-ft diameter column was adequate for Example 13-1 at a pressure of 700 kPa. Using the same data as in Problem 14-D4, estimate the packing cost. The number of equilibrium contacts will be the same as in Problem 14-D1.

F.   *Problems Requiring Other Resources*

F1.   Look up the current cost index in *Chemical Engineering* magazine. Use this to update the cost equations in Table 14-1 and the ordinates in Figures 14-1 and 14-2.

chapter 15
ABSORPTION AND STRIPPING

Up to now we have talked almost entirely about distillation. There are other unit operations that are very useful in the processing of chemicals or in pollution control. Absorption is the unit operation where one or more components of a gas stream are removed by being taken up (absorbed) in a nonvolatile liquid (solvent). In this case the liquid solvent must be added as a *separating agent*.

Stripping is the opposite to absorption. In this unit operation, one or more components of a liquid stream are removed by being vaporized into an insoluble gas stream. Here the gas stream (stripping agent) must be added as a separating agent.

What was the separating agent for distillation? Heat.

Absorption can be either physical or chemical. In *physical absorption* the gas is removed because it has greater solubility in the solvent than other gases. An example is the removal of butane and pentane from a refinery gas mixture ($C_4 - C_5$) with a heavy oil. In *chemical absorption* the gas to be removed reacts with the solvent and remains in solution. An example is the removal of CO_2 or H_2S by reaction with NaOH or with monoethanolamine (MEA). The reaction can be either irreversible (as with NaOH) or reversible (with MEA). For irreversible reactions the resulting liquid must be disposed of, whereas in reversible reactions the solvent can be regenerated (in stripper or distillation columns). Thus reversible reactions are often preferred. Chemical absorption systems are discussed in more detail by Astarita *et al* (1983), Kohl (1987), and Kohl and Riesenfeld (1985).

Chemical absorption usually has a much more favorable equilibrium relationship than physical absorption (solubility of most gases is usually very low) and is therefore often preferred. However, the Murphree efficiency is often quite low (10% is not unusual), and this must be taken into account.

Both absorption and stripping can be operated as equilibrium stage

472

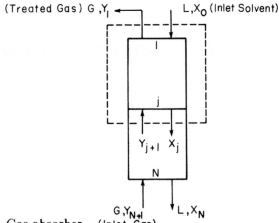

(Treated Gas) G, Y_1 ← L, X_0 (Inlet Solvent)

Y_{j+1} X_j

G, Y_{N+1} L, X_N

Figure 15-1.   Gas absorber. (Inlet Gas)

operations with contact of liquid and vapor. Since distillation is also an equilibrium stage operation with contact of liquid and vapor, we would expect the equipment to be quite similar. This is indeed the case; both absorption and stripping are operated in packed and plate towers. Plate towers can be designed by following an adaption of the McCabe-Thiele method. Packed towers can be designed by use of HETP or preferably by mass transfer considerations (see Chapter 19).

In both absorption and stripping a separate phase is added as the separating agent. Thus the columns are simpler than those for distillation in that reboilers and condensers are normally not used. Figure 15-1 is a schematic of a typical absorption column. In this column solute B entering with insoluble carrier gas C in stream Y_{N+1} is absorbed into the nonvolatile solvent A.

Figure 15-2.   Gas treatment plant.

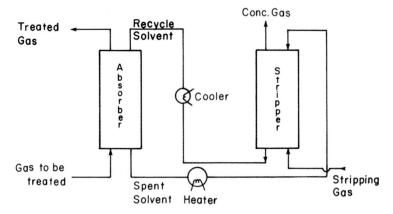

A gas treatment plant often has both absorption and stripping columns as shown in Figure 15-2. In this operation the solvent is continually recycled. The heat exchanger heats the saturated solvent, changing the equilibrium characteristics of the system so that the solvent can be stripped. A very common type of gas treatment plant is used for the removal of CO_2 and/or H_2S from refinery gas or natural gas. In this case MEA or other amine solvents in water are used as the solvent, and steam is used as the stripping gas (for more details see Kohl and Riesenfeld, 1985).

15.1. ABSORPTION AND STRIPPING EQUILIBRIA

For absorption and stripping in three component systems we often assume that

1. Carrier gas is insoluble.

2. Solvent is nonvolatile.

3. The system is isothermal and isobaric.

The Gibbs phase rule is

$$F = C - P + 2 = 3(A, B, \text{ and } C) - 2(\text{vapor and liquid}) + 2 = 3$$

If we set T and p constant, there is one remaining degree of freedom. The equilibrium data are usually represented either by plotting solute composition in vapor versus solute composition in liquid or by giving a Henry's law constant. Henry's law is

$$p_B = H_B x_B \qquad (15\text{-}1)$$

where H_B is Henry's law constant, in atm/mole frac, $H = H(p,T,\text{composition})$; x_B is the mole fraction B in the liquid; and p_B is the partial pressure of B in the vapor.

Henry's law is valid only at low concentrations of B. Since partial pressure is defined as

$$y_B \equiv \frac{p_B}{p_{tot}} \qquad (15\text{-}2)$$

Henry's law becomes

$$y_B = \frac{H_B}{p_{tot}} x_B \qquad (15\text{-}3)$$

This will plot as a straight line if H_B is a constant. If the component is pure, $y_B = 1$ and $p_B = p_{tot}$. Equilibrium data for absorption are given by Hwang (1981), Kohl (1987), Kohl and Riesenfeld (1985), Perry *et al.* (1963, pp. 14-2 to 14-12), Perry and Chilton (1973, p. 14-3), Perry and Green (1984, pp. 3-101 to 3-103), and Siedell and Linke (1952). For example, the values given for CO_2, CO, and H_2S are shown in Table 15-1 (Perry *et al.*, 1963). The large H values in Table 15-1 show that CO_2 and H_2S are very sparingly soluble in water. Since H is roughly independent of p_{tot}, this means that more gas is absorbed at higher pressure. This phenomenon is commonly taken advantage of to make carbonated beverages. When the bottle or can is opened the pressure drops and the gas desorbs, forming little bubbles.

The Henry's law constants depend upon temperature and usually fol-

Table 15-1. Henry's Law Constants, H, for CO_2, CO, and H_2S in Water. H is in atm/mole frac.

T ° C	CO_2	CO	H_2S
0	728	35,200	26,800
5	876	39,600	31,500
10	1040	44,200	36,700
15	1220	48,900	42,300
20	1420	53,600	48,300
25	1640	58,000	54,500
30	1860	62,000	60,900
35	2090	65,900	67,600
40	2330	69,600	74,500
45	2570	72,900	81,400
50	2830	76,100	88,400
60	3410	82,100	103,000
70	----	84,500	119,000
80	----	84,500	135,000
90	----	84,600	144,000
100	----	84,600	148,000

Source: Perry *et al.* (1963), pp. 14-4 and 14-6.

Table 15-2. Absorption of Ammonia in Water.

Weight NH₃ per 100 weight H₂O	Partial pressure of NH₃, mm Hg						
	0 °C	10 °C	20 °C	30 °C	40 °C	50 °C	60 °C
100	947						
90	785						
80	636	987	1450	----	----	3300	
70	500	780	1170	----	----	2760	
60	380	600	945	----	----	2130	
50	275	439	686	----	----	1520	
40	190	301	470	----	719	1065	
30	119	190	298	----	454	692	
25	89.5	144	227	----	352	534	825
20	64	103.5	166	----	260	395	596
15	42.7	70.1	114	----	179	273	405
10	25.1	41.8	69.6	----	110	167	247
7.5	17.7	29.9	50.0	----	79.7	120	179
5	11.2	19.1	31.7	----	51.0	76.5	115
4	----	16.1	24.9	----	40.1	60.8	91.1
3	----	11.3	18.2	23.5	29.6	45	67.1
2.5	----	----	15.0	19.4	24.4	(37.6)*	(55.7)
2	----	----	12.0	15.3	19.3	(30.0)	(44.5)
1.6	----	----	----	12.0	15.3	(24.1)	(35.5)
1.2	----	----	----	9.1	11.5	(18.3)	(26.7)
1.0	----	----	----	7.4	----	(15.4)	(22.2)
0.5	----	----	----	3.4			

* Extrapolated values.
Source: Perry et al. (1963). Copyright 1963. Reprinted with permission of McGraw-Hill.

low an Arrhenius relationship. Thus,

$$H = H_0 \exp\left(\frac{-E}{RT}\right) \qquad (15\text{-}4)$$

A plot of log H versus $1/T$ will often give a straight line.

The effect of concentration is shown in Table 15-2, where the absorption of ammonia in water (Perry et al., 1963) is illustrated. Note that the solubilities are nonlinear and $H = p_{NH_3}/x$ is not a constant. This behavior is fairly general for soluble gases.

We will convert equilibrium data to the concentration units required for calculations. If mole or mass ratios are used, equilibrium *must* be converted into ratios.

15.2. OPERATING LINES FOR ABSORPTION

The McCabe-Thiele diagram is most useful when the operating line is straight. This requires that

The energy balances be automatically satisfied
Liquid flow rate/vapor flow rate = constant

In order for energy balances to be automatically satisfied, we must assume that

1. The heat of absorption is negligible
2. Operation is isothermal

These two assumptions will guarantee satisfaction of the enthalpy balances. When the gas and liquid streams are both fairly dilute, the assumptions will probably be satisfied.

We also desire a straight operating line. This will be automatically true if we define

$$L/G = \frac{\text{moles nonvolatile solvent/hr}}{\text{moles insoluble carrier gas/hr}}$$

and if we assume that:

3. Solvent is nonvolatile
4. Carrier gas is insoluble

Assumptions 3 and 4 are often very closely satisfied. The results of these last two assumptions are that the mass balance for solvent becomes

$$L_N = L_j = L_0 = L = \text{constant} \tag{15-5}$$

while the mass balance for the carrier gas is

$$G_{N+1} = G_j = G_1 = G = \text{constant} \tag{15-6}$$

Note that we cannot use overall flow rates of gas and liquid in concentrated mixtures because a significant amount of solute may be absorbed which would change gas and liquid flow rates and give a curved operating line. For very dilute solutions ($<$ 1% solute), overall flow rates can be used, and mass or mole fractions can be used for operating equations and equilibria. Since we want to use L = moles nonvolatile solvent (S)/hr and G = moles insoluble carrier gas (C)/hr, we must define our compositions in such a way that we can write a mass balance for solute B. How do we do this?

After some manipulation we find that the correct way to define our compositions is as *mole ratios.* Define

$$Y = \frac{\text{moles B in gas}}{\text{moles pure carrier gas C}} \quad \text{and} \quad X = \frac{\text{moles B in liquid}}{\text{moles pure solvent S}} \quad (15\text{-}7a)$$

The mole ratios Y and X are related to our usual mole fractions by

$$Y = \frac{y}{1-y} \quad \text{and} \quad X = \frac{x}{1-x} \quad (15\text{-}7b)$$

Note that both Y and X can be greater than 1.0. With the mole ratio units, we have

$$Y_j G = \left[\frac{\text{moles B in gas stream j}}{\text{moles carrier gas}}\right] \left[\frac{\text{moles carrier gas}}{\text{hr}}\right]$$

$$= \frac{\text{moles B in gas stream j}}{\text{hr}}$$

and

$$X_j L = \left[\frac{\text{moles B in liquid stream j}}{\text{moles solvent}}\right] \left[\frac{\text{moles solvent}}{\text{hr}}\right]$$

$$= \frac{\text{moles B in liquid stream j}}{\text{hr}}$$

Thus we can easily write the steady-state mass balance, input = output, in these units. The mass balance around the top of the column using the mass balance envelope shown in Figure 15-1 is

$$Y_{j+1} \, G + X_0 L = X_j L + Y_1 G \qquad (15\text{-}8)$$

or

$$\text{Moles B in/hr} = \text{moles B out/hr}$$

Solving for Y_{j+1} we obtain

$$Y_{j+1} = \frac{L}{G} X_j + \left[Y_1 - \frac{L}{G} X_0 \right] \qquad (15\text{-}9)$$

This is a straight line with slope L/G and intercept $\left(Y_1 - \dfrac{L}{G} X_0 \right)$. It is our *operating line* for absorption. Thus if we plot ratios Y vs X we have a McCabe-Thiele type of graph as shown in Figure 15-3.

The steps in this procedure are:

1. Plot Y vs X equilibrium data (convert from fractions to ratios).

2. Values of X_0, Y_{N+1}, Y_1 and L/G are known. Point (X_0, Y_1) is on operating line, since it represents passing streams.

3. Slope is L/G. Plot operating line.

4. Starting at stage 1, step off stages: equilibrium, operating, equilibrium, etc.

Note that the operating line is *above* the equilibrium line, because solute is being transferred from the gas to the liquid. In distillation we had material (the more volatile component) transferred from liquid to gas, and the operating line was below the equilibrium curve. If we had plotted the less volatile component, that operating line would be above the equilibrium curve in distillation.

Equilibrium data must be converted to ratio units, Y vs X. These values can be greater than 1.0, since $Y = y/(1 - y)$ and $X = x/(1 - x)$. The $Y = X$ line has no significance in absorption. As usual the stages are counted at the equilibrium curve. A minimum L/G ratio can be defined as shown in Figure 15-3. If the system is not isothermal, the operating line will not be affected, but the equilibrium line will be. Then the McCabe-Thiele method must be modified to include changing equilibrium curves.

For very dilute systems we can use mole fractions, since total flows are approximately constant. This is illustrated in the section on the Kremser equation.

Example 15-1. Graphical Absorption Analysis

A gas stream is 90 mole % N_2 and 10 mole % CO_2. We wish to

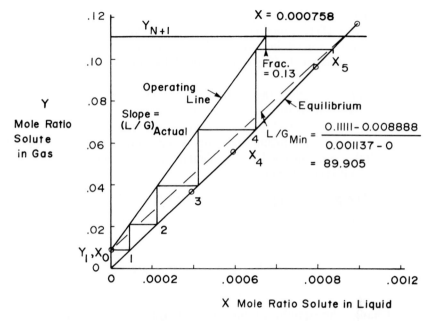

Figure 15-3.   McCabe-Thiele diagram for absorption, Example 15-1.

absorb the CO_2 into water. The inlet water is pure and is at 5°C. Because of cooling coils, operation can be assumed to be isothermal. Operation is at 10 atm. If the liquid flow rate is 1.5 times the minimum liquid flow rate, how many equilibrium stages are required to absorb 92% of the CO_2? Choose a basis of 1 mole/hr of entering gas.

Solution

A. Define. See the sketch. We need to find the minimum liquid flow rate, the value of the outlet gas concentration, and the number of equilibrium stages required.

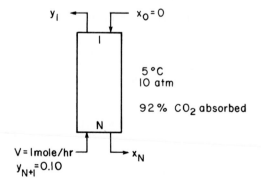

B. Explore. First we need equilibrium data. These are available in Table 15-1. Since concentrations are fairly high, the problem should be solved in mole ratios. Thus we need to convert all compositions including equilibrium data to mole ratios.

C. Plan. Derive the equilibrium equation from Henry's law. Convert compositions from mole fractions to mole ratios using Eqs. (15-7). Calculate Y_1 by a percent recovery analysis. Plot mole ratio equilibrium data on a YX diagram, and determine $(L/G)_{min}$ and hence L_{min}. Calculate actual L/G, plot operating line, and step off stages.

The problem appears to be straightforward.

D. Do It. *Equilibrium:*

$$y = \frac{H}{P_{tot}} x = \frac{876}{10} x = 87.6 \, x$$

Change the equilibrium data to mole ratios with a table as shown below. (The equation can also be converted, but it is easier to avoid a mistake with a table.)

x	$X = \dfrac{x}{1-x}$	$y = 87.6 \, x$	$Y = \dfrac{y}{1-y}$
0	0	0	0
0.0001	0.0001	0.00876	0.00884
0.0004	0.0004	0.0350	0.0363
0.0006	0.0006	0.0526	0.0555
0.0008	0.0008	0.0701	0.0754
0.0010	0.0010	0.0876	0.0960
0.0012	0.0012	0.10512	0.1175

Note that $x = X$ in this concentration range, but $y \neq Y$. The inlet gas mole ratio is

$$Y_{N+1} = \frac{y_{N+1}}{1 - y_{N+1}} = \frac{0.1}{0.9} = 0.1111 \quad \frac{\text{moles } CO_2}{\text{moles } N_2}$$

$$G = (1 \text{ mole total gas/hr})(1 - y_{N+1}) = 0.9 \quad \frac{\text{moles } N_2}{\text{hr}}$$

Percent Recovery Analysis: 8% of CO_2 exits.

(0.1 mole in)(0.08 recovered) = 0.008 mole CO_2 out
Thus,

$$Y_1 = \frac{\text{moles } CO_2}{\text{moles } N_2} = \frac{0.008 \text{ mole } CO_2}{0.9 \text{ mole } N_2} = 0.008888$$

Operating Line:

$$Y_{j+1} = \frac{L}{G} X_j + \left(Y_1 - \frac{L}{G} X_0\right)$$

Goes through point $(Y_1, X_0) = (0.008888, 0)$.

$(L/G)_{min}$ is found as the slope of the operating line from point (Y_1, X_0) to the intersection with the equilibrium curve at Y_{N+1}. This is shown on Figure 15-3.

$$(L/G)_{min} = 89.905 \text{ so } L_{min} = (89.905)(0.9) = 80.914 \frac{\text{moles water}}{\text{hr}}$$

$$L_{actual} = 1.5 L_{min} = 121.37 \quad \text{and} \quad (L/G)_{actual} = 134.86$$

Plot operating line from (Y_1, X_0) with this slope (practice to be sure you can do this when axes are different). Step off stages on the diagram. Need 4.13 equilibrium stages as shown in Figure 15-3.

The fraction was calculated as

$$\text{Frac} = \frac{X_{out} - X_4}{X_5 - X_4} = \frac{0.000758 - 0.00071}{0.00108 - 0.00071} = 0.13$$

E. Check. The overall mass balances are satisfied by the outlet concentrations. The significant figures carried in this example are excessive compared with the equilibrium data. Thus they should be rounded off when reported. The concentrations used were quite high for Henry's law. Thus, it would be wise to check the equilibrium.

F. Generalize. Note that the gas concentration is considerably greater than the liquid concentration. This situation is common for physical absorption (solubility is low). Chemical absorption is used to obtain more favorable equilibrium. The liquid flow rate required for physical absorption is excessive. Thus, in practice, this type of operation uses chemical absorption.

If we had assumed that total gas and liquid flow rates were constant (dilute solutions), the result would be in error. An estimate of this error can be obtained by estimating $(L/G)_{min}$. The minimum operating line goes from $(y_1, x_0) = (0.00881, 0)$ to $(y_{N+1}, x_{equil,N+1})$. $y_{N+1} = 0.1$ and $x_{equil,N+1} = y_{N+1}/87.6 = 0.1/87.6 = 0.0011415$.

Then

$$(L/G)_{min,dilute} = \frac{y_{N+1} - y_1}{x_{equil,N+1} - x_0} = \frac{0.1 - 0.00881}{0.0011415 - 1} = 79.886$$

This is in error by more than 10%.

15.3. STRIPPING ANALYSIS

Since stripping is very similar to absorption we expect a similar result. The mass balance for the column shown in Figure 15-4 is the same as for absorption and the operating line is still

$$Y_{j+1} = \frac{L}{G} X_j + (Y_1 - \frac{L}{G} X_0)$$

For stripping we know X_0, X_N, Y_{N+1}, and L/G. Since (X_N, Y_{N+1}) is a point on the operating line, we can plot the operating line and step off stages. This is illustrated in Figure 15-5.

Figure 15-4.   Stripping column.

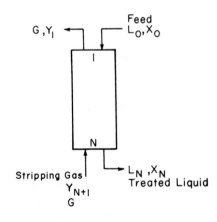

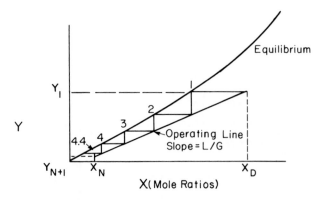

Figure 15-5.   McCabe-Thiele diagram for stripping.

Note that the operating line is below the equilibrium curve because solute is transferred from liquid to gas. This is therefore similar to the stripping section of a distillation column. A maximum L/G ratio can be defined; this corresponds to the minimum amount of stripping gas. Start from the known point (Y_{N+1}, X_N) and draw a line to the intersection of $X = X_0$ and the equilibrium curve. Alternatively, there may be a tangent pinch point. For a stripper, $Y_1 > Y_{N+1}$, while the reverse is true in absorption. Thus the top of the column is on the right side in Figure 15-5 but on the left side in Figure 15-3. Stripping often has large temperature changes, so the method used here may have to be modified.

Murphree efficiencies can be used on these diagrams if they are defined as

$$\frac{Y_j - Y_{j+1}}{Y_j^* - Y_{j+1}} = E_{MV} \tag{15-10}$$

For dilute systems the more common definition of Murphree vapor efficiency in mole fractions would be used. Efficiencies for absorption and stripping are often quite low.

Usually the best way to determine efficiencies is to measure them on commercial-scale equipment. In the absence of such data a rough prediction of the overall efficiency can be obtained from O'Connell's correlation shown in Figure 15-6 (O'Connell, 1946). Although originally done for bubble-cap systems, the results can be used for a first estimate for sieve and valve trays. More detailed estimates can be made using a mass transfer analysis (see Chapter 19). Correlations for very dilute strippers are given by Hwang (1981).

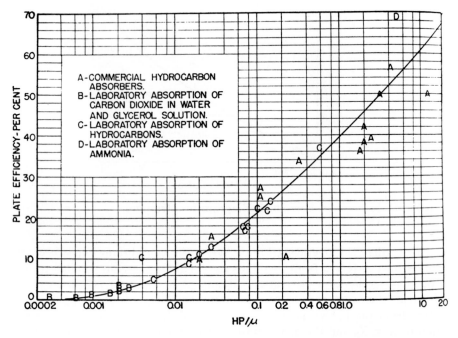

Figure 15-6. O'Connell's correlation for overall efficiency of bubble-cap absorbers. Reprinted from O'Connell, *Trans. AIChE, 42,* 741 (1946). Copyright 1946, AIChE.

15.4. COLUMN DIAMETER

For absorption and stripping, the column diameter is designed the same way as for a staged (Chapter 12) or packed (Chapter 13) distillation column. However, note that the gas flow rate, G, must now be converted to the total gas flow rate, V. The carrier gas flow rate, G, is

$$G, \text{ moles carrier gas/hr} = (1 - y_j)V_j \qquad (15\text{-}11)$$

Since $y_j = \dfrac{Y_j}{1+Y_j}$, this is

$$G = \left(\frac{1}{1+Y_j}\right)V_j \qquad (15\text{-}12)$$

or

$$(1+Y_j)G = V_j \qquad (15\text{-}13)$$

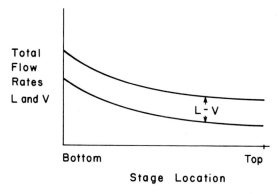

Figure 15-7.   Total flow rates in absorber.

The total liquid flow rate, L_j, can be determined from an overall mass balance. Using the balance envelope shown in Figure 15-1, we obtain

$$L_j - V_{j+1} = L_0 - V_1 = \text{constant for any } j \qquad (15\text{-}14)$$

Note that the difference between total flow rates of passing streams is constant. This is illustrated in Figure 15-7.

Both total flows V_j and L_j will be largest where Y_j and X_j are largest. This is at the bottom of the column for absorption, and therefore you design the diameter at the bottom of the column. In strippers, flow rates are highest at the top of the column, so you design the diameter for the top of the column. Specific design details for absorbers and strippers are discussed by Zenz (1979).

15.5. ANALYTICAL SOLUTION: KREMSER EQUATION

When the solution is quite dilute (say less than 1% solute in both gas and liquid), the total liquid and gas flow rates will not change significantly since little solute is transferred. Then the entire analysis can be done with mole or mass fractions and total flow rates. In this case the column shown in Figure 15-1 will look like the one in Figure 15-8, where streams have been relabeled. The operating equation is derived by writing a mass balance around stage j and solving for y_{j+1}. The result,

$$y_{j+1} = \frac{L}{V}x_j + \left(y_1 - \frac{L}{V}x_0\right) \qquad (15\text{-}15)$$

is essentially the same as Eq. (15-9) except that the units are different.

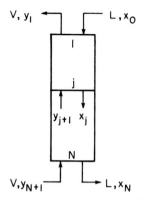

Figure 15-8.   Dilute absorber. L and V are total flow rates, y and x are mass or mole fractions.

To use Eq. (15-15) in a McCabe-Thiele diagram, we assume:

1. L/V (total flows) is constant
2. Isothermal system
3. Isobaric system
4. Negligible heat of absorption

These are reasonable assumptions for dilute absorbers and strippers. The solutions on a plot of y versus x (mole or mass fraction) will look like Figure 15-2 for absorbers and like Figure 15-4 for strippers. The operating line slope will be L/V. Figures 15-9 and 15-10 show two special cases for absorbers.

If one additional assumption is valid, the stage-by-stage problem can be solved analytically. This additional assumption is:

5. Equilibrium line is straight.

$$y_j = mx_j + b \qquad (15\text{-}16)$$

This assumption is reasonable for very dilute solutions and agrees with Henry's law, Eq. (15-3), if $m = H_A/p_{tot}$ and $b=0$.

An analytical solution for absorption is easily derived for the special case shown in Figure 15-9, where the operating and equilibrium lines are parallel. Now the distance between operating and equilibrium lines, Δy, is constant. To go from outlet to inlet concentrations with N stages, we

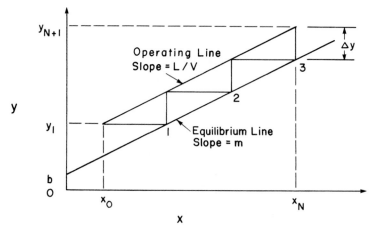

Figure 15-9. McCabe-Thiele diagram for dilute absorber with parallel equilibrium and operating lines.

have

$$N \, \Delta y = y_{N+1} - y_1 \qquad (15\text{-}17)$$

since each stage causes the same change in vapor composition. Δy can be obtained by subtracting the equilibrium equation (15-16) from the operating equation (15-15).

Figure 15-10. McCabe-Thiele diagram for dilute absorber. $(L/V) <$ m.

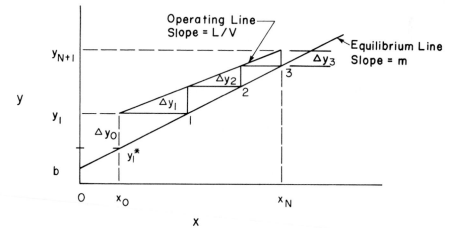

$$(\Delta y)_j = y_{j+1} - y_j = (\frac{L}{V} - m)x_j + (y_1 - \frac{L}{V}x_0 - b) \qquad (15\text{-}18)$$

For the special case shown in Figure 15-9 $L/V = m$ and Eq. (15-18) becomes

$$\Delta y = (\Delta y)_j = y_1 - \frac{L}{V} x_0 - b = \text{constant} \qquad (15\text{-}19)$$

Combining Eqs. (15-17) and (15-19), we get

$$N = \frac{y_{N+1} - y_1}{(y_1 - \frac{L}{V} x_0 - b)} \qquad \text{for} \quad \frac{L}{mV} = 1 \qquad (15\text{-}20)$$

Equation (15-20) is a special case of the Kremser equation. When this equation is applicable, absorption and stripping problems can be solved quite simply and accurately without the need for a stage-by-stage calculation.

Figure 15-9 and the resulting Eq. (15-20) were for a special case. The more general case is shown in Figure 15-10. Now Δy_j varies from stage to stage. The Δy values can be determined from Eq. (15-18). Equation (15-18) is easier to use if we replace x_j with the equilibrium equation (15-16),

$$x_j = \frac{y_j - b}{m} \qquad (15\text{-}21)$$

Then

$$(\Delta y)_j = (\frac{L}{mV} - 1) y_j + (y_1 - \frac{L}{mV} b - \frac{L}{V} x_0) \qquad (15\text{-}22a)$$

$$(\Delta y)_{j+1} = (\frac{L}{mV} - 1) y_{j+1} + (y_1 - \frac{L}{mV} b - \frac{L}{V} x_0) \qquad (15\text{-}22b)$$

Subtracting Eq. (15-22a) from (15-22b),

$$(\Delta y)_{j+1} - (\Delta y)_j = (\frac{L}{mV} - 1)(y_{j+1} - y_j) = (\frac{L}{mV} - 1)(\Delta y)_j$$

and solving for $(\Delta y)_{j+1}$

$$(\Delta y)_{j+1} = \frac{L}{mV}(\Delta y)_j \qquad (15\text{-}23)$$

Equation (15-23) relates the change in vapor composition from stage to stage to (L/mV), which is known as the *absorption factor*. If either the operating or equilibrium line is curved, this simple relationship no longer holds and a simple analytical solution does not exist.

The difference between inlet and outlet gas concentrations must be the sum of the Δy_j values shown in Figure 15-10. Thus,

$$\Delta y_1 + \Delta y_2 + \cdots + \Delta y_N = y_{N+1} - y_1 \tag{15-24}$$

Applying Eq. (15-23)

$$\Delta y_1 (1 + \frac{L}{mV} + (\frac{L}{mV})^2 + \cdots + (\frac{L}{mV})^{N-1}) = y_{N+1} - y_1 \tag{15-25}$$

The summation in Eq. (15-25) can be calculated. The general formula is

$$\sum_{i=0}^{k} aA^i = \frac{a(1 - A^{k+1})}{(1 - A)} \quad \text{for} \quad |A| < 1$$

Then Eq. (15-25) is

$$\frac{y_{N+1} - y_1}{\Delta y_1} = \frac{1 - (\frac{L}{mV})^N}{1 - \frac{L}{mV}} \tag{15-26}$$

If $L/mV > 1$, then divide both sides of Eq. (15-25) by $(L/mV)^{N-1}$ and do the summation in terms of mV/L. The resulting equation will still be Eq. (15-26). From Eq. (15-23), $\Delta y_1 = \Delta y_0 L/mV$ where $\Delta y_0 = y_1 - y_1^*$ is shown in Figure 15-10. The vapor composition y_1^* is the value that would be in equilibrium with the inlet liquid, x_0. Thus,

$$y_1^* = mx_0 + b \tag{15-27}$$

Removal of Δy_1 from Eq. (15-26) gives

$$\frac{y_{N+1} - y_1}{y_1 - y_1^*} = \frac{\frac{L}{mV} - (\frac{L}{mV})^{N+1}}{1 - \frac{L}{mV}} \tag{15-28}$$

Equation (15-28) is one form of the Kremser equation (Kremser, 1930 Souders and Brown, 1932). A large variety of alternative forms can be developed by algebraic manipulation. For instance, if we add 1.0 to both sides of Eq. (15-28) and rearrange, we have

$$\frac{y_{N+1} - y_1^*}{y_1 - y_1^*} = \frac{1 - (\frac{L}{mV})^{N+1}}{1 - \frac{L}{mV}} \tag{15-29}$$

which can be solved for N. After manipulation, this result is

$$N = \frac{\ln[(1 - \frac{mV}{L})(\frac{y_{N+1} - y_1^*}{y_1 - y_1^*}) + \frac{mV}{L}]}{\ln(\frac{L}{mV})} \tag{15-30}$$

where $L/mV \neq 1$. Equations (15-29) and (15-30) are also known as forms of the Kremser equation. Alternative derivations of the Kremser equation are given by Brian (1972), King (1980), and Mickley *et al.* (1957).

A variety of forms of the Kremser equation can be developed. Several alternative forms in terms of the gas-phase composition are

$$\frac{y_{N+1} - y_1}{y_{N+1} - y_1^*} = \frac{(L/mV) - (L/mV)^{N+1}}{1 - (L/mV)^{N+1}} \tag{15-31}$$

$$\frac{y_{N+1} - y_{N+1}^*}{y_1 - y_1^*} = (\frac{L}{mV})^N \tag{15-32}$$

$$N = \frac{\ln[(y_{N+1} - y_{N+1}^*)/(y_1 - y_1^*)]}{\ln(L/mV)} \tag{15-33}$$

$$N = \frac{\ln[(y_{N+1} - y_{N+1}^*)/(y_1 - y_1^*)]}{\ln[(y_{N+1} - y_1)/(y_{N+1}^* - y_1^*)]} \tag{15-34}$$

where

$$y_{N+1}^* = mx_n + b \quad \text{and} \quad y_1^* = mx_0 + b \tag{15-35}$$

Alternative forms in terms of the liquid phase composition are

$$N = \frac{\ln[(1 - \frac{L}{mV})(\frac{x_0 - x_N^*}{x_N - x_N^*}) + \frac{L}{mV}]}{\ln(mV/L)} \tag{15-36}$$

$$N = \frac{\ln[(x_N - x_N^*)/(x_0 - x_0^*)]}{\ln(mV/L)} \tag{15-37}$$

$$N = \frac{\ln[(x_N - x_N^*)/(x_0 - x_0^*)]}{\ln[(x_0^* - x_N^*)/(x_0 - x_N)]} \tag{15-38}$$

$$\frac{x_N - x_N^*}{x_0 - x_N^*} = \frac{1 - (mV/L)}{1 - (mV/L)^{N+1}} \tag{15-39}$$

$$\frac{x_N - x_N^*}{x_0 - x_0^*} = (\frac{L}{mV})^N \tag{15-40}$$

where

$$x_N^* = \frac{y_{N+1} - b}{m} \quad \text{and} \quad x_0^* = \frac{y_1 - b}{m} \tag{15-41}$$

A form including a constant Murphree vapor efficiency is

$$N = -\frac{\ln\{[1 - mV/L][(y_{N+1} - y_1^*)/(y_1 - y_1^*)] + mV/L\}}{\ln[1 + E_{MV}(mV/L - 1)]} \tag{15-42}$$

Forms for systems with three phases, with two phases flowing cocurrently and countercurrent to the third phase, were developed by Wankat (1980). Forms of the Kremser equation for columns with multiple sections are developed by Brian (1972, Chapt. 3) and by King (1980, pp. 371-376).

Which form of the Kremser equation to use depends upon the problem statement. When the assumptions required for the derivation are valid, the Kremser equation has several advantages over the stage-by-stage calculation procedure. If the number of stages is large, the Kremser equation is much more convenient to use, and it is easy to program on a computer or calculator. When the number of stages is specified, the McCabe-Thiele stage-by-stage procedure is trial-and-error, but the use of the Kremser equation is not. Because calculations can be done faster, the effects of varying y_1, x_0, L/V, m etc. are easy to determine. The major disadvantage of the Kremser equation is that it is accurate only for dilute solutions where L/V is constant, equilibrium is linear, and the system is isothermal.

Example 15-2. Kremser Equation

A plate tower providing six equilibrium stages is employed for stripping ammonia from a waste water stream by means of countercurrent air at atmospheric pressure and 80°F. Calculate the concentration of ammonia in the exit water if the inlet liquid concentration is 0.1 mole % ammonia in water, the inlet air is free of ammonia, and 30 standard cubic feed (scf) of air are fed to the tower per pound of waste water.

Solution

A. Define. The column is sketched in the figure.

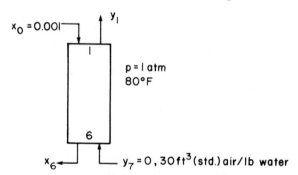

We wish to find the exit water concentration, x_6.

B. Explore. Since the concentrations are quite low we can use the Kremser equation. Equilbrium data are available in several sources. From King (1971, p. 273) we find $y_{NH_3} = 1.414\ x_{NH_3}$ at 80°F.

C. Plan. We have to convert flow to molar units. Since we want a concentration of liquid, forms (15-39) or (15-40) of the Kremser equation will be convenient. We will use Eq. (15-39).

D. Do It. We can calculate ratio V/L,

$$\frac{V}{L} = \frac{30 \text{ scf air}}{1 \text{ lb water}} \times \frac{1 \text{ lb mole air}}{379 \text{ scf air}} \times \frac{18 \text{ lb water}}{1 \text{ lb mole water}}$$

$$= 1.43 \text{ moles air/mole water}$$

Note that the individual flow rates are not needed.

The Kremser equation [form (15-39)] is

$$\frac{x_N - x_N^*}{x_0 - x_N^*} = \frac{1 - \dfrac{mV}{L}}{1 - \left(\dfrac{mV}{L}\right)^{N+1}}$$

where $x_N = x_6$ is unknown, $x_0 = 0.001$, $m = 1.414$, $b = 0$,
$$x_N^* = y_7/m = 0, \quad V/L = 1.43, \quad N = 6$$
Rearranging,

$$x_N = \frac{1 - mV/L}{1 - (mV/L)^{N+1}} x_0$$

$$x_N = \frac{1 - (1.414)(1.43)}{1 - [(1.414)(1.43)]^7} (0.001)$$

$$= 7.45 \times 10^{-6} \text{ mole fraction}$$

Most of the ammonia is stripped out by the air.

E. Check. We can check with a different form of the Kremser equation or by solving the results graphically; both give the same result. We should also check that the major assumptions of the Kremser equation (constant flow rates, linear equilibrium, and isothermal) are satisfied. In this dilute system they are.

F. Generalize. This problem is trial-and-error when it is solved graphically. Also, the Kremser equation is very easy to set up on a computer or calculator. Thus, *when it is applicable,* the Kremser equation is very convenient.

15.6. DILUTE MULTISOLUTE ABSORBERS AND STRIPPERS

Up to this point we have been restricted to cases where there is a single solute to recover. Both the stage-by-stage McCabe-Thiele procedures and the Kremser equation can be used for multisolute absorption and stripping if certain assumptions are valid. The single-solute analysis by the Kremser equation required systems that (1) are isothermal, (2) are isobaric, (3) have a negligible heat of absorption, and (4) have constant flow rates. These assumptions are again required.

To see what other assumptions are required consider the Gibbs phase rule for a system with three solutes plus a solvent and a carrier gas. The phase rule is

$$F = C - P + 2 = 5 - 2 + 2 = 5$$

Five degrees of freedom is a large number. In order to represent equilibrium as a single curve or in a linear form like Eq. (15-16), four of these degrees of freedom must be specified. Constant temperature and pressure utilize two degrees of freedom. The other two degrees of freedom can be specified by assuming that (5) solutes are independent of each other, in other words, that equilibrium for any solute does not depend on the amounts of other solutes present. This assumption requires dilute solutions. In addition, the analysis must be done in terms of mole or mass *fractions* and total flow rates, for which dilute solutions are also required. An analysis using ratio units will not work because the ratio calculation

$$Y_i = \frac{y_i}{1 - y_1 - y_2 - y_3}$$

involves other solute concentrations that will be unknown.

The practical effect of assumption 5 is that we can solve the multisolute problem once for each solute, treating each problem as a single-component problem. This is true for both the stage-by-stage solution method and the Kremser equation. Thus for the absorber shown in Figure 15-11 we solve three single-solute problems.

Each additional solute increases the degrees of freedom for the absorber by two. These two degrees of freedom are required to specify the inlet gas and inlet liquid compositions. For the usual design problem, as shown in Figure 15-11, the inlet gas and inlet liquid compositions and flow rates will be specified. With temperature and pressure also

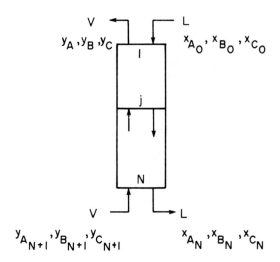

Figure 15-11.   Dilute multisolute absorber.

specified, one degree of freedom is left. This is usually used to specify one of the outlet solute concentrations such as $y_{B,1}$. The design problem for solute B is now fully specified. To solve for the number of stages, we can plot the equilibrium data, which are of the form

$$y_B = f_B(x_B) \tag{15-43}$$

on a McCabe-Thiele diagram. The operating equation

$$y_{B,j+1} = \frac{L}{V} x_{B,j} + \left(y_{B,1} - \frac{L}{V} x_{B,0}\right) \tag{15-44}$$

is the same as Eq. (15-15) and can also be plotted on the McCabe-Thiele diagram. Then the number of stages is stepped off as usual. This is shown in Figure 15-12A.

Once the number of stages has been found from the solute B calculation, the concentrations of solutes A and C can be determined by solving two fully specified simulation problems. That is, the number of stages is known and the outlet compositions have to be calculated. Simulation problems require a trial-and-error procedure when a stage-by-stage calculation is used. One way to do this calculation for component A is:

1. Plot the A equilibrium curve, $y_A = f_A(x_A)$.

2. Guess $y_{A,1}$ for solute A.

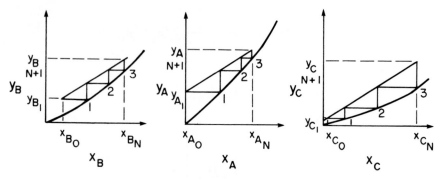

Figure 15-12.   McCabe-Thiele solution for dilute three-solute absorber. Solute B is specified. Solutes A and C are trial-and-error.

3. Plot the A operating line,

$$y_{A,j+1} = \frac{L}{V} x_{A,j} + \left(y_{A,1} - \frac{L}{V} x_{A,0}\right)$$

Slope $= L/V$, which is same as for solute B. Point $(y_{A,1}, x_{A,0})$ is on the operating line. This is shown in Figure 15-12B.

4. Step off stages up to $y_{A,N+1}$ (see Figure 15-12).

5. Check: Is number of stages same as calculated? If yes, you have the answer. If no, return to step 2.

The procedure for solute C is the same.

The three diagrams shown in Figure 15-12 are often plotted on the same yx graph. This saves paper but tends to be confusing.

If the equilibrium function for each solute is linear as in Eq. (15-16), the Kremser equation can be used. First the design problem for solute B is solved by using Eq. (15-30), (15-33), or (15-34) to find N. Then separately solve the two simulation problems (for solutes A and C) using equations such as (15-28), (15-29), (15-31), or (15-32). Remember to use m_A and m_C when you use the Kremser equation. Note that when the Kremser equation can be used, the simulation problems are *not* trial-and-error.

These solution methods are restricted to very dilute solutions. In more concentrated solutions, flow rates are not constant, solutes may not have independent equilibria, and temperature effects become important. When this is true, more complicated computer solution methods involving simultaneous mass and energy balances plus equilibrium are required. These methods are briefly discussed in the next section.

15.7. MATRIX SOLUTION

For more concentrated solutions, absorbers and strippers are usually not isothermal, total flow rates are not constant, and solutes may not be independent. The matrix methods discussed for multicomponent distillation (reread Section 8.5) can be adapted for absorption and stripping.

Absorbers are equivalent to very wide boiling feeds. Thus, in contrast with distillation, a wide-boiling feed (sum rates) flow chart like Figure 15-13 should be used. The flow rate loop is now solved first, since flow rates are never constant in absorbers. The energy balance, which requires the most information, is used to calculate new temperatures, since this is done last. The initial steps are very similar to those for distillation, and usually the same physical properties package is used.

The mass balances and equilibrium equations are very similar to those for distillation and the column is again numbered from the bottom up as shown in Figure 15-14. To fit into the matrix form, streams V_0 and L_{N+1} are relabeled as feeds to stages 1 and N, respectively. The stages from 2 to $N - 1$ have the same general shape as for distillation (Figure 8-8). Thus the mass balances and the manipulations [Eqs. (8-18) to (8-23)] are the same for distillation and absorption.

For stage 1 the mass balance becomes

$$B_1 \ell_1 + C_1 \ell_2 = D_1 \tag{15-45}$$

where

$$B_1 = 1 + \frac{V_1 K_1}{L_1} , \quad C_1 = -1 , \quad D_1 = F_1 z_1 = V_0 y_0 \tag{15-46}$$

and the component flow rates are $\ell_1 = L_1 x_1$ and $\ell_2 = L_2 x_2$. These equations are repeated for each component.

For stage N the mass balance is

$$A_n \ell_{N-1} + B_N \ell_N = D_N \tag{15-47}$$

498

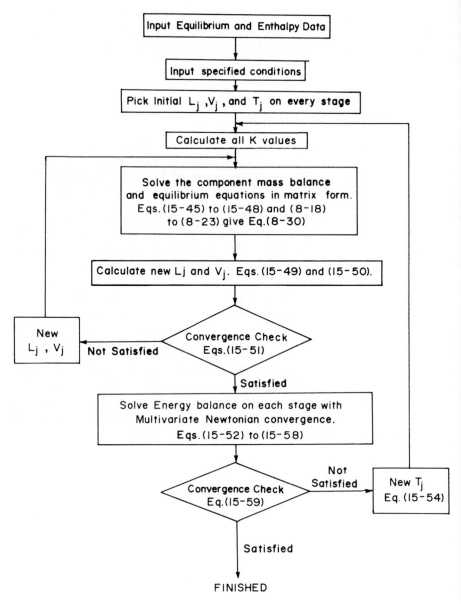

Figure 15-13.   Sum rates convergence procedure for absorption and stripping.

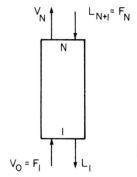

Figure 15-14.   Absorber nomenclature for matrix analysis.

where

$$A_N = -\frac{V_{N-1}K_{N-1}}{L_{N-1}} \ , \ \ B_N = 1 + \frac{V_N K_N}{L_N} \tag{15-48}$$

$$D_N = F_N z_N = L_{N+1} x_{N+1}$$

This equation differs from the one for distillation, since stage N is an equilibrium stage in absorption, not a total condenser.

Combining Eqs. (15-45), (8-22), and (15-47) results in a tridiagonal matrix, Eq. (8-30), with all terms defined in Eqs. (15-46), (8-23), and (15-48). There is one matrix for each component. These tridiagonal matrices can each be inverted with the Thomas algorithm (Table 8-1). The results are liquid component flow rates, $\ell_{i,j}$, that are valid for the assumed L_j, V_j, and T_j.

The next step is to use the summation equations to find new total flow rates L_j and V_j. The new liquid flow rate is conveniently determined as

$$L_{j,new} = \sum_{i=1}^{C} \ell_{i,j} \tag{15-49}$$

The vapor flow rates are determined by summing the component vapor flow rates,

$$V_{j,new} = \sum_{i=1}^{C} \left[\left(\frac{K_{i,j}V_j}{L_j} \right)_{old} \ell_{i,j} \right] \tag{15-50}$$

Convergence can be checked with

$$\left| \frac{L_{j,old} - L_{j,new}}{L_{j,old}} \right| < \epsilon \quad \text{and} \quad \left| \frac{V_{j,old} - V_{j,new}}{V_{j,old}} \right| < \epsilon \qquad (15\text{-}51)$$

for all stages. For computer calculations, an ϵ of 10^{-4} or 10^{-5} can be used. If convergence has not been reached, new liquid and vapor flow rates are determined, and we return to the component mass balances (see Figure 15-13). Direct substitution ($L_j = L_{j,new}$, $V_j = V_{j,new}$) is usually adequate.

Once the flow rate loop has converged, the energy balances are used to solve for the temperatures on each stage. This can be done in several different ways (Henley and Seader, 1981; Holland, 1975; King, 1980; Smith, 1963). We will discuss only a multivariate Newtonian convergence procedure. For the general stage j, the energy balance was given as Eq. (8-40). This can be rewritten as

$$E_j(T_{j-1}, T_j, T_{j+1}) = \qquad (15\text{-}52)$$

$$L_j h_j + V_j H_j - L_{j+1} h_{j+1} - V_{j-1} H_{j-1} - F_j h_{Fj} - q_j = 0$$

The multivariate Newtonian approach is an extension of the single variable Newtonian convergence procedure. The change in the energy balance is

$$(E_j)_{k+1} - (E_j)_k = \qquad (15\text{-}53)$$

$$\frac{\partial E_j}{\partial T_{j-1}}(\Delta T_{j-1}) + \frac{\partial E_j}{\partial T_j}(\Delta T_j) + \frac{\partial E_j}{\partial T_{j+1}}(\Delta T_{j+1})$$

where k is the trial number. The ΔT values are defined as

$$\Delta T_{j-1} = (T_{j-1})_{k+1} - (T_{j-1})_k, \quad \Delta T_j = (T_j)_{k+1} - (T_j)_k$$

$$\Delta T_{j+1} = (T_{j+1})_{k+1} - (T_{j+1})_k \qquad (15\text{-}54)$$

The partial derivatives can be determined from Eq. (15-52) by determining which terms in the equation are direct functions of the stage temperatures T_{j-1}, T_j, and T_{j+1}. These partial derivatives are

$$\frac{\partial E_j}{\partial T_{j-1}} = -V_{j-1}\frac{\partial H_{j-1}}{\partial T_{j-1}} = -V_{j-1}C_{PV,j-1} = A_j \qquad (15\text{-}55a)$$

$$\frac{\partial E_j}{\partial T_j} = L_j\frac{\partial h_j}{\partial T_j} + V_j\frac{\partial H_j}{\partial T_j} = L_j c_{PL,j} + V_j C_{PV,j} = B_j \qquad (15\text{-}55b)$$

$$\frac{\partial E_j}{\partial T_{j+1}} = -L_{j+1}\frac{\partial h_{j+1}}{\partial T_{j+1}} = -L_{j+1}c_{PL,j+1} = C_j \qquad (15\text{-}55c)$$

where we have identified the terms as A, B, and C terms for a matrix. The total stream heat capacities can be determined from individual component heat capacities. For ideal mixtures this is

$$c_{PL,j} = \left(\sum_{i=1}^{C} c_{PL,i}\, x_i\right)_j, \quad C_{PV,j} = \left(\sum_{i=1}^{C} C_{PV,i}\, y_i\right)_j \qquad (15\text{-}56)$$

For the next trial we hope to have $(E_j)_{k+1} = 0$. If we define

$$-(E_j)_k = D_j \qquad (15\text{-}57)$$

then the equations for ΔT_j can be written as

$$\begin{bmatrix} B_1 & C_1 & & & & \cdot \\ A_2 & B_2 & C_2 & & & \cdot \\ 0 & A_3 & B_3 & C_3 & & \cdot \\ \cdot & \cdot & \cdot & & \cdot & \cdot & \cdot \\ & & & A_{N-1} & B_{N-1} & C_{N-1} \\ & & & 0 & A_N & B_N \end{bmatrix} \begin{bmatrix} \Delta T_1 \\ \Delta T_2 \\ \Delta T_3 \\ \cdot \\ \Delta T_{N-1} \\ \Delta T_N \end{bmatrix} = \begin{bmatrix} D_1 \\ D_2 \\ D_3 \\ \cdot \\ D_{N-1} \\ D_N \end{bmatrix} \qquad (15\text{-}58)$$

Equation (15-58) can be inverted using any computer inversion program or the Thomas algorithm shown in Table 8-1. The result will be all the ΔT_j values.

Convergence can be checked from the ΔT_j. If

$$|\Delta T_j| < \epsilon_T \qquad (15\text{-}59)$$

for all stages, then convergence has been achieved. The problem is finished! If Eq. (15-59) is not satisfied, determine new temperatures from Eq. (15-54) and return to calculate new K values and redo the component mass balances (see Figure 15-13). A reasonable range for ϵ_T for computer solution is 10^{-2} to 10^{-3}.

15.8. IRREVERSIBLE ABSORPTION

Absorption with an irreversible chemical reaction is often used in small facilities for removing obnoxious chemicals. For example, NaOH is used to remove both CO_2 and H_2S; it reacts with the acid gas in solution and forms a nonvolatile salt. This is convenient in small facilities, because the absorber is usually small and simple and no regeneration facilities are required. However, the cost of the reactant (NaOH) can make operation expensive. In large-scale systems it is usually cheaper to use a solvent that can be regenerated.

Consider a simple absorber where the gas to be treated contains carrier gas C and solute B. The solvent contains a nonvolatile solvent S and a reagent R that will react irreversibly with B according to the irreversible reaction

$$R + B \rightarrow RB \tag{15-60}$$

The resulting product RB is nonvolatile. At equilibrium, x_B (in the free form) $= 0$ since the reaction is irreversible and any B that dissolves will form product RB. Thus, $y_B = 0$ at equilibrium. As long as there is any reagent R present, the equilibrium expression is $y_B = 0$. From the stoichiometry of the reaction shown in Eq. (15-60), there will be reagent available as long as $Lx_{R,0} > Vy_{B,N+1}$.

For a dilute countercurrent absorber, the mass balance was given by Eq. (15-15), where x is the total mole fraction of B in the liquid (as free B and as bound RB). The operating and equilibrium diagrams can be plotted on a McCabe-Thiele diagram as shown in Figure 15-15. One equilibrium stage will give $y_1 = 0$, which is more than sufficient. Unfortunately, the stage efficiency is often very low because of low mass transfer rates of the solute into the liquid. If the Murphree vapor efficiency

$$E_{MV} = \frac{y_{in} - y_{out}}{y_{in} - y_{out}^*} \tag{15-61}$$

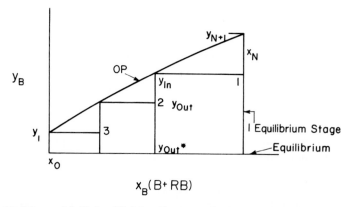

Figure 15-15.   McCabe-Thiele diagram for countercurrent irreversible
absorption.

is used in Figure 15-15, the number of real stages can be stepped off.
Murphree vapor efficiencies less than 30% are common.

Since only one equilibrium stage is required, alternatives to countercurrent cascades may be preferable. A cocurrent cascade is shown in
Figure 15-16A. Packed columns would normally be used for the
cocurrent cascade. The advantage of the cocurrent cascade is that it
cannot flood, so smaller diameter columns with higher vapor velocities
can be used. The higher vapor velocities give higher mass transfer rates
(see Chapter 19), and less packing will be required. The mass balance
for the cocurrent system using the mass balance envelope shown in Figure 15-16A is

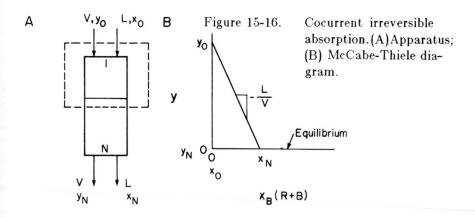

Figure 15-16.   Cocurrent irreversible
absorption. (A) Apparatus;
(B) McCabe-Thiele dia-
gram.

$$y_0 V + L x_0 = y V + L x \qquad (15\text{-}62)$$

Solving for y, we obtain the operating equation

$$y = -\frac{L}{V} x + \frac{y_0 V - L x_0}{V} \qquad (15\text{-}63)$$

This is a straight line with a slope of $-L/V$, and is plotted in Figure 15-16B. At equilibrium, $y_N = 0$, and x_N can be found from the operating equation as shown in Figure 15-16B. When equilibrium is not attained, the system can be designed with the mass transfer analysis discussed in Chapter 19. Cocurrent absorbers are used commercially for irreversible absorption.

15.9. SUMMARY-OBJECTIVES

In this chapter we studied absorption and stripping. The objectives for this chapter are:

1. Explain what absorption and stripping do and describe a complete gas treatment plant.

2. Use the McCabe-Thiele method to analyze absorption and stripping systems for both concentrated and dilute systems.

3. Design the column diameter for an absorber or stripper for staged columns (Chapter 12) and packed columns (Chapter 13).

4. Derive the Kremser equation for dilute systems.

5. Use the Kremser equation for dilute absorption and stripping problems.

6. Solve problems for dilute multicomponent absorbers and strippers both graphically and analytically.

7. Use the matrix solution method for nonisothermal multicomponent absorption or stripping.

8. Discuss how irreversible absorption differs from reversible systems, and design countercurrent irreversible absorbers.

REFERENCES

Astarita, G., D.W. Savage, and A. Bisio, *Gas Treating with Chemical Solvents*, Wiley, New York, 1983.

Brian, P.L.T., *Staged Cascades in Chemical Processes,* Prentice-Hall, Englewood Cliffs, NJ, 1972.

Henley, E.J. and J.D. Seader, *Equilibrium Stage Separation Operations in Chemical Engineering,* Wiley, New York, 1981.

Holland, C.D., *Fundamentals and Modeling of Separation Processes. Absorption, Distillation, Evaporation and Extraction,* Prentice-Hall, Englewood Cliffs, NJ, 1975.

Hwang, S.T., "Tray and Packing Efficiencies at Extremely Low Concentrations," in N.N. Li (Ed.), *Recent Developments in Separation Science,* Vol. 6, CRC Press, Boca Raton, FL, 1981, pp. 137-148.

King, C.J., *Separation Processes,* McGraw-Hill, New York, 1971.

King, C.J., *Separation Processes,* 2nd ed., McGraw Hill, New York, 1980.

Kohl, A.L., "Absorption and Stripping," in R.W. Rousseau (Ed.), *Handbook of Separation Process Technology,* Wiley, New York, 1987, Chapter 6.

Kohl, A.L. and F.C. Riesenfeld, *Gas Purification,* 4th ed., Gulf Pub. Co., Houston, TX, 1985.

Kremser, A., *Nat. Petrol. News,* (May 30, 1930), p. 43.

Maxwell, J.B., *D ta Book on Hydrocarbons,* Van Nostrand, Princeton, NJ, 1950.

Mickley, H.S., T.K. Sherwood, and C.E. Reed, *Applied Mathematics in Chemical Engineering,* McGraw-Hill, New York, 1957.

O'Connell, H.E., "Plate Efficiency of Fractionating Columns and Absorbers," *Trans. AIChE, 42,* 741 (1946).

Perry, R.H., C.H. Chilton, and S.O. Kirkpatrick (Eds.), *Chemical Engineer's Handbook,* 4th ed., McGraw-Hill, New York, 1963.

Perry, R.H. and C.H. Chilton (Eds.), *Chemical Engineer's Handbook,* 5th ed., McGraw-Hill, New York, 1973.

Perry, R.H. and D. Green (Eds.), *Perry's Chemical Engineers' Handbook,* 6th ed., McGraw-Hill, New York, 1984.

Sherwood, T.K., R.L. Pigford, and C.R. Wilke, *Mass Transfer,* McGraw-Hill, New York, 1975.

Siedell and Linke, *Solubilities of Inoganic and Organic Compounds,* Van Nostrand, Princeton, NJ, 1952.

Souders, M. and G.G. Brown, "Fundamental Design of High Pressure Equipment Involving Paraffin Hydrocarbons. IV. Fundamental Design of Absorbing and Stripping Columns for Complex Vapors," *Ind. Eng. Chem., 24,* 519 (1932).

Smith, B.D., *Design of Equilibrium Stage Processes,* McGraw-Hill, New York, 1963.

Wankat, P.C., "Calculations for Separations with Three Phases: 1. Staged Systems," *Ind. Eng. Chem. Fundam., 19,* 358 (1980).

Zenz, F.A., "Design of Gas Absorption Towers," in P.A. Schweitzer (Ed.), *Handbook of Separation Techniques for Chemical Engineers,* McGraw-Hill, New York, 1979, Section 3.2.

HOMEWORK

A. *Discussion Problems*

A1. How can the direction of mass transfer be reversed as it is in a complete gas plant? What controls whether a column is a stripper or an absorber?

A2. Why is the Murphree efficiency often lower in chemical absorption than in physical absorption? (What additional resistances are present?)

A3. Outline the method of determining the column diameter for an absorber or stripper in:

   a. A plate column (review Chapter 12)
   b. A packed column (review Chapter 13)

A4. As the system becomes dilute, $L/G \to L/V$, $Y \to y$, and $X \to x$. At what concentration levels could you safely work in terms of fractions and total flows instead of ratios and flows of solvent and carrier gas? What variable will this depend on? Explore numerically. See also Problem 15-C4.

A5. How can the gas plant in Figure 15-2 be made thermodynamically more efficient with heat exchange? Refer to Chapter 14 for heat exchange in distillation.

A6. Explain the significance of the curves in Figure 15-7. Sketch a plot of Lx and Vy versus stage location.

A7. Equation (15-16) has an extra b term that doesn't appear in Henry's law. In Example 15-2, b = 0. When might it be useful to have a nonzero b term?

A8. Explain how the single assumption that "solutes are independent of each other" can specify more than one degree of freedom.

A9. What are the advantages of using the Kremser equation for multisolute absorption and stripping instead of a graphical solution? When can't the Kremser equation be used? See also Problem 15-C6.

A10. Develop your key relations chart for this chapter.

B. *Generation of Alternatives*

B1. The Kremser equation can be used for more than just determining the number of stages. List as many types of problems (where a different variable is solved for) as you can. What variables would be specified? How would you solve the equation?

B2. Many other configurations of absorbers and strippers can be devised (for example, there could be two feeds). Generate as many as possible.

C. *Derivations*

C1. Derive Eq. (15-30) starting with Eq. (15-29).

C2. Derive Eq. (15-39). Follow the procedure used to derive Eqs. (15-28) and (15-29) except work in terms of Δx. Thus start by writing the equilibrium and operating equations in form $x = \cdots$.

C3. Derive Eq. (15-14).

C4. Plot a graph of ratio units (Y or X) versus fraction units (y or x) to use for converting units.

C5. Derive an operating equation similar to Eq. (15-9), but draw your balance envelope around the bottom of the column. Show that the result is equivalent to Eq. (15-9).

C6. Develop an analytical method for stage-by-stage analysis of an isothermal multisolute absorber similar to the graphical analysis shown in Figure 15-11. Do this for systems with nonlinear isotherms. Develop a computer flow chart for this trial-and-error problem.

C7. Derive Eq. (15-50).

C8. Complete the derivation of Eq. (15-58) by doing the steps outlined in the derivation.

D. *Problems*

D1. Does CO_2 absorption in water follow an Arrhenius relationship? If it does, determine H_0 and E in Eq. (15-4).

D2. We wish to remove solute Q from a heavy oil stream by stripping the oil with air. The column will operate at 60°C and 2.2 atm pressure. At this temperature and pressure the oil can be considered to be nonvolatile and the air is insoluble in the oil. The inlet liquid flows at 100 moles/hr (total flow rate) and is 20 mole % Q and 80 mole % heavy oil. We wish to reduce the concentration of Q in the oil to 1 mole % Q. The inlet air stream contains no Q and flows at 100 moles/hr. Equilibrium data at 60°C and 2.2 atm have been obtained in the laboratory and can be expressed as: Mole fraction Q in vapor = 1.5 × (mole fraction Q in liquid). Find the outlet gas composition and the number of equilibrium stages required for this separation.

D3. Component A in a water stream is to be stripped out using an air stream in a countercurrent staged stripper. Inlet air is pure, and flow rate is G = 500 lb/hr. Inlet liquid stream has a mass ratio of X = 0.10 and a flow rate of pure water of 500 lb/hr. The desired outlet mass ratio is X = 0.005. Assume that water is nonvolatile and air is insoluble. Find the number of equilibrium stages and the outlet gas mass ratio. Equilibrium data can be represented by Y = 1.5X. Also find $(L/G)_{max}$.

D4. A vent gas stream in your chemical plant is 15.0 wt % Z; the rest is air. The local pollution authorities feel that Z is a minor pollutant and require a maximum concentration of 4.0 wt %. You have decided to build an absorption tower using water as the absorbent. The inlet water is pure and at 30°C. The operation is essentially isothermal. At 30°C your laboratory has found that the equilibrium data can be approximated by y = 0.5x (where y and x are weight fractions of Z in vapor and liquid, respectively).

a. Find the minimum ratio of water to air $(L/G)_{min}$ (This is on a Z-free basis.)

b. With an $L/G = 1.22 (L/G)_{min}$ find the total number of equilibrium stages and the outlet liquid concentration.

Assume that air is not soluble in water and that water is nonvolatile.

D5. A packed column 3 in in diameter with 10 ft of Intalox saddle packing is being run in the laboratory. P is being stripped from nC_9 using methane gas. The methane can be assumed to be insoluble and the nC_9 is nonvolatile. Operation is isothermal. The laboratory test results are:

$$X_{in} = 0.40 \ \frac{lb \ P}{lb \ n-C_9} \ , \quad Y_{out} = 0.50 \ \frac{lb \ P}{lb \ methane}$$

$$X_{out} = 0.06 \ \frac{lb \ P}{lb \ n-C_9} \ , \quad Y_{in} = 0.02 \ \frac{lb \ P}{lb \ methane}$$

Equilbrium data can be approximated as $Y = 1.5X$. Find the HETP for the packing.

D6. We wish to design a stripping column to remove carbon dioxide from water. This is done by heating the water and passing it countercurrent to a nitrogen stream in a staged stripper. Operation is isothermal and isobaric at 60°C and 1 atm pressure. The water contains 9.2×10^{-6} mole fraction CO_2 and flows at 100,000 lb/hr. Nitrogen (N_2) enters the column as pure nitrogen and flows at 2500 ft^3/hr. Nitrogen is at 1 atm and 60°C. We desire an outlet water concentration that is 2×10^{-7} mole fraction CO_2. Ignore nitrogen solubility in water and ignore the volatility of the water. Equilibrium data are in Table 15-1. Use a Murphree vapor efficiency of 40%. Find outlet vapor composition and number of real stages needed.

D7. You wish to absorb solute Q from an air stream into a solvent. The inlet solvent is pure. The entering air stream is 0.0015 mole frac Q, and the exiting air stream is to be 0.00003 mole frac Q. The outlet liquid stream is 0.00065 mole frac Q. At the temperature and pressure of operation, equilibrium can be represented by Henry's law. In terms of mole fractions of Q, equilibrium is $y = 2.21x$. How many stages are required? *Use one form of the Kremser equation.*

D8. A stripping column with four equilibrium stages is used to treat a liquid with 0.207 mole frac impurity in a nonvolatile solvent. Total flow rate of the liquid stream is 1003.8 moles/hr. We desire an outlet liquid composition of 0.005 mole frac impurity. The inlet gas is 0.005 mole frac impurity. The equilibrium data in mole fractions is $y = 3.0x$. What gas flow rate should be used?

D9. A heavy oil stream at 115°F is used in an absorber to remove dilute quantities of impurity Z from an air stream. The heavy oil is then recycled back to the process where Z is removed. The process is being run on a pilot plant basis, and information for scale-up is desired. The current absorber is a 30-plate sieve-tray column. Pilot plant data are as follows:

Liquid flow rate = 5.0 moles/hr

Gas flow rate = 2.5 moles/hr

$y_{Z,in} = 0.08$, $y_{Z,out} = 0.0002$, $x_{Z,in} = 0.0002$

Equilibrium for Z is given as $y_Z = 0.7x_Z$. Find the overall efficiency for component Z. Assume that liquid and gas flow rates are roughly constant.

D10. A stripping tower with four equilibrium stages is being used to remove ammonia from waste water using air as the stripping agent. Operation is at 80°F and 1 atm. The inlet air is pure air, and the inlet water contains 0.02 mole fraction ammonia. The column operates at $L/V = 0.65$. Equilibrium data in mole fractions are given as $y = 1.414x$. Find the outlet concentrations.

D11. An absorption column for laboratory use has been carefully constructed so that it has exactly 4.0 equilibrium stages and is being used to measure equilibrium data. Water is used as the solvent to absorb ammonia from air. The system operates isothermally at 80°F and 1 atm. The inlet water is pure distilled water. The ratio of $L/V = 1.2$, inlet gas concentration is 0.01 mole frac ammonia, and the measured outlet gas concentration is 0.0027 mole frac ammonia. Assuming that equilibrium is of the form $y = mx$, calculate the value of m for ammonia. Check your result.

D12. Read the section on cross flow in Chapter 16 before proceeding. We wish to strip CO_2 from a liquid solvent using air as the carrier gas. Since the air and CO_2 mixtures will be vented and since cross flow has a lower pressure drop, we will use a cross-flow system. The inlet liquid is 20.4 wt % CO_2, and the total inlet liquid flow rate is 1000 kg/hr. We desire an outlet liquid composition that is 2.5 wt % (0.025 wt frac) CO_2. The gas flow to each stage is 25,190 kg air/hr. In the last stage a special purified air is used that has no CO_2. In all other stages the inlet air is 0.0012 wt frac CO_2. Find the number of equilibrium stages required. In weight fraction units, equilibrium is y = 0.04x. Note: Use unequal axes for your McCabe-Thiele diagram.

D13. A complete gas treatment plant often consists of both an absorber to remove the solute and a stripper to regenerate the solvent. Some of the treated gas is heated and used in the stripper. This is called stream B. In a particular application we wish to remove obnoxious impurity A from the inlet gas. The absorber operates at 1.5 atm and 24°C where equilibrium is given as y = 0.5x (units are mole fractions). The stripper operates at 1.0 atm and 92°C where equilibrium is y = 3.0 x   (units are mole fractions). The total gas flow rate is 1400 moles/day, and the gas is 15 mole % A. The nonsoluble carrier is air. We desire a treated gas concentration of 0.5 mole % A. The liquid flow rate into the absorber is 800 moles/day and the liquid is 0.5 mole % A.

   a. Calculate the number of stages in the absorber and the liquid concentration leaving.

   b. If the stripper is an already existing column with four equilibrium stages, calculate the gas flow rate of stream B (concentration is 0.5 mole % A) and the outlet gas concentration from the stripper.

D14. We want to remove traces of propane and n-butane from a hydrogen stream by adsorbing them into a heavy oil. The feed is 150 kg moles/hr of a gas that is 0.0017 mole frac propane and 0.0006 mole frac butane. We desire to recover 98.8% of the butane. Assume that H_2 is insoluble. The heavy oil enters as a pure stream (which is approximately C_{10} and can be assumed to be nonvolatile). Liquid flow rate is 300 kg moles/hr. Operation is at 700 kPa and 20 ° C. K values can be obtained from Figures 2-11 and 2-12 or from Eq. (2-12). Find the equilibrium number of stages required and the compositions of gas and liquid streams leaving the absorber.

D15. An air stream at 26 ° C and 101.3 kPa contains component E. The gas-phase concentration is y_E = 0.0047. We wish to absorb E into a water stream at 26 ° C in the countercurrent absorber shown in the figure. Since the gas stream is essentially saturated

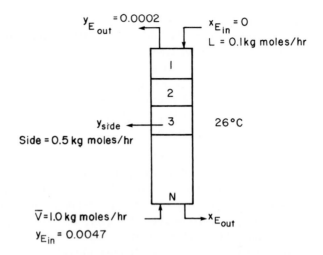

with water there will be essentially no transfer of water and the operation will be isothermal. Equilibrium data at $26°$ C are given as $y_E = 0.0789x_E$. The total gas flow rate entering is 1 kg mole/hr and the liquid flow rate is 0.1 kg moles/hr. The side stream is withdrawn at 0.5 kg mole/hr from the third stage. The final outlet vapor composition should be 0.0002 mole frac E. Use a graph with different scales. Assume that total flow rates are constant and solve in mole fraction units.

  a. What is the mole fraction of E in the side stream?

  b. What is the outlet E mole fraction leaving with the water?

  c. How many equilibrium stages are required?

D16. Repeat 15-D15, but solve in ratio units.

D17. We wish to absorb ammonia from air into water. Equilibrium data are given as $y_{NH_3} = 1.414x_{NH_3}$ in mole fractions. The countercurrent column has three equilibrium stages. The entering air stream has a total flow rate of 10 kg moles/hr and is 0.0083 mole frac NH_3. The inlet water stream contains 0.0002 mole frac ammonia. We desire an outlet gas stream with 0.0005 mole frac ammonia. Use the Kremser equation and find the required liquid flow rate, L.

F. *Problems Requiring Other Resources*

F1. Laboratory tests are being made prior to the design of an absorption column to absorb bromine (Br_2) from air into water. Tests were made in a laboratory-packed column that is 6 in in diameter, has 5 ft of packing, and is packed with saddles. The column was operated at 20°C and 5 atm total pressure, and the following data were obtained:

Inlet solvent is pure water.
Inlet gas is 0.02 mole frac bromine in air.
Exit gas is 0.002 mole frac bromine in air.
Exit liquid is 0.001 mole frac bromine in water.

What is the L/G ratio for this system? (Base your answer on flows of pure carrier gas and pure solvent.) What is the HETP obtained at these experimental conditions? Henry's law constant data are given in Perry's (4th ed) on pages 14-2 to 14-12. Note: Use mole ratio units. Assume that water is nonvolatile and air is insoluble.

F2.   A liquid stream from an absorber is to be regenerated in a stripping column. The column will operate at 80°C and 1 atm. The stream to be treated is a water stream containing 0.001 mole frac bromine (Br_2). The liquid flow rate is 100 moles/hr. To strip this water an air stream saturated with water at 80°C will be used. The inlet air contains no bromine and flows at a rate of 0.1975 moles air/hr. If the outlet water should contain 0.00005 mole frac bromine, determine the number of equilibrium stages required and the outlet gas concentration. Why might this column be difficult to operate? Note that increasing the temperature and decreasing the pressure makes stripping possible. Note also that since the inlet air stream is saturated with water vapor, no additional water will evaporate and the water can be treated as nonvolatile. Utilize mole ratios and solve this problem graphically.

F3.   (Difficult) An absorber with three equilibrium stages is operating at 1 atm. The feed is 10 moles/hr of a 60 mole % ethane, 40 mole % n-pentane mixture and enters at 30 °F. The solvent used is pure n-octane at 70 °F, solvent flow rate is 20 moles/hr. We desire to find all the outlet compositions and temperatures. The column is insulated. For a first guess assume that all stages are at 70 °F. As a first guess on flow rates, assume:

$$L_4 = 20.0 \qquad V_3 = 6.00 \text{ moles/hr}$$
$$L_3 = 20.2 \qquad V_2 = 6.18$$
$$L_2 = 20.8 \qquad V_1 = 6.66$$
$$L_1 = 24.0 \qquad V_0 = 10.0$$

Then go through *one* iteration of the sum rates convergence procedure (Figure 15-13) using direct substitution to estimate new flow rates on each stage. You could use these new flow rates for a second iteration, but instead of doing a second iteration of the flow loop use a paired simultaneous convergence routine. To do this, use the new values for liquid and vapor flow rates to find compositions on each stage. Then calculate enthalpies and use the multivariable Newtonian method to calculate new temperatures on each stage. You will then be ready to recalculate K values and solve the mass balances for the second iteration. However, for purposes of this assigment stop after the new temperatures have been estimated. Use a DePriester chart or Eq. (2-12) for K values. Pure-component enthalpies are given in Maxwell (1950) and on pages 629 and 630 of Smith (1963). Assume ideal solution behavior to find the enthalpy of each stream.

F4. (Difficult) An absorber with three equilibrium stages is operating at 1 atm. The feed is 100 moles/hr of a 60 mole % ethane, 40 mole % n-pentane mixture and the feed enters at 40 °F. The solvent used is pure n-octane at 90 °F and solvent flow rate is 150 moles/hr. We desire to find all the outlet compositions. The column is insulated. For a first guess assume that all stages are 80 °F. As a first guess on flow rates, assume:

$$L_4 = 150.0 \qquad V_3 = 60.35 \text{ moles/hr}$$
$$L_3 = 152.0 \qquad V_2 = 63.0$$
$$L_2 = 159.0 \qquad V_1 = 67.8$$
$$L_1 = 191.25 \qquad V_0 = 100.0$$

Then go through *one* iteration of the sum rates convergence procedure using direct substitution to estimate new flow rates on each stage. You could use these new flow rates a second iteration, but instead use a paired simultaneous convergence routine. To do this, use the new values for liquid and vapor flow rates to find compositions on each stage and then calculate enthalpies and use a direct substitution method to calculate new temperatures on each stage. For purposes of this assignment, stop after the new temperatures have been estimated. Use a DePriester chart or Eq. (2-12) for K values. Pure-component enthalpies are given in Maxwell (1950) and on pages 629 and 630 of Smith (1963). Assume ideal solution behavior to find the enthalpy of each stream.

chapter 16
IMMISCIBLE EXTRACTION

16.1. EXTRACTION PROCESSES AND EQUIPMENT

Extraction is a process where one or more solutes are removed from a liquid by transferring the solute(s) into a second liquid phase. The two liquid phases must be immiscible (that is, insoluble in each other) or partially immiscible. In this chapter we will discuss extraction equipment and immiscible extraction, while in Chapter 18 we will discuss partially miscible extraction. In both cases the separation is based on different solubilities of the solute in the two phases. Since vaporization is not required, extraction can be done at low temperature and is a gentle process suitable for unstable molecules.

Extraction is a common laboratory and commercial unit operation. For example, in commercial penicillin manufacture, after the fermentation broth is sent to a centrifuge to remove cell particles the pencillin is extracted from the broth. Then the solvent and the pencillin are separated from each other by one of several techniques. In petroleum processing, aromatic hydrocarbons such as benzene, toluene, and xylenes are separated from the paraffins by extraction with a solvent such as sulfolane. The mixture of sulfolane and aromatics is sent to a distillation column, where the sulfolane is the bottoms product, and is recycled back to the extractor. Flow charts for acetic acid recovery from water and for the separation of aromatics from aliphatics are given by Robbins (1979). Reviews of industrial applications of extraction are presented by Bailes *et al.* (1976), Lo *et al.* (1983), and Ritchy and Ashbrook (1979).

As these commercial examples illustrate, the complete extraction process includes the extraction unit and the solvent recovery process. This is shown schematically in Figure 16-1. In many applications the downstream solvent recovery step (usually distillation) is more expensive than the actual extraction step. A variety of extraction cascades including single stages, countercurrent cascades, and cross-flow cascades can be used; we will discuss these later.

The variety of equipment used for extraction is much greater than

515

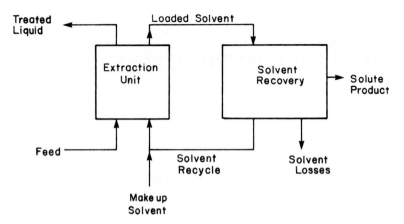

Figure 16-1. Complete extraction process.

for distillation, absorption, and stripping. Efficient contacting and separating of two liquid phases is considerably more difficult than contacting and separating a vapor and a liquid. In addition to plate and packed columns, many specialized pieces of equipment have been developed. Some of these are illustrated in Figure 16-2. Details of the different types of equipment are provided by Lo (1979), Lo *et al.* (1983), Reissinger and Schroeter (1978), Perry and Chilton (1973, pp. 21-4 to 21-29), Perry and Green (1984, pp. 21-55 to 21-83), and Jamrack (1963). A summary of the features of the various types of extractors is presented in Table 16-1. Decision methods for choosing the type of extractor to use are presented by King (1980), Lo (1979), Reissinger and Schroeter (1978), and Robbins (1979).

The number of equilibrium contacts required can be determined using the same stage-by-stage procedures for all the extractors. Determining stage efficiencies and hydrodynamic characteristics is more difficult. The more complicated systems are designed by specialists (e.g., see Lo *et al.*, 1983; or Skelland and Tedder, 1987).

16.2. COUNTERCURRENT EXTRACTION

The most common type of extraction cascade is the countercurrent system shown schematically in Figure 16-3. In this cascade the two phases flow in opposite directions. Each stage is assumed to be an equilibrium stage so that the two phases leaving the stage are in equilibrium.

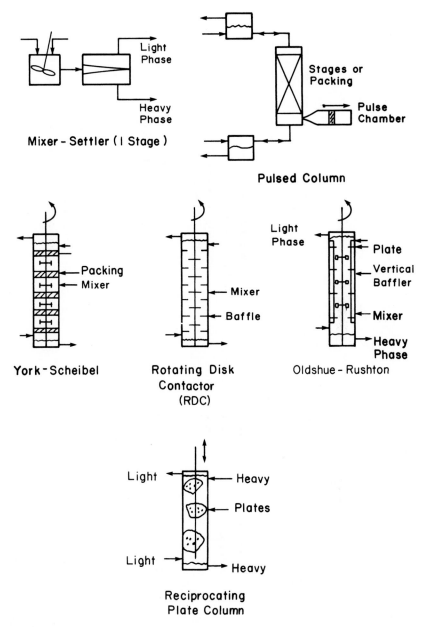

Figure 16-2.   Extraction equipment.

Table 16-1. Extractor Types

Extractor	Examples	General Features
Unagitated columns	Plate columns; packed columns; spray columns	Low capital cost Low operating and maintenance cost Simplicity in construction Handles corrosive material
Mixer settlers	Mixer-settler; vertical mixer-settler; Morris contactor; static mixers	Higher-stage efficiency Handles wide solvent ratios High capacity Good flexibility Reliable scale-up Handles liquids with high viscosity Requires large area Expensive if large number of stages required
Pulsed columns	Perforated plate; packed	Low HETP No internal moving parts Many stages possible

Rotary-agitation columns	Rotating disk contactor; Oldshue-Rushton; Scheibel (several types); Kuhni; Asymmetric rotating disk	Reasonable capacity Reasonable HETP Many stages possible Reasonable construction cost Low operating & maintenance cost
Reciprocating-plate columns	Karr column; segmental passages; counter-moving plates	High throughput Low HETP Great versatility and flexibility Simplicity in contruction Handles liquids containing suspended solids Handles mixtures with emulsifying tendencies
Centrifugal extractors and separators	Podbielniak; Quadronic; Alfa-Laval; Westfalia; Robatel	Short contacting time for unstable material Limited space required Handles easily emulsified material Handles systems with little liquid density difference

Source: Adapted from Lo (1979).

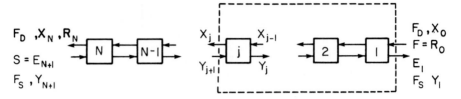

Figure 16-3.   Mass balance envelope for countercurrent cascade.

The solute, A, is initially dissolved in diluent, D, in the feed. Solute is extracted with solvent, S. Streams with high concentrations of diluent are called *raffinate,* while streams with high concentrations of solvent are the *extract.* The nomenclature in both weight fraction and weight ratio units is given in Table 16-2.

16.2.1. McCabe-Thiele Method

The McCabe-Thiele analysis for immiscible extraction is very similar to the analysis for absorption and stripping discussed in Chapter 15. It was first developed by Evans (1934) and is reviewed by Jamrack (1963) and Robbins (1979). In order to use a McCabe-Thiele type of analysis we must be able to plot a single equilibrium curve, have the energy balances automatically satisfied, and have one operating line for each section.

For equilibrium conditions, the Gibbs phase rule is: $F = C - P + 2$. There are three components (solute, solvent and diluent) and two phases. Thus there are three degrees of freedom. In order to plot equilibrium data as a single curve, we must reduce this to one degree of freedom. The following two assumptions are usually made:

1. The system is isothermal.

2. The system is isobaric.

To have the energy balances automatically satisfied, we must also assume

3. The heat of mixing is negligible.

These three assumptions are usually true for dilute systems. The operating line will be straight, which makes it easy to work with, and the solvent and diluent mass balances will be automatically satisfied if the following assumption is valid:

Table 16-2. Nomenclature for Extraction

Weight Fraction Units:

$x_{A,j}$	Raffinate wt fraction of solute leaving stage j
$x_{D,j}$	Raffinate wt fraction of diluent leaving stage j
$x_{S,j}$	Raffinate wt fraction of solvent leaving stage j
x_{A_F}, x_{A_M}	Wt. fraction solute in feed and mixed streams (Chapt. 18)
$y_{A,j}$	Extract wt fraction of solute leaving stage j
$y_{D,j}$	Extract wt fraction of diluent leaving stage j
$y_{S,j}$	Extract wt fraction of solvent leaving stage j
y_{A_S}	Wt fraction solute in solvent stream
z_A, z_B	Wt fraction in feed to fractional extractor
R_j	Raffinate flow rate leaving stage j, kg/hr
E_j	Extract flow rate leaving stage j, kg/hr
$S = E_{N+1}$	Solvent flow rate entering extractor (not necessarily pure), kg/hr
$F = R_o$	Feed rate entering extractor, kg/hr
M	Flow rate of mixed streams, kg/hr (Chapt. 18)

Weight Ratio Units (Solvent and Diluent Immiscible):

X_j	Weight ratio solute in diluent leaving stage j, kg A/kg D
Y_j	Weight ratio solute in solvent leaving stage j, kg A/kg S
F_D	Flow rate of diluent, kg diluent/hr
F_S	Flow rate of solvent, kg solvent/hr

Names:

A	Solute. Material being extracted
D	Diluent. Chemical solute is dissolved in the feed
S	Solvent. Separating agent added for the separation

4a. Diluent and solvent are totally immiscible.

When the fourth assumption is valid, then

$$F_D = \text{diluent flow rate, kg D/hr} = \text{constant} \qquad (16\text{-}1)$$

$$F_S = \text{solvent flow rate, kg S/hr} = \text{constant} \qquad (16\text{-}2)$$

These are flow rates of diluent only and solvent only and do *not* include the total raffinate and extract streams. Equations (16-1) and (16-2) are the diluent and solvent mass balances, and they are automatically satisfied when the phases are immiscible.

When the diluent and solvent are immiscible, weight ratio units are related to weight fractions as

$$X = \frac{x}{1-x} \quad \text{and} \quad Y = \frac{y}{1-y} \qquad (16\text{-}3)$$

where X is kg solute/kg diluent and y is kg solute/kg solvent. Note that these equations require that the phases be immiscible.

The operating equation can be derived with reference to the mass balance envelope shown in Figure 16-3. In weight ratio units, the steady-state mass balance is

$$F_S\, Y_1 + F_D\, X_j = F_S\, Y_{j+1} + F_D\, X_0 \qquad (16\text{-}4)$$

Solving for Y_{j+1} we obtain the operating equation,

$$Y_{j+1} = \left(\frac{F_D}{F_S}\right)X_j + \left(Y_1 - \frac{F_D}{F_S}\, X_0\right) \qquad (16\text{-}5)$$

When plotted on a McCabe-Thiele diagram of Y versus X, this is a straight line with slope F_D/F_S and Y intercept $\left(Y_1 - \dfrac{F_D}{F_S}\, X_0\right)$, as illustrated in Example 16-1. Note that since F_D and F_S are constant, the operating line is straight. For the usual design problem, F_D/F_S will be known as well be X_0, X_N and Y_{N+1}. Since X_N and Y_{N+1} are the concentrations of passing streams, they represent the coordinates of a point on the operating line.

Equilibrium data for dilute extraction are usually represented as a distribution ratio, K_d,

$$K_d = \frac{y_A}{x_A} \tag{16-6}$$

in weight fractions or mole fractions. For very dilute systems K_d will be constant, while at higher concentrations K_d often becomes a function of concentration. Values of K_d are tabulated in Perry and Green (1984, pp. 15-9 to 15-13), Hartland (1970, Chapt. 6), and Francis (1972). A brief listing is given in Table 16-3. The equilibrium "constant" is temperature- and pH-dependent. The temperature dependence is illustrated in Table 16-3 for the distribution of acetic acid between water (the diluent) and benzene (the solvent). Note that there is an optimum temperature at which K_d is a maximum. Benzene is not a good solvent for acetic acid because the K_d values are low, but water would be a good solvent if benzene were the diluent. As shown in Table 16-3, 1-butanol is a much better solvent then benzene for acetic acid. In addition, benzene would probably not be used as a solvent because it is carcinogenic. The use of extraction to fractionate components requires that the selectivity, $\alpha_{21} = K_{D2}/K_{D1}$, be large. An example where fractional extraction is feasible is the separation of ethylbenzene and xylenes illustrated in Table 16-3. The ethylbenzene - p-xylene separation will be the most difficult of these, but is nonetheless feasible.

For the McCabe-Thiele diagram, if flow rates are given as in Eqs. (16-1) and (16-2) the equilibrium data must be expressed as weight or mole ratios. Equations (16-3) can be used to transform the equilibrium data, which is easy to do in tabular form (see Example 16-1). Usually equilibrium data are reported in fractions, not the ratio units given in the second part of Table 16-3.

With the equilibrium data and the operating equation known, the McCabe-Thiele diagram is plotted as shown in Figure 16-4. First the point (Y_{N+1}, X_N) is plotted, and the operating line passes through this point with a slope of F_D/F_S. Y_1 can be found from the operating line at the inlet raffinate concentration X_o. Then stages are easily stepped off.

Example 16-1. Countercurrent Immiscible Extraction

An 11.5 wt % mixture of acetic acid in water is to be extracted with 1-butanol at 1 atm pressure and 26.7° C. We desire outlet concentrations of 0.5 wt % in the water and 9.6 wt % in the

Table 16-3. Distribution Coefficients for Immiscible Extraction

Solute (A)	Solvent	Diluent	$T, °C$	$K_d = y_A/x_A$
Equilibrium in Weight Fraction Units (Perry and Green, 1984)				
Acetic acid	Benzene	Water	25	0.0328
Acetic acid	Benzene	Water	30	0.0984
Acetic acid	Benzene	Water	40	0.1022
Acetic acid	Benzene	Water	50	0.0588
Acetic acid	Benzene	Water	60	0.0637
Acetic acid	1-Butanol	Water	26.7	1.613
Furfural	Methyli sobutyl ketone	Water	25	7.10
Ethyl benzene	β, β'-Thiodipropionitrile	n-Hexane	25	0.100
m-Xylene	β, β'-Thiodipropionitrile	n-Hexane	25	0.050
o-Xylene	β, β'-Thiodipropionitrile	n-Hexane	25	0.150
p-Xylene	β, β'-Thiodipropionitrile	n-Hexane	25	0.080
Equilibrium in Mass Ratio Units (Brian, 1972)				
Linoleic acid ($C_{17}H_{31}COOH$)	Heptane	Methylcellosolve + 10 vol % water		2.17
Abietic acid ($C_{19}H_{29}COOH$)	Heptane	Methylcellosolve + 10 vol % water		1.57
Oleic acid	Heptane	Methylcellosolve + 10 vol % water		4.14

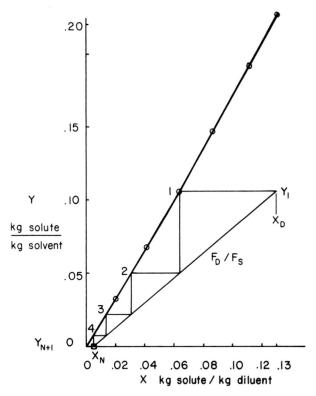

Figure 16-4.   McCabe-Thiele diagram for dilute extraction, Example 16-1.

butanol. Inlet butanol is pure. Find the number of equilibrium stages required and the ratio of water to 1-butanol.

Solution

A. Define. The sketch is shown. We wish to find N and the ratio of water flow rate to butanol flow rate.

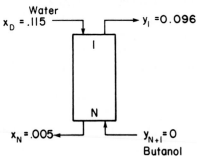

B. Explore. Equilibrium is given in Table 16-3, y = 1.613x. Since concentrations are relatively high, we will use mass ratio units. We also assume that water and butanol are immiscible.

C. Plan. Solve first using a McCabe-Thiele diagram in mass ratio units. First convert the equilibrium data to mass ratios using Eqs. (16-3). Find values of all known streams in mass ratio units. Plot passing streams and obtain the operating line. Then step off stages.

D. Do It. Graphical Solution: The inlet and outlet streams in mass ratio units can be found from Eqs. (16-3).

$$Y_{N+1} = \frac{y_{N+1}}{1 - y_{N+1}} = 0$$

$$X_N = \frac{x_N}{1 - x_N} = \frac{0.005}{1 - 0.005} = 0.00503$$

$$Y_1 = \frac{y_1}{1 - y_1} = \frac{0.096}{1 - 0.096} = 0.106$$

$$X_0 = \frac{x_0}{1 - x_0} = \frac{0.115}{1 - 0.115} = 0.13$$

The equilibrium data, y = 1.613x, are converted to ratio units in the following table.

x	y	$X = \dfrac{x}{1 - x}$	$Y = \dfrac{y}{1 - y}$
0	0	0	0
0.02	0.0323	0.0204	0.0334
0.04	0.0645	0.0417	0.0689
0.06	0.0968	0.0638	0.1072
0.08	0.1290	0.087	0.1481
0.10	0.1613	0.1111	0.1923
0.115	0.1855	0.1299	0.2277
0.14	0.2258	0.1628	0.2917

The equilibrium values of Y vs X are plotted in Figure 16-4. Note that the equilibrium line is slightly curved. The pairs of passing streams (Y_{N+1}, X_N) and (Y_1, X_0) are on the operating line and are also shown in Figure 16-4. The slope of the operating line is

$$\frac{F_D}{F_S} = \frac{Y_1 - Y_{N+1}}{X_0 - X_N} = \frac{0.106 - 0}{0.13 - 0.005} = 0.848$$

which is the ratio of water flow rate (without acid) to the butanol flow rate (without acid).

Stages are stepped off as shown in Figure 16-4. Four equilibrium stages are sufficient.

E. Check. The equilibrium parameter and the assumption of immiscible phases should be checked for these high concentrations. The most common mistakes are to solve in fraction units (which is wrong because R/E is not constant) or to convert inlet and outlet concentrations to ratio units but not change the equilibrium data.

F. Generalize. Note that the nonlinear transformation of variables going from fractions to ratios, Eq. (16-3), converts a straight equilibrium line into a curve. For $K_d > 1$ the curve slopes upward as shown in Figure 16-4, while for $K_d < 1$ the curve slopes downward. The curvature becomes quite noticeable for very large or very small K_d.

The McCabe-Thiele diagram shown in Figure 16-4 is very similar to McCabe-Thiele diagrams for stripping. This is true because the processes are analogous unit operations in that both contact two phases and solute is transferred from the X phase to the Y phase. In both analyses we defined constant flows in each phase and used ratio units. The analogy breaks down when we consider stage efficiencies and sizing the column diameter, since mass transfer characteristics and flow hydrodynamics are very different for extraction and stripping.

In stripping there was a maximum L/G. For extraction, a maximum value of F_D/F_S can be determined in the same way. This gives the minimum solvent flow rate, $F_{S\,min}$, for which the desired separation can be obtained with an infinite number of stages.

For simulation problems the number of stages is specified but the outlet raffinate concentration is unknown. A trial-and-error procedure is required in this case. This procedure is essentially the same as the simulation procedure used for absorption or stripping.

When there is some partial miscibility of diluent and solvent, the McCabe-Thiele analysis can still be used if the following alternative assumption is valid.

4b. The concentration of solvent in the raffinate and the concentration of diluent in the extract are both constant.

The flow rates of the diluent and solvent streams are now defined as

$$F_D = \frac{kg \ (D \ + \ S) \ in \ raffinate}{hr} = constant \qquad (16\text{-}7a)$$

$$F_S = \frac{kg \ (S \ + \ D) \ in \ extract}{hr} = constant \qquad (16\text{-}7b)$$

The ratio units are defined as

$$X = \frac{kg \ A \ in \ raffinate}{kg \ (D \ + \ S) \ in \ raffinate} \qquad (16\text{-}8a)$$

$$Y = \frac{kg \ A \ in \ extract}{kg \ (S \ + \ D) \ in \ extract} \qquad (16\text{-}8b)$$

The calculation procedure now follows Eqs. (16-4) to (16-6) and Figure 16-4. When the phases are partially miscible and assumption 4b is not valid, the methods developed in Chapter 18 should be used.

If the system is very dilute, total flows will be constant and the McCabe-Thiele diagram can be plotted in fractions. Since adding large amounts of solute often makes the phases partially miscible, the assumption of complete immiscibility is often valid only for very dilute systems, for which we can use mole or weight fractions and total extract and raffinate flow rates.

Dilute multicomponent extraction can be analyzed on a McCabe-Thiele diagram if we add two more assumptions.

5. Total extract and total raffinate flow rates are constant.

6. Each solute is independent.

When these assumptions are valid, the entire problem can be solved in mole or weight fractions. Then for each solute the mass balance for the balance envelope shown in Figure 16-3 is

$$y_{i,j+1} = \frac{R}{E} x_{i,j} + \left(y_{i,1} - \frac{R}{E} x_{i,0}\right) \qquad (16\text{-}9)$$

where i represents the solute and terms are defined in Table 16-2. Since R/E is assumed to be constant, this is a straight line on a McCabe-Thiele diagram plotted in fractions. The operating lines for each solute have the same slopes but different y intercepts. The equilibrium curves for each solute are independent because of assumption 6. Then we can solve for each solute independently. This solution procedure is similar to the one for dilute multicomponent absorption and stripping. We first solve for the number of stages, N, using the solute that has a specified outlet raffinate concentration. Then with N known, a trial-and-error procedure is used to find x_{N+1} and y_1 for each of the other solutes.

Solvent selection is critical for the development of an economical extraction system. The solvent should be highly selective for the desired solute and not very selective for undesired solutes. It should have a high capacity for the desired solute. Both the selectivity, $\alpha = K_{desired}/K_{undesired}$, and $K_{desired}$ should be as large as possible. The solvent should also be easy to separate from the diluent. This implies either a totally immiscible system or a partially immiscible system where separation by other means such as distillation is easy. In addition, the solvent should be nontoxic, noncorrosive, readily available, chemically stable, and cheap. Many of the guidelines for solvent selection are the same as for extractive distillation (see Chapter 10). King (1981), Lo *et al.* (1983), and Treybal (1963) discuss solvent selection in detail.

16.2.2. Kremser Method

If one additional assumption can be made, the Kremser equation can be used for dilute extraction. Obviously, this assumption is

7. Equilibrium is linear.

In this case, equilibrium has the form

$$y_i = m_i x_i + b_i \qquad (16\text{-}10)$$

The dilute extraction model now satisfies all the assumptions used to derive the Kremser equations in Chapter 15, so they can be used directly. Since we have used different symbols for flow rates, we replace L/V with R/E. Then Eq. (15-20) becomes

$$N = \frac{y_{N+1} - y_1}{y_1 - \dfrac{R}{E} x_o - b} \quad \text{for} \quad \frac{R}{mE} = 1 \qquad (16\text{-}11)$$

while Eqs. (15-29) (inverted) and (15-30) become

$$\frac{y_1 - y_1^*}{y_{N+1} - y_1^*} = \frac{1 - \dfrac{R}{mE}}{1 - \left(\dfrac{R}{mE}\right)^{N+1}} \qquad (16\text{-}12)$$

and

$$N = \frac{\ln\left[\left(1 - \dfrac{mE}{R}\right)\left(\dfrac{y_{N+1} - y_1^*}{y_1 - y_1^*}\right) + \dfrac{mE}{R}\right]}{\ln\left(\dfrac{R}{mE}\right)} \qquad (16\text{-}13)$$

Other forms of the Kremser equation can also be written with this substitution. When the Kremser equation is used, simulation problems are no longer trial-and-error. Application of the Kremser equation to extraction is considered in detail by Hartland (1970), who also gives linear fits to equilibrium data over various concentration ranges.

16.3. FRACTIONAL EXTRACTION

Very often extraction is used to separate solutes from each other. In this situation we can use fractional extraction with two solvents as illustrated in Figure 16-5. In fractional extraction the two solvents are chosen so that solute A prefers solvent 1 and concentrates at the top of the column, while solute B prefers solvent 2 and concentrates at the bottom of the column. In Figure 16-5 solvent 2 is labeled as diluent so that we can use the nomenclature of Table 16-2. The column sections in Figure 16-5 are often separate so that each section can be at a different pH or temperature. This will make the equilibrium curve different for the two sections.

A common problem in fractional extraction is the *center cut*. In Figure 16-6 solute B is the desired solute while A represents a series of solutes that are more strongly extracted by solvent 1 and C represents a series of solutes that are less strongly extracted by solvent 1. Center cuts are common when pharmaceuticals are produced by fermentation, since a host of undesired chemicals are also produced.

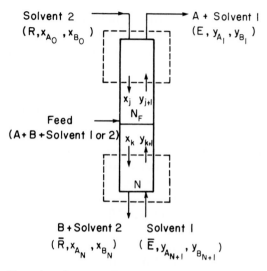

Figure 16-5.   Fractional extraction.

Before looking at the analysis of fractional extraction it will be helpful to develop a simple criterion to predict whether a solute will go up or down in a given column. At equilibrium the solutes distribute between the two liquid phases. The ratio of *solute* flow rates in the column shown in Figure 16-5 is

$$\frac{\text{Solute A flow up column}}{\text{Solute A flow down column}} = \frac{y_{A,j}E_j}{x_{A,j}R_j} = \frac{K_{D,A,j}E_j}{R_j} \qquad (16\text{-}14)$$

Figure 16-6.   Center-cut extraction.

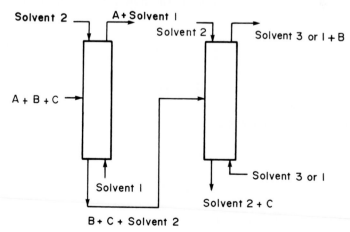

where we have used the equilibrium expression, Eq. (16-6). If $(K_{D,A}E/R)_j > 1$, the net movement of solute A is up at stage j, while if $(K_{D,A}E/R)_j < 1$, the net movement of solute A is down at stage j. In Figure 16-5 we want

$$\left(\frac{K_{D,A}\,E}{R}\right)_j > 1 \quad \text{and} \quad \left(\frac{K_{D,B}\,E}{R}\right)_j < 1 \tag{16-15}$$

in both sections of the column. By adjusting $K_{D,i}$ (say, by changing the solvents used or temperature) and/or the two solvent flow rates (E and R), we can change the direction of movement of a solute. This was done to solute B in Figure 16-6; B goes down in the first column and up in the second column. Ranges of $(K_D E/R)$ can be derived that will make the fractional extractor work (see Problem 16-A10). Since it is quite expensive to have a large number of equilibrium stages in a commercial extractor, the ratios in Eq. (16-15) should be significantly different.

The analysis of fractional extraction is straightforward for dilute mixtures when the solutes are independent and total flow rates in each section are constant. The external mass balances for the fractional extraction cascade shown in Figure 16-5 are

$$R + \bar{E} + F = E + \bar{R} \tag{16-16a}$$

$$Rx_{A,0} + \bar{E}y_{A,N+1} + Fz_{A,F} = Ey_{A,1} + \bar{R}x_{A,N} \tag{16-16b}$$

$$Rx_{B,0} + \bar{E}y_{B,N+1} + Fz_{B,F} = Ey_{B,1} + \bar{R}x_{B,N} \tag{16-16c}$$

If the feed is contained in solvent 1, the flow rates in the two sections are related by the expressions

$$E = \bar{E} + F \tag{16-17a}$$

$$\bar{R} = R \tag{16-17b}$$

while if the feed is in solvent 2,

$$E = \bar{E} \tag{16-18a}$$

$$\bar{R} = R + F \tag{16-18b}$$

The solute operating equations for the top section using the top mass balance envelope in Figure 16-5 is Eq. (16-9), which was derived earlier. For the bottom section the mass balances are represented by

$$y_{i,k+1} = \frac{\overline{R}}{\overline{E}} x_k - (\frac{\overline{R}}{\overline{E}} x_{i,N} - y_{i,N+1})$$

(16-19)

Since the feed is usually dissolved in one of the solvents, the phase flow rates are usually slightly different in the two sections. A McCabe-Thiele diagram can be plotted for each solute. Each diagram will have two operating lines and one equilibrium line, as shown in Figure 16-7.

Figures 16-7A and B show the characteristics of both absorber and stripper diagrams. The solutes are being "absorbed" in the top section

Figure 16-7. McCabe-Thiele diagram for fractional extraction. (A) Solute A; (B) Solute B.

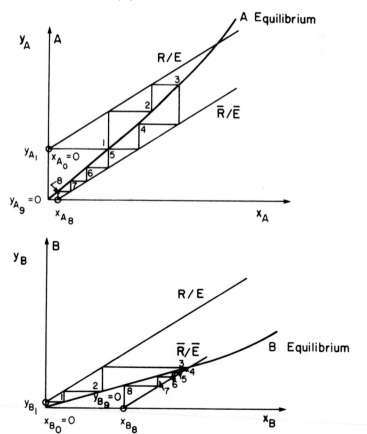

(that is, the solute concentration is increasing as we go down the column) while the solute is being "stripped" in the bottom section (solute concentration is increasing as we go up the column). Thus the solute is most concentrated at the feed stage and diluted at both ends (because we add lots of extra solvent). The two operating lines will intersect at a feed line which is at the concentration of the solute in the feed. This feed concentration is usually much greater than the solute concentration on the feed stage. If there is no solvent in the feed, then the effective feed concentrations (y_i or x_i) are very large and the operating lines are almost parallel.

The top section of the column in Figures 16-5 and 16-7 is removing ("absorbing") component B from A. The more stages in this section, the purer the A product will be (smaller $y_{B,1}$), and the higher the recovery of B in the B product will be. The bottom section of the column is removing ("stripping") component A from B. Extra stages in the bottom section increase the purity of the B product (reduce $x_{A,N}$) and increase the recovery of A in the A product.

Specifications would typically include temperature, pressure, feed composition (A, B, and solvents), feed flow rate, and both solvent compositions. Some of the ways of specifying the *four* remaining degrees of freedom are illustrated below.

Case 1. Specify $y_{A,1}$, R, $\overline{E}$, and N_F. Calculate $x_{A,N}$ from Eq. (16-16b), calculate $\overline{R}$ from Eq. (16-17b) or (16-18b), and calculate E from Eq. (16-17a) or (16-18a). Since pairs of passing streams and slopes are known, both operating lines can be plotted for solute A. The total number of stages can easily be determined from the solute A diagram. Solution for solute B is trial-and-error.

Case 2. Specify $\overline{E}$, R, N_F, and N. This is trial-and-error for each solute, but the two solute problems are not coupled and can be solved separately.

Case 3. Specify R, $\overline{E}$, $y_{A,1}$, and $x_{B,N}$. After doing external mass balances, you can plot the operating lines as in Figure 16-7, but the problem is still trial-and-error. The feed stage must be varied until the total number of stages is the same for both solutes. Small changes in compositions or flow rates will probably be required to get an exact fit.

Because they are inherently trial-and-error, fractional extraction problems are naturals for computer solution.

Brian (1972, Chapt. 3) explores fractional extraction in detail. He illustrates the use of extract reflux and derives forms of the Kremser equation for multisection columns. An abbreviated treatment of the

Kremser equation for fractional extraction is also presented by King (1980).

16.4. CROSS FLOW AND TWO-DIMENSIONAL SYSTEMS

Although countercurrent cascades are the most common, other cascades can be employed. One type occasionally used is the cross-flow cascade shown in Figure 16-8. Each stage is assumed to be an equilibrium stage. In this cascade, fresh extract streams are added to each stage and extract products are removed. For single-solute systems the same four assumptions made for countercurrent cascades are required for the McCabe-Thiele analysis. For dilute multicomponent systems, assumptions 5 and 6 are again required.

To derive an operating equation, we use a mass balance envelope around a single stage as shown around stage j in Figure 16-8. For concentrated systems, ratio units are again used and the resulting steady-state mass balance is

$$F_D \, X_{j-1} + F_{S,j} \, Y_{j,in} = F_D \, X_j + F_{S,j} \, Y_j \tag{16-20}$$

Solving for the outlet extract mass ratio, Y_j, we get

$$Y_j = -\frac{F_D}{F_{S,j}} \, X_j + \left(\frac{F_D}{F_{S,j}} \, X_{j-1} + Y_{j,in}\right) \tag{16-21}$$

Each stage will have a different operating equation. On a McCabe-Thiele diagram plotted as Y versus X, this is a straight line of slope $-F_D/F_{S,j}$ and Y intercept $(X_{j-1} \, F_D/F_{S,j} + Y_{j,in})$. In these equations $Y_{j,in}$ is the mass ratio of solute in the extract entering stage j and $F_{S,j}$ is the flow rate of solvent entering stage j. The designer can specify all

Figure 16-8.   Cross-flow cascade.

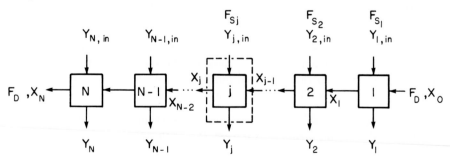

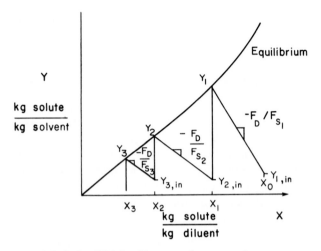

Figure 16-9.   McCabe-Thiele diagram for cross flow.

values of F_{Sj} and $Y_{j,in}$ as well as X_0, F_D, and either X_N or N. If the calculation is started at the first stage ($j = 1$), $X_{j-1} = X_0$ is known and the operating equation can be plotted. Since the stage is an equilibrium stage, X_1 and Y_1 are in equilibrium in addition to being related by operating equation (16-18). Thus the intersection of the operating line and the equilibrium curve is at Y_1 and X_1 (see Figure 16-9). The raffinate input to stage 2 is X_1. Thus the point $(Y_{2,in}, X_1)$ on the operating line is known and the operating line for stage 2 can be plotted. The procedure is repeated until X_N is reached or N stages have been stepped off.

In general, each stage can have different solvent flow rates, F_{S_j}, and different inlet solvent mass ratios, $Y_{j,in}$, as illustrated in Figure 16-9. This figure also shows that the point with the inlet concentrations $(Y_{j,in}, X_{j-1})$ is on the operating line for each stage, which is easily proved with Eq. (16-16). That is, if we let $Y_j = Y_{j,in}$ and $X_j = X_{j-1}$, Eq. (16-18) is satisfied. Thus the point representing the inlet concentrations is on the operating line.

The operating lines in Figure 16-9 are similar to those we found for binary flash distillation. Both single-stage systems and cross-flow systems are arranged so that the two outlet streams are in equilibrium *and* on the operating line. This is not true of countercurrent systems.

The analysis for dilute multicomponent systems follows as a logical extension of the independent solution for each solute as was discussed for counter-current systems. Total flow rates and mole or weight fractions are used in these calculations. Note that simulation problems do *not* require trial-and-error solution for cross-flow systems.

Example 16-2. Cross-Flow Immiscible Extraction

The 11.5 wt % aqueous solution of acetic acid processed in a countercurrent column in Example 16-1 is to be treated in a four-stage cross-flow system. The same amount of solvent (pure 1-butanol) is to be used, but now it will be divided equally among the four stages. Operation is at 1 atm and 26.7 °C. Find the outlet concentrations of all streams.

Solution

A. Define. The process sketch is shown. We wish to find X_4, Y_1, Y_2, Y_3, and Y_4.

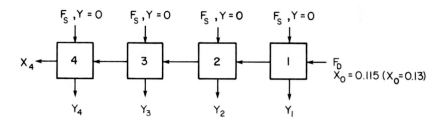

B. Explore. Equilibrium data are given in Table 16-3 and were converted to mass ratios in Example 16-1. This seems to be a straightforward cross-flow calculation.

C. Plan. Plot the equilibrium curve and plot the operating lines following Eq. (16-18).

D. Do It. Since F_D/F_S was 0.848 in Example 16-1 and $F_{S \, cross-flow} = 0.25 \, F_{S \, countercurrent}$, $F_D/F_S = 4(0.848) = 3.39$. For stage 1 the point $(Y = 0, X = 0.13)$ is on the operating line. Slope $= -F_D/F_S = -3.39$. This is plotted in Figure 16-10. The result for stage 1 gives X_1. Then $(Y_1 = 0, X_1)$ is on the operating line for stage 2. This procedure can be continued for stages 3 and 4. The result is $X_4 = 0.026$, $Y_1 = 0.148$, $Y_2 = 0.098$, $Y_3 = 0.065$, $Y_4 = 0.043$.

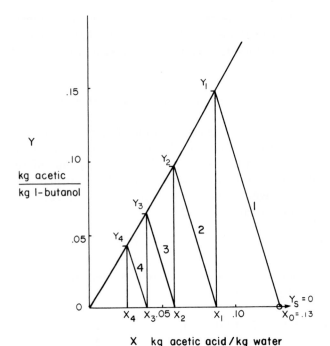

Figure 16-10.   Graphical solution for Example 16-2.

E. Check. An overall mass balance can be checked. If we set F_D = 1, then F_S = 1/3.39 for each stage. Overall mass balance is

$$(F_D)(X_0 - X_4) + \sum_{j=1}^{4} (F_S Y_{in} - F_S Y_{j,out}) = 0$$

Using the results obtained, the left hand side of this equation equals 0.0004. Thus, the overall mass balance is satisfied within the accuracy of the graph.

F. Generalize. With the same amount of solvent and the same number of stages, the countercurrent cascade produced a much cleaner water stream (0.005 vs 0.026). In addition, the counter-current system produced only one concentrated solvent stream instead of four streams of different concentrations. This result explains why countercurrent cascades are much more common than cross-flow cascades.

Cross-flow systems have also been explored for stripping (Wnek and Snow, 1972). The analysis procedure for staged cross-flow systems would be very similar to the extraction calculations developed here (see Problem 15-D12 for an example).

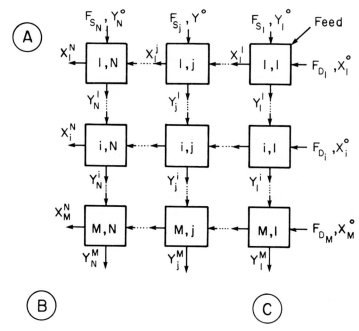

Figure 16-11.   Two-dimensional cascade.

In countercurrent systems the solvent is reused in each stage while in cross-flow systems it is not. Because solvent is reused, countercurrent systems can obtain more separation with the same total amount of solvent and the same number of stages. They can also obtain both high purity (x_{N+1} small) and high yield (high recovery of solute). Cross-flow systems can obtain either high purity or concentrated solvent streams but not both. For extraction they may have an advantage when flooding or slow settling of the two phases is a problem. For absorption and stripping the pressure drop may be significantly lower in a cross-flow system. Cascades that combine cross-flow and countercurrent operation have been studied by Thibodeaux *et al.* (1977).

Another cascade variation is the two-dimensional cascade shown in Figure 16-11; it can be considered as a series of cross-flow cascades connected together. The top row of the cascade looks like Figure 16-8, and its graphical solution can be obtained as in Figure 16-9. This gives the inputs (y_1^1, y_2^1, etc.) for the second row, which can again be solved as a cross-flow cascade. This procedure is then repeated.

The advantage of the two-dimensional cascade is that it can separate a multisolute mixture in a single steady-state system. For example, suppose the feed in Figure 16-11 consists of three solutes, A, B, and C. If solute A greatly prefers the diluent, it will remain in the diluent phase and exit at location A. If solute C prefers the solvent, it will be extracted out and exit at point C. Solute B, which distributes between the two phases, will exit somewhere near location B. The steady-state version of the two-dimensional cascade has been studied by Wankat (1972a) and Hudson and Wankat (1973). Unsteady versions in which a pulse of feed is input have been studied by Meltzer (1958) and Wankat (1972b). The disadvantages of the two-dimensional cascade are that it uses a large number of stages and large amounts of solvent. A more economical alternative regenerates the solvent inside the cascade by changing temperature or pH (Wankat *et al.*, 1976).

Two-dimensional systems have also been extensively studied for adsorption, chromatography, and electrophoresis. Some commercial units have been constructed for these applications. Two-dimensional filters and magnetic separators are in common use. The entire area of two-dimensional separations was reviewed by Wankat (1984-85).

16.5. SUMMARY-OBJECTIVES

In this chapter we looked at extraction processes and extended the McCabe-Thiele method and the Kremser equation to immiscible extraction. The objectives for this chapter are:

1. Explain what extraction is and outline the types of equipment used for extraction.

2. Use the McCabe-Thiele method to solve both concentrated and dilute immiscible extraction problems.

3. Use the Kremser equation for dilute extraction problems.

4. Explain the operation of fractional extraction and analyze fractional extraction with a McCabe-Thiele analysis.

5. Describe cross-flow and two-dimensional cascades and solve immiscible extraction problems using these cascades.

REFERENCES

Bailes, P.J., C. Hanson, and M.A. Hughes, "Liquid-Liquid Extraction: Nonmetallic Materials," *Chem. Eng., 83* (10), 115 (May 10, 1976).

Brain, P.L.T., *Staged Cascades in Chemical Processing*, Prentice-Hall, Englewood Cliffs, NJ, 1972.

Evans, T.W., "Counter-current and Multiple Extraction," *Ind. Eng. Chem., 23,* 860 (1934).

Francis, A.W., *Handbook for Components in Solvent Extraction,* Gordon and Breach, New York, 1972.

Hartland, S., *Counter-Current Extraction,* Pergamon, London, 1970.

Hudson, B.L. and P.C. Wankat, "Two-Dimensional Cross-Flow Extraction," *Separation Sci., 8,* 599 (1973).

Jamrack, W.D., *Rare Metal Extraction by Chemical Engineering Methods,* Pergamon, New York, 1963.

King, C.J., *Separation Processes,* 2nd ed., McGraw-Hill, New York, 1980.

Lo, T.C., "Commercial Liquid-Liquid Extraction Equipment," in P. Schweitzer (Ed.), *Handbook of Separation Techniques for Chemical Engineers,* McGraw-Hill, New York, 1979, pp. 1-283 to 1-342.

Lo, T.C., M.H.I. Baird, and C. Hanson (Eds.), *Handbook of Solvent Extraction,* Wiley, New York, 1983.

Meltzer, H.L., "Three-Phase Counter-Current Distribution: Theory and Application to the Study of Strandin," *J. Biol. Chem., 233* (6), 1327 (1958).

Perry, R.H. and C.H. Chilton (Eds.), *Chemical Engineers' Handbook,* 5th ed., McGraw-Hill, New York, 1973.

Reissinger, K.H. and J. Schroeter, "Modern Liquid-Liquid Extractors: Review and Selection Criteria," in *Alternatives to Distillation,* Inst. Chem. Eng., No. 54, 1978, pp. 33-48.

Ritchy, G.M. and A.W. Ashbrook, "Hydrometallurgical Extraction," in P. Schweitzer (Ed.), *Handbook of Separation Techniques for Chemical Engineers,* McGraw-Hill, New York, 1979, pp. 2-105 to 2-130.

Robbins, L.A., "Liquid-Liquid Extraction," in P. Schweitzer (Ed.), *Handbook of Separation Techniques for Chemical Engineers,* McGraw-Hill, New York, 1979, pp. 1-255 to 1-282.

Skelland, A.H.P. and D.W. Tedder, "Extraction-Organic Chemicals Processing," in R.W. Rousseau (Ed.), *Handbook of Separation Process Technology,* Wiley, New York, 1987, Chapter 7.

Thibodeaux, L.J., D.R. Daner, A. Kimura, J.D. Millican, and R.I. Parikh, "Mass Transfer Units in Single and Multiple Stage Packed Bed, Cross-Flow Devices," *Ind. Eng. Chem. Process Des. Develop., 16,* 325 (1977).

Treybal, R.E., *Liquid Extraction,* 2nd ed., McGraw-Hill, New York, 1963.

Wankat, P.C., "Two-Dimensional Cross-Flow Cascades," *Separation Sci., 7,* 233 (1972a).

Wankat, P.C., "Two-Dimensional Development in Staged Systems," *Separation Sci., 7,* 345 (1972b).

Wankat, P.C., "Two-Dimensional Separation Processes," *Separation Sci. Technol., 19,* 801 (1984-85).

Wankat, P.C., A.R. Middleton, and B.L. Hudson, "Steady State Continuous Multicomponent Separations in Regenerated Two-Dimensional Cascades," *Ind. Eng. Chem. Fundam., 15,* 309 (1976).

Wnek, W.J. and R.H. Snow, "Design of Cross-Flow Cooling Towers and Ammonia Stripping Towers," *Ind. Eng. Chem. Process Des. Develop., 11,* 343 (1972).

HOMEWORK

A. *Discussion Problems*

A1. What is the designer trying to do in the extraction equipment shown in Figure 16-2 and listed in Table 16-1? Why are there so many types of extraction equipment and only two major types of equipment for vapor-liquid contact?

A2. What are some of the properties you would look for in a good solvent?

A3. Why aren't Eqs. (16-3) valid when diluent and solvent are partially miscible? What is the correct equation to use when the phases are partially miscible? Show that the units balance in Eqs. (16-4) and (16-5).

A4. Compare the advantages and disadvantages of the McCabe-Thiele and Kremser design procedures.

A5. List the steps in the trial-and-error procedure for a McCabe-Thiele analysis of countercurrent extraction when the number of stages is specified.

A6. Explain the similarities (analogies) between extraction and stripping (or absorption). How do the separations differ?

A7. Develop your one-page key relations chart for this chapter.

A8. a. Draw a three-stage countercurrent mixer-settler system for extraction.

b. Draw a three-stage cross-flow mixer-settler system for extraction.

A9. Compare $K_D E/R$ to KV/L used in distillation, absorption, and stripping. Do these quantities have the same significance?

A10. What are the appropriate ranges of $K_{D,j}E/R$ for solutes A, B, C in the two columns in Figure 16-6?

A11. Without looking at the text, define: raffinate, extract, diluent, solute, cross-flow cascade, two-dimensional cascade, and fractional extraction.

A12. In fractional extraction what happens to solute C if;

a. $(\dfrac{K_{D,c}E}{R})_{top} > 1$ and $(\dfrac{K_{D,c}E}{R})_{bottom} < 1$?

b. $(\dfrac{K_{D,c}E}{R})_{top} < 1$ and $(\dfrac{K_{D,c}E}{R})_{bottom} > 1$?

c. How would you adjust the extractor so that the conditions in parts a or b would occur?

B. *Generation of Alternatives*

B1. For fractional extraction, list possible problems other than the three in the text. Outline the solution to these problems.

B2. How would you couple together cross-flow and countercurrent cascades? What might be the advantages of this arrangement?

C. *Derivations*

C1. For the McCabe-Thiele solution we arbitrarily decided to plot Y versus X. We could also have plotted X as ordinate versus Y as abscissa. In this case the graph would be similar to the one for absorption. Derive the operating line and plot the McCabe-Thiele diagram.

C2. Derive Eq. (16-11) starting with a McCabe-Thiele diagram (follow the procedure of Chapter 15).

C3. Derive Eq. (16-12) starting with a McCabe-Thiele diagram (follow the procedure of Chapter 15).

C4. Derive Eq. (16-13) from Eq. (16-12).

C5. Derive Eq. (16-19).

C6. Prove that the two operating lines in fractional extraction (Figure 16-7) intersect at a feed line.

C7. For fractional extraction outline in detail a solution procedure for (a) case 1, (b) case 2, and (c) case 3.

C8. For the cross-flow cascade, show that the point $(Y_{j,in}, X_{j-1})$ is on the operating line. Also show that if the two entering feeds are combined as a mixed feed the operating equation is the same as for flash distillation.

C9. Develop the solution method for a dilute multicomponent extraction in a cross-flow cascade. Sketch the McCabe-Thiele diagrams.

C10. Single-stage systems $(N = 1)$ can be designed as countercurrent systems, Figure 16-4, or as cross-flow systems, Figure 16-9. Develop the methods for both these designs. Which is easier? If the system is dilute, how can the Kremser equation be used?

C11. Sketch the McCabe-Thiele graphical solution for a 3 row by 3 column (9-stage) two-dimensional cascade. Show what happens to solutes that:
a. Prefer the diluent phase

b. Prefer the solvent phase

c. Distribute between the two phases

C12. Develop a computer flow chart for fractional extraction calculations. How will your program differ for cases 1, 2, and 3?

D. *Problems*

D1. An extraction column is used to remove an impurity P from toluene with water as the solvent. The toluene and water may be considered to be completely immiscible. The toluene enters the column with 20 wt % P, and it should leave with 1.0 wt % P. The water used as solvent enters the column as pure water. Feed flow rate is 200 lb/hr. Solvent rate is 300 lb/hr. Over the required range of operation, equilibrium data can be satisfactorily represented as

wt frac P in water = 0.75 (wt frac P in toluene)

+ 10.0 (wt frac P in toluene)^2

Find the number of equilibrium stages required. If the overall efficiency is 16%, how many real stages are needed?

D2. A water solution containing 0.005 mole frac benzoic acid is to be extracted using pure benzene as the solvent. If the feed rate is 100 moles/hr and the solvent rate is 10 moles/hr, find the number of equilibrium stages required to reduce the water concentration to 0.0001 mole frac benzoic acid. Operation is isothermal at 6° C, where equilibrium data can be represented as

Mole fraction benzoic acid in water = 0.0446 × (mole frac benzoic acid in benzene)

D3. We wish to extract glop out of water into toluene. The water enters with 15 wt % glop and we want it to leave at 2 wt % glop. Flow rate of the entering steam F_D (water + glop) is 1000 kg/hr. We have available enough pure toluene to input 1466.67 kg/hr (stream $F_{S,1}$, $Y_{N+1} = 0$). We also can feed $F_{S,2} = 671.73$ kg/hr of a toluene that is 2 wt % glop. Toluene and water are completely immiscible. At the temperature and pressure of the system the equilibrium data are y = 0.5x in weight fractions. Find the optimum feed location for feed $F_{S,2}$ and the total number of stages required.

D4. We have a mixture of acetic acid in water and wish to extract this with 3-heptanol at 25° C. Equilibrium is

$$\frac{\text{Wt frac acetic acid in solvent}}{\text{Wt frac acetic acid in water}} = 0.828$$

The inlet water solution flows at 550 lb/hr and is 0.0097 wt frac acetic acid. We desire an outlet water concentration of 0.00046 wt frac acetic acid. The solvent flow rate is 700 lb/hr. The entering solvent contains 0.0003 wt frac acetic acid. Find the outlet solvent concentration and the number of equilibrium stages required (use the Kremser equation). Is this an economical way to extract acetic acid?

D5. We have an extraction column with 30 equilibrium stages. We are extracting acetic acid from water into 3-heptanol at 25° C. Equilibrium is given in Problem 16-D4. The aqueous feed flows at a rate of 500 kg/hr. The feed is 0.011 wt frac acetic acid, and the exit water should be 0.00037 wt frac acetic acid. The inlet 3-heptanol contains 0.0002 wt frac acetic acid. What solvent flow rate is required? Assume that total flow rates are constant.

D6. We have a mixture of linoleic and oleic acids dissolved in methylcellosolve and 10% water. Feed is 0.003 wt frac linoleic acid and 0.0025 wt frac oleic acid. Feed flow rate is 1500 kg/hr. A simple countercurrent extractor will be used with 750 kg/hr of pure heptane as solvent. We desire a 99% recovery of the oleic acid in the extract product. Equilibrium data are given in Table 16-3. Find N and the recovery of linoleic acid in the extract product.

D7. The fractional extraction system shown in Figure 16-5 is separating abietic acid from other acids. Solvent 1, heptane, enters at $\overline{E} = 1000$ kg/hr and is pure. Solvent 2, methylcellosolve + 10% water, is pure and has a flow rate of $R = 2500$ kg/hr. Feed is 5 wt % abietic acid in solvent 2 and flows at 1 kg/hr. There are only traces of other acids in the feed. We desire to recover 95% of the abietic acid in the bottom raffinate stream. Feed is on stage 6. Assume that the solvents are completely immiscible and that the system can be considered to be very dilute. Equilibrium data are given in Table 16-3. Find N.

D8. You are to compare a countercurrent extraction system to a cross-flow system. 200 moles/hr of a toluene acid solution that is 0.05 mole frac acid is to be extracted with water. The water is recycled from the acid-water distillation column and contains 0.002 mole frac acid. A total water flow rate of 30 moles/hr is used. Operation is at 1 atm and 25° C. For practical purposes, toluene and water are immiscible.

At 25 ° C the equilibrium data can be represented as

$$\frac{\text{Moles acid}}{\text{Moles acid } + \text{ toluene}} = 0.0846 \frac{\text{moles acid}}{\text{moles acid } + \text{ water}}$$

a. If the outlet toluene stream is to contain 0.001 mole frac acid, how many stages are required for a countercurrent cascade?

b. If the water is split up equally among five stages of a cross-flow cascade, what is the outlet toluene concentration?

c. Which cascade is more effective?

D9. The system shown in the figure is extracting acetic acid from water using benzene as the solvent. The temperature shift is used to regenerate the solvent and return the acid to the water phase.

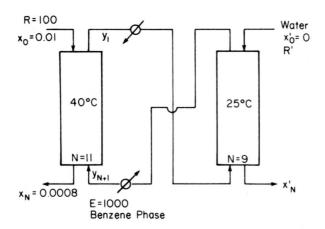

a. Determine y_1 and y_{N+1} (units are wt. fracs.) for the column at 40°C.

b. Determine R' and x_N' for the column at 25 ° C.

c. Is this a practical way to concentrate the acid?

Data are in Table 16-3. Note: A similar scheme is used commercially for citric acid concentration using a more selective solvent.

E. *More Complex Problems*

E1. We have a liquid feed that is 48 wt % m-xylene and 52 wt % o-xylene, which are to be separated in a fractional extractor (Figure 16-5) at 25°C and 101.3 kPa. Solvent 1 is β,β'-thiodipropionitrile, and solvent 2 is n-hexane. Equilibrium data are in Table 16-3. For each kilogram of feed, 200 kg of solvent 1 and 20 kg of solvent 2 are used. Both solvents are pure when they enter the cascade. We desire a 92% recovery of o-xylene in solvent 1 and a 94% recovery of m-xylene in n-hexane. Find outlet composition, N, and N_f. Adjust the recovery of m-xylene if necessary to solve this problem.

F.   *Problems Requiring Other Resources*

F1.   The article by Lo (1979) is an excellent review of commercial liquid-liquid extraction equipment. Read it and write a critique (maximum of two pages, typed double-spaced).

F2.   In one of the references (e.g., Bailes *et al.*, 1976; Jamrack, 1963; Lo, 1979; Lo *et al.*, 1983; Ritchy *et al.*, 1979; or Robbins, 1979) look up the flow chart for a complete extraction process. Write a short report on this process and compare it to the general process shown in Figure 16-1. State the type of extractor used and give reasons for its use. Your report should also answer the following: What other types of extractors could be used? What solvent is used? Why? How is the solvent recovered? Can you think of any possible improvements?

F3.   To see what real extraction equipment looks like, tour your school's undergraduate laboratory or a local plant and look at the extraction equipment. What category would you place it in in Table 16-1? If possible, watch the equipment in operation.

F4.   If the systems are partially miscible, flow rates will vary. Both cross-flow and two-dimensional cascades can conveniently be designed with a modification of the Rachford-Rice equation [Eq. (3-29)]. Derive the modified form of the Rachford-Rice equation for cross-flow and two-dimensional cascades. Outline the solution method that would be used. Compare your method with the solution developed by Hudson and Wankat (1973).

chapter 17
EXTENDING McCABE-THIELE ANALYSIS AND THE KREMSER EQUATION TO OTHER SEPARATIONS

Up to this point we have seen that the stage-by-stage analysis originally developed for distillation can be extended to absorption, stripping, and extraction. In this chapter we will first look at the McCabe-Thiele procedure in a very general, abstract way. Then we will briefly apply this general technique to several separation processes. We will also apply the Kremser equation to these separation methods when it is applicable. Remember that stage-by-stage calculations can easily be done analytically whenever a McCabe-Thiele analysis is valid. Thus this section and chapter as a whole are also applicable to computer calculations.

17.1. GENERALIZED McCABE-THIELE PROCEDURE

The McCabe-Thiele procedure has been applied to flash distillation, continuous countercurrent distillation, batch distillation, absorption, stripping, and extraction. What are the common factors for the McCabe-Thiele analysis in all these cases?

All the McCabe-Thiele graphs are plots of concentration in one phase versus concentration in the other phase. In all cases there is a single equilibrium curve, and there is one operating line for each column section. It is very desirable for this operating line to be straight. In addition, although it isn't evident on the graph, we want to satisfy the energy balance and mass balances for all other species.

In order to obtain a single equilibrium curve, we have to specify enough variables that only one degree of freedom remains. For binary distillation this can be done by specifying constant pressure. For absorption, stripping, and extraction we specified that pressure and temperature were constant, and if there were several solutes we assumed that they were independent. In general, we will specify that pressure and/or temperature are constant, and for multisolute systems we will assume that the solutes are independent.

549

To have a straight operating line for the more volatile component in distillation we assumed that constant molal overflow was valid, which meant that in each section total flows were constant. If this was not true, special units (latent heat units) were defined so that flow rates would be constant. For absorption, stripping, and extraction we could make the assumption that total flows were constant if the systems were very dilute. For more concentrated systems we assumed that there was one chemical species in each phase that did not transfer into the other phase; then the flow of this species (carrier gas, solvent, or diluent) was constant. In general, we have to assume either that total flows are constant or that flows of nontransferred species are constant.

These assumptions control the concentration units used to plot the McCabe-Thiele diagram. If total flows are constant, the solute mass balance is written in terms of *fractions,* and fractions are plotted on the McCabe-Thiele diagram. If flows of nontransferred species are constant, *ratio* units must be used, and ratios are plotted on the McCabe-Thiele diagram.

The McCabe-Thiele operating line satisfies the mass balance for only the more volatile component or the solute. In binary distillation the constant molal overflow assumption forces total vapor and liquid flow rates to be constant and therefore the overall mass balance will be satisfied. In absorption, when constant carrier and solvent flows are assumed, the mass balances for these two chemicals are automatically satisfied. In general, if overall flow rates are assumed constant, we are satisfying the overall mass balance. If the flow rates of nontransferred species are constant, we are satisfying the balances for these species.

The energy balance is automatically satisfied in distillation when the constant molal overflow assumption is valid. In absorption, stripping, and extraction, the energy balances were satisfied by assuming constant temperature and a negligible heat of absorption, stripping or mixing. In general, we will assume constant temperature and a negligible heat involved in contacting the two phases.

The Kremser equation was used for absorption, stripping, and extraction in Chapters 15 and 16. When total flows, pressure, and temperature are constant and the heat of contacting the phases is negligible, we can use the Kremser equation if the equilibrium expression is linear. When these assumptions are valid, the Kremser equation can be used for other separations.

Unfortunately, the assumptions required to use a McCabe-Thiele analysis or the Kremser equation may not be valid for a given separation. If the assumptions are not valid, the results of the analysis could be garbage. To determine the validity of the assumptions, the engineer has to examine each specific case in detail. The more dilute the solute, the more likely it is that the assumptions will be valid.

In the remainder of this chapter these principles will be applied to generalize the McCabe-Thiele approach and the Kremser equation for a variety of unit operations. A listing of various applications is given in Table 17-1. It may be helpful to study this table in detail to refresh your memory.

17.2. WASHING

When solid particles are being processed in liquid slurries, the solids entrain liquid with them. The removal of any solute contained in this entrained liquid is called *washing*. To be specific, consider an operation mining sand from the ocean. The wet sand contains salt, and this salt can be removed by washing with pure water. The entrained liquid is called underflow liquid, because the solids are normally removed from the bottom of a settler as shown in Figure 17-1A. Washing is done by mixing solid (sand) and wash liquor (water) together in a mixer and sending the mixture to a settler. The solids and entrained underflow liquid exit from the bottom of the settler, and clear overflow liquid without solids is removed from the top (see Figure 17-1A). In washing, the solute (salt) is not held up or attached to the inert solid (sand). The salt is assumed to be at the same concentration in the underflow liquid as it is in the overflow liquid. Thus it can be removed by displacing it with clear water. The separation can be done in single-stage, cross-flow, and countercurrent cascades.

The equilibrium condition for a washer is that solute concentration is the same in both the underflow and overflow liquid streams. This statement does not say anything about the solid, which changes the relative underflow and overflow flow rates but does not affect concentrations. Thus the equilibrium equation is

$$y = x \tag{17-1}$$

where y = mass fraction solute in the overflow liquid and x = mass fraction solute in the underflow liquid.

For the general countercurrent cascade shown in Figure 17-1B, it is easy to write a steady state-mass balance for the indicated mass balance envelope:

Table 17-1. Applications of McCabe-Thiele and Kremser Procedures

Unit Operation	Feed	Const. Flow
Absorption (Chapt.15)	Gas phase	L = kg solvent/hr G = kg carrier gas/hr
Dilute abs. (Chapt. 15)	Gas phase	Total flow rates
Stripping (Chapt. 15)	Liquid	L = kg solvent/hr G = kg carrier gas/hr
Dilute stripping (Chapt. 15)	Liquid	Total flow rates
Dilute extraction (Chapt. 16)	Raffinate	F_D = kg diluent/hr F_S = kg solvent/hr
Very dilute extract (Chapt. 16)	Raffinate	E = total extract kg/hr R = total raffinate
Washing (Chapt. 17)	Solids + underflow liquid	U = underflow liquid, kg/hr O = overflow liquid, kg/hr
Leaching (Chapt. 17)	Solids & solutes	F_{solv} = kg solvent/hr F_{solid} = kg insoluble solids/hr
Adsorption and ion exchange	Gas or liquid	S = adsorbent solids, kg/hr G = fluid, kg/hr
Three-phase (Chapt. 17)	One or two phases	L, W, V

Conc. Units	Comments
Y,X, mass ratios	Op line above equil line
y,x, fractions	McCabe-Thiele or Kremser ($y = mx + b$)
Y,X, mass ratios	Op line below equil line Heat effects may be be important
y,x, fractions	McCabe-Thiele or Kremser
$Y = \dfrac{\text{kg solute}}{\text{kg solvent}}$ $X = \dfrac{\text{kg solute}}{\text{kg diluent}}$	McCabe-Thiele Op line below equil line
y = wt frac in extract x = wt frac in raffinate	McCabe-Thiele or Kremser ($y = mx + b$)
y = wt frac in overflow x = wt frac in underflow	Equil $y = x$ Op line under equil line or use Kremser
$Y = \dfrac{\text{kg solute}}{\text{kg solvent}}$ $X = \dfrac{\text{kg solute}}{\text{kg solid}}$	Op line under equil line Kremser ($y = mx + b$)
$Y = \dfrac{\text{kg solute}}{\text{kg solvent}}$ $q = \dfrac{\text{kg solute}}{\text{kg adsorbent}}$	Op line above equil line for adsorption and below for desorption. Kremser for linear equil
y, fraction x' (Eq. 17-15)	Pseudo-equilibrium Kremser for linear equil

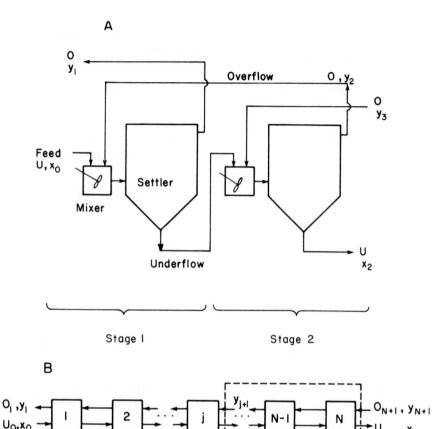

Figure 17-1.   Countercurrent washing. (A) Two-stage mixer-settler system; (B) general system.

$$O_{j+1} y_{j+1} + U_N x_N = O_{N+1} y_{N+1} + U_j x_j \qquad (17\text{-}2)$$

where O_j and U_j are the total overflow and underflow liquid flow rates in kg/hr leaving stage j. The units for Eq. (17-2) are kg solute/hr. To develop the operating equation, we solve Eq. (17-2) for y_{j+1}:

$$y_{j+1} = \frac{U_j}{O_{j+1}} x_j + \frac{O_{N+1}}{O_{j+1}} y_{N+1} - \frac{U_N}{O_{j+1}} x_N \qquad (17\text{-}3)$$

In order for this to plot as a straight line the underflow liquid and overflow liquid flow rates must be constant.

The underflow liquid flow rate can be calculated from the volume of liquid entrained with the solids. Let ϵ be the porosity of the solids in the underflow. That is,

$$\epsilon = \frac{\text{volume voids}}{\text{total volume}} = \frac{\text{volume liquid}}{\text{total volume}} \qquad (17\text{-}4)$$

and then

$$1 - \epsilon = \frac{\text{volume solids}}{\text{total volume}} \qquad (17\text{-}5)$$

We can now calculate the underflow liquid flow rate, U_j. Often the specifications will give the flow rate of dry solids or the flow rate of wet solids. Suppose we are given the flow rate of dry solids. Then we can calculate the volume of solids per hour.

$$\left(\frac{\text{kg dry solids}}{\text{hr}}\right)\left(\frac{1}{\rho_s \text{ kg/m}^3}\right) = \text{m}^3 \text{ dry solids/hr}$$

The total volume of underflow is

$$\left(\frac{\text{kg dry solids}}{\text{hr}}\right)\left(\frac{1}{\rho_s \dfrac{\text{kg}}{\text{m}^3}}\right)\frac{1}{(1-\epsilon)\dfrac{\text{m}^3 \text{ solids}}{\text{m}^3 \text{ underflow}}} = \frac{\text{m}^3 \text{ total underflow}}{\text{hr}}$$

Then the volume of underflow liquid in m^3 liquid per hour is

$$\left(\frac{\text{kg dry solids}}{\text{hr}}\right)\left(\frac{1}{\rho_s \dfrac{\text{kg}}{\text{m}^3}}\right)\frac{1}{(1-\epsilon)\left(\dfrac{\text{m}^3 \text{ solids}}{\text{m}^3 \text{ underflow}}\right)}\left(\epsilon \frac{\text{m}^3 \text{ liquid}}{\text{m}^3 \text{ total underflow}}\right)$$

and finally the kg/hr of underflow liquid, U_j, is

$$U_j = (\text{rate dry solids, } \frac{\text{kg}}{\text{hr}})\left(\frac{\epsilon}{1-\epsilon}\right)\frac{\rho_f}{\rho_s} \qquad (17\text{-}6)$$

In these equations ρ_f is the fluid density in kg/m^3 and ρ_s is the density of dry solids in kg/m^3.

If the solids rate, ϵ, ρ_f, and ρ_s are all constant, then from Eq. (17-6), $U_j = U = $ constant. If U is constant, then an overall mass balance shows that the overflow rate, O_j, must also be constant. Thus to have constant flow rates we assume:

1. No solids in the overflow. This ensures that the solids flow rate will be constant.

2. ρ_f and ρ_s constant. Constant ρ_f implies that the solute has little effect on fluid density or that the solution is dilute.

3. ϵ constant. Thus the volume of liquid entrained from stage to stage is constant.

When these assumptions are valid, O and U are constant, and the operating equation simplifies to

$$y_{j+1} = \frac{U}{O} x_j + \left(y_{N+1} - \frac{U}{O} x_N\right) \tag{17-7}$$

which obviously represents a straight line on a McCabe-Thiele plot. Note that this equation is similar to all the other McCabe-Thiele operating equations we have developed. Only the nomenclature has changed.

An alternative way of stating the problem would be to specify the volume of *wet* solids processed per hour. Then the underflow volume is

Underflow liquid volume, m^3/hr = (Volume wet solids/hr) ϵ

and

$$U = (\text{Volume wet solids/hr}) \ \epsilon \ \rho_f \tag{17-8}$$

If densities and ϵ are constant, volumetric flow rates are constant. Now the washing problem can be solved using volumetric flow rates and concentrations in kg solute/m^3 (see Problem 17-D4).

Note that we could have just assumed that overflow and underflow rates are constant and derived Eq. (17-7). However, it is much more informative to show the three assumptions required to make overflow and underflow rates constant. These assumptions show that this analysis for washing is likely to be invalid if the settlers are not removing all the solid, if for some reason the amount of liquid entrained changes, or if the fluid density changes markedly. The first two problems will not occur in well-designed systems. The third is easy to check with density data.

The McCabe-Thiele diagram can now be plotted as shown in Figure 17-2. This McCabe-Thiele diagram is unique, since temperature and

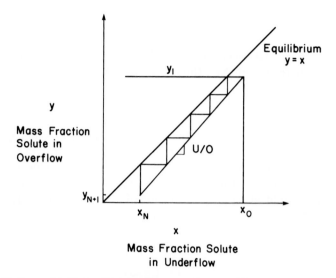

Figure 17-2.   McCabe-Thiele diagram for washing.

pressure do not affect the equilibrium. Temperature will affect the rate of attaining equilibrium and hence the efficiency, because at low temperatures more viscous solutions will be difficult to wash off the solid.

The analysis for washing can be extended to a variety of modifications. These include simulation problems, use of efficiencies, calculation of maximum U/O ratios, and calculations for cross-flow systems. The Kremser equation can also be applied to countercurrent washing with no additional assumptions. This adaptation is a straightforward translation of nomenclature and is illustrated in Example 17-1. Brian (1972) discusses application of the Kremser equation to washing in considerable detail.

Washing is also commonly done by collecting the solids on a filter and then washing the filter cake. This approach, which is often used for crystals and precipitates that may be too small to settle quickly and is usually a batch operation, is discussed by Mullin (1972).

Example 17-1. Washing

In the production of sodium hydroxide by the lime soda process, a slurry of calcium carbonate particles in a dilute sodium hydroxide solution results. A four-stage countercurrent washing system is used. The underflow entrains approximately 3 kg liquid/kg calcium carbonate solids. The inlet water is pure water. If 8 kg wash water/kg calcium carbonate solids is used, predict the recovery of NaOH in the wash liquor.

Solution

A. Define. Recovery is defined as $1 - x_{out}/x_{in}$. Thus recovery can be determined even though x_{in} is unknown.

B, C. Explore and Plan. If we pick a basis of 1 kg calcium carbonate/hr, then $O = 8$ kg wash water/hr and $U = 3$ kg/hr. This problem can be solved with the Kremser equation if we translate variables. To translate: Since $y = $ overflow liquid weight fraction, we set $O = V$. Then $U = L$. This translation keeps $y = mx$ as the equilibrium expression. It is convenient to use the Kremser equation in terms of x. For instance, Eq. (15-39) becomes

$$\frac{x_N - x_N^*}{x_0 - x_N^*} = \frac{1 - m\,O/U}{1 - (m\,O/U)^{N+1}} \qquad (17\text{-}9)$$

D. Do It. Equilibrium is $y = x$. Thus $m = 1$. Since inlet wash water is pure, $y_{N+1} = 0$. Then $x_N^* = y_{N+1}/m = 0$, $m\,O/U = (1)(8)/3 = 8/3$, and $N = 4$. Then Eq. (17-9) is

$$\frac{x_N}{x_0} = \frac{1 - 8/3}{1 - (8/3)^5} = 0.01245$$

and   Recovery $= 1 - x_N/x_0 = 0.98755$

E. Check. This solution can be checked with a McCabe-Thiele diagram. Since the x_N value desired is known, the check can be done without trial and error.

F. Generalize. Recoveries for linear equilibrium can be determined without knowing the inlet concentrations. This can be useful for the leaching of natural products because the inlet concentration fluctuates. The translation of variables shown here can be applied to other forms of the Kremser equation.

17.3. LEACHING

Leaching or solid-liquid extraction is a process in which a soluble solute is removed from a solid matrix using a solvent to dissolve the solute. The most familiar example is making coffee from ground coffee beans or tea from tea leaves. The complex mixture of chemicals that give the coffee or tea its odor, taste, and physiological effects are leached from the solids by the hot water. Instant coffee and tea are made by leaching ground coffee beans or tea leaves with hot water and then drying the liquid to produce a solid. There are many other commercial applications of leaching such as leaching soybeans to recover soybean oil, leaching ores to recover a variety of minerals, and leaching plant leaves to extract a variety of pharmaceuticals (Rickles, 1965; Schwartzberg, 1980, 1987).

The equipment and operation of washing and leaching systems may be very similiar. In both cases a solid and a liquid must be contacted, allowed to equilibrate, and then separated from each other. Thus the mixer-settler type of equipment shown in Figure 17-1 is also commonly used for leaching easy-to-handle solids. A variety of other specialized equipment has been developed to move the solid and liquid counter-currently during leaching. Prabhudesai (1979) and Schwartzberg (1980, 1987) present good introductions to this leaching equipment.

In leaching, the solute is initially part of the solid and dissolves into the liquid. In washing, the solute is retained in the pores of the solid and the solid itself does not dissolve. In leaching, the equilibrium equa-tion is not $y = x$, and the total solid flow rate is not constant. Since diffusion rates in a solid are low, mass transfer rates are low. Thus, equilibrium may take days for large pieces such as pickles where it is desirable to leach out excess salt. A rigorous analysis of leaching requires that the changing solid and liquid flow rates be included. This situation is very similiar to partially miscible extraction and is included in Chapter 18. In this section we will look at simple cases where a modified McCabe-Thiele or Kremser equation can be used.

A countercurrent cascade for leaching is shown in Figure 17-3A. We will consider the (idealized) case where entrainment of liquid with the solid underflow can be ignored. The assumptions are:

1. System is isothermal.

2. System is isobaric.

3. No solvent dissolves into solid.

4. No solvent entrained with the solid.

5. There is an insoluble solid backbone or matrix.

6. Heat of mixing of solute in solvent is negligible.

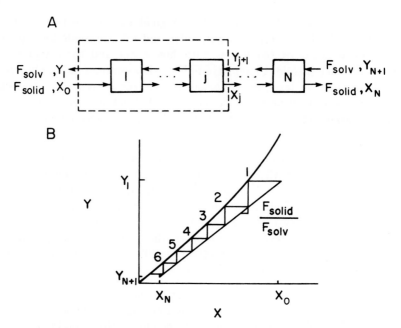

Figure 17-3.   Countercurrent leaching. (A) Cascade; (B) McCabe-Thiele diagram.

7. Stages are equilibrium stages.

With these assumptions the energy balance is automatically satisfied. A straight operating line is easily derived using the mass balance envelope shown in Figure 17-3A. Defining,

$$Y = \frac{\text{kg solute}}{\text{kg solvent}} \; , \quad X = \frac{\text{kg solute}}{\text{kg insoluble solid}}$$

$$F_{solv} = \text{kg solvent/hr} \; , \quad F_{solid} = \text{kg insoluble solid/hr} \qquad (17\text{-}10)$$

the operating equation is

$$Y_{j+1} = \frac{F_{solid}}{F_{solv}} X_j + Y_1 - \frac{F_{solid}}{F_{solv}} X_0 \qquad (17\text{-}11)$$

This represents a straight line as plotted in Figure 17-3B. The equilibrium curve is now the equilibrium of the solute between the solvent and solid phases. The equilibrium data must be measured experimentally. If the equilibrium line is straight, the Kremser equation can be applied.

In the previous analysis, assumptions 4 and 7 are often faulty. There is always entrainment of liquid in the underflow (for the same reason that there is an underflow liquid in washing). Since diffusion in solids is very slow, equilibrium is seldom attained in real processes. The combined effects of entrainment and nonequilibrium stages are often included by determining an "effective equilibrium constant." This effective equilibrium depends on flow conditions and residence times and is valid only for the conditions at which it was measured. Thus, the effective equilibrium constant is not a very fundamental quantity. However, it is easy to measure and use. The McCabe-Thiele diagram will look the same as Figure 17-3B. Further simplification is obtained by assuming that the effective equilibrium is linear, $y = m_E x$.

17.4. SUPERCRITICAL FLUID EXTRACTION

There has been an increasing amount of interest in the use of supercritical fluids for extracting compounds from solids or liquids. In this section we will briefly consider the properties of supercritical fluids that make them interesting for extraction. Then a typical process for supercritical fluid (SCF) extraction will be explored and several applications will be discussed.

First, what is a supercritical fluid? Figure 17-4A shows a typical pressure-temperature diagram for a single component. Above the critical temperature T_c, it is impossible to liquefy the compound. The critical pressure p_c is the pressure required to liquefy the compound at the critical temperature. The critical temperatures and pressures for a large number of compounds have been determined (Paulaitis et al., 1983; Reid, et al., 1977). SCFs of interest include carbon dioxide ($p_c = 72.8$ atm, $T_c = 31\,°C$, $\rho = 0.47$ g/mL), propane ($p_c = 41.9$ atm, $T_c = 97\,°C$, $\rho = 0.22$ g/mL), and water ($p_c = 217.7$ atm, $T_c=374\,°C$, $\rho = 0.32$ g/mL). An SCF behaves like a gas in that it will expand to fill the confines of the container.

As can be seen from this very short list, SCFs have densities much greater than those of typical gases and less than those of liquids by roughly a factor of 2 to 3. The viscosities of SCFs are about one-tenth those of liquids. This leads to low pressure drops. The diffusivities are in between those of liquids and gases and are roughly 10 times those of liquids. This plus the lack of a phase boundary leads to very high mass transfer rates and low HETP values in packed beds. The SCFs can often dissolve almost the same amount of solute as a good solvent. Extraction can often be carried out at low temperatures, particularly

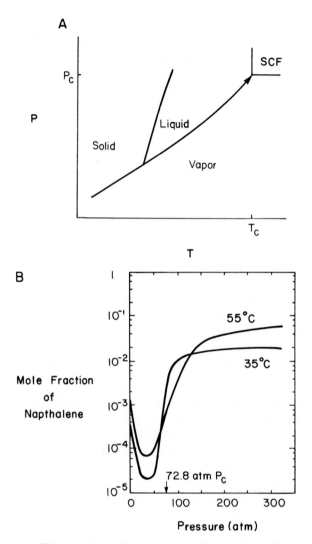

Figure 17-4.   Thermodynamics of supercritical fluid (SCF) extraction. (A) Pressure-temperature diagram for pure component; (B) solubility of naphthalene in CO_2.

when CO_2 is the SCF. This is particularly advantageous for extraction of foods and pharmaceuticals. Many SCFs are also completely natural and thus are totally acceptable as additives in foods and pharmaceuticals. This is a major advantage to the use of supercritical CO_2.

The solubility of a solute in an SCF is a complex function of temperature and pressure. This is commonly illustrated with the solubility

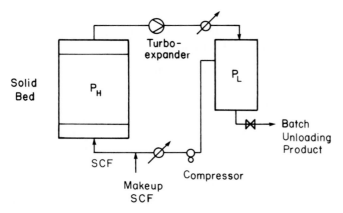

Figure 17-5.   Batch supercritical fluid extraction. Regeneration is by
pressure swing.

of naphthalene in CO_2, which is illustrated in Figure 17-4B (Hoyer,
1985; Paulaitis *et al.*, 1983). As pressure is increased the solubility first
decreases and then increases. At both high and low pressures the
naphthalene is more soluble at high temperatures than at low temperatures. This is the expected behavior, because the vapor pressure of
naphthalene increases with increasing temperature. Immediately above
the critical pressure the solute is more soluble at the lower temperature.
This is a *retrograde* phenomenon. If naphthalene solubility is plotted
versus CO_2 density, the retrograde behavior does not appear. In addition to having high solubilities, the SCF should be selective for the
desired solutes. Solute-solvent interactions can affect the solubility and
the selectivity of the SCF, and therefore entrainers are often added to
the SCF to increase solubility and selectivity.

Figure 17-4B also shows how the solute can be recovered from the
CO_2. If the pressure is dropped, the naphthalene solubility plummets
and naphthalene will drop out as a finely divided solid. A typical process using pressure reduction is shown in Figure 17-5. Note that this
will probably be a batch process if solids are being processed because of
the difficulty in feeding and withdrawing solids at the high pressures of
supercritical extraction. Regeneration can also be acheived by changing
the temperature. In some cases this may be preferable because it will
decrease the cost of compression.

Several current applications have been widely publicized. Kerr-
McGee developed its ROSE (Residuum Oil Supercritical Extraction)
process in the 1950s. When oil prices went up, the process attracted
considerable attention, since it has much lower operating costs than
competing processes. The ROSE process uses an SCF such as propane

to extract more useful hydrocarbons from the residue left after distillation. This process utilizes the high temperatures and pressures expected for residuum treatment to lead naturally to SCF extraction.

Much of the commercial interest has been in the food and pharmaceutical industries. Here, the major driving force is the desire to have completely "natural" processes, which cannot contain any residual hydrocarbon or chlorinated solvents. Supercritical carbon dioxide has been the SCF of choice because it is natural, nontoxic, and cheap, is completely acceptable as a food or pharmaceutical ingredient, and often has good selectivity and capacity. Currently, supercritical CO_2 is used to extract caffeine from green coffee beans to make decaffeinated coffee. Supercritical CO_2 is also used to extract flavor compounds from hops to make a hop extract that is used in beer production. The leaching processes that were replaced were adequate in all ways except that they used solvents that were undesirable in the final product.

A variety of other SCF extraction processes have been explored (Hoyer, 1985; Paulaitis et al., 1983). These include extraction of oils from seeds such as soybeans, removal of excess oil from potato chips, fruit juice extraction, extraction of oxgenated organics such as ethanol from water, removal of lignite from wood, desorption of solutes from activated carbon, and treatment of hazardous wastes.

The main problems in applying SCF extraction on a large scale have been scaling up for the high pressure required. The high-pressure equipment becomes quite heavy and expensive. In addition, methods for charging and discharging solids continuously have not been well developed for these high-pressure applications. Another problem has been the lack of design data for supercritical extraction. Supercritical extraction is not expected to be a cheap process. Typical costs are probably in the range of 5 to 10 cents per pound of material processed (Worthy, 1981). Thus, the most likely applications are extractions for which existing separation methods have at least one serious drawback and for which SCF extraction does not have major processing disadvantages.

17.5. THREE-PHASE SYSTEMS

Equilibrium stage separations with three phases present are something of a curiosity, but they do occasionally occur. Examples are steam distillation with a liquid water layer, extraction with two immiscible solvents and a diluent so that there are three phases, and slurry adsorption. Slurry adsorption is a process in which an adsorbent such as activated carbon is slurried in a liquid and then contacted countercurrently to the

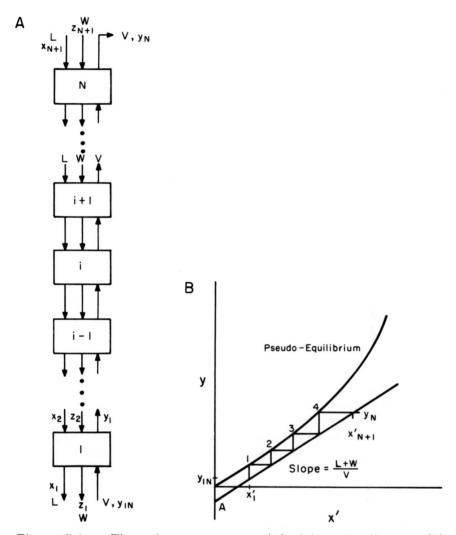

Figure 17-6.   Three-phase separators. (A) Schematic diagram; (B) pseudo-equilibrium curve analysis. From Wankat, 1980. Reprinted with permission from *Ind. Eng. Chem. Fundam., 19,* 358 (1980). Copyright 1980, American Chemical Society.

gas phase (Frost, 1974). The liquid can be selected so that it is an absorption solvent for the solute.

The analysis of a staged three-phase separator is very straightforward if the stages are equilibrium stages (Wankat, 1980). A countercurrent cascade for a general three-phase system is shown in Figure 17-6A. Streams L and W flow cocurrently and are countercurrent to stream V. Since each stage is assumed to be an equilibrium stage, streams L, W, and V leaving each stage are in equilibrium. The average mole fraction of the two cocurrent phases is

$$x_i' = \frac{L\,x_i + W z_i}{L + W} = \frac{\dfrac{L}{W}\,x_i + z_i}{\dfrac{L}{W} + 1} \tag{17-15}$$

Using a mass balance envelope around the top of the column, we obtain the operating equation

$$y_i = \frac{L + W}{V}\,x_{i+1}' + \left(y_N - \frac{L + W}{V}\,x_{N+1}'\right) \tag{17-16}$$

This will plot as a straight line on a plot of y vs x' if the total flow rates are constant. If flow rates are not constant, the analysis can be done in ratio units. The pseudo-equilibrium curve can be plotted by picking an arbitrary y value, determining x and z values in equilibrium with it, and then calculating x'. Note that the pseudo-equilibrium curve is for a specific L/W ratio. The modified McCabe-Thiele diagram is shown in Figure 17-6B.

If the y-x and y-z equilibria are linear, then the y-x equilibrium must also be linear. Then Figure 17-6B will consist of stages stepped off between two straight lines. This is exactly the same situation as in Figure 15-10. The Kremser equation can now be derived by the same method used in Chapter 15. For linear equilibria of the form

$$y = K_1 x + b_1 , \quad y = K_2 z + b_2 \tag{17-17}$$

the resulting Kremser equation is

$$\frac{y_N - y_{in}}{y_{N+1} - y_{in}} = \frac{\alpha' - 1}{1 - (\alpha')^{N+1}} \tag{17-18}$$

or

$$N = \frac{\ln\left[(1 - \alpha)\left(\dfrac{y_{in} - y_{N+1}^*}{y_N - y_{N+1}^*}\right) + \alpha\right]}{\ln(\alpha')} \tag{17-19}$$

where the terms are defined as

$$\alpha = \frac{V}{\dfrac{W}{K_2} + \dfrac{L}{K_1}} \quad , \quad \alpha' = \frac{1}{\alpha} \qquad (17\text{-}20a,b)$$

$$y_{N+1}^* = \frac{\dfrac{L}{K_1}(K_1 \, x_{N+1} + b_1) + \dfrac{W}{K_2}(K_2 \, z_{N+1} + b_2)}{\dfrac{W}{K_2} + \dfrac{L}{K_1}} \qquad (17\text{-}20c)$$

These equations are similiar to Eqs. (15-28) and (15-30) except for the changes in nomenclature due to the columns in Figures 15-8 and 17-6A being numbered differently. When $W = 0$, Eqs. (17-18) and (17-19) reduce to the usual forms of the Kremser equation.

17.6. APPLICATION TO OTHER SEPARATIONS

The McCabe-Thiele and Kremser methods can be applied to analyze other separation processes. Adsorption and ion exchange are occasionally operated in countercurrent columns. The application of the McCabe-Thiele procedure in this case is discussed in Wankat (1988). In some situations crystallization can be analyzed as an equilibrium stage separation. This analysis is also discussed in Wankat (1988).

The McCabe-Thiele and Kremser methods have also been applied to analyze less common separation methods, such as continuous countercurrent chromatography (see Wankat, 1986, for a review). A modification of the McCabe-Thiele method has also been applied to parametric pumping, which is a cyclic adsorption and ion exchange process (see Grevillot and Tondeur 1976, 1977). A similar modification can be used to analyze cycling zone adsorption (see Wankat, 1986).

17.7. SUMMARY - OBJECTIVES

In this chapter we looked at the general applicability of the McCabe-Thiele and Kremser analysis procedures. The methods are reviewed in Table 17-1. The objectives for this chapter are:

1. Explain in general terms how the McCabe-Thiele and Kremser analyses can be applied to other separation schemes. Delineate when these procedures are applicable.

2. Explain what washing is and apply the McCabe-Thiele and Kremser procedures to washing problems.

3. Explain what leaching is and apply both McCabe-Thiele and Kremser methods to leaching problems.

4. Explain how supercritical extraction works, and discuss its advantages and disadvantages.

5. Use both the McCabe-Thiele and Kremser approaches for three-phase separators.

REFERENCES

Brian, P.L.T., *Staged Cascades in Chemical Processing,* Prentice-Hall, Englewood Cliffs, NJ, 1972.

Frost, A.C., "Slurry Bed Adsorption with Molecular Sieves," *Chem. Eng. Prog. 70,* (5), 70 (1974).

Hoyer, G.G., "Extraction with Supercritical Fluids: Why, How, and So What," *Chemtech,* July 1985, p. 440.

Grevillot, G. and D. Tondeur, "Equilibrium Staged Parametric Pumping. I. Single Transfer Step per Half Cycle and Total Reflux - The Analogy with Distillation," *AIChE J., 22,* 1055 (1976).

Grevillot, G. and D. Tondeur, "Equilibrium Staged Parametric Pumping. II. Multiple Transfer Steps per Half Cycle and Reservoir Staging," *AIChE J., 23,* 840 (1977).

Mullin, J.W., *Crystallization,* Butterworths, London, and CRC Press, Boca Raton, FL, 1972, pp. 290-293.

Paulaitis, M.E., J.M.L. Penniger, R.D. Gray, Jr., and P. Davidson, *Chemical Engineering at Supercritical Fluid Conditions,* Butterworths, Boston, 1983.

Prabhudesai, R.K., "Leaching," in P.A. Schweitzer (Ed.), *Handbook of Separation Techniques for Chemical Engineers,* McGraw-Hill, New York, 1979, Section 5.1.

Reid, R.C., J.M. Prausnitz, and T.K. Sherwood, *The Properties of Gases and Liquids,* 3rd ed., McGraw-Hill, New York, 1977.

Rickles, R.H., "Liquid-Solid Extraction," *Chem. Eng., 72,* 157 (March 15, 1965).

Schwartzberg, H.G., "Continuous Counter-current Extraction in the Food Industry," *Chem. Eng. Prog., 76* (4), 67 (April 1980).

Schwartzberg, H.G., "Leaching-Organic Materials," in R.W. Rousseau (Ed.), *Handbook of Separation Process Technology,* McGraw-Hill, New York, 1987, Chapt. 10.

Treybal, R.E., *Mass Transfer Operations,* 3rd ed., McGraw-Hill, New York, 1980, Chapt. 13.

Wankat, P.C., "Calculations for Separations with Three Phases. 1. Staged Systems," *Ind. Eng. Chem. Fundam., 19,* 358 (1980).

Wankat, P.C., *Large-Scale Adsorption and Chromatography,* CRC Press, Boca Raton, FL, 1986.

Wankat, P.C., *Mass Transfer Limited Separations,* Elsevier, New York (in press, 1988).

Worthy, W., "Supercritical Fluids Offer Improved Separations," *C&EN,* Aug. 3, 1981, p. 16.

HOMEWORK

A. *Discussion Problems*

A1. Develop your key relations chart for this chapter. Remember that a key relations chart is *not* a core dump but is selective.

A2. In your own words, describe the McCabe-Thiele analytical procedure in general terms that could be applied to any separation.

A3. How do the ideas of a general McCabe-Thiele procedure and the concept of unit operation relate to each other?

A4. How does the solid enter into washing calculations? Where does solids flow rate implicitly appear in Figure 17-2?

A5. In your own words, describe the Kremser analytical procedure in general terms that can be applied to any separation.

A6. Referring to Table 17-1, list similarities and differences between absorption, stripping, extraction, washing, and leaching.

A7. Show how Figure 17-5 could be modified to use a temperature swing instead of a pressure swing. What might be the advantage and disadvantage of doing this?

C. *Derivations*

C1. Explain the derivation of Eq. (17-6). Write out the units for this equation.

C2. Adapt the Kremser equation to leaching.

C3. Derive the operating equations and sketch the McCabe-Thiele analytical procedure for cross-flow and single-stage washing systems.

C4. Derive Eqs. (17-11) and (17-14).

C5. Derive Eqs. (17-18) and (17-19).

C6. Show that Eqs. (17-18) and (17-19) reduce to the usual form of the Kremser equation when $W = 0$.

D. *Problems*

D1. We wish to wash a solids stream. A countercurrent washing system with seven stages is used. The inlet solids flow is 10,000 kg/hr of dry solids. Each kilogram of dry solids also entrains 4 liters of solution. Thus, total inlet feed is 10,000 kg solids plus 40,000 liters of solution. The inlet concentration of this entrained solution is 2 wt % NaOH and 98 wt % water. Pure water at a flow rate of 25,000 kg/hr is used for washing. Assume that the liquid density is constant at 1 kg/liter. What is the outlet concentration of the NaOH in the overflow liquid?

D2. You are working on a new glass factory near the ocean. The sand is to be mined wet from the beach. However, the wet sand carries with it seawater entrained between the sand grains. Several studies have shown that 40% by volume seawater is consistently carried with the sand. The seawater is 0.035 wt frac salt, which must be removed by a washing process.

Densities: Water, 1.0 g/cm^3 (assume constant); Dry sand, 1.8 g/cm^3 (including air in voids); Dry sand without air, $1.8/0.6 = 3.0$ g/cm^3.

a. We desire a final wet sand product in which the entrained water has 0.002 wt frac salt. For each 1000 cm^3 of wet sand fed we will use 0.5 kg of pure wash water. In a countercurrent washing process, how many stages are required? What is the outlet concentration of the wash water?

b. In a cross-flow process we wish to use seven stages with 0.2 kg of pure wash water added to each stage for each 1000 cm^3 of wet sand fed. What is the outlet concentration of the water entrained with the sand?

D3. We wish to wash an alumina solids to remove NaOH from the entrained liquid. The underflow from the settler tank is 20 vol % solid and 80 vol % liquid. The two solid feeds to the system are also 20 vol % solids. In one of these feeds, NaOH concentration in the liquid is 5 wt %. This feed's solid flow rate (on a dry basis) is 1000 kg/hr. The second feed has a NaOH concentration in the liquid of 2 wt %, and its solids flow rate (on a dry basis) is 2000 kg/hr. We desire the final NaOH concentration in the underflow liquid to be 0.6 wt % (0.006 wt frac) NaOH. A countercurrent operation is used. The inlet washing water is pure and flows at 4000 kg/hr. Find the optimum feed location for the intermediate feed and the number of equilibrium stages required.

Data: ρ_w = 1.0 kg/liter (constant), $\rho_{alumina}$ = 2.5 kg/liter (dry crushed)

D4. We wish to wash a dilute acid from crushed rock. 100 m^3 of wet rock must be washed every day. Upon settling, the porosity is 0.4. Thus there is 40 m^3/day of underflow solution. The initial liquid entrained with the rock has a concentration of 1.0 kg acid/m^3, and the wash liquid contains no acid. If we use a cross-flow washer with three equilibrium stages and an overflow rate of 20 m^3/day in each stage (total amount of wash water is 60 m^3/day), what is the outlet concentration of the underflow liquid? Assume that porosity and densities are constant.

D5. You are mining wet sand from a beach by the ocean and want to wash out the salt. The sand carries with it 38 vol % liquid (ϵ = 0.38). The original seawater is 0.035 wt frac salt. You are expected to process 10,000 kg/hr of dry sand (the feed sand also contains saltwater). An 8-stage countercurrent cascade of mixer settlers is available. The density of the liquid solutions can be assumed constant at 1.0 g/cm^3. The density of dry sand without air is 3.0 g/cm^3, while the bulk density of dry sand containing air is $(3.0)(1 - 0.38) = 1.86$ g/cm^3.

a.   If the entering wash water rate is 1500 kg/hr, what is the outlet concentration of salt in the underflow liquid? The entering wash water contains no salt. Use a form of the Kremser equation.

b.   What is the total weight of entering underflow feed (sand + water) per hour?

D6.   You are working on a new glass factory near the ocean. The sand is to be mined wet from the beach. However, the wet sand carries with it seawater entrained between the sand grains. Data are given in Problem 17-D2. The salt must be removed by a washing process. A cross-flow process will be employed, with 0.2 kg of wash water added to each stage for each 1000 cm³ of wet sand fed. The wash water outlet from the last stage will be used as the wash water inlet for stage 3. Wash water outlet from stage N−1 will be used as wash inlet for stage 2, and wash water outlet from stage N−2 as wash water inlet to stage 1. All other stages have pure wash water inlet (see figure). We desire an outlet concentration of less than 0.002 wt frac salt in the entrained liquid. What is the minimum number of stages required to obtain this concentration? (This is *not* a minimum wash water flow rate problem. This specification means you don't have to obtain exactly 0.002 with an integer number of stages.) Note: This problem is *not* trial-and-error.

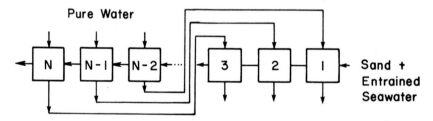

D7.   In the leaching of sugar from sugar cane, water is used as the solvent. Typically about 11 stages are used in a countercurrent Rotocel or other leaching system. On a volumetric basis liquid flow rate/solid flow rate = 0.95. The effective equilibrium constant is $m_E = 1.18$, where m_E = (concentration, g/liter, in liquid)/(concentration, g/liter, in solid) (Schwartzberg, 1980). If pure water is used as the inlet solvent, predict the recovery of sugar in the solvent.

D8.   We are testing out a system to leach caffeine, $C_8H_{10}N_4O_2$, out of coffee beans using a new solvent. Over the range of concentrations studied, the concentration of caffeine in the beans (measured as g caffeine per kg insoluble solid) is related to the weight ratio in the liquid by $Y = 0.8X$, where Y = weight ratio caffeine in the liquid, g caffeine/kg solvent; and X = g caffeine/kg insoluble solid. This is an effective equilibrium constant. The solvent we are studying appears to be unique in that only caffeine dissolves in it and none of the other water-soluble components in coffee dissolve. The solvent does not adsorb or crystallize onto the solids. The system uses three mixer-centrifuges, each of which acts as an equilibrium stage. The mixer-centrifuges are arranged so that the solid and fluid phases flow countercurrently to each other. Solvent entering the system is pure. We desire the final value of X to be $0.05X_{initial}$. How many kilograms of solvent are required for each kilogram of fresh coffee?

D9.   The use of slurry adsorbents has received some industrial attention because it allows for countercurrent movement of the solid phase. Your manager wants you to design a slurry adsorbent system for removing methane from a hydrogen gas stream. The actual separation process is a complex combination of adsorption and absorption, but the total equilibrium can be represented by a simple equation. At $5°$C, equilibrium can be represented as

Weight fraction CH_4 in gas = $1.2 \times$ (weight fraction CH_4 in slurry)

At $5°$C, no hydrogen could be detected in the slurry and the heat of sorption was negligible. We wish to separate a gas feed at $5°$C that contains 100 lb/hr of hydrogen and 30 lb/hr of methane. An outlet gas concentration of 0.05 wt frac methane is desired. The entering slurry will contain no methane and flows at a rate of 120 lb/hr. Find the number of equilibrium stages required for this separation and the mass fraction methane leaving with the slurry. Note: Because of the way the data are presented, this is NOT a three-phase contacting problem.

D10. A countercurrent leaching system is recovering oil from soybeans. The system has five stages. On a volumetric basis, liquid flow rate/solids flow rate = 1.36. 97.5% of the oil entering with the nonsoluble solids is recovered with the solvent. Solvent used is pure. Determine the effective equilibrium constant, m_E, where m_E is (kg/m^3 of solute in solvent)/(kg/m^3 of solute in solid) and is given by the equation $y = m_E x$.

D11. Slurry adsorption is to be used to remove benzene from an air stream V. The benzene is adsorbed onto activated carbon, phase L, which is in water, phase W. The vapor flow rate is 80 moles/hr, W = 45 moles/hr, and L = 70 moles/hr. The inlet water and carbon are both pure, $z_{in} = x_{in} = 0$. The inlet gas is 0.008 mole frac benzene, while an outlet gas containing 0.002 mole frac benzene is desired. At 24°C and 1 atm pressure, equilibrium could be fit by $y = 0.17x$ and $y = 269z$. Find the number of equilibrium stages required.

F. *Problems Requiring Other Resources*

F1. Read one of the articles by Prabhudesai (1979) or Schwartzberg (1980, 1987) and write a one- or two-page critique on the equipment used for leaching.

chapter 18
EXTRACTION OF
PARTIALLY MISCIBLE SYSTEMS

Liquid-liquid extraction was introduced in Chapter 16, and equipment was briefly discussed there. All extraction systems are partially miscible to some extent. When partial miscibility is very low, as for toluene and water, we can treat the system as if it were completely immiscible and use McCabe-Thiele analysis or the Kremser equation shown in Chapter 16. When partial miscibility becomes appreciable, it can no longer be ignored, and a calculation procedure that allows for variable flow rates must be used. In this case a different type of stage-by-stage analysis, which is very convenient for ternary systems, can be used. For multicomponent systems, computer calculations are required.

Leaching, or solid-liquid extraction, was introduced in Chapter 17. When a large part of the solid dissolves, the calculation procedures are essentially the same as for partially miscible liquid-liquid extraction. These leaching calculations will be discussed at the end of the chapter.

18.1. EXTRACTION EQUILIBRIA

Extraction systems are noted for the wide variety of equilibrium behavior that can occur in them. In the partially miscible range utilized for extraction, two liquid phases will be formed. At equilibrium the temperatures and pressures of the two phases will be equal and the compositions of the two phases will be related. The number of independent variables that can be arbitrarily specified (i.e., the degrees of freedom) for a system at equilibrium can be determined from the Gibbs phase rule

$$F = C - P + 2$$

which for a ternary extraction system is $F = 3 - 2 + 2 = 3$ degrees of freedom. In an extraction system, temperature and pressure are almost always constant so only one degree of freedom remains. Thus if we

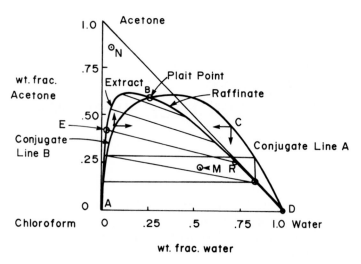

Figure 18-1.   Equilibrium for water-chloroform-acetone at 25°C and 1 atm.

specify the composition of one component in either phase, all other compositions will be set at equilibrium.

Extraction equilibrium data are easily shown graphically as either right triangular diagrams or equilateral triangular diagrams. Figure 18-1 shows the data listed in Table 18-1 (Alders, 1959) for the system water-chloroform-acetone at 25°C on a right triangular diagram. We have chosen chloroform as solvent, water as diluent, and acetone as solute. Curved line AEBRD represents the *solubility envelope* for this system. Any point below this line represents a two-phase mixture that will

Table 18-1. Equilibrium Data for the System Water-Chloroform-Acetone at 1 atm and 25°C

Water Phase, wt %			Chloroform Phase, wt %		
x_D Water	x_S Chloroform	x_A Acetone	y_D Water	y_S Chloroform	y_A Acetone
82.97	1.23	15.80	1.3	70.0	28.7
73.11	1.29	25.60	2.2	55.7	42.1
62.29	1.71	36.00	4.4	42.9	52.7
45.6	5.1	49.3	10.3	28.4	61.3
34.5	9.8	55.7	18.6	20.4	61.0

Note: Water and chloroform phases on the same line are in equilibrium with each other.

separate at equilibrium into a saturated extract phase and a saturated raffinate phase. Line AEB is the saturated extract line, while line BRD is the saturated raffinate line. Point B is called the *plait point* where extract and raffinate phases are identical. Remember that the extract phase is the phase with the higher concentration of solvent. Tie line ER connects extract and raffinate phases that are in equilibrium.

Point N in Figure 18-1 is a single phase because the ternary system is miscible at these concentrations. Point M represents a mixture of two phases, since it is in the immiscible range for this ternary system. At equilibrium the mixture represented by M will separate into a saturated raffinate phase and a saturated extract phase in equilibrium with each other. Either of the conjugate lines shown in Figure 18-1 can be used to draw tie lines. Consider tie line ER, which was found by drawing a horizontal line from point E to the conjugate line (point C) and then a vertical line from point C to the saturated raffinate curve (point R). Points E and R are in equilibrium, so they are on the ends of a tie line. These lines are shown in Figure 18-2 for different equilibrium data. This procedure is analogous to the use of an auxiliary line on an enthalpy-composition diagram illustrated in Figure 2-5.

To find the raffinate and extract phases that result when mixture M separates into two phases, we need a tie line through point M. This requires a simple eyeball trial-and-error calculation. Guess the location of the end point of the tie line on the saturated extract or raffinate curve, construct a tie line through this point, and check to see if the line passes through point M. If the first guess does not pass through M, repeat the process until you find a tie line that does. This is not too difficult because tie lines that are close to each other are approximately parallel.

The solubility envelope, tie lines, and conjugate lines shown on the triangular diagrams are derived from experimental equilibrium data. To

Figure 18-2.   Construction of tie line using conjugate line.

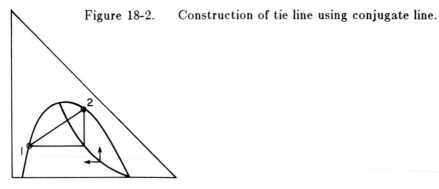

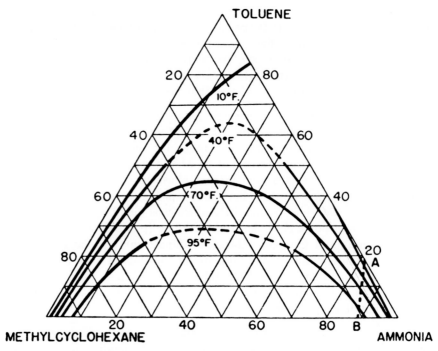

Figure 18-3.   Effect   of   temperature   on   equilibrium   of
                methylcyclohexane-toluene-ammonia   system,   From
                Fenske *et al., AIChE Journal, 1,* 335 (1955). Copyright
                1955, AIChE.

obtain these data a mixture can be made up and allowed to separate in
a separatory funnel. Then the concentrations of extract and raffinate
phase in equilibrium with each other are measured. This measurement
will give the location of one point on the saturated extract line, one
point on the saturated raffinate line, and the tie line connecting these
two points. One point on the conjugate line can be constructed from
this tie line by reversing the procedure used to construct a tie line when
the conjugate line was known.

The equilibrium data represented by Figure 18-1 are often called a
type I system, since there is *one pair* of immiscible binary compounds.
It is also possible to have systems with zero, two, and three immiscible
binary pairs (Alders, 1959; Ashton *et al.,* 1983). It is possible to go from
a type I to a type II system as temperature decreases. This is shown in
Figure 18-3 (Fenske *et al.,* 1955) for the methylcyclohexane-toluene-
ammonia system. At 10 ° F this is a type II system.

We will use right triangular diagrams exclusively in the remainder of
this chapter, because they are easy to read, they don't require special

paper, the scales of the axes can be varied, and portions of the diagram can be enlarged. Although equilateral diagrams have none of these advantages, they are used extensively in the literature for reporting extraction data; therefore it is important to be able to read and use this type of extraction diagram.

Equilibrium data can be correlated and estimated with thermo-dynamic models that calculate activity coefficients. These calculations are similar to those for vapor-liquid equilibrium. See Chapter 2 for references.

18.2. MIXING CALCULATIONS AND THE LEVER-ARM RULE

Triangular diagrams can be used for mixing calculations. In Figure 18-4A a simple mixing operation is shown, where streams F_1 and F_2 are mixed to form stream M. Streams F_1, F_2, and M can be either single-phase or two-phase. Operation of the mixer is assumed to be isother-mal. For ternary systems there are three independent mass balances. With right triangular diagrams it is convenient to use the diluent bal-ance, the solute balance, and the overall mass balance. The solvent mass balance will be automatically satisfied if the three independent bal-ances are satisfied. The nomenclature is the same as in Chapter 16 in fraction units (see Table 16-2).

For the mixing operation in Figure 18-4A the flow rates F_1 and F_2 would be given as well as the concentration of the two feeds: x_{A,F_1}, x_{D,F_1}, x_{A,F_2}, x_{D,F_2}. The three independent mass balances used to

Figure 18-4. Mixing operation. (A) Equipment; (B) triangular diagram.

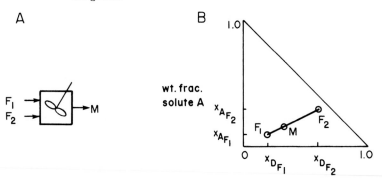

solve for M, $x_{A,M}$ and $x_{D,M}$ are,

$$F_1 + F_2 = M \qquad (18\text{-}1)$$

$$F_1 x_{A,F_1} + F_2 x_{A,F_2} = M x_{A,M} \qquad (18\text{-}2a)$$

$$F_1 x_{D,F_1} + F_2 x_{D,F_2} = M x_{D,M} \qquad (18\text{-}2b)$$

The concentrations of the mixed stream M are

$$x_{A,M} = \frac{F_1 x_{A,F_1} + F_2 x_{A,F_2}}{F_1 + F_2} \qquad (18\text{-}3a)$$

$$x_{D,M} = \frac{F_1 x_{D,F_1} + F_2 x_{D,F_2}}{F_1 + F_2} \qquad (18\text{-}3b)$$

We will now show that points F_1, F_2, and M are collinear as shown in Figure 18-4B. We will first use Eq. (18-1) to remove the mixed stream flow rate M from Eqs. (18-2). Next, we solve the resulting equations for the ratio F_1/F_2 and then set these two equations equal to each other. The manipulations are as follows:

$$F_1 x_{A,F_1} + F_2 x_{A,F_2} = (F_1 + F_2) x_{A,M}$$

$$F_1 x_{D,F_1} + F_2 x_{D,F_2} = (F_1 + F_2) x_{D,M}$$

Then

$$\frac{F_1}{F_2} = \frac{x_{A,M} - x_{A,F_2}}{x_{A,F_1} - x_{A,M}} \qquad (18\text{-}4a)$$

$$\frac{F_1}{F_2} = \frac{x_{D,M} - x_{D,F_2}}{x_{D,F_1} - x_{D,M}} \qquad (18\text{-}4b)$$

Finally, setting these equations equal to each other and rearranging, we have

$$\text{slope from point M to } F_2 = \frac{x_{A,M} - x_{A,F_2}}{x_{D,M} - x_{D,F_2}} \qquad (18\text{-}5)$$

$$= \frac{x_{A,F_1} - x_{A,M}}{x_{D,F_1} - x_{D,M}} = \text{slope from point M to } F_1$$

Equation (18-5) is the three-point form of a straight line. Equation (18-5) states that the three points $(x_{A,M}, x_{D,M})$, (x_{A,F_2}, x_{D,F_2}) and (x_{A,F_1}, x_{D,F_1}) lie on a straight line. The manipulations used to derive Eq. (18-5) are very similar to those used to develop difference points for countercurrent calculations, and we will return to them shortly.

It will often prove convenient to be able to determine the location of the mixing point on the line between F_1 and F_2 without having to solve the mass balances analytically. This can be done using Eqs. (18-3) or (18-4), which relate the ratio of the feed rates to differences in the ordinate and abscissa, respectively. With F_1/F_2, x_{A,F_1} and x_{A,F_2} known, Eq. (18-3) can be used to find $x_{A,M}$. This calculation can also be done graphically by requiring that

$$\frac{F_1}{F_2} = \frac{\text{distance from } x_{A,M} \text{ to } x_{A,F_2}}{\text{distance from } x_{A,M} \text{ to } x_{A,F_1}}$$

With a ruler or calipers, the value of $x_{A,M}$ can be found quickly. Equation (18-4) can be used in a similar way to find $x_{D,M}$.

In Figure 18-5 similar triangles F_1AM and MBF_2 have been drawn. Since the triangles are similar,

$$\frac{\text{Distance from } F_1 \text{ to } M}{\text{Distance from } F_1 \text{ to } A} = \frac{\text{distance from } M \text{ to } F_2}{\text{distance from } M \text{ to } B} \qquad (18\text{-}6a)$$

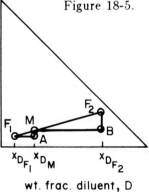

Figure 18-5. Development of lever-arm rule with similar triangles.

wt. frac. solute A

wt. frac. diluent, D

Rearranging this formulation, we have

$$\frac{\overline{MF_2}}{\overline{F_1M}} = \frac{\overline{MB}}{\overline{F_1A}} = \frac{x_{D,M} - x_{D,F_2}}{x_{D,F_1} - x_{D,M}} \qquad (18\text{-}6b)$$

where the bar denotes distance. According to Eq. (18-4b), the right-hand side of this equation is equal to F_1/F_2. Thus we have shown that

$$\frac{F_1}{F_2} = \frac{\overline{MF_2}}{\overline{F_1M}} \qquad (18\text{-}7)$$

Equation (18-7) is the lever-arm rule, which was first introduced in Chapter 2. It may be helpful to review that material now, including Figure 2-10. By measuring along the straight line between F_1 and F_2 we can find point M so that the lever-arm rule is satisfied. When you use the lever-arm rule, you don't need the individual values of the flow rates F_1 and F_2 to find the location of M.

Using Eq. (18-7), point M might be found by trial and error. Since this is a cumbersome procedure, it is worthwhile to develop the lever-arm rule in a different form. These alternative forms are (see Problem 18-C1)

$$\frac{F_1}{M} = \frac{\overline{F_2M}}{\overline{F_1F_2}}, \qquad \frac{F_2}{M} = \frac{\overline{F_1M}}{\overline{F_1F_2}} \qquad (18\text{-}8)$$

In this form the lever-arm rule is useful for finding the location of a stream M that is the sum of the two streams F_1 and F_2.

18.3. SINGLE-STAGE AND CROSS-FLOW SYSTEMS

Single-stage extraction systems can easily be solved with the tools we have developed. A batch extractor would consist of a single vessel equipped with a mixer. The two feeds would be charged to the vessel, mixed, and then allowed to settle into the two product phases. A continuous single-stage system requires a mixer and a settler as shown in Figure 16-2. Here the feed and solvent are fed continuously to the mixer, and the raffinate and extract products are continuously withdrawn from the settler. Figure 18-6 shows this schematically. The calculational procedures for batch and continuous operation are the same,

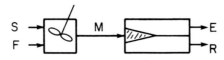

Figure 18-6.   Continuous mixer-settler.

the only real difference being that in batch operation S, F, M, E, and R are measured as total weight of material, whereas in continuous operation they are flow rates.

Usually the solvent and feed streams will be completely specified in addition to temperature and pressure. Thus the known variables are S, F, $y_{A,S}$, $y_{D,S}$, $x_{A,F}$, $x_{D,F}$, T, and p. The values of E, R, $y_{A,E}$, $y_{D,E}$, $x_{A,R}$, and $x_{D,R}$ are usually desired. If we make the usual assumption that the mixer-settler combination acts as one equilibrium stage, then streams E and R are in equilibrium with each other.

The calculation method proceeds as follows. (1) Plot the locations of S and F on the triangular equilibrium diagram. (2) Draw a straight line between S and F and use the lever-arm rule or Eqs. (18-3) to find the location of the mixed stream M. Now we know that stream M settles into two phases in equilibrium with each other. Therefore, (3) construct a tie line through point M to find the compositions of the extract and raffinate streams. (4) Find the ratio E/R using mass balances. We will follow this method to solve the following example.

Example 18-1. Single-Stage Extraction

A solvent stream containing 10% by weight acetone and 90% by weight chloroform is used to extract acetone from a feed containing 55 wt % acetone and 5 wt % chloroform with the remainder being water. The feed rate is 250 kg/hr, while the solvent rate is 400 kg/hr. Operation is at 25°C and atmospheric pressure. Find the extract and raffinate compositions and flow rates when one equilibrium stage is used for the separation.

Solution

A. Define.   The equipment sketch is the same as Figure 18-6 with S = 400, $y_{A,S}$ = 0.1, $y_{S,S}$ = 0.9, $y_{D,S}$ = 0 and F = 250, $x_{A,F}$ = 0.55, $x_{S,F}$ = 0.05, $x_{D,F}$ = 0.40. Find $x_{A,R}$, $x_{D,R}$, $y_{A,E}$, $y_{D,E}$, R, and E.

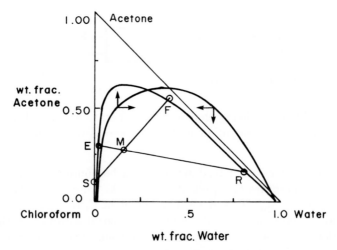

Figure 18-7.   Solution for single-stage extraction, Example 18-1.

B. Explore. Equilibrium data are obviously required. They can be obtained from Table 18-1 and Figure 18-1. The remainder of the problem is straightforward.

C. Plan. Plot streams F and S. Find mixing point M from the lever-arm rule or from Eq. (18-4). Then a tie line through M gives locations of streams E and R. Flow rates can be found from mass balances.

D. Do It. The graphical solution is shown in Figure 18-7. After locating streams F and S, M is on the line SF and can be found from the lever arm rule,

$$\frac{F}{M} = \frac{\overline{SM}}{\overline{FS}} = \frac{250}{250+400} = 0.385$$

or from Eq. (18-3),

$$x_{AM} = \frac{F x_{AF} + S y_{A,S}}{F + S} = \frac{(250)(0.55) + 400(0.1)}{650} = 0.273$$

A tie line through M is then constructed by trial and error, and the extract and raffinate locations are obtained. Concentrations are

$$y_{A,E} = 0.30, \qquad y_{D,E} = 0.02, \qquad x_{A,R} = 0.16, \qquad x_{D,R} = 0.83$$

The flow rates can be determined from the mass balances

$$M = E + R \quad \text{and} \quad Mx_{A,M} = Ey_{A,E} + Rx_{A,R}$$

Solving for R, we obtain

$$R = M \frac{x_{A,M} - y_{A,E}}{x_{A,R} - y_{A,E}} \tag{18-9}$$

or $R = (650) \left| \dfrac{0.273 - 0.30}{0.16 - .30} \right| = 125.36 \text{ kg/hr}$

and $E = M - R = 650 - 125.36 = 524.64$.

The lever arm rule can also be used but tends to be slightly less accurate.

E. Check. We can check the solute or diluent mass balances. For example,

$$Sy_{A,S} + Fx_{A,F} = Ey_{A,E} + Rx_{A,R}$$

which is

$$(400)(0.1) + (250)(0.55) = (524.64)(0.30) + (125.36)(0.16)$$

or $177.5 \sim 177.45$, which is well within the accuracy of the calculation. The diluent mass balance also checks.

F. Generalize. This procedure is similar to the one we used for binary flash distillation in Figure 3-3. Thus there is an analogy between distillation calculations on enthalpy-composition diagrams (Ponchon-Savarit diagrams) and extraction calculations on triangular diagrams.

From this example it is evident that a single extraction stage is sufficient to remove a considerable amount of acetone from water. However, quite a bit of solvent was needed for this operation, the resulting extract phase is not very concentrated, and the raffinate phase is not as dilute as it could be.

The separation achieved with one equilibrium stage can easily be enhanced with a cross-flow system as shown in Figure 18-8A. Assume

586

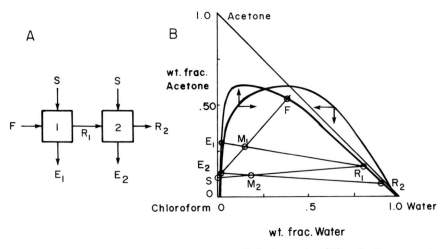

Figure 18-8.   Cross flow extraction. (A) Cascade; (B) solution on triangular diagram.

that a cross-flow stage is added to the problem given in Example 18-1
and another 400 kg/hr of solvent (stream S_2 with 10% acetone, 90%
chloroform) is used in stage 2. The concentrations of E_2 and R_2 are
easily found by doing a second mixing calculation with streams S_2 and
R_1.   During this mixing calculation, R_{j-1} (the feed to stage j) is
different for each stage. A tie line through the new mixing point M_2
gives the location of streams E_2 and R_2. This is illustrated in Figure
18-8B. Note that $x_{A,R_2} < x_{A,R_1}$ as desired.

In Chapter 16 we found that cross-flow systems are less efficient than
countercurrent systems (see Problem 18-D7). In the next section the calculations for countercurrent cascades will be developed.

18.4. COUNTERCURRENT EXTRACTION CASCADES

18.4.1. External Mass Balances

A countercurrent cascade allows for more complete removal of the
solute, and the solvent is reused so less is needed. A schematic diagram
of a countercurrent cascade is shown in Figure 18-9. All calculations
will assume that the column is isothermal and isobaric and is operating
at steady state. In the usual design problem, the column temperature
and pressure, the flow rates and compositions of streams F and S, and
the desired composition (or percent removal) of solute in the raffinate
product are specified. The designer is required to determine the number

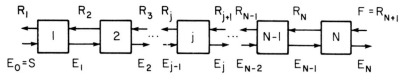

Figure 18-9.   Countercurrent extraction cascade.

of equilibrium stages needed for the specified separation and the flow rates and compositions of the outlet raffinate and extract streams. Thus the known variables are T, p, R_{N+1}, E_0, $x_{A,N+1}$, $x_{D,N+1}$, $y_{A,0}$, $y_{D,0}$, and $x_{A,1}$, and the unknown quantities are E_N, R_1, $x_{D,1}$, $y_{A,N}$, $y_{D,N}$ and N.

For an isothermal ternary extraction problem, the outlet compositions and flow rates can be calculated from external mass balances used in conjunction with the equilibrium relationship. The mass balances around the entire cascade are

$$E_0 + R_{N+1} = R_1 + E_N \tag{18-10a}$$

$$E_0 y_{A,0} + R_{N+1} x_{A,N+1} = R_1 x_{A,1} + E_N y_{A,N} \tag{18-10b}$$

$$E_0 y_{D,0} + R_{N+1} x_{D,N+1} = R_1 x_{D,1} + E_N y_{D,N} \tag{18-10c}$$

Since five variables are unknown (actually, there are seven, but $x_{S,1}$ and $y_{S,N}$ are easily found once $x_{A,1}$, $x_{D,1}$, $y_{A,N}$, and $y_{D,N}$ are known), a total of five independent equations are needed.

To find two additional relationships, note that streams R_1 and E_N are both leaving equilibrium stages. Thus the compositions of stream R_1 must be related in such a way that R_1 is on the saturated raffinate curve. This gives a relationship between $x_{A,1}$ and $x_{D,1}$. Similarly, since stream E_N must be a saturated extract, $y_{A,N}$ and $y_{D,N}$ are related. If the saturated extract and saturated raffinate relationships are known in analytical form, these two equations can be added to the three mass balances, and the resulting five equations can be solved simultaneously for the five unknowns.

The procedure can also be carried out conveniently on a triangular diagram. Let us represent the cascade shown in Figure 18-9 as a mixing tank followed by a black box separation scheme that produces the desired extract and raffinate as shown in Figure 18-10A. In Figure 18-10A, streams E_N and R_1 are *not* in equilibrium as they were in Figure 18-6, but stream E_N is a saturated extract and stream R_1 is a saturated raffinate. The external mass balances for Figure 18-10A are

A.

B.

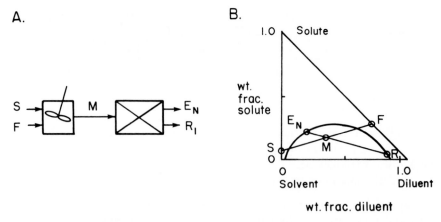

wt. frac. diluent

Figure 18-10.   External mass-balance calculation. (A) Mixer-separation representation; (B) solution on triangular diagram.

$$E_0 + R_{N+1} = M = E_N + R_1 \qquad (18\text{-}11a)$$

$$E_0 y_{A,0} + R_{N+1} x_{A,N+1} = M x_{A,M} = E_N y_{A,N} + R_1 x_{A,1} \qquad (18\text{-}11b)$$

$$E_0 y_{D,0} + R_{N+1} x_{D,N+1} = M x_{D,M} = E_N y_{D,N} + R_1 x_{D,1} \qquad (18\text{-}11c)$$

The coordinates of point M can be found from Eqs. (18-11):

$$x_{A,M} = \frac{E_0 y_{A,0} + R_{N+1} x_{A,N+1}}{E_0 + R_{N+1}} \qquad (18\text{-}12a)$$

$$x_{D,M} = \frac{E_0 y_{D,0} + R_{N+1} x_{D,N+1}}{E_0 + R_{N+1}} \qquad (18\text{-}12b)$$

Since Eqs. (18-11) are the same type of mass balances as for a mixer, the points representing streams E_N, R_1, and M lie on a straight line and the flow rates are related to the length of line segments by the lever-arm rule. You can prove this by developing the three-point form of a straight line (see Problem 18-C2). We also know that E_N must lie on the saturated extract line and R_1 must lie on the saturated raffinate line. Since $x_{A,1}$ is known, the location of R_1 on the saturated raffinate line can be found. A straight line from R_1 extended through M (found from $x_{A,M}$ and $x_{D,M}$ or from the lever-arm rule) will intersect the saturated extract stream at the value of E_N. This construction is illustrated in Figure 18-

10B. Note that this procedure is very similar to the one used for single equilibrium stages, but the line R_1ME_N is *not* a tie line. Mass balances can then be used to solve for the flow rates E_N and R_1.

The external mass balance and the equilibrium diagram in Figure 18-10B can be used to determine the effect of variation in the feed or solvent concentrations, the raffinate concentration, or the ratio F/S on the resulting separation. For example, if the amount of solvent is increased, the ratio of F/S will decrease. The mixing point M will move toward point S, and the resulting extract will contain less solute.

18.4.2. Difference Points and Stage-by-Stage Calculations

To determine the number of stages or flow rates and compositions inside the cascade, stage-by-stage calculations are needed. But first we use the external mass balances to find concentrations $y_{A,N}$ and $y_{D,N}$ and flow rates E_N and R_1. Starting at stage 1 (Figure 18-9) we note that streams R_1 and E_1 both leave equilibrium stage 1. Therefore, these two streams are in equilibrium and the concentration of stream E_1 can be found from an equilibrium tie line.

Streams E_1 and R_2 pass each other in the diagram and are called *passing streams*. These streams can be related to each other by mass balances around stage 1 and the raffinate end of the extraction train. The unknown variables for these mass balances are concentrations $x_{A,2}$, $x_{D,2}$, and $x_{S,2}$ and flow rates E_1 and R_2. Concentration $x_{S,2}$ can be determined from the stoichiometric relation $x_{S,2} = 1.0 - x_{A,2} - x_{D,2}$. Taking this equation into account, there are four unknowns (E_1, R_2, $x_{A,2}$, and $x_{D,2}$) but only three independent mass balances. What is the fourth relation that must be used?

To develop a fourth relation we must realize that stream R_2 is a saturated raffinate stream. Thus it will be located on the saturated raffinate line, and $x_{A,2}$ and $x_{D,2}$ are related by the relationship describing the saturated raffinate line. With four equations and four unknowns we can now solve for the variables E_1, R_2, $x_{A,2}$, and $x_{D,2}$.

To continue along the column, we repeat the procedure for stage 2. Since streams E_2 and R_2 are in equilibrium, a tie line will give the concentration of stream E_2. Streams E_2 and R_3 are passing streams; thus, they are related by mass balances. It will prove to be convenient if we write the mass balances around stages 1 and 2 instead of around stage 2 alone. The fourth required relationship is that stream R_3 must be a saturated raffinate stream. The stage-by-stage calculation procedure is then continued for stages 3, 4, etc. When the calculated solute concentration in the extract is greater than or equal to the specified concentration, that is, $y_{A,j_{calc}} > y_{A,N \, specified}$, the problem is finished.

These stage-by-stage calculations can be done analytically and can be programmed for computer solution if equations are available for the tie lines and the saturated extract and saturated raffinate curves. If the equations are not readily available, either the equilibrium data must be fitted to an analytical form or a data matrix with a suitable interpolation routine must be developed. Graphical techniques can be employed and have the advantage of giving a visual interpretation of the process.

In a graphical procedure for countercurrent systems the equilibrium calculations can easily be handled by constructing tie lines. The relation between $x_{A,j}$ and $x_{D,j}$ is already shown as the saturated raffinate curve. All that remains is to develop a method for representing the mass balances graphically.

Referring to Figure 18-9, we can do a mass balance around the first stage. After rearrangement, this is

$$E_0 - R_1 = E_1 - R_2$$

If we now do mass balances around each stage and rearrange each balance as the difference between passing streams, we obtain

$$\Delta = E_0 - R_1 = \cdots = E_j - R_{j+1} = \cdots = E_N - R_{N+1} \qquad (18\text{-}13a)$$

Thus the difference in flow rates of passing streams is constant even though both the extract and raffinate flow rates are varying. The same difference calculation can be repeated for solute A,

$$\Delta x_{A,\Delta} = E_0 y_{A,0} - R_1 x_{A_1} = \cdots = E_j y_{A,j} - R_{j+1} x_{A,j+1}$$

$$= \cdots = E_N y_{A,N} - R_{N+1} x_{A,N+1} \qquad (18\text{-}13b)$$

and for diluent D,

$$\Delta x_{D,\Delta} = E_0 y_{D,0} - R_1 x_{D,1} = \cdots = E_j y_{D,j} - R_{j+1} x_{D,j+1}$$

$$= \cdots = E_N y_{D,N} - R_{N+1} x_{D,N+1} \qquad (18\text{-}13c)$$

The differences in flow rates (which is the net flow) of solute and diluent are constant.

Equations (18-13) define a difference or Δ (delta) point. The coordinates of this point are easily found from Eqs. (18-13b) and (18-13c).

$$x_{A,\Delta} = \frac{E_0 y_{A,0} - R_1 x_{A,1}}{\Delta} = \frac{E_N y_{A,N} - R_{N+1} x_{A,N+1}}{\Delta} \quad (18\text{-}14a)$$

$$x_{D,\Delta} = \frac{E_0 y_{D,0} - R_1 x_{D,1}}{\Delta} = \frac{E_N y_{D,N} - R_{N+1} x_{D,N+1}}{\Delta} \quad (18\text{-}14b)$$

where Δ is given by Eq. (18-13a). $x_{A,\Delta}$ and $x_{D,\Delta}$ are the coordinates of the difference point and are not compositions that occur in the column. Note that $x_{A,\Delta}$ and $x_{D,\Delta}$ can be negative.

The difference point can be treated as a stream for mixing calculations. Thus Eqs. (18-13a,b,c) show that the following points are collinear.

$$\Delta(x_{A,\Delta}, x_{D,\Delta}), E_0(y_{A,0}, y_{D,0}), R_1(x_{A,1}, x_{D,1})$$

$$\Delta(x_{A,\Delta}, x_{D,\Delta}), E_1(y_{A,1}, y_{D,1}), R_2(x_{A,2}, x_{D,2})$$

$$\Delta(x_{A,\Delta}, x_{D,\Delta}), E_j(y_{A,j}, y_{D,j}), R_{j+1}(x_{A,j+1}, x_{D,j+1})$$

$$\Delta(x_{A,\Delta}, x_{D,\Delta}), E_N(y_{A,N}, y_{D,N}), R_{N+1}(x_{A,N+1}, x_{D,N+1})$$

The existence of these straight lines and the applicability of the lever-arm rule can be proved formally (see Problem 18-C3).

Since all pairs of passing streams lie on a straight line through the Δ point, the Δ point is used to determine operating lines for the mass balances. A difference point in each section replaces the single operating line used on a McCabe-Thiele diagram. The procedure for stepping off stages will be illustrated after we discuss finding the location of the Δ point.

There are three methods for finding the location of Δ:

1. Graphical Construction. Since the points Δ, E_0, and R_1; and Δ, E_N, and R_{N+1} are on straight lines, we can draw these two straight lines. The point of intersection must be the Δ point. For the typical design problem (see Figure 18-9), points R_{N+1}, E_0, and R_1 are easily plotted. E_N can be found from external balances (Figure 18-10B). Then Δ is found as shown in Figure 18-11.

2. Coordinates. The coordinates of the difference point were found in

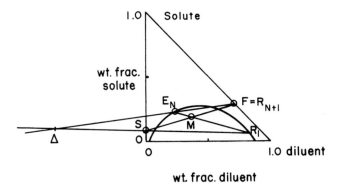

wt. frac. diluent

Figure 18-11.　Location of difference point.

Eqs. (18-14). These coordinates can be used to find the location of Δ. It may be convenient to draw one of the straight lines in method 1 (such as line $\overline{E_0R_1}$) and use one of the coordinates (such as $x_{D,\Delta}$) to find Δ. This procedure is useful since accurate graphical determination of Δ can be difficult.

3. Lever-Arm Rule. The general form of the lever-arm rule for two passing streams is (see Problem 18-C3)

$$\frac{R_{j+1}}{E_j} = \frac{\overline{E_j\Delta}}{\overline{R_{j+1}\Delta}} \tag{18-15}$$

The lever-arm rule can be used to find the Δ point. For instance, if flow rates R_1 and E_0 are known, then Δ can be found on the straight line through points R_1 and E_0 at a distance that satisfies Eq. (18-15) with $j = 0$.

Consider again the stage-by-stage calculation routine that was outlined previously. We start with the known concentration of raffinate product stream R_1 and use an equilibrium tie line (stage 1) to find the location of saturated extract stream E_1 (see Figure 18-12). The points representing Δ, E_1, and R_2 are collinear, since E_1 and R_2 are passing streams. If the location of Δ is known, the straight line from Δ to E_1 can be drawn and then be extended to the saturated raffinate curve. This has to be the location of stream R_2 (see Figure 18-12). Thus the difference point allows us to solve the three simultaneous mass balances by drawing a single straight line. This straight line is called the *operating line*. The procedure may now be continued by constructing a tie line (representing stage 2) to find the location of stream E_2, which is in

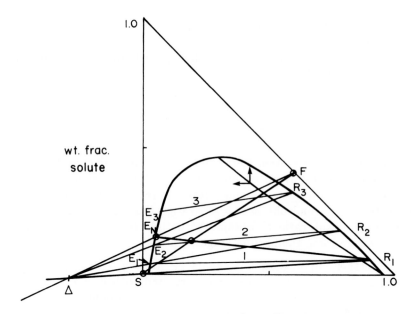

Figure 18-12.   Stage-by-stage solution.

equilibrium with stream R_2. Then the mass balances are again solved simultaneously by drawing a straight line from Δ through E_2 to the saturated raffinate line, which locates R_3. This process of alternating between equilibrium and mass balances is continued until the desired separation is achieved. The stages are counted along the tie lines, which represent extract and raffinate streams in equilibrium. To obtain an accurate solution, a large piece of graph paper is needed and care must be exercised in constructing the diagram.

Before continuing you should carefully reread the preceding paragraph; it contains the essence of the stage-by-stage calculation method.

In Figure 18-12 we see that two equilibrium stages do not quite provide sufficient separation, and three equilibrium stages provide more separation than is needed. In a case like this, an approximate fractional number of stages can be reported.

$$\text{Fraction} = \frac{\overline{E_2 E_N}}{\overline{E_2 E_3}} \sim 0.19$$

This fraction can be measured along the curved saturated extract line. It should be stressed that the resulting number of stages, 2.19, is only

594

approximate. The fractional number of stages is useful when the actual stages are not equilibrium stages. Thus if a sieve-plate column with an overall plate efficiency of 25% were being used, the actual number of plates required would be

$$\frac{2.19}{0.25} = 8.76 \text{ or } 9 \text{ plates}$$

If a mixer-settler system were used, where each mixer-settler combination is approximately an equilibrium stage, then we would have three choices: (1) Use three stages and obtain more separation than desired; (2) use two stages and obtain less separation than desired; or (3) change the feed-to-solvent ratio to obtain the desired separation with exactly two or exactly three equilibrium stages.

One further important point should be stressed with respect to Figure 18-12. If two equilibrium stages were used with $F/S = 2$ as in the original problem statement, the saturated extract would *not* be located at the value E_2 shown on the graph and saturated raffinate would *not* be at the value R_1 shown. The streams R_1 and E_2 do *not* satisfy the external mass balance for this system. The values R_1 and E_N do, but R_1 and E_2 do not. The exact compositions of the product streams for a two-stage system require a trial-and-error solution (see Section 18.8).

What do we do if flow rate E_0 is less than R_1? The easiest solution is to define Δ so that it is now positive.

$$\Delta = R_1 - E_0 = R_{j+1} - E_j = R_{N+1} - E_N \tag{18-16}$$

Δ is still equal to the difference between the flow rates of any pair of passing streams, but it is now raffinate minus extract. The corresponding lever-arm rule for any pair of passing streams is still Eq. (18-15), but

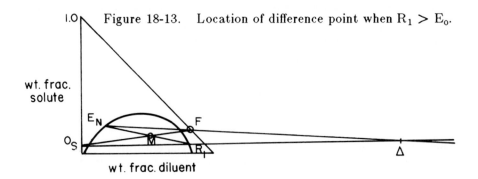

Figure 18-13. Location of difference point when $R_1 > E_0$.

the Δ point will be on the opposite side of the triangular diagram. This situation is shown in Figure 18-13. The stage-by-stage calculation procedure is unchanged when the location of Δ is on the right side of the diagram.

18.4.3. Complete Extraction Problem

At this point you should be ready to solve a complete extraction problem.

Example 18-2. Countercurrent Extraction

A solution of acetic acid (A) in water (D) is to be extracted using isopropyl ether as the solvent (S). The feed is 1000 kg/hr of a solution containing 35 wt % acid and 65 wt % water. The solvent used comes from a solvent recovery plant and is essentially pure isopropyl ether. Inlet solvent flow rate is 1475 kg/hr. The exiting raffinate stream should contain 10 wt % acetic acid. Operation is at 20 °C and 1 atm. Find the outlet concentrations and the number of equilibrium stages required for this separation. The equilibrium data are given by Treybal (1968) and are reproduced in Table 18-2.

Table 18-2. Equilibrium Data for Water-Acetic Acid-Isopropyl Ether at 20 °C and 1 atm

Water Layer, wt %			Isopropyl Ether Layer, wt %		
Acetic Acid	Water	Isopropyl Ether	Acetic Acid	Water	Isopropyl Ether
x_A	x_D	x_S	y_A	y_D	y_S
0.69	98.1	1.2	0.18	0.5	99.3
1.41	97.1	1.5	0.37	0.7	98.9
2.89	95.5	1.6	0.79	0.8	98.4
6.42	91.7	1.9	1.93	1.0	97.1
13.30	84.4	2.3	4.82	1.9	93.3
25.50	71.1	3.4	11.40	3.9	84.7
36.70	58.9	4.4	21.60	6.9	71.5
44.30	45.1	10.6	31.10	10.8	58.1
46.40	37.1	16.5	36.20	15.1	48.7

Points on the same horizontal line are in equilibrium.
Source: Treybal (1968).

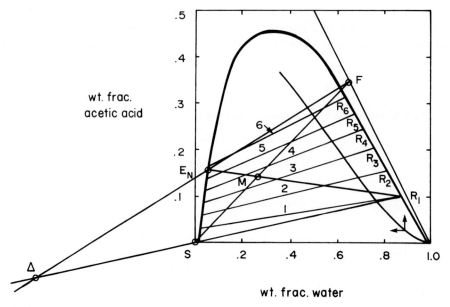

Figure 18-14.   Solution to Example 18-3.

Solution

A. Define. The extraction will be a countercurrent system as shown in Figure 18-9. $F = 1000$, $x_{A,F} = 0.35$, $x_{D,F} = 0.65$, $S = 1475$, $y_{A,S} = 0$, $y_{D,S} = 0$, $x_{A,1} = 0.1$. Find $x_{D,1}$, $y_{A,N}$, $y_{D,N}$ and N.

B, C. Explore and Plan. This looks like a straightforward design problem. Use the method illustrated in Figures 18-11 and 18-13 to find Δ. Then step off stages as illustrated in Figure 18-12.

D. Do It. The solution is shown in Figure 18-14.

1. Plot equilibrium data and construct conjugate line.

2. Plot locations of streams $E_0 = S$, $R_{N+1} = F$, and R_1.

3. Find mixing point M on line through points S and F at $x_{A,M}$ value calculated from Eq. (18-12).

$$x_{A,M} = \frac{(1475)(0) + (1000)(0.35)}{1475 + 1000} = 0.1414$$

4. Line $R_1 M$ gives point E_N.

5. Find Δ point as intersection of straight lines E_0R_1 and E_NR_{N+1}

6. Step off stages, using the procedure shown in Figure 18-12. To keep the diagram less crowded, the operating lines, $\Delta E_j R_{j+1}$, are not shown. You can use a straight edge on Figure 18-14 to check the operating lines. A total of 5.8 stages are required.

E. Check. Small errors in plotting the data or in drawing the operating and tie lines can cause fairly large errors in the number of stages required. If greater accuracy is required, a much larger scale and more finely divided graph paper can be used, or a McCabe-Thiele diagram (see the next section) or computer methods can be used.

F. Generalize. Note that the portion of the triangular diagram above 50% acetic acid is not needed. In many cases the diagram can be expanded (using unequal axes) to increase accuracy.

This completes the basic procedure for designing new columns. In the remainder of this chapter we will look at the relations between McCabe-Thiele and triangular diagrams, computer calculations, determination of minimum solvent rate, simulation problems, two-feed columns, and extract reflux.

18.5. RELATIONSHIP BETWEEN McCABE-THIELE AND TRIANGULAR DIAGRAMS

Stepping off a lot of stages on a triangular diagram can be difficult and inaccurate. More accurate calculations can be done with a McCabe-Thiele diagram. Since total flow rates are *not* constant, the triangular diagram and the Δ point are used to plot a curved operating line on the McCabe-Thiele diagram. This construction is illustrated in Figure 18-15 for a single point. For any arbitrary operating line (which must go through Δ), the values of the extract and raffinate concentrations of passing streams $(y_{A,op}, x_{A,op})$ are easily determined. These concentrations must represent a point on the operating line in the McCabe-Thiele diagram. Thus the values of $y_{A,op}$ and $x_{A,op}$ are transferred to the diagram. Since the raffinate value is an x, the $y = x$ line is used to find x. A very similar procedure was used in Figure 2-6B to relate McCabe-Thiele to enthalpy-composition diagrams.

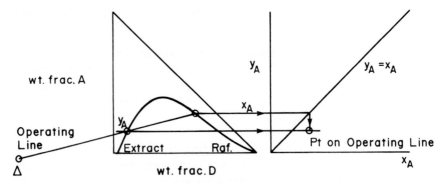

Figure 18-15. Use of triangular diagram to plot operating line on McCabe-Thiele diagram.

When this construction is repeated for a number of arbitrary operating lines, a curved operating line is generated on the McCabe-Thiele diagram. The equilibrium data, y_A vs x_A, can also be plotted. Then stages can be stepped off on the diagram. This is shown in Figure 18-16 for Example 18-2. The equilibrium data were obtained from Table 18-2. The Δ point on the triangular diagram was used to find the operating

Figure 18-16. Use of triangular and McCabe-Thiele diagrams to solve Example 18-3.

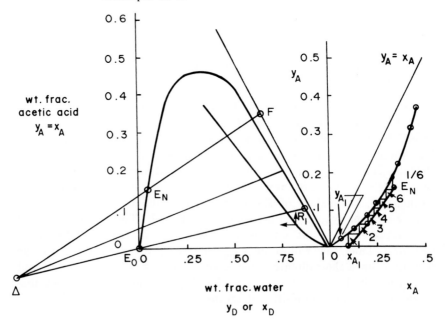

line. A total of 6 1/6 stages was determined, which is reasonably close to the 5.8 found in Example 18-2.

Note that in this example the operating line is close to straight. Thus R_{j+1}/E_j is approximately constant. This occurs when the change in solubility of the solvent in the raffinate streams is approximately the same as the change in solubility of the diluent in the extract streams. When the changes in partial miscibility of the extract and raffinate phases are very unequal, R_{j+1}/E_j will vary significantly and the operating line will show more curvature. When the feed is not presaturated with solvent and the entering solvent is not presaturated with diluent, there can be a large change in flow rates on stages 1 and N. This is not evident in Figure 18-16, since the extract and raffinate phases are close to immiscible for the ranges of concentration shown.

The McCabe-Thiele diagrams are useful for more complicated extraction columns.

18.6. COMPUTER CALCULATIONS

Partially miscible ternary extraction calculations can be adapted to programmable calculators or computers. The following steps can be used for a stage-by-stage calculation procedure that mimics the graphical procedure.

1. Determine an equation for the saturated raffinate curve and one for the saturated extract curve. These should fit the solubility data over the ranges of interest for the extraction. Equations in the forms

$$x_D = f_{raf}(x_A), \quad y_D = f_{ext}(y_A) \tag{18-17a,b}$$

   are convenient.

2. Determine an equation to fit the tie-line data. Since the solubility envelope equations are known separately, a single equation such as

$$y_A = f_{tie}(x_A) \tag{18-18}$$

   is sufficient. This equation fits the equilibrium data on the McCabe-Thiele diagram.

3. Solve the external mass balances Eqs. (18-10) simultaneously with the saturated extract and raffinate curve equations. This will

mimic the construction shown in Figure 18-10B. Mass balance equations (18-10) can be rearranged to the equation for a straight line (see Problem 18-C3).

$$y_{D,N} = (y_{A,N} - x_{A,M})\left(\frac{x_{D,M} - x_{D,1}}{x_{A,M} - x_{A,1}}\right) + x_{D,M} \tag{18-19}$$

where $x_{A,M}$ and $x_{D,M}$ are given by Eq. (18-12). The value of $x_{D,1}$ can be found from Eq. (18-17b) from the known raffinate concentration $x_{A,1}$. Now Eqs. (18-19) and (18-17b) are solved simultaneously to find the outlet extract concentrations $(y_{A,N}, y_{D,N})$. Determine flow rates R_1 and E_N from Eqs. (18-11), and calculate $x_{A,\Delta}$ and $x_{D,\Delta}$ from Eqs. (18-14).

4. Stream $E_j(y_{A,j}, y_{D,j})$ is in equilibrium with stream R_j. Calculate $y_{A,j}$ from Eq. (18-18) and then $y_{D,j}$ from Eq. (18-17b). For the first stage, set $j = 1$ and calculate $y_{A,1}$ from the known value $x_{A,1}$. This step is equivalent to drawing tie lines.

5. Streams $R_{j+1}(x_{A,j+1}, x_{D,j+1})$ and $E_j(y_{A,j}, y_{D,j})$ are passing streams. Since stream $E_j(y_{A,j}, y_{D,j})$ is known, calculate $x_{A,j+1}$ and $x_{D,j+1}$ from the simultaneous solution of the mass balances, Eqs. (18-13), and the saturated raffinate expression, Eq. (18-17a). We want to find the point of intersection of the operating line and the saturated raffinate curve. The operating equation can be written as (see Problem 18-C3)

$$x_{D,j+1} = (x_{A,j+1} - x_{A,\Delta})\left(\frac{x_{D,\Delta} - y_{D,j}}{x_{A,\Delta} - y_{A,j}}\right) + x_{D,\Delta} \tag{18-20}$$

The values of $y_{A,j}$ and $x_{A,j}$ are known from step 4, while $x_{A,\Delta}$ and $x_{D,\Delta}$ were found in step 3. Solve Eq. (18-20) simultaneously with the equation for the saturated raffinate, Eq. (18-17a). This calculation mimics drawing the operating line through the Δ point.

6. Repeat steps 4 and 5 until $y_{A,j} \geq y_{A,N}$. The required number of stages is equal to $j-1$; and a fractional number of stages can be estimated.

$$\text{Fraction} = \frac{y_{A,N} - y_{A,j-1}}{y_{A,j} - y_{A,j-1}} \tag{18-21}$$

An alternative is to report that j equilibrium stages are more than enough to achieve the desired separation.

If the equilibrium data are accurately represented by Eqs. (18-17) and (18-18), this method will be as accurate as a very careful graphical analysis. The limit to accuracy is the accuracy of the equilibrium data. When only one calculation is required, the graphical procedures will be quicker. When many calculations are required, a calculator or computer routine will be advantageous. In the remainder of this chapter we will use graphical methods to look at alternative operating methods. The analyses for these methods can be programmed on a calculator or computer using modifications of the method outlined in this section.

18.7. EFFECT OF SOLVENT RATE

As the solvent rate is increased, the separation should become easier, and the outlet extract stream, E_N, should become more diluted. The effect of increasing S/F is shown in Figure 18-17. As S/F increases, the mixing point moves toward the solvent and the solute concentration in stream E_N decreases. The difference point starts on the right-hand side of the diagram $(R_{j+1} > E_j)$ and moves away from the diagram. When $R_{j+1} = E_j$, Δ is at infinity. A further increase in S/F puts Δ on the left-hand side of the diagram $(R_{j+1} < E_j)$. It now moves toward the diagram as S/F continues to increase. Some of these S/F ratios will be too low, and even a column with an infinite number of stages will not be able to do the desired separation.

It is often of interest to calculate the minimum amount of solvent that can be used and still obtain the desired separation. The minimum

Figure 18-17. Effect of increasing S/F. $(S/F)_4 > (S/F)_3 > (S/F)_2 > (S/F)_1$

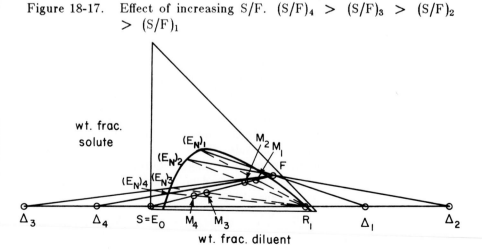

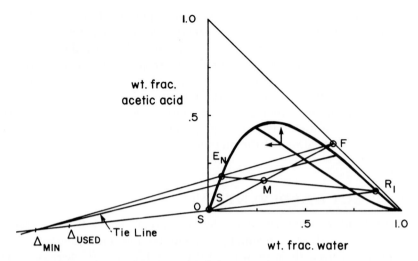

Figure 18-18.   Determination of minimum solvent rate.

solvent rate (or minimum S/F) is the rate at which the desired separation can be achieved with an infinite number of stages. If less solvent is
used, the desired separation is impossible; while if more solvent is used,
the separation can be achieved with a finite number of stages. The
corresponding Δ value, Δ_{min}, thus represents the dividing point between
impossible cases and possible solutions. This situation is analogous to
minimum reflux in distillation.

To determine Δ_{min} and hence $(S/F)_{min}$, we note that to have an
infinite number of stages, tie lines and operating lines must coincide (be
parallel) somewhere on the diagram. This will require an infinite
number of stages. The construction to determine Δ_{min} is shown in Figure 18-18 for Example 18-2, and is outlined here:

1. Draw and extend line R_1S.

2. Draw a series of arbitrary tie lines in the range between points F
    and R_1. Extend these tie lines until they intersect line R_1S.

3. Δ_{min} is located at the point of intersection of a tie line that is
    closest to the diagram on the left-hand side or furthest from the
    diagram if on the right-hand side. Thus Δ_{min} is at the largest S/F
    that requires an infinite number of stages. Often the tie line that
    when extended goes through the feed point is the desired tie line.
    This is not the case in Figure 18-18.

4. Draw the line $\Delta_{min}F$. The intersection of this line with the
    saturated extract curve is $E_{N,min}$.

5. Draw the line from $E_{N,min}$ to R_1. Intersection of this line with the line from S to F gives M_{min}.

6. From the lever-arm rule, $(S/F)_{min} = \overline{FM}/\overline{SM}$. In Figure 18-18, $(S/F)_{min} = 1.296$ and $S_{min} = 1296$ kg/hr. The actual solvent rate for Example 18-2 is 1475 kg/hr, so the ratio $S/S_{min} = 1.138$. Use of more solvent in Figures 18-14 or 18-16 would decrease the required number of stages.

On a McCabe-Thiele diagram the behavior will appear simpler. At low S/F the operating line will intersect the equilibrium curve. As S/F increases, the operating line will eventually just touch the equilibrium curve (minimum solvent rate). Then as S/F increases further, the operating line will move away from the equilibrium curve. Unfortunately, since the operating line is curved, it is difficult to find an accurate value of S/F from the McCabe-Thiele diagram. An approximate value can easily be estimated.

18.8. SIMULATION PROBLEMS

When the extraction system already exists, we wish to determine how much separation can be achieved. For countercurrent systems a stage-by-stage analysis will be by trial and error. Since this was also the case in distillation, absorption, stripping, and immiscible extraction, the need for a trial-and-error calculation is not surprising.

For a simulation problem the specified variables are usually (see Figure 18-9) E_0, $y_{A,0}$, $y_{D,0}$, R_{N+1}, $x_{A,N+1}$, $x_{D,N+1}$, T, p, and N (equilibrium contacts). Points E_0 and R_{N+1} are easily plotted, and the mixing point M can be determined. Since $x_{A,1}$ is not known, stream R_1 cannot be found. Thus the mixing point alone does not allow us to find point E_N. Without the locations of R_1 and E_N, Δ cannot be found. We are unable to get started!

The solution to this impasse is to use a trial-and-error procedure such as the the following:

1. Guess $x_{A,1}$ and plot point R_1.

2. Draw line R_1M and find point E_N.

3. Find the Δ point.

4. Step off N stages starting at R_1.

5. Check: Is E_N (step 2) = E_N (step 4)? If not, repeat the calculation.

This type of trial-and-error problem is ideal for computer calculations.

Once we have the separation, we can determine the flow rates on every stage with the aid of mass balances or the lever-arm rule. From these flow rates we can determine whether the extractor will operate as expected. Flooding behavior in extractors depends on the type of equipment used. (See Laddha and Degaleesan, 1978, or Perry and Green, 1984, Section 21.)

18.9. TWO-FEED EXTRACTION COLUMN

In practice we may find ourselves in the situation where we have two feeds of different composition that both have to be treated in an extraction system. In this case it will be more economical to design a column with two separate feeds instead of mixing the feeds together before sending them to the column. Philosophically this is sound, since mixing would destroy the partial separation that already exists. A schematic diagram of a two-feed column is shown in Figure 18-19A.

First, we do the external balances:

$$E_0 + F_2 + R_{N+1} = M = E_N + R_1 \tag{18-22a}$$

$$E_0 y_{A,0} + F_2 x_{A,F_2} + R_N x_{A,N+1} = M x_{A,M} = E_N y_{A,N} + R_1 x_{A,1} \tag{18-22b}$$

$$E_0 y_{D,0} + F_2 x_{D,F_2} + R_{N+1} x_{D,N+1} = M x_{D,M} = E_N y_{D,N} + R_1 x_{D,1} \tag{18-22c}$$

From these equations the coordinates of point M are easily derived:

$$x_{A,M} = \frac{E_0 y_{A,0} + F_2 x_{A,F_2} + R_{N+1} x_{A,N+1}}{E_0 + F_2 + R_{N+1}} \tag{18-23a}$$

$$x_{D,m} = \frac{E_0 y_{D,0} + F_2 x_{D,F_2} + R_{N+1} x_{D,N+1}}{E_0 + F_2 + R_{N+1}} \tag{18-23b}$$

We can now plot point M. Equations (18-22) also show that there is a straight line between points M, E_N, and R_1. Since we can find R_1 on the saturated raffinate curve, we can draw the line $ME_N R_1$ to locate point E_N. Flow rates E_N and R_1 are determined by solving Eqs. (18-22a) and (18-22b) simultaneously.

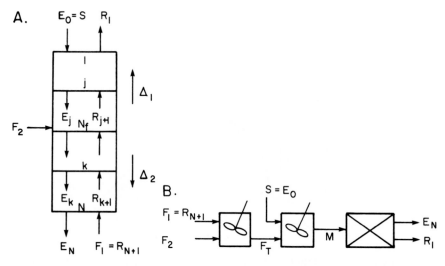

Figure 18-19. Two-feed extraction. (A) Equipment; (B) two-mixer/separator representation.

$$E_N = M\left[\frac{x_{A,M} - x_{A,1}}{y_{A,N} - x_{A,1}}\right] \tag{18-24a}$$

$$R_1 = M - E_N = M\left[\frac{y_{A,N} - x_{A,M}}{y_{A,N} - x_{A,1}}\right] \tag{18-24b}$$

An alternative method for doing the external mass balances is to represent the column as shown in Figure 18-19B (see also Problem 18-C5). Then the mass balances are

$$F_1 + F_2 = F_T \tag{18-25a}$$

$$F_T + S = M \tag{18-25b}$$

$$M = E_N + R_1 \tag{18-25c}$$

From these equations, points F_T, M, and finally E_N can be found. This construction is shown in Figure 18-20.

In the top of the column we can write a mass balance around stages 1 through j and then rearrange it to form a difference between passing streams,

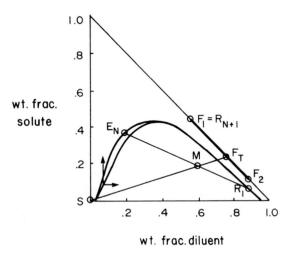

Figure 18-20. Construction for external balance for two-feed extractor.

$$R_{j+1} - E_j = R_1 - E_0 \equiv \Delta_1 \tag{18-26}$$

For the bottom of the column we write a mass balance around stages N through $k + 1$. Rearranging this as the difference between passing streams, we have

$$E_k - R_{k+1} = E_N - R_{N+1} \equiv \Delta_2 \tag{18-27}$$

The difference flows Δ_1 and Δ_2 have been defined as net flows out of the column. If $E_0 > R_1$, the flow Δ_1 will be negative.

The relation between Δ_1 and Δ_2 can be obtained from the external overall mass balance equation (18-22a). Rearranging this equation, we have

$$F_2 = (E_N - R_{N+1}) + (R_1 - E_0) \tag{18-28}$$

or

$$F_2 = \Delta_1 + \Delta_2 \tag{18-29}$$

In general, Δ_1 will not be equal to Δ_2; thus, two difference points are required.

Equations (18-22), (18-23), and (18-25) can all be considered as mixing equations. Thus we know that we can develop the three-point form of a straight line to show that points representing streams E_0, R_1, and Δ_1; E_N, R_{N+1}, and Δ_2; and F_2, Δ_1, and Δ_2 form three straight lines. This fact will prove useful in locating the two difference points.

To find the required number of stages, we must locate the two difference points. There are three methods we can use to find Δ_1 and Δ_2:

1. Use the coordinates of one of the Δ points. These coordinates can be determined as in Eqs. (18-13) and (18-14). The results are (see Problem 18-C6)

$$x_{A,\Delta_1} = \frac{R_1 x_{A,1} - E_0 y_{A,0}}{\Delta_1}, \quad x_{D,\Delta_1} = \frac{R_1 x_{D,1} - E_0 y_{D,0}}{\Delta_1} \qquad (18\text{-}30)$$

$$x_{A,\Delta_2} = \frac{E_N y_{A,N} - R_{N+1} x_{A,N+1}}{\Delta_2} \qquad (18\text{-}31a)$$

$$x_{D,\Delta_2} = \frac{E_N y_{D,N} - R_{N+1} x_{D,N+1}}{\Delta_2} \qquad (18\text{-}31b)$$

where flow rates Δ_1 and Δ_2 are given by Eqs. (18-26) and (18-27).

2. Use the lever-arm rule. The lever-arm rule is

$$\frac{R_1}{E_0} = \frac{\overline{\Delta_1 E_0}}{\overline{\Delta_1 R_1}} \qquad (18\text{-}32a)$$

Equation (18-32a) can be used to find Δ_1 on line $E_0 R_1$ if R_1/E_0 is known. Once Δ_1 has been found, Δ_2 is at the intersection of the line $\Delta_1 F_2$ [Eq. (18-29)] and $E_N R_{N+1}$ [Eq. (18-27)]. Alternatively, if R_{N+1}/E_N is known, Δ_2 can be found from

$$\frac{R_{N+1}}{E_N} = \frac{\overline{\Delta_2 E_N}}{\overline{\Delta_2 R_{N+1}}} \qquad (18\text{-}32b)$$

Now Δ_1 is at the intersection of lines $\Delta_2 F_2$ [Eq. (18-29)] and $E_0 R_1$ [Eq. (18-26)].

3. Use a graphical construction. The external mass balance, Eq. (18-22a), can be rearranged to

$$F_1 + F_2 = E_N + R_1 - E_0$$

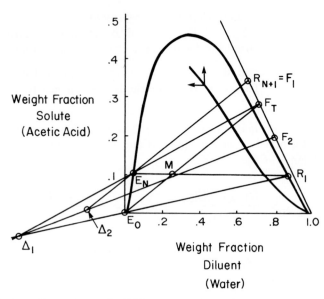

Figure 18-21.   Location of difference points for two-feed extractor, Example 18-3.

Replacing $F_1 + F_2$ with F_T and $R_1 - E_0$ with Δ_1 this is

$$F_T = E_N + \Delta_1 \qquad (18\text{-}33)$$

Thus there is a straight line connecting the points representing streams F_T, Δ_1, and E_N. Then Δ_1 is found at the intersection of the lines $E_0 R_1$ and $F_T E_N$. Once Δ_1 has been found, Δ_2 is located at the intersection of the lines $E_N R_{N+1}$ and $\Delta_1 F_2$. Difference points may be on opposite sides of the diagram.

The location of the difference points in a two-feed column is illustrated in Figure 18-21.

Once the difference points have been located, we can step off stages. This procedure again consists of alternating between equilibrium (represented by tie lines) and mass balances (represented by operating lines). For mass balances in the section above the feed stage for feed F_2, difference point Δ_1 is used. For mass balances in the column section below the feed stage for feed F_2, difference point Δ_2 is used. If the feed stage location N_f is specified in advance, we switch from one difference point to the other (or one operating line to the other) at the specified stage. If the optimum feed stage location is desired, the feed stage should be the one that results in a minimum total number of stages. When the tie line representing the feed stage intersects line $\Delta_1 F_2 \Delta_2$ (which represents the *feed line*), you have located the optimum feed stage. The truth of this statement is not readily evident, but some manipulation with the triangular diagrams will convince you.

The procedure for stepping off stages and determining the optimum feed plate location is illustrated in Example 18-3.

Example 18-3. Two-Feed Extractor

Pure isopropyl ether (S) at a flow rate of 1800 kg/hr is used to extract acetic acid (A) from water (D). Feed 1 (R_{N+1}) has a flow rate of 600 kg/hr and weight fractions $x_{A,N+1} = 0.35$ and $x_{D,N+1} = 0.65$. Feed 2 (F_2) has a flow rate of 400 kg/hr and weight fractions $x_{A,F_2} = 0.20$ and $x_{D,F_2} = 0.80$. We desire an outlet raffinate weight fraction of $x_{A_1} = 0.10$. Find the outlet extract composition, the optimum feed stage and the total number of equilibrium stages required.

Solution

A. Define. This system is the same as in Figure 18-19A. We wish to find N_f, N, $y_{A,N}$, and $y_{D,N}$.

B. Explore. Equilibrium data are given in Table 18-2. It appears that the two-mixer representation in Figure 18-19B can be used to find M and E_N. Then Δ_1 and Δ_2 can be found from case 1, 2, or 3.

C. Plan. Steps are as follows.

1. Plot equilibrium data.

2. Find F_T.

3. Find M and then E_N.

4. Find Δ_1 and then Δ_2 graphically (case 3).

5. Starting at R_1, step off stages using Δ_1 until a tie line intersects the feed line ($\Delta_1 \Delta_2 F_2$). This gives the optimum feed stage.

6. Finish stepping off stages using Δ_2.

D. Do It.

1. Equilibrium data are plotted in Figure 18-21.

2. $F_T = F_1 + F_2 = R_{N+1} + F_2 = 1000$ kg/hr.
F_T is on a straight line through R_{N+1} and F_2. From the lever-arm rule,

$$0.6 = \frac{F_1}{F_T} = \frac{\overline{F_2 F_T}}{\overline{F_1 F_2}}$$

or, modifying Eq. (18-4a),

$$x_{A,F_T} = \frac{x_{A,N+1}\, F_1 + x_{A,F_2}\, F_2}{F_1 + F_2} = 0.29$$

F_T can be plotted as shown in Figure 18-21.

3. $F_T + S = M = 2800$
M is on a straight line from S to F_T. From the lever-arm rule,

$$\frac{1000}{2800} = \frac{F_T}{M} = \frac{\overline{SM}}{\overline{F_T S}}$$

or from a modification of Eq. (18-4a),

$$x_{A,M} = \frac{S\, y_{A,S} + F_T\, x_{A,F_T}}{S + F_T} = 0.104$$

M is plotted in Figure 18-21. E_N is found by drawing line $R_1 M$ to the saturated extract curve. From the graph: $y_{A,N} = 0.1$, $y_{D,N} = 0.025$.

4. Δ_1 is at the intersection of lines $E_0 R_1$ and $F_T E_N$. This is shown in Figure 18-21. Δ_2 is at the intersection of lines $E_N R_{N+1}$ and $\Delta_1 F_2$. This is also shown in Figure 18-21.

5. Points Δ_1, Δ_2, E_N, F_T, F_1, F_2, R_1 are all shown in Figure 18-22 without the construction lines. Stages are stepped off starting at R_1. Δ_1 is used for the operating line. The tie line for stage 2 intersects the feed line $\Delta_1 \Delta_2 F_2$. Thus stage 2 is the optimum feed stage.

6. Δ_2 is used for the operating line above the feed. 3 stages provide more than enough separation.

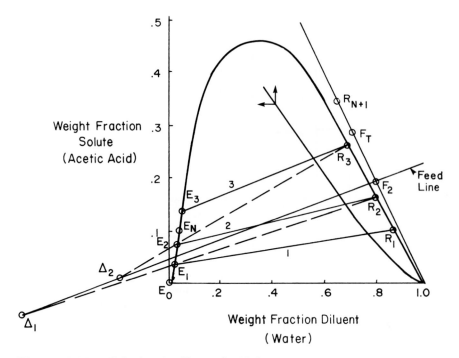

Figure 18-22.   Solution for Example 18-3.

E. Check. The location of M can also be found from Eqs. (18-23), which is probably easier. The location of the Δ points can be checked using the calculations of case 1 or 2.

F. Generalize. The procedure for finding the optimum feed stage is general and can also be used when there is extract reflux. This example illustrates that Eq. (18-4) can be generalized to any mixing operation. Using Eqs. (18-23) to find M and Eqs. (18-31) to find Δ_1 and Δ_2 is probably easier than the graphical construction illustrated here, and they are certainly easier to program. This problem can also be solved on a McCabe-Thiele diagram. The two difference points are used to generate two operating lines.

18.10. EXTRACT REFLUX

One problem with the simple countercurrent cascade and the two-feed system shown in Figure 18-19A is that the concentration of solute in the outlet extract stream is limited to a relatively low value because the outlet extract stream E_N is a passing stream with feed R_{N+1}. Thus the maximum concentration of A in E_N will occur when the column is

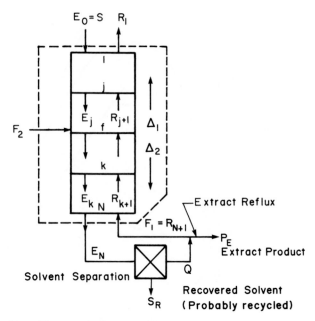

Figure 18-23.   Extract reflux system.

operated at minimum solvent rate. This limitation can often be over-come by using extract reflux.

The generalized idea of reflux is to take a product stream, change its phase, and return a portion of the stream to the same location in the column. This is done with reflux in distillation where the outlet vapor is condensed to a liquid and the liquid phase is returned to the column. It is also used with the boilup in distillation where the bottoms liquid is boiled and the vapor is returned to the column. To use this idea in extraction, part of the exiting extract stream is changed into a raffinate phase by removing solvent, and this raffinate is returned to the column. The solvent is usually removed by distillation, although other separa-tions such as flash distillation, reverse osmosis, ultrafiltration, or another extraction column could be used. A system with extract reflux is shown in Figure 18-23. Note that if we cut this diagram at the dashed line it looks like the two-feed column shown in Figure 18-19A. Thus we should expect that the analysis procedure for extract reflux would be similar to that for a two-feed column. Extract reflux is used most often with type II systems, where solute A is only partially miscible with the extracting solvent.

The solvent removal system shown in Figure 18-23 does not neces-sarily produce pure solvent or a saturated raffinate. The raffinate

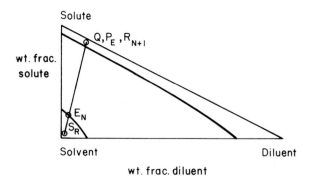

Figure 18-24. Solvent removal to produce extract reflux.

stream produced, Q, is then split into the reflux, R_{N+1}, and the product, P_E. These streams have the same composition. Mass balances around the solvent separator are

$$E_N = Q + S_R \qquad (18\text{-}34a)$$

$$E_N y_{A_N} = Q x_{A,N+1} + S_R y_{A,S_R} \qquad (18\text{-}34b)$$

$$E_N y_{D_N} = Q x_{D,N+1} + S_R y_{D,S_R} \qquad (18\text{-}34c)$$

Usually flow rates E_N, Q, S_R and compositions $y_{A,N}$ and $y_{D,N}$ are unknown. We do know that stream E_N is a saturated extract; thus, $y_{A,N}$ and $y_{D,N}$ are on the saturated extract curve represented by Eq. (18-17b). Equations (18-34) and (18-17b) are a series of four equations with five unknowns. If we divide Eqs. (18-34) by S_R, we can consider them a set of four equations with the four unknowns E_N/S_R, Q/S_R, $y_{A,N}$, and $y_{D,N}$. Graphically, this set of equations can be solved as shown in Figure 18-24. The ratios E_N/S_R and Q/S_R can be determined from the lever-arm rule.

The procedure for stage-by-stage calculations for extract reflux is similar to the procedure already developed for a two-feed column. However, the exact method used to find the mixing points and difference points depends upon what variables are specified in the problem statement. Once the difference points have been found, the procedure for stepping off stages and determining the optimum feed stage is the same as for the two-feed column.

The simplest extract reflux problem results when the specified variables are F_2, x_{A,F_2}, x_{D,F_2}, $y_{A,0}$, $y_{D,0}$, $x_{A,1}$, R_{N+1}, $x_{A,N+1}$, $x_{D,N+1}$, y_{A,S_R},

A

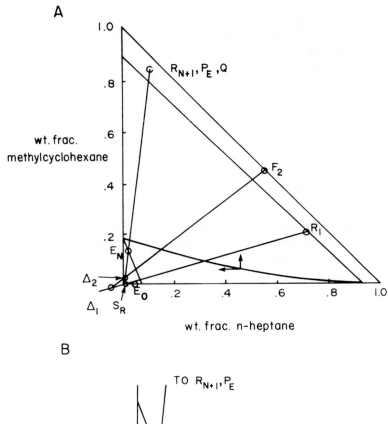

wt. frac.
methylcyclohexane

wt. frac. n-heptane

B

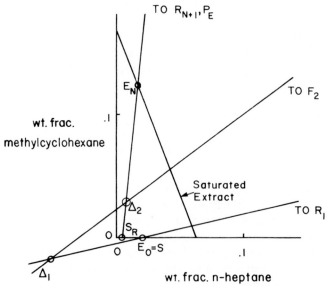

wt. frac.
methylcyclohexane

wt. frac. n-heptane

Figure 18-25.   Solution for Example 18-4.   (A) Entire diagram; (B)
expansion of solvent vertex.

and y_{D,S_R}, that is, when the feed, solvent, recovered solvents, reflux, and exiting raffinate are specified. The concentrations of the recovered solvent will depend upon the separation scheme used for solvent recovery. With this set of variables, the procedure for finding the mixing and difference points is straightforward.

1. Plot locations of F_2, E_0, $F_1 = R_{N+1}$, R_1, and S_R.

2. Locate E_N using construction shown in Figure 18-24.

3. Solve the external mass balances around the column (dashed line in Figure 18-23). These mass balances were given in Eqs. (18-22). The three unknowns are flow rates E_0, R_1, and E_N.

4. The difference points are defined by Eqs. (18-26) and (18-27) and can now be found in the same way as for a two-feed column. Since all flow rates are now known, this is easily done using Eqs. (18-30) and (18-31). The graphical methods used for two-feed columns can also be used.

The construction is illustrated in Figures 18-25A and B. Usually the two Δ points will be on the same side of the diagram as illustrated in these figures. Obviously, the figures are crowded, and care must be exercised to avoid mistakes.

This set of specifications with flow rates F_2 and R_{N+1} (or equivalently the ratio R_{N+1}/F_2) given is the most convenient set for solution. A similar procedure can be used if solvent flow rate $E_0 = S$ is given but outlet raffinate composition is unknown. This technique is also useful if the concentrations of recovered solvent, y_{A,S_R} and y_{D,S_R}, are unknown but E_0 and $x_{A,1}$ are given in addition to the other variables.

18.10.1. Extract Reflux Ratio

Although specifying the flow rate R_{N+1} or ratio R_{N+1}/F_2 is convenient for calculations, it may not be the most convenient specification for design or operation. We often wish to specify the external reflux ratio, R_{N+1}/P_E, or the internal reflux ratio, R_{N+1}/E_N. The most common set of specifications would be

$$F_2, x_{A,F_2}, x_{D,F_2}, y_{A,0}, y_{D,0}, x_{A,1}, x_{A,N+1}, x_{D,N+1}, y_{A,S_R}, y_{D,S_R}, \text{ and } R_{N+1}/P_E$$

This means the feed, solvent, recovered solvent, reflux and exit raffinate concentrations, and external reflux ratio are specified. We could also specify the above set of variables and substitute R_{N+1}/E_N for R_{N+1}/P_E.

First, do steps 1 and 2 of the procedure just outlined. Now notice that step 3, where you find E_0, R_1, and E_N, cannot be accomplished. Thus the external mass balances and the difference points cannot be found by this procedure. Fortunately, there is a simple alternative.

When the internal reflux ratio, R_{N+1}/E_N, is specified, it is easy to determine the coordinates of Δ_2. Rearranging Eqs. (18-31) we obtain

$$x_{A,\Delta_2} = \frac{y_{A,N} - x_{A,N+1}\, R_{N+1}/E_N}{1 - R_{N+1}/E_N} \qquad (18\text{-}35a)$$

$$x_{D,\Delta_2} = \frac{y_{D,N} - x_{D,N+1}\, R_{N+1}/E_N}{1 - R_{N+1}/E_N} \qquad (18\text{-}35b)$$

We can now plot point Δ_2 and graphically find Δ_1 at the intersection of straight lines $R_1\,E_0$ and $F_2\,\Delta_2$. Alternatively, we can write the mass balances around the top of the column and through stage k.

$$F_2 + E_0 = R_1 + \Delta_2 \qquad (18\text{-}36a)$$

$$F_2\, x_{A,F_2} + E_0\, y_{A,0} = R_1\, x_{A,1} + \Delta_2\, x_{A,\Delta_2} \qquad (18\text{-}36b)$$

$$F_2\, x_{D,F_2} + E_0\, y_{D,0} = R_1\, x_{D,1} + \Delta_2\, x_{D,\Delta_2} \qquad (18\text{-}36c)$$

Solve for the three unknowns E_0, R_1, and Δ_2. Then use Eqs. (18-30) to calculate the coordinates of Δ_1. Stages can now be stepped off.

Example 18-4. Extract Reflux

Methylcyclohexane is being extracted from n-heptane using aniline as the solvent. Equilibrium data obtained by Varteressian and Fenske (1937) are given in Table 18-3. Feed F_2 is 100 kg/hr of a 45 wt % methylcyclohexane, 55 wt % n-heptane mixture. A raffinate that is 20 wt % methylcyclohexane is desired. The solvent used is 98 wt % aniline and 2 wt % n-heptane. The recovered solvent from the solvent separator is 99.7 wt % aniline and 0.3 wt % n-heptane. The extract product stream, P_E, is 85 wt % methylcyclohexane and 10 wt % n-heptane. The internal reflux ratio, R_{N+1}/E_N, is 0.115. Find the locations of Δ_1 and Δ_2 and the flow rates E_0, R_{N+1}, R_1, and E_N.

Table 18-3. Equilibrium Data for Methylcyclohexane, n-Heptane-Aniline

| Hydrocarbon Phase, wt % | | Aniline-Rich Phase, wt % | |
| Methylcyclohexane | n-Heptane | Methylcyclohexane | n-Heptane |
x_A	x_D	y_A	y_D
0.0	92.6	0.0	6.2
9.2	83.1	0.8	6.0
18.6	73.4	2.7	5.3
22.0	69.8	3.0	5.1
33.8	57.6	4.6	4.5
40.9	50.4	6.0	4.0
46.0	45.0	7.4	3.6
59.0	30.7	9.2	2.8
67.2	22.8	11.3	2.1
71.6	18.2	12.7	1.6
73.6	16.0	13.1	1.4
83.3	5.4	15.6	0.6
88.1	0.0	16.9	0.0

Source: Varteressian and Fenske (1937).

Solution

A. Define. A sketch of the system is shown.

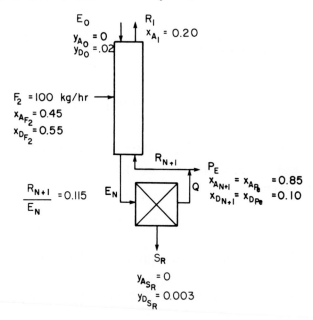

B. Explore. This is an extract reflux problem with the internal extract reflux ratio given. Thus follow the procedure outlined in the text.

C. Plan. We can proceed as follows.

1. Plot the equilibrium data.

2. Plot known points: E_0, F_2, S_R, R_{N+1} (P_E, Q), R_1 (saturated raffinate).

3. Find the location of E_N from mass balances around the solvent separator (see Figure 18-24). E_N is a saturated extract steam.

4. Find the coordinates of Δ_2 from Eq. (18-35).

5. Find Δ_1 at the intersection of lines R_1E_0 and $F_2\Delta_2$ or by solving Eqs. (18-36) and using Eq. (18-30).

6. Use mass balances to find flow rates.

D. Do It. Steps 1 to 3 are straightforward (see Figures 18-25A and B). Point E_N is at $y_{A,N} = 0.122$, $y_{D,N} = 0.019$. To find Δ_2 use Eq. (18-35):

$$x_{A,\Delta_2} = \frac{y_{A,N} - \dfrac{R_{N+1}}{E_N} x_{A,N+1}}{1 - R_{N+1}/E_N} = \frac{0.122 - (0.115)(0.85)}{1 - 0.115} = 0.0274$$

$$x_{D,\Delta_2} = \frac{0.019 - (0.1)(0.115)}{1 - 0.115} = 0.00847$$

How to locate Δ_1 is shown in Figures 18-25A and B.

To find the flow rates we use the mass balances. Equations (18-36) become

$$100 + E_0 = R_1 + \Delta_2$$

$$(100)(0.45) + E_0(0) = R_1(0.20) + \Delta_2(0.0274)$$

$$(100)(0.55) + E_0(0.02) = R_1(0.715) + \Delta_2(0.00847)$$

Solving simultaneously, we obtain $E_0 = 960.76$, $R_1 = 92.33$, and

$\Delta_2 = 968.43$ kg/hr. To find E_N and R_{N+1}, solve

$$968.43 = \Delta_2 = E_N - R_{N+1}$$

$$R_{N+1}/E_N = 0.115$$

simultaneously. The result is $R_{N+1} = 125.84$ and $E_N = 1094.27$ kg/hr.

E. Check. The flow rates and compositions do satisfy the external mass balances. In addition, the points are all consistent when we calculate them using other methods. For example, Eq. (18-30a) can be used to calculate x_{A,Δ_1}.

$$\Delta_1 = R_1 - E_0 = 92.33 - 960.76 = -868.43$$

$$x_{A,\Delta_1} = \frac{R_1 x_{A,1} - E_0 y_{A_0}}{\Delta_1} = \frac{(92.33)(0.20) - (960.76)(0)}{-868.43} = -0.021$$

which agrees with the graphical construction.

F. Generalize. A couple of comments about this example are in order. First, we note that the solvent vertex becomes quite crowded. This affects the accuracy of our answer. Some relief may be obtained by plotting weight fraction solute versus weight fraction solvent, but the solvent vertex will still be crowded. We could avoid this difficulty by using mass-ratio diagrams (Wankat, 1982). Calculation of the number of stages is similar to Figure 18-22 but is not shown in Figure 18-25.

Example 18-4 showed the situation where the internal reflux ratio was specified. If the external reflux ratio, R_{N+1}/P_E, is specified instead of R_{N+1}/E_N, some additional manipulations are necessary to determine the locations of the difference points. This is the subject of Problem 18-C10 and 18-D12.

18.11. LEACHING

Leaching (solid-liquid extraction) and liquid-liquid extraction will have very different hydrodynamic and mass transfer characteristics. However, equilibrium staged analysis is almost identical for the two proesses

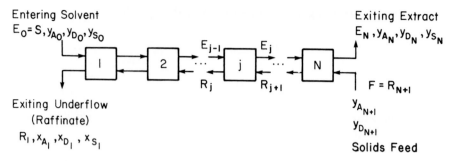

Figure 18-26.   Countercurrent leaching nomenclature.

because it is not affected by these differences. In this section we will briefly consider the analysis of leaching systems using triangular diagrams. In leaching the flow rates will generally not be constant. The variation in flow rates can be included by doing the analysis on a triangular diagram. A very similar technique uses ratio units in a Ponchon-Savarit diagram (Prabhudesai, 1979; McCabe and Smith, 1985; Treybal, 1980).

A schematic of a countercurrent leaching system is shown in Figure 18-26. Since leaching (liquid-solid extraction) is quite similiar to liquid-liquid extraction, the same nomenclature is used (see Table 16-2). Even if flow rates E and R vary, it is easy to show that the differences in total and component flow rates for passing streams are constant. Thus we can define the difference point from these differences. This was done in Eqs. (18-13) and (18-14) for the liquid-liquid extraction cascade of Figure 18-9. Since the cascades are the same (compare Figures 18-9 and 18-26), the results for leaching are the same as for liquid-liquid extraction.

The calculation procedure for countercurrent leaching operations is exactly the same as for liquid-liquid extraction:

1.   Plot the equilibrium data.

2.   Plot the locations of known points.

3.   Find mixing point M.

4.   Locate E_N.

5.   Find the Δ point.

6.   Step off stages.

This procedure is illustrated in Example 18-5.

Table 18-4. Test Data for Extraction of Oil from Meal with Benzene

Mass Fraction Oil (Solute) in Solution	Mass Fraction Underflow (Raffinate)		
y_A	x_A	x_D	x_S
0	0	0.670	0.333
0.1	0.0336	0.664	0.302
0.2	0.0682	0.660	0.272
0.3	0.1039	0.6541	0.242
0.4	0.1419	0.6451	0.213
0.5	0.1817	0.6366	0.1817
0.6	0.0224	0.6268	0.1492
0.7	0.268	0.6172	0.1148

Source: Prabhudesai (1979).

The equilibrium data for leaching must be obtained experimentally since it will depend on the exact nature of the solids, which may change from source to source. If there is no entrainment, the overflow (extract) stream will often contain no inert solids (diluent). However, the raffinate stream will contain solvent. Test data for the extraction of oil from meal with benzene are given in Table 18-4 (Prabhudesai, 1979). In these data, inert solids are not extracted into the benzene. The data are plotted on a triangular diagram in Figure 18-27 (see Example 18-5). The conjugate line is constructed in the same way as for extraction.

Example 18-5. Leaching Calculations on a Triangular Diagram

We wish to treat 1000 kg/hr (wet basis) of meal that contains 0.20 wt frac oil and no benzene. The inlet solvent is pure benzene and flows at 662 kg/hr. We desire an underflow product that is 0.04 wt frac oil. Temperature and pressure are constant, and the equilibrium data are given in Table 18-4. Find the outlet extract concentration and the number of equilibrium stages needed in a countercurrent leaching system.

Solution

A. Define. The system is similiar to that of Figure 18-26 with streams E_0 and R_{N+1} specified. In addition, weight fraction $x_{A,1}$ = 0.04 and is a saturated raffinate. We wish to find the composition of stream E_N and the number of equilibrium stages required.

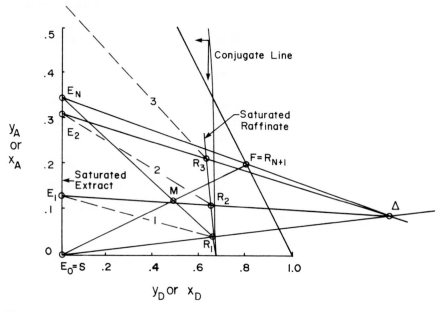

Figure 18-27.   Solution to leaching problem, Example 18-5.

B. Explore. This looks like a straightforward leaching problem, which can be solved like the corresponding extraction problem.

C. Plan. We will plot the equilibrium diagram on a scale that allows the Δ point to fit on the graph. Then we will plot points E_0, R_{N+1}, and R_1. We will use the lever-arm rule to find point M and then find point E_N. Next we find Δ and finally we step off the stages.

D. Do It. The diagram in Figure 18-27 shows the equilibrium data and the points that have been plotted. Point M was found along the line $E_0 R_{N+1}$ from Eq. (18-12).

$$x_{A,M} = \frac{E_0 y_{A,0} + F x_{A,N+1}}{E_0 + F} = \frac{0 + (1000)(0.2)}{662 + 1000} = 0.120$$

Then points E_N and Δ were found as shown in Figure 18-27. Finally, stages were stepped off in exactly the same way as for a triangular diagram for extraction. The exit extract concentration is 0.305 wt frac oil, and 3 stages are more than enough. 2 stages are not quite enough. Approximately 2.1 stages are needed.

E. Check. The outlet extract concentration can be checked with an overall mass balance. The number of stages could be checked by solving the problem with another method.

F. Generalize. Since the leaching example was quite similiar to liquid-liquid extraction, we might guess that the other calculation procedures developed for extraction would also be valid for leaching. Since this is true, there is little reason to reinvent the wheel and rederive all the methods. Some of these applications are explored in Problems 18-C8, D14, D15, and D16.

McCabe-Thiele diagrams (Chapt. 17) can also be used for leaching, since flow rates are often close to constant. If flow rates are not constant, curved operating lines can be constructed on the McCabe-Thiele diagram with the triangular diagram.

18.12. SUMMARY - OBJECTIVES

In this chapter, methods for ternary partially miscible extraction systems are explored. At the end of this chapter you should be able to satisfy the following objectives:

1. Plot extraction equilibria on a triangular diagram. Find the saturated extract, saturated raffinate, and conjugate lines.

2. Find the mixing point and solve single-stage and cross-flow extraction problems.

3. For countercurrent systems, do the external mass balances, find the difference points, and step off the equilibrium stages.

4. Use the difference points to plot the operating line(s) on a McCabe-Thiele diagram.

5. Set up a computer or programmable calculator program for ternary extraction.

6. Solve problems that involve simulation, two feeds, or extract reflux.

7. Apply the methods to leaching problems.

REFERENCES

Alders, L. *Liquid-Liquid Extraction,* 2nd ed., Elsevier, Amsterdam, 1959.

624

Ashton, N.F., C. McDermott, and A. Brench, "Chemistry of Extraction of Nonreacting Solutes," in T.C. Lo, M.H.I. Baird, and C. Hanson (Eds.), *Handbook of Solvent Extraction,* Wiley, New York, 1983, Chapt. 1.

Fenske, M.R., R.H. McCormick, H. Lawroski, and R.G. Geier, "Extraction of Petroleum Fractions by Ammonia Solvents," *AIChE J.,* 1, 335 (1955).

King, C.J., *Separation Processes,* 2nd ed., McGraw-Hill, New York, 1981.

Laddha, G.S. and T.E. Degaleesan, *Transport Phenomena in Liquid Extraction,* McGraw-Hill, New York, 1978.

McCabe, W.L., J.C. Smith, and P. Harriott, *Unit Operations of Chemical Engineering,* 4th ed., McGraw-Hill, New York, 1985.

Perry, R.H. and D. Green (Eds.), *Perry's Chemical Engineer's Handbook,* 6th ed., McGraw-Hill, New York, 1984.

Prabhudesai, R.K., "Leaching," in P.A. Schweitzer (Ed.), *Handbook of Separation Techniques for Chemical Engineers,* McGraw-Hill, New York, 1979, Section 5.1.

Smith, B.D., *Design of Equilibrium Stage Processes,* McGraw-Hill, New York, 1963.

Treybal, R.E., *Mass Transfer Operations,* 3rd ed., McGraw-Hill, New York, 1980.

Varteressian, K.A. and M.R. Fenske, "The System Methylcyclohexane - Aniline - N-Heptane," *Ind. Eng. Chem., 29,* 270 (1937).

Wankat, P.C., "Advanced Graphical Extraction Calculations," in J.M. Calo and E.J. Henley (Eds.), *Stagewise and Mass Transfer Operations,* Vol. 3, *Extraction and Leaching,* AIChEMI, Series B, AICHE, New York, 1982, p. 17.

HOMEWORK

A.   *Discussion Problems*

A1.   Answer the following questions without referring to the text.

    a.   Define solubility envelope.

    b.   Define plait point.

    c.   Define extract.

    d.   Explain the difference between type I and type II systems.

A2.   Construct a conjugate line on Figure 18-3 at $10\,^\circ$ C.

A3.   When $E_0 > R_1$ the difference point is on the left side of the diagram; and when $R_1 > E_0$ the difference point is on the right of the diagram. What happens when $R_1 = E_0$? Answer this question using a logical argument.

A4.   For Figure 18-12 suppose we desired to obtain the desired raffinate concentration with exactly two equilibrium stages. This can be accomplished by changing the amount of solvent used. Would we want to increase or decrease the amount of solvent? Explain the effect this change will have on M, E_N, Δ, and the number of stages required.

A5.   You have the mixer in Figure 18-4A and know F_1 and M and the values x_{A,F_1}, x_{D,F_1}, $x_{A,M}$ and $x_{D,M}$. Explain how you would use the lever-arm rule and a triangular diagram to find x_{A,F_2} and x_{D,F_2}.

A6.   All the calculations in this chapter use right triangle diagrams. What differences would result if equilateral triangle diagrams were used to solve the extraction problem?

A7.   How will the diagram change if we switch axes and plot weight fraction solvent versus weight fraction diluent?

A8.   If the extract and raffinate phases are totally immiscible, can extraction problems still be solved using triangular diagrams? Explain how, and describe what the equilibrium diagram will look like.

A9.   What can be done to increase the removal of solute (i.e., decrease $x_{A,1}$)?

A10. What can be done to increase the concentration of solute in the extract (i.e., increase $y_{A,N}$)?

A11. What situation in extraction is analogous to minimum reflux in distillation?

A12. What situation in extraction is superficially analogous to total reflux in distillation? How does it differ?

A13. Write your key relations chart for this chapter.

A14. Study Figure 18-17. Explain what happens as S/F increases. What happens to M? What happens to E_N? What happens to Δ? How do you find Δ_{min} if it lies on the left-hand side? How do you find Δ_{min} if it lies on the right-hand side?

A15. Explain why extract reflux has more potential for increasing separation in type II systems than in type I systems.

A16. Explain why raffinate reflux is not usually helpful.

A17. Show how to use the lever-arm rule to find the location of stream Q in Figure 18-24 if E_N, $y_{A,N}$, S_R, y_{A,S_R}, and y_{D,S_R} are given.

A18. What happens to the required number of stages in a problem with extract reflux if F_2 is doubled but all concentrations and the ratios E_0/F_2 and R_{N+1}/F_2 remain unchanged?

A19. Can the calculations for extract reflux be done on an equilateral triangular diagram? If so, how will it differ from a right triangular diagram? If not, why not?

A20. Compare the minimum extract reflux ratio to the minimum reflux ratio in distillation. How are they similar, and how do they differ?

A21. The exact and relative locations of points on Figures 18-25A and B may or may not be significant. This is explored in the following questions.

    a. Must S_R be on the solute axis?

    b. Must $E_0 = S$ be on the solute axis?

    c. Must F_2 be on the hypotenuse?

    d. Must R_1 be on the saturated raffinate curve?

    e. Must E_N be on the saturated extract curve?

    f. Must Δ_2 lie between Δ_1 and F_2?

    g. Must Δ_2 lie between E_N and S_R?

    h. Could E_0 and S_R be the same composition?

A22. Three analysis procedures were developed for extraction in Chapters 16 and 18: McCabe-Thiele, Kremser, and triangular diagrams. If you have an extraction problem, how do you decide which method to use? (In other words, explain when each is applicable.)

A23. Would you expect stage efficiencies to be higher or lower in leaching than in liquid-liquid extraction? Explain.

C. *Derivations*

C1. Derive Eqs. (18-8).

C2. Prove that the locations of streams M, E_N, and R_1 in Figure 18-10A lie on a straight line as shown in Figure 18-10B.

C3. Define Δ and the coordinates of Δ from Eqs. (18-11) and (18-12). Prove that points Δ, E_j, and R_{j+1} (passing streams) lie on a straight line by developing the three-point form of a straight line. While doing this, prove that the lever-arm rule is valid. Then derive Eq. (18-20).

C4. Derive Eq. (18-19).

C5. The mixing diagram shown in Figure 18-19B for the external mass balances assumed that F_1 and F_2 were mixed first. However, we could also draw the column as shown.

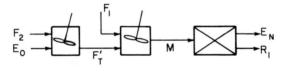

Write the balances for this arrangement. Plot the appropriate mixing points on a right triangular diagram for Example 18-3. Show that M is the same point. What is the relationship between F_T' and the difference points. Derive an equation that relates F_T' to one of the difference points. Show that Δ_1 and Δ_2 are the same as in Example 18-3.

C6. Derive Eqs. (18-28) and (18-29).

C7. Derive a form of Eqs. (18-29) that can be used to find Δ_2 if the ratio R_{N+1}/E_N is known but individual flow rates are not known.

C8. Develop the procedures for single-stage and cross-flow systems for leaching.

C9. Derive Eq. (18-9).

C10. Extract reflux problems when the specified variables are F_2, x_{A,F_2}, x_{D,F_2}, $y_{A,0}$, $y_{D,0}$, $x_{A,1}$, $x_{A,N+1}$, $x_{D,N+1}$, y_{A,S_R}, y_{D,S_R}, and R_{N+1}/P_E are more difficult than for other cases. Briefly, one proceeds as follows: Find the ratio E_N/S_R from the solution of Eqs. (18-34) and (18-17b) or from the lever-arm rule using the construction shown in Figure 18-26. Then, in the balance around the solvent recovery unit, $E_N = S_R + R_{N+1} + P_E$, let $E_N = (E_N/S_R) S_R$ and solve for S_R/P_E. Then determine R_{N+1}/S_R from the balance equation. Now divide the numerator and denominator of Eqs. (18-31) by S_R and calculate the coordinates of Δ_2. With the location of Δ_2 known, Δ_1 is at the intersection of straight lines $R_1 E_0$ and $\Delta_2 F$. Flow rates can now be found from mass balances. Develop this procedure in detail.

D. *Problems*

D1. The equilibrium data for extraction of methylcyclohexane (A) from n-heptane (D) into aniline (S) is given in Table 18-3. We have 100 kg/hr of a feed that is 60% methylcyclohexane and 40% n-heptane and 50 kg/hr of a feed that is 20% methylcyclohexane and 80% n-heptane. These two feeds are mixed with 200 kg/hr of pure aniline in a single equilibrium stage.

    a.   What are the extract and raffinate compositions leaving the stage?

    b.   What is the flow rate of the extract product?

D2. The equilibrium data for acetic acid-water-isopropyl ether are given in Table 18-2. In Example 18-2 the feed was 1000 kg/hr of a solution containing 35 wt % acid and 65 wt % water. Pure isopropyl ether is used. We desire an exit raffinate that contains 10 wt % acetic acid. If an existing apparatus has 184 equilibrium stages, estimate the exiting extract concentration and the solvent flow rate. Operation is at 20 °C and 1 atm.

D3. A feed containing 38 wt % acetone and 62 wt % water is to be sent at 250 kg/hr to an extraction column to recover the acetone using 1,1,2-trichloroethane $(C_2H_3Cl_3)$ as the solvent. 85.5 kg/hr of pure $C_2H_3Cl_3$ is used as the solvent stream. We desire a raffinate product that contains 8% acetone. Equilibrium data are given by McCabe and Smith (1976) and are reproduced in Tables 18-5 and 18-6. Solve this problem on a right triangular diagram plotting weight fraction $C_2H_3Cl_3$ (solvent) as ordinate and weight fraction acetone (solute) as the abscissa. There are two reasons for asking you to change the axes: (1) Solutions often appear like this in the literature and (2) it serves as a check that you understand the material and are not merely memorizing the features of the diagram.

Table 18-5. Equilibrium Data - Limiting Solubility Curve

$C_2H_3Cl_3$, wt %	Water, wt %	Acetone, wt %
94.73	0.26	5.01
79.58	0.76	19.66
67.52	1.44	31.04
54.88	2.98	42.14
38.31	6.84	54.85
24.04	15.37	60.59
15.39	26.28	58.33
6.77	41.35	51.88
1.75	61.11	37.17
0.92	74.54	24.54
0.65	87.63	11.72
0.44	99.56	0.00

Source: McCabe and Smith (1976).

Table 18-6. Equilibrium Data - Tie Lines

Weight % in Water Layer			Weight % in Trichloroethane Layer		
$C_2H_3Cl_3$	Water	Acetone	$C_2H_3Cl_3$	Water	Acetone
0.52	93.52	5.96	90.93	0.32	8.75
0.73	82.23	17.04	73.76	1.10	25.14
1.02	72.06	26.92	59.21	2.27	38.52
1.17	67.95	30.88	53.92	3.11	42.97
1.60	62.67	35.73	47.53	4.26	48.21
2.10	57.00	40.90	40.00	6.05	53.95
3.75	50.20	46.05	33.70	8.90	57.40
6.52	41.70	51.78	26.26	13.40	60.34

Source: McCabe and Smith (1976).

    a. Plot the solubility envelope and the conjugate line for this system.

    b. Find the exiting extract concentration and flow rate.

    c. Find the number of equilibrium stages required for this separation. Since the direction stages are counted is arbitrarily, start at the exiting extract stream.

D4. The equilibrium data for the system water-acetic acid-isopropyl ether are given in Table 18-2. We have a feed that is 30 wt % acetic acid and 70 wt % water. Feed flow rate is 150 kg/hr. A single equilibrium stage will be used. The feed is to be treated with pure isopropyl ether.

    a. If the solvent rate is 100 kg/hr, find the outlet extract and raffinate compositions.

    b. If a raffinate composition of $x_A = 0.1$ is desired, how much solvent is needed?

D5.   A countercurrent system is to be used for the water-acetic acid-isopropyl ether extraction (see Table 18-2). Feed is 39 wt % acetic acid and 59 wt % water. Feed flow rate is 1000 kg/hr. Solvent added contains 1 wt % acetic acid but no water. Total flow rate of added solvent is 1500 kg/hr. We desire a raffinate that is 7 wt % acetic acid. What are the weight fractions of E_N and R_1? What are the flow rates of E_N and R_1? How many stages are required?

D6.   A countercurrent system with three equilibrium stages is to be used for water-acetic acid-isopropyl ether extraction (see Table 18-2). Feed is 40 wt % acetic acid and 60 wt % water. Feed flow rate is 2000 kg/hr. Solvent added contains 1 wt % acetic acid but no water. We desire a raffinate that is 5 wt % acetic acid. What solvent flow rate is required? What are the weight fractions of E_N and R_1? What are the flow rates of E_N and R_1?

D7.   A feed that is 40 wt % acetic acid and 60 wt % water is fed to a three-stage cross-flow cascade. Feed flow rate is 2000 kg/hr. 2500 kg/hr of solvent containing 99 wt % isopropyl ether and 1% acetic acid is added to each stage. Operation is at 20°C and 1 atm. Equilibrium data are in Table 18-2.

    a. Determine the weight fractions of the outlet raffinate and the three outlet extract streams.

    b. Determine the flow rate of the outlet raffinate.

D8.   We wish to remove acetic acid from water using isopropyl ether as solvent. The operation is at 20°C and 1 atm (see Table 18-2). The feed is 0.45 wt frac acetic acid and 0.55 wt frac water. Feed flow rate is 2000 kg/hr. A countercurrent system is used. Pure solvent (no acetic acid and no water) is used. We desire an extract stream that is 0.20 wt frac acetic acid and a raffinate that is 0.20 wt frac acetic acid.

    a. How much solvent is required?

    b. How many equilibrium stages are needed?

D9.   We are extracting acetic acid from water with isopropyl ether at 20°C and 1 atm pressure. Equilibrium data are in Table 18-2. The column has three equilibrium stages. The entering feed rate is 1000 kg/hr. The feed is 40 wt % acetic acid and 60 wt %

water. The exit ⌐ extract stream has a flow rate of 2500 kg/hr and is 20 wt % acetic acid. The entering extract stream (which is *not* pure isopropyl ether) contains no water. Find:

a. The exit raffinate concentration.

b. The required entering extract stream concentration.

c. Flow rates of exiting raffinate and entering extract streams.

Trial and error is *not* needed.

D10. 1000 kg/hr of a feed containing 48 wt % acetone, 50 wt % water, and 2 wt % $C_2H_3Cl_3$ is to be sent to an extraction column to recover the acetone using 1,1,2-trichloroethane ($C_2H_3Cl_3$) as the solvent. 1424 kg/hr of 98 wt % $C_2H_3Cl_3$ and 2 wt % acetone is used as the solvent stream. Two equilibrium stages are to be used. Data are given in Tables 18-5 and 18-6. Solve this problem on a right triangular diagram plotting weight fraction $C_2H_3Cl_3$ (solvent) as ordinate and weight fraction acetone (solute) as the abscissa. There are two reasons for asking you to change axes: (1) Solutions often appear like this in the literature and (2) it serves as a check that you understand the material and are not merely memorizing the general features of the diagrams. Find exiting extract and raffinate concentrations and flow rates. Do this problem carefully and accurately.

D11. 200 kg/hr of a feed containing 60 wt % acetone and 40 wt % water and 200 kg/hr of a second feed containing 20 wt % acetone and 80 wt % water are sent to an extraction column to recover acetone using $C_2H_3Cl_3$ as solvent. Pure solvent at a flow of 124.25 kg/hr is used. Outlet raffinate composition of 5 wt % acetone is desired. Find the optimum location for the feed plate and the total number of stages required. Plot weight fraction $C_2H_3Cl_3$ as ordinate and weight fraction acetone as abscissa. Data are given in Tables 18-5 and 18-6.

D12. Methylcyclohexane is being extracted from n-heptane using aniline as the solvent. Equilibrium data are given in Table 18-3. We wish to use a system with extract reflux. Feed is 500 kg/hr of a 45 wt % n-heptane and 55 wt % methylcyclohexane mixture. A raffinate that is 16 wt % methylcyclohexane is desired. The solvent recovered from the solvent separator is 2 wt % methylcyclohexane and 1 wt % n-heptane. Stream E_0 is pure aniline. The extract product stream, P_E, is 80 wt % methylcyclohexane and 18 wt % n-heptane. The external reflux ratio, R_{N+1}/P_E, is 14.89. Find the difference points. Use results of Problem 18-C10.

D13. A simple countercurrent extraction is removing methylcyclohexane from n-heptane using aniline as the solvent. The feed is 60 wt % methylcyclohexane and 40 wt % n-heptane. The outlet raffinate is 60 wt % n-heptane and has a flow rate of $R_1 = 1000$ kg/hr. The inlet solvent is 5 wt % n-heptane and contains no methylcyclohexane. The inlet solvent flow rate is $E_0 = 5000$ kg/hr. Assume that all stages are equilibrium stages. Equilibrium data are in Table 18-3.

    a. Determine the composition of outlet extract stream E_N.

    b. What is the flow rate of feed required?

D14. Repeat Example 18-5 except for a single-stage system and unknown underflow product concentration.

D15. Repeat Example 18-5 except for a three-stage countercurrent system and unknown underflow product concentration.

D16. Repeat Example 18-5 except for a three-stage cross-flow system, with pure solvent at the rate of 421 kg/hr added to each stage and unknown underflow product concentration.

E. *More Complex Problems*

E1. Repeat Example 18-2, but do it with a calculator or computer instead of graphically. Note that the equilibrium data for the saturated raffinate and extract curves, Eqs. (18-17a,b) can be estimated as straight lines over the concentration range of interest. The tie lines, Eq. (18-18), can be estimated as a quadratic function (see y_A vs x_A diagram, Figure 18-16). Since Eqs. (18-17a,b) are linear, the simultaneous solution of Eqs. (18-19) and (18-17b) and the simultaneous solution of Eqs. (18-20) and (18-17a) are straightforward.

E2. An extraction column with extract reflux (see Figure 18-23) is recovering methylcyclopentane (A) from n-hexane (D) with aniline (S). Feed $F_2 = 100$, $x_{A,F_2} = 0.41$, $x_{D,F_2} = 0.59$. Entering solvent $y_{A,0} = 0$ and $y_{D,0} = 0.04$. Outlet raffinate $x_{A,1} = 0.035$. Recovered solvent has $y_{A,S_R} = 0.01$ and $y_{D,S_R} = 0.005$. Product P_E and reflux has composition $x_{A,N+1} = 0.68$ and $x_{D,N+1} = 0.25$. Reflux flow rate $R_{N+1} = 1500$. All compositions are mole fractions. Determine the optimum feed stage location and the total number of stages required. The equilibrium data at $25\,°C$ are given in Table 18-7.

Table 18-7. Equilibrium Data for Methylcyclopentane
(A), n-Hexane (D) and Aniline at 25 ° C.

x_A	x_D	y_A	y_D
0	0.082	0	0.93
0.016	0.076	0.090	0.838
0.022	0.073	0.187	0.736
0.080	0.050	0.430	0.480
0.117	0.037	0.618	0.278
0.209	0.007	0.795	0.075
0.244	0	0.856	0

Source: Darwent and Winkler (1943).

chapter 19
MASS TRANSFER ANALYSIS

Up to now we have used an equilibrium stage analysis procedure even in packed columns where there are no stages. The major advantage of this procedure is that it does not require determination of the mass transfer on the stage.

In packed columns, it is conceptually incorrect to use the staged model even though it works if the correct HETP is used. In this chapter we will develop a physically more realistic model for packed columns that is based on mass transfer between the phases. After developing the model for distillation, we will discuss mass transfer correlations that allows us to predict the required coefficients for common packings. Next, we will repeat the analysis for both dilute and concentrated absorbers and strippers and analyze cocurrent absorbers. Then a simple model for mass transfer on a stage will be developed, and the estimation of stage efficiency will be considered. Finally, the equations for combined heat and mass transfer will be developed with particular application to cooling towers.

It is assumed that readers have some previous knowledge of basic mass transfer concepts.

19.1. BASICS OF MASS TRANSFER

The basic mass transfer equation in words is:

$$\text{Mass transfer rate} = \tag{19-1}$$

$$(\text{area}) \times (\text{mass transfer coeff.}) \times (\text{driving force})$$

In this equation the mass transfer rate will have units such as lb moles/hr. The area is the area across which mass transfer occurs expressed in ft^2. The driving force is the concentration difference that

Table 19-1. Definitions of Mass Transfer Coefficients and HTUs

Basic Eq.	Units on k	HTU Eq.	Reference
$N_A = k_y \Delta y_A$	lb moles/ft^2–hr or g moles/cm^2–s	$H = \dfrac{V}{k_y \, a \, A_c}$	This book Bennett & Myers (1982) Bolles & Fair (1982) Hines & Maddox (1985) Treybal (1980) Greenkorn & Kessler (1972)
$\bar{N}_A = \bar{k}_y \, \Delta \bar{y}_A$	lb/hr–ft^2	$\bar{H} = \dfrac{G}{\bar{k}_y \, a \, A_c}$	Transfer in mass units
$N_A = k \Delta c$	cm/s or ft/hr	$H = \dfrac{V}{k a c A_c}$	Cussler (1984)
$N_A = k' \, \Delta p$	$k' = k_y/P_{tot}$ lb moles/ft^2–hr–atm g moles/cm^2–s–atm	$H = \dfrac{V}{k' a P_{tot} A_c}$	Sherwood et al. (1975) McCabe et al. (1984) Greenkorn and Kessler (1972)

drives the mass transfer and can be represented as a difference in mole fractions, a difference in partial pressures, a difference in concentrations in kg moles/liter and so forth. The mass transfer coefficient includes the effect of diffusivity and flow conditions, and its units depend upon the units used for the other terms. We will typically use a driving force in mole fractions and will have lb moles/hr-ft^2 for the units of the mass transfer coefficient. Other definitions of the mass transfer coefficient are outlined in Table 19-1.

For equilibrium staged separations we are interested in mass transfer from one phase to another. This is illustrated schematically in Figure 19-1 for the transfer of component A from the liquid to a vapor phase. x_I and y_I are the interfacial mole fractions. For dilute absorbers and strippers and for distillation where there is equimolar counter-transfer of the more volatile and less volatile components, the mass transfer equation becomes

$$\text{Rate} = A_I \, k_y \, (y_{AI} - y_A) \qquad (19\text{-}2a)$$

or

$$\text{Rate} = A_I \, k_x \, (x_A - x_{AI}) \qquad (19\text{-}2b)$$

where k_y and k_x are the individual mass transfer coefficients for the vapor and liquid phases, respectively. These equations define k_y and k_x. Unfortunately, there are two major problems with these equations when they are applied to vapor-liquid and liquid-liquid contactors. First, the interfacial area A_I between the two phases is very difficult to measure. This problem is usually avoided by writing Eqs. (19-2) as

$$\text{Rate/Volume} = k_y a (y_{AI} - y_A) \qquad (19\text{-}3a)$$

$$\text{Rate/Volume} = k_x a (x_A - x_{AI}) \qquad (19\text{-}3b)$$

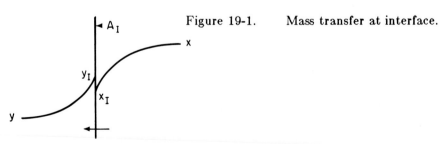

Figure 19-1.   Mass transfer at interface.

where a is the interfacial area per unit volume of the column. Since a is no easier to measure than A_I, we usually measure and correlate the products k_ya and k_xa. Typical units for ka are lb moles/hr-ft^3.

The second problem is that the interfacial mole fractions are also very difficult to measure. To avoid this problem, mass transfer calculations often use a driving force defined in terms of hypothetical equilibrium mole fractions.

$$\text{Rate/Volume} = K_y \, a \, (y_A^* - y_A) \qquad (19\text{-}4a)$$

$$\text{Rate/Volume} = K_x \, a \, (x_A - x_A^*) \qquad (19\text{-}4b)$$

These equations define the overall mass transfer coefficients K_y and K_x. y_A^* is the vapor mole fraction which would be in equilibrium with the bulk liquid of mole fraction x_A, and x_A^* is the liquid mole fraction in equilibrium with the bulk vapor of mole fraction y_A.

To obtain the relationship between the overall and individual coefficients, we begin by assuming that there is no resistance to mass transfer at the interface. This assumption implies that x_I and y_I must be in equilibrium. The mole fraction difference in Eq. (19-4a) can be written as

$$(y_A^* - y_A) = (y_A^* - y_{AI}) + (y_{AI} - y_A) = m(x_A - x_{AI}) + (y_{AI} - y_A) \qquad (19\text{-}5a)$$

where m is now the average slope of the equilibrium curve at x_A and x_{AI}.

$$m = \left(\frac{\partial y_A^*}{\partial x_A}\right)_{\text{avg}} \qquad (19\text{-}5b)$$

Combining Eq. (19-5a) with Eqs. (19-3) and (19-4a), we obtain

$$\frac{\text{Rate/Volume}}{K_ya} = \left[\frac{\text{rate}}{\text{volume}}\right]\left[\frac{m}{k_xa} + \frac{1}{k_ya}\right]$$

which leads to the result

$$\frac{1}{K_ya} = \frac{m}{k_xa} + \frac{1}{k_ya} \qquad (19\text{-}6a)$$

Similar manipulations starting with Eq. (19-4b) lead to

$$\frac{1}{K_x a} = \frac{1}{k_x a} + \frac{1}{m k_y a} \qquad (19\text{-}6b)$$

This *sum-of-resistances* model shows that the overall coefficients will not be constant even if k_x and k_y are constant if the equilibrium is curved and m varies. Equations (19-6) also show the effect of equilibrium on the controlling resistance. If m is small, then from Eq. (19-6a) $K_y \sim k_y$ and the gas-phase resistance controls. If m is large, then Eq. (19-6b) gives $K_x \sim k_x$ and the liquid-phase resistance controls. If there is a resistance at the interface (for example, if a surface-active agent is present), then this resistance must be added to Eqs. (19-6).

19.2 DISTILLATION

Consider the packed distillation tower shown in Figure 19-2. Only binary distillations with constant molal overflow will be considered. Let A be the more volatile component and B the less volatile component. In addition to making L/V constant and satisfying the energy balances, constant molal overflow automatically requires equimolal counterdiffusion, $N_A = -N_B$. Thus CMO simplifies the mass balances, eliminates the need to solve the energy balances, and simplifies the mass

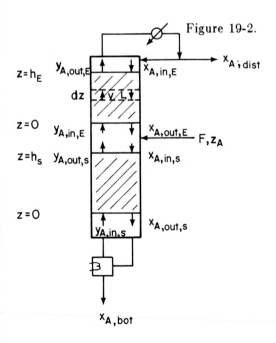

Figure 19-2.   Packed distillation column.

transfer equations. We will also assume perfect plug flow of the liquid and vapor. This means that there is no eddy mixing to reduce the separation.

The mass transfer can be written in terms of the individual coefficients as in Eqs. (19-3) or overall coefficients as in Eqs.(19-4). For the differential height dz in the rectifying section, the mass transfer rate is

$$N_A \, a \, A_c \, dz = k_y \, a \, (y_{AI} - y_A) \, A_c \, dz \qquad (19\text{-}7)$$

where N_A is the flux of A in lb moles/ft^2-hr and A_c is the column cross-sectional area in ft^2. This equation then has units of lb moles/hr. The mass transfer rate must also be equal to the changes in the amount of the more volatile component in the liquid and vapor phases.

$$N_A \, a \, A_c \, dz = V \, dy_A = L dx_A \qquad (19\text{-}8)$$

where L and V are constant molal flow rates. Combining Eqs. (19-7) and (19-8), we obtain

$$dz = \frac{V}{k_y \, a \, A_c \, (y_{AI} - y_A)} dy_A \qquad (19\text{-}9)$$

Integrating this from z=0 to z=h, where h is the total height of packing in a section, we obtain

$$h = \frac{V}{k_y \, a \, A_c} \int_{y_{A,in}}^{y_{A,out}} \frac{dy_A}{y_{AI} - y_A} \qquad (19\text{-}10)$$

We have assumed that the term $V/k_y a A_c$ is constant. The limits of integration for y_A in each section are shown in Figure 19-2. Equation (19-10) is often written as

$$h = H_G \, n_G \qquad (19\text{-}11)$$

where the height of a gas-phase transfer unit H_G is

$$H_G = \frac{V}{k_y \, a \, A_c} \qquad (19\text{-}12a)$$

and the number of gas-phase transfer units n_G is

$$n_G = \int_{y_{A,in}}^{y_{A,out}} \frac{dy_A}{y_{AI} - y_A} \qquad (19\text{-}12b)$$

The height of transfer unit terms are commonly known as HTUs and the number of transfer units as NTUs. Thus, the model is often called the HTU-NTU model.

An exactly similar analysis can be done in the liquid phase by starting with Eq. (19-3b). The result for each section is

$$h = \frac{L}{k_x \, a \, A_c} \int_{x_{A,out}}^{x_{A,in}} \frac{dx_A}{x_A - x_{AI}} \qquad (19\text{-}13)$$

which is usually written as

$$h = H_L \, n_L \qquad (19\text{-}14)$$

where

$$H_L = \frac{L}{k_x \, a \, A_c} \qquad (19\text{-}15a)$$

and

$$n_L = \int_{x_{A,out}}^{x_{A,in}} \frac{dx_A}{x_A - x_{AI}} \qquad (19\text{-}15b)$$

In order to do the integration to calculate n_G and n_L we must relate the interfacial mole fractions y_{AI} and x_{AI} to the bulk mole fractions y_A and x_A. To do this we start by setting Eqs. (19-3a) and (19-3b) equal to each other. After simple rearrangement, this is

$$\frac{y_{AI} - y_A}{x_{AI} - x_A} = -\frac{k_x \, a}{k_y \, a} = -\frac{L}{V} \frac{H_G}{H_L} \qquad (19\text{-}16)$$

where the last equality on the right comes from the definitions of H_G and H_L. The left-hand side of this equation can be identified as the slope of a line from the point representing the interfacial mole fractions (y_{AI}, x_{AI}) to the point representing the bulk mole fractions (y_A, x_A). Since there is no interfacial resistance, the interfacial mole fractions are

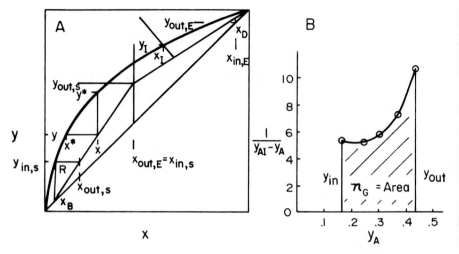

Figure 19-3.   Analysis of number of transfer units. (A) Determination of equilibrium or interfacial values. (B) Graphical
integration of Eq. (19-12b). Shown for stripping section
of Example 19-1.

in equilibrium. The bulk mole fractions are easily related by a mass balance through segment dz around either the top or the bottom of the
column. This *operating line* in the rectifying section is

$$y_A = \frac{L}{V} x + (1 - \frac{L}{V})x_{A,\text{dist}} \qquad (19\text{-}17)$$

In the stripping section the operating line that relates y_A to x_A is

$$y_A = \frac{\overline{L}}{\overline{V}} x - (\frac{\overline{L}}{\overline{V}} - 1) x_{A,\text{bot}} \qquad (19\text{-}18)$$

Note that these operating equations are exactly the same as the operating equations for staged systems. Thus the operating lines will intersect
at the feed line.

We can now use a modified McCabe-Thiele diagram to determine x_{AI}
and y_{AI}. From any point (y_A, x_A) on the operating line, draw a line of
slope $-k_x a/k_y a$. The intersection of this line with the equilibrium curve
gives the interfacial mole fractions that correspond to y_A and x_A (see
Figure 19-3A). After this calculation is done for a series of points, we
can plot $1/(y_{AI} - y_A)$ vs y_A as shown in Figure 19-3B. The area under
the curve is n_G. n_L is determined by plotting $1/(x_A - x_{AI})$. The areas

can be determined from graphical integration or numerical integration such as Simpson's rule [see Eq. (11-12) and Example 19-1].

It will be most accurate to do the calculations for the stripping and enriching sections separately. For example, in the stripping section,

$$H_{G,S} = \frac{\overline{V}}{k_y \, a \, A_c} \tag{19-19a}$$

and

$$n_{G,S} = \int_{y_{A,in,S}}^{y_{A,out,S}} \frac{dy_A}{y_{AI} - y_A} \tag{19-19b}$$

In the determination of n_G for the stripping section, $y_{A,in,S}$ is the vapor mole fraction leaving the reboiler. This is illustrated in Figure 19-2 for a partial reboiler. Mole fraction $y_{A,out,S}$ is the mole fraction leaving the stripping section. This mole fraction can be estimated at the intersection of the operating lines. This is shown in Figure 19-3A. Note that this estimate makes $y_{A,out,S} = y_{A,in,E}$.

Calculating the interfacial mole fractions adds an extra step to the calculation. Since it is often desirable to avoid this step, the overall mass transfer coefficients in Eqs. (19-4) are often used. In terms of the overall driving force the mass transfer rate corresponding to Eq. (19-7) is

$$N_A \, a \, A_c \, dz = K_y \, a \, (y_A^* - y_A) \, A_c \, dz \tag{19-20}$$

Setting this equation equal to Eq. (19-8), we obtain

$$dz = \frac{V}{K_y \, a \, A_c \, (y_A^* - y_A)} \, dy_A \tag{19-21}$$

Integration of this equation over a section of the column gives

$$h = \frac{V}{K_y \, a \, A_c} \int_{y_{A,in}}^{y_{A,out}} \frac{dy_A}{y_A^* - y_A} \tag{19-22}$$

This equation is usually written as

$$h = H_{OG} \, n_{OG} \tag{19-23}$$

where the height of an overall gas phase transfer unit is

$$H_{OG} = \frac{V}{K_y \, a \, A_c} \qquad (19\text{-}24a)$$

and the number of overall gas phase transfer units is

$$n_{OG} = \int_{y_{A,in}}^{y_{A,out}} \frac{dy_A}{y_A^* - y_A} \qquad (19\text{-}24b)$$

Exactly the same steps can be done in terms of the liquid mole fractions. The result is

$$h = H_{OL} \, n_{OL} \qquad (19\text{-}25)$$

where

$$H_{OL} = \frac{L}{K_x \, a \, A_c} \qquad (19\text{-}26a)$$

and

$$n_{OL} = \int_{x_{A,in}}^{x_{A,out}} \frac{dx_A}{x_A - x_A^*} \qquad (19\text{-}26b)$$

The advantage of this formulation is that $y_A^* - y_A$ is easily found from the vertical line shown in Figure 19-3A. The value $x_A - x_A^*$ can be found from the horizontal line shown in the figure. The number of transfer units, n_{OG} or n_{OL}, is then easily determined. Calculation of n_{OG} is similiar to the calculation of n_G illustrated in Figure 19-3B. The disadvantage of using the overall coefficients is that the height of an overall transfer unit, H_{OG} or H_{OL}, is much less likely to be constant than H_G or H_L. This is easy to illustrate, since we can calculate the overall HTU from the individual HTUs. For example, substituting Eq. (19-6a) into Eq. (19-23) and rearranging, we obtain

$$H_{OG} = \frac{mV}{L} H_L + H_G \qquad (19\text{-}27a)$$

H_{OL} can be found by substituting Eq. (19-6b) into Eq. (19-26a):

$$H_{OL} = H_L + \frac{L}{mV} H_G \qquad (19\text{-}27b)$$

Obviously, H_{OG} and H_{OL} are related:

$$H_{OL} = \frac{L}{mV} H_{OG} \qquad (19\text{-}27c)$$

If H_G and H_L are constant, H_{OG} and H_{OL} cannot be exactly constant, since m, the slope of the equilibrium curve, varies in the column. The various NTU values must be related, since h in Eqs. (19-11), (19-14), (19-23), and (19-25) is obviously the same, but the HTU values vary. These relationships are derived in Problem 19-C1.

This approach can easily be extended to the more complex continuous columns discussed in Chapters 6 and 10 and to the batch columns discussed in Chapter 11. Any of these situations can be analyzed by plotting the appropriate operating lines and then proceeding with the HTU-NTU analysis.

Example 19-1. Distillation in a Packed Column

We wish to repeat Example 5-3 (distillation of ethanol and water) except that a column packed with 2-in metal pall rings will be used. F = 1000 kg moles/hr, z = 0.2, T_F = 80 °F, x_D = 0.8, x_B = 0.02, L/D = 5/3, and p = 1 atm. Use a vapor flow rate that is nominally 75% of flooding. In the enriching section estimate H_G = 1.33 ft and H_L = 0.83 ft, and in the stripping section H_G = 0.93 ft and H_L = 0.35 ft (see Example 19-2).

Solution

A. Define. Determine the height of packing in the stripping and enriching sections.

B and C. Explore and Plan. The solution obtained in Example 5-3 can be used to plot the operating lines and the feed line. These are exactly the same as in Figure 5-13. Since the ethanol-water equilibrium is very nonlinear, the design will be more accurate if an individual mass transfer coefficient is used. Thus, use Eqs. (19-11) and (19-12) for the enriching and stripping sections separately. The term $(y_{AI} - y_A)$ can be determined as illustrated

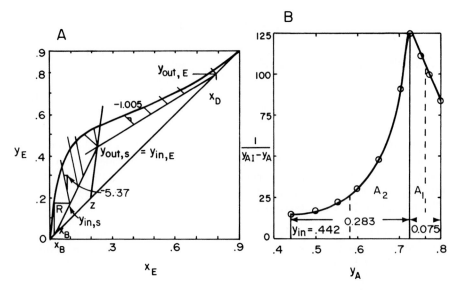

Figure 19-4.   Solution to Example 19-1. (A) Determination of y_E; (B) graphical integration for enriching section.

in Figure 19-3A. n_G can be found for each section as shown in Figure 19-3B.

D. Do It. The equilibrium and operating lines from Example 5-3 are plotted in Figure 19-4. In the *stripping* section, Eq. (19-16) gives a slope of

$$\text{Slope} = -\frac{\bar{L}}{\bar{V}}\frac{H_G}{H_L} = -2.04\left[\frac{0.93}{0.35}\right] = -5.37$$

where $\bar{L}/\bar{V} = 2.04$ from Example 5-13 or from mass balances. Lines of slope −5.37 are drawn from arbitrary points on the stripping section operating line to the equilibrium curve. This is illustrated in Figure 19-4A. Values of y_A are on the operating line, while y_{AI} values are on the equilibrium line. The following table was generated.

y_{AI}	y_A	$y_{AI} - y_A$	$1/(y_{AI} - y_A)$
0.354	0.17	0.184	5.44
0.44	0.25	0.19	5.26
0.477	0.306	0.171	5.85
0.512	0.375	0.137	7.30
0.535	0.442	0.093	10.75

From this table $1/(y_{AI} - y_A)$ vs y_A is easily plotted, as shown in Figure 19-3B. n_G is the area under this curve from $y_{A,in,S} = 0.17$ to $y_{A,out,S} = 0.442$. $y_{A,in,S}$ is the vapor mole fraction leaving the partial reboiler. Determination of $y_{A,in,S}$ is shown in Figure 19-4. $y_{A,out,S}$ is the vapor mole fraction at the intersection of the operating lines. The area in Figure 19-3B can be estimated from Simpson's rule (although the area will be overestimated since the minimum in the curve is not included).

$$n_{G,S} = \frac{y_{A,out} - y_{A,in}}{6}\left[f(y_{A,in}) + 4\ f(y_{A,avg}) + f(y_{A,out})\right] \quad (19\text{-}28a)$$

where

$$f = \frac{1}{y_{AI} - y_A} \quad , \quad y_{A,avg} = \frac{y_{A,in} + y_{A,out}}{2} \quad (19\text{-}28b)$$

Note that the end points and the middle point were calculated. For the stripping section, this is

$$n_{G,S} = \frac{0.272}{6}[5.44 + 4(5.85) + 10.75] = 1.79$$

And the height of packing in the stripping section is

$$h_S = H_{G,S}\ n_{G,S} = (0.93)(1.79) = 1.66 \text{ ft}$$

In the *enriching* section the slope is

$$\text{Slope} = -\frac{L}{V}\frac{H_G}{H_L} = -\left(\frac{5}{8}\right)\left(\frac{1.33}{0.83}\right) = -1.005$$

Arbitrary lines of this slope are shown on Figure 19-4. The following table was generated.

y_{AI}	y_A	$y_{AI} - y_A$	$1/(y_{AI} - y_A)$
0.505	0.442	0.063	15.9
0.557	0.5	0.057	17.5
0.594	0.55	0.044	22.7
0.632	0.60	0.032	31.25
0.671	0.65	0.021	47.6
0.711	0.7	0.011	90.9
0.733	0.725	0.008	125
0.759	0.75	0.009	111.1
0.785	0.775	0.01	100
0.812	0.8	0.012	83.3

The plot of $1/(y_{AI} - y_A)$ vs y_A is shown in Figure 19-4B. An approximate area can be found using Simpson's rule, Eqs. (19-28a) and (19-28b), for areas A_1 and A_2. For area A_1 the initial point is selected as the maximum point, and the dividing point $y_A = 0.7625$ was calculated.

$$A_1 = \frac{0.075}{6}[125 + 4(107) + 83.3] = 7.95$$

$$A_2 = \frac{0.283}{6}[15.9 + 4(26.5) + 125] = 11.65$$

$n_{G,E}$ is the total area = 19.6. Then the height is

$$h_E = H_{G,E} \, n_{G,E} = (1.33)(19.6) = 26.1 \text{ ft}$$

E. Check. The operating and equilibrium curves were checked in Example 5-3. The areas can be checked by counting squares in Figures 19-3B and 19-4B. More accuracy for area A_1 could be obtained by dividing this region into two parts. The HTU values will be estimated and checked in Example 19-2.

F. Generalize. The method illustrated here can obviously be used in other distillation systems. Since the curve for n_G can be very nonlinear, it is a good idea to plot the curve as shown in Figure 19-4B before doing the numerical integration.

19.3. RELATIONSHIP BETWEEN HETP AND HTU

In simple cases the HTU-NTU approach and the HETP approach discussed in Chapter 13 can be related. If the operating and equilibrium curves are straight and parallel, $mV/L = 1$, we have the situation shown in Figure 19-5A. The equilibrium equation is

$$y = mx + b \tag{19-29a}$$

while a general equation for the straight operating line is

$$y = \frac{L}{V}x + y_{out} - \frac{L}{V}x_{in} \tag{19-29b}$$

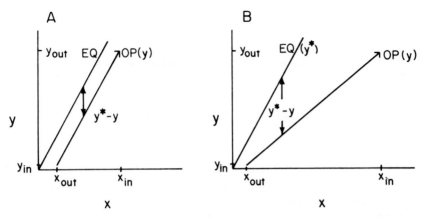

Figure 19-5.   Calculation of $y^* - y$ with linear equilibrium and operating lines. (A) $mV/L = 1$;   (B) $mV/L \neq 1$.

Now the integral in the definition of n_{OG} can easily be evaluated analytically. The difference between the equilibrium and operating lines, $y^* - y$, is

$$y^* - y = mx + b - \left[\frac{L}{V} x + y_{out} - \frac{L}{V} x_{in} \right] \qquad (19\text{-}30)$$

When $m = L/V$, this becomes

$$y^* - y = \frac{L}{V} x_{in} + b - y_{out}$$

Then Eq. (19-24b) becomes

$$n_{OG} = \int_{y_{in}}^{y_{out}} \frac{dy}{y^* - y} = \int_{y_{in}}^{y_{out}} \frac{dy}{\dfrac{L}{V} x_{in} + b - y_{out}}$$

which is easily integrated.

$$n_{OG} = \frac{y_{in} - y_{out}}{y_{out} - \dfrac{L}{V} x_{in} - b} \quad , \quad \text{for} \quad \frac{mV}{L} = 1 \qquad (19\text{-}31)$$

Since $h = H_{OG} n_{OG} = N \times (\text{HETP})$, we can solve for HETP.

$$\text{HETP} = \frac{n_{OG}}{N} H_{OG} \tag{19-32}$$

N can be obtained from Eq. (15-20). Comparison of Eqs. (19-31) and (15-20) shows that $N = n_{OG}$ when $mV/L = 1$. Thus,

$$\text{HETP} = H_{OG}, \qquad \text{for} \quad \frac{mV}{L} = 1 \tag{19-33}$$

If the operating and equilibrium lines are straight but not parallel, then we have the situation shown in Figure 19-5B. The difference between equilibrium and operating lines is still given by Eq. (19-30), but the terms with x do not cancel out. By substituting in x from the operating equation, $y^* - y$ in Eq. (19-24) can be determined as a linear function of y. After integration and considerable algebraic manipulation, n_{OG} is found to be

$$n_{OG} = \left[\frac{1}{1 - mV/L}\right] \ln \left[(1 - \frac{mV}{L})(\frac{y_{in} - y_{out}^*}{y_{out} - y_{out}^*}) + \frac{mV}{L}\right] \tag{19-34a}$$

where

$$y_{out}^* = mx_{in} + b \tag{19-35a}$$

This analysis can also be done in terms of liquid mole fractions. The result is

$$n_{OL} = \left[\frac{1}{1 - L/mV}\right] \ln \left[(1 - L/mV)\left[\frac{x_{A,in} - x_{A,out}^*}{x_{A,out} - x_{A,out}^*}\right] + \frac{L}{mV}\right] \tag{19-34b}$$

where

$$x_{A,out}^* = (y_{in} - b)/m \tag{19-35b}$$

Equations (19-34) are known as the Colburn equations. The value of HETP can now be determined from Eq. (19-32), where N is found from the Kremser equation (15-39) and n_{OG} from Eq. (19-34a). This result is

$$\text{HETP} = \frac{H_{OG} \ln (mV/L)}{mV/L - 1} \tag{19-36}$$

The use of this result is illustrated in Example 19-2.

Although it was derived for a straight operating line and straight equilibrium lines, Eq. (19-36) will be approximately valid for curved equilibrium or operating lines. HETP should be determined separately for each section of the column, since mV/L is not usually the same in the enriching and stripping sections. For maximum accuracy the HETP can be calculated for each stage in the column (Sherwood *et al.*, 1975).

19.4. MASS TRANSFER CORRELATIONS FOR PACKED TOWERS

In order to use the HTU-NTU analysis procedure we must be able to predict the mass transfer coefficients or the HTU values. There has been considerable effort expended in correlating these terms. Care must be exercised in using these correlations since HTU values in the literature may be defined differently. The definitions given here are based on using mole fractions in the basic transfer equations (see Table 19-1). If concentrations or partial pressures are used, the mass transfer coefficients will have different units, which will lead to different definitions for HTU although the HTU will still have units of height. In working with these correlations, terms must be expressed in appropriate units.

19.4.1. Detailed Correlations

We will use the correlation of Bolles and Fair (1982), for which HTUs are defined in the same way as here. The Bolles-Fair correlation is based on the previous correlation of Cornell *et al.* (1960a,b) and a data bank of 545 observations and includes distillation, absorption, and stripping.

The correlation for H_G is

$$H_G = \frac{\psi (D_{col}')^{b_1} (h_p/10)^{1/3} (Sc_V)^{1/2}}{\left[(3600) W_L \, (\mu_L/\mu_w)^{0.16} (\rho_L/\rho_w)^{-1.25} (\sigma_L/\sigma_w)^{-0.8} \right]^{b_2}} \tag{19-37}$$

where ψ is a packing parameter that is given in Figure 19-6 (Bolles and Fair, 1982) for common packings, and other special terms are defined in Table 19-2. Viscosity, density, surface tension, and diffusivities should be defined in consistent units so that the Schmidt number and the ratios of liquid to water properties are dimensionless. The packing height h_p is the height of each packed bed; thus, the stripping and enriching sections should be considered separately.

652

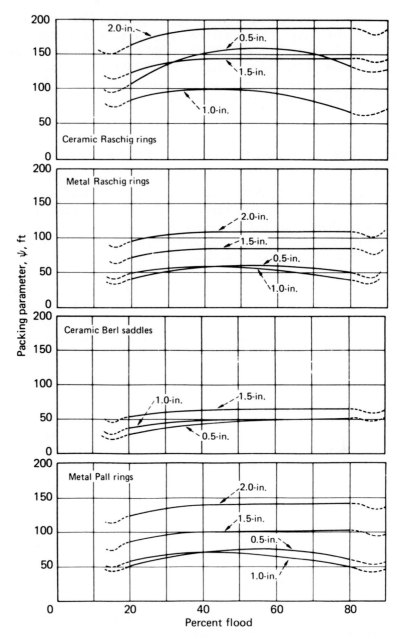

Figure 19-6.   Packing parameter ψ for H_G calculation (Bolles and Fair, 1982). Excerpted by special permission from *Chemical Engineering*, 89 (14), 109 (July 12, 1982). Copyright 1982, McGraw-Hill, Inc., New York, N.Y. 10020.

Table 19-2. Terms for Eqs. (19-37) and (19-38)

D'_{col}	= lesser of column diameter in ft or 2
b_1	= 1.24 (rings) or 1.11 (saddles)
h_p	= height of each packed bed in ft
H_G, H_L	= HTU in ft
Sc_v	= Schmidt no. for vapor = $\mu_v/\rho_v D_v$
W_L	= Weight mass flux of liquid, lb/s–ft^2
b_2	= 0.6 (rings) or 0.5 (saddles)
Sc_L	= Schmidt no. for liquid = $\mu_L/\rho_L D_L$

The correlation for H_L is

$$H_L = \phi\, C_{fL}\, (h_p/10)^{0.15}\, (Sc_L)^{1/2} \tag{19-38}$$

In this equation ϕ is a packing parameter shown in Figure 19-7, and C_{fL} is a vapor load coefficient shown in Figure 19-8 (Bolles and Fair, 1982). The value of u_{flood} in Figure 19-8 is from the packed bed flooding correlation in Chapter 13.

The calculated H_G and H_L values can vary from location to location in the column. When this occurs, an integrated mean value should be used. The overall HTU values can be obtained from Eqs. (19-27). Even if H_G and H_L are constant, H_{OG} and H_{OL} will vary owing to the curvature of the equilibrium curve.

Bolles and Fair (1982) show that there is considerable scatter in modeled HETP data versus experimental HETP data. HETP was calculated from Eq. (19-36). For 95% confidence in the results, Bolles and Fair suggest a safety factor of 1.70 in the determination of HETP. They note that this large a safety factor is usually not used, since there are often a number of hidden safety factors such as not including end effects and using nonoptimum operating conditions. However, if a tight design is used, then the 1.70 safety factor is required.

Example 19-2. Estimation of H_G and H_L

Estimate the values of H_G and H_L for the distillation in Examples 5-3 and 19-1 using 2-in metal pall rings.

654

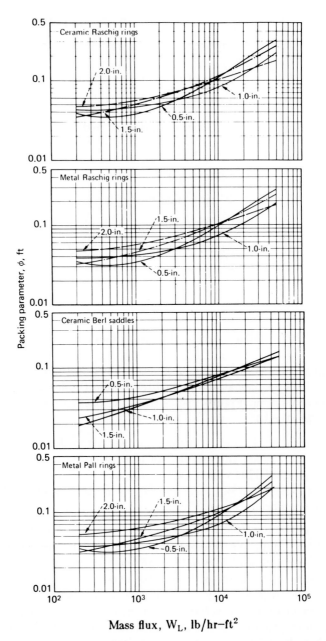

Figure 19-7.   Packing parameter ϕ for H_L calculation (Bolles and Fair, 1982). Excerpted by special permission from *Chemical Engineering*, 89 (14), 109 (July 12, 1982). Copyright 1982, McGraw-Hill, Inc., New York, N.Y. 10020.

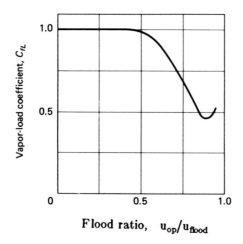

Flood ratio, u_{op}/u_{flood}

Figure 19-8.   Vapor load coefficient C_{fL} for H_L calculation (Bolles and Fair, 1982). Excerpted by special permission from *Chemical Engineering*, 89 (14), 109 (July 12, 1982). Copyright 1982, McGraw-Hill, Inc., New York, N.Y. 10020.

Solution

A. Define. We want to find H_G and H_L in both the stripping and enriching sections. This will be done as if we had completed Example 5-3 but not Example 19-1. Thus, we know the number of equilibrium stages required but we have not estimated packing heights.

B and C. Explore and Plan. We will use the Bolles and Fair (1982) correlation shown in Eqs. (19-37) and (19-38) and Figures 19-6 to 19-8. Obviously, we need to estimate the physical properties required in this correlation. We will do this for the stripping and enriching sections separately. This estimation is easiest if a physical properties package is available. We will illustrate the estimation using values in the literature. The packing height, h_p, must be estimated for each section. These heights will be estimated from the number of stages in each section multiplied by an estimated HETP. Flow rates will be found from mass balances and then will be converted to weight units. A diameter calculation will be done to determine the actual percent flooding.

D. Do It. We will do calculations at the top of the column and assume that these values are reasonably accurate throughout the enriching section. Estimation of properties at the bottom of the column will be used for the stripping section. External balances give D = 230.8 kg moles/hr, B = 769.2 kg moles/hr.

Flooding at Top.

$$x_D = y_1 = 0.8$$

$$\overline{MW}_v = y_1 \, MW_E + (1 - y_1)MW_w = 40.4$$

From the ideal gas law,

$$\rho_v = \frac{P \, (MW_v)}{RT} = 1.393 \times 10^{-3} \, g/cm^3$$

where T = 78.4 °C from Figure 5-14.

Liquid Density. 80 mole % ethanol is 91.1 wt %. From Perry and Green (1984), $\rho_L = 0.7976$ g/mL at 40 °C and $\rho_L = 0.82386$ at 0 °C. By linear interpolation, $\rho_L = 0.772$ g/mL at 78.4 °C. At 78.4 °C, $\rho_w = 0.973$ g/mL.

For the flooding curve in Figure 13-4 the abscissa is

$$\frac{L'}{G'}\left[\frac{\rho_v}{\rho_L}\right]^{1/2} = \frac{L}{V}\left[\frac{\rho_v}{\rho_L}\right]^{1/2} = \left[\frac{5}{8}\right]\left[\frac{1.393 \times 10^{-3}}{0.77}\right]^{1/2} = 0.27$$

Ordinate (flooding) = 0.197, which gives

$$G'_{flood} = \left[\frac{(ordinate)(\rho_G)(\rho_L)(g_c)}{F \, \psi \, \mu^{0.2}}\right]^{1/2}$$

or

$$G'_{flood} = \left[\frac{(0.197)(1.393 \times 10^{-3})(62.4)^2(0.77)(32.2)}{20\left[\frac{0.973}{0.772}\right](0.52)^{0.2}}\right]^{1/2} = 1.16 \, \frac{lb}{s\text{--}ft^2}$$

The $(62.4)^2$ converts ρ_L and ρ_G to lb/ft³. μ_L is estimated as 0.52. Then,

$$G'_{actual} = 0.75 \, G'_{flood} = 0.87 \, \frac{lb}{s\text{--}ft^2}$$

The molar vapor flow rate is

$$V = (L/D + 1)D = 615.4 \text{ kg moles/hr}$$

which allows us to find the column cross-sectional area.

$$\text{Area} = \frac{V}{G'_{actual}} \left[\frac{1 \text{ lb mole}}{0.46 \text{ kg mole}} \right] \left[\frac{40.4 \text{ lb}}{\text{lb mole}} \right] \left[\frac{1 \text{ hr}}{3600 \text{ s}} \right] = 17.2 \text{ ft}^2$$

$$\text{Dia} = \left[\frac{4 \text{ Area}}{\pi} \right]^{1/2} = 4.68 \text{ ft}$$

Round this off to 5.0 ft. Area = 19.6 ft^2. This roundoff reduces the % flooding.

$$\text{Actual frac. flood} = 0.75 \ (17.2/19.6) = 0.66$$

A repeat of the calculation at the bottom of the column shows that the column will flood first at the top since the molecular weight is much higher.

Estimation at Top.

Liquid Diffusivities. From Reid *et al.* (1977, p. 577), for very dilute systems $D_{EW} = 1.25 \times 10^{-5}$ cm^2/s and $D_{WE} = 1.132 \times 10^{-5}$ at 25 °C. The effect of temperature can be estimated since the ratio $D_L \, \mu_L/T \sim$ constant. At the top we want D_L at 78.4 °C = 351.5 K.

$$D_L(78.4) = \frac{351.5}{\mu_L(78.4)} \left[\frac{D_L(25)\mu(25)}{298} \right]$$

Estimating viscosities from Perry and Green (1984) using 95% ethanol:

$$D_L(78.4) = \frac{351.5}{0.47 \text{ cP}} \frac{(1.132 \times 10^{-5})(1.28 \text{ cP})}{298}$$

$$D_L = 3.64 \times 10^{-5} \text{ cm}^2/\text{s}$$

The liquid surface tension can be estimated from data in the *Handbook of Chemistry and Physics*.

$$\sigma_L(78.4) = 18.2 \text{ dynes/cm}$$

$$\sigma_W(78.4) = 62.9 \text{ dynes/cm}$$

For vapors the Schmidt number can be estimated from kinetic theory (Sherwood *et al.* 1975, pp. 17-24). The equation is

$$Sc_v = \left[\frac{\mu_v}{\rho_v D}\right] = 1.18 \frac{\Omega_D}{\Omega_v} \left[\frac{MW_A}{MW_A + MW_B}\right]^{1/2} \left[\frac{\sigma_{AB}}{\sigma_B}\right]^2$$

where the collision integrals Ω_D, Ω_v and the Lennard-Jones force constants σ_{AB} and σ_B are discussed by Sherwood *et al.* (1975). At the top of the column the result is $Sc_v = 0.355$.

The liquid flow rate at the top is

$$L = (L/D)D = 384.6 \text{ kg moles/hr.}$$

The liquid flux W_L is

$$W_L = \frac{(L \frac{\text{kg moles}}{\text{hr}})(\frac{1 \text{ lb mole}}{0.46 \text{ kg mole}})(\frac{40.4 \text{ lb}}{\text{lb mole}})(\frac{1 \text{ hr}}{3600 \text{ s}})}{\text{Area}}$$

$$= 0.479 \frac{\text{lb}}{\text{s-ft}^2}$$

In Eq. (19-37) $D'_{col} = 2$, $\psi = 141$ from Figure 19-6 at 66% flood, $b_1 = 1.24$ and $b_2 = 0.6$. We can estimate h_p as (No. stages) × (HETP), where an average HETP is about 2 ft. Then

$$h_p = (11)(2) = 22 \text{ ft}$$

Equation (19-37) is then

$$H_{G,E} = \frac{(141)(2)^{1.24}(2.2)^{1/3}(0.355)^{1/2}}{[(3600)(.479)(.47/.36)^{0.16}(.772/.973)^{-1.25}(18.2/62.9)^{-0.8}]^{0.6}}$$

$$H_{G,E} = 1.33 \text{ ft}$$

For H_L we calculate W_L as 1724 lb/hr-ft^2 and $\phi = 0.07$ from Figure 19-7. $C_{fl} = 0.81$ from Figure 19-8. Then from Eq. (19-38),

$$H_{L,E} = (0.07)(0.81)(2.2)^{.15}\left[\frac{0.0047}{(0.772)(3.64 \times 10^{-5})}\right]^{1/2} = 0.83 \text{ ft}$$

Note that μ_L in Sc_L is in poise (0.01 P = 1 cP).

These calculations can be repeated for the bottom of the column. The results are: $H_{G,S} = 0.93$ ft and $H_{L,S} = 0.35$ ft.

E. Check. One check can be made by estimating HETP using Eq. (19-36). At the top of the column the slope of the equilibrium curve is m ~ 0.63. This will vary throughout the column. Then from Eq. (19-27a), at the top

$$H_{OG} = \frac{m}{L/V}\,H_L + H_G = \frac{0.63}{5/8}\,(0.827) + 1.33 = 2.16 \text{ ft}$$

Note that H_{OG} will vary in the enriching section since m varies. From Eq. (19-36),

$$\text{HETP} = \frac{2.16}{0.63(8/5)-1}\,\ln\,[0.63(8/5)] = 2.15 \text{ ft}$$

This is close to our estimated HETP, so our results are reasonable. The packing heights calculated in Example 19-1, $h_S = 1.66$ and $h_E = 26.1$, differ from our initial estimates. A second iteration can be done to correct H_G and H_L. For example,

$$H_{G,E,cor} = \left[\frac{2.61}{2.2}\right]^{1/3} H_{G,E,initial} = 1.41 \text{ ft}$$

which is a 6% correction. Changing H_G and H_L will change the slopes of the lines used to calculate y_{AI}; thus, n_G will also change.

F. Generalize. This calculation is long and involved because of the need to estimate physical properties. This part of the problem is greatly simplified if a physical properties package is available on the computer. In this example m is close to one. Thus both terms in Eqs. (19-6) are significant and neither resistance controls. Thus, H_G and H_L are the same order of magnitude.

19.4.2. Simple Correlations

The detailed correlation is fairly complex to use if a physical properties package is not available. Simplified correlations are available but will not be as accurate (Bennett and Myers, 1982; Greenkorn and Kessler, 1972; Perry and Green, 1984; Sherwood et al., 1975; Treybal, 1955): For H_G the following empirical form has been used (Bennett and Myers, 1982; Greenkorn and Kessler, 1972; Treybal, 1955):

$$H_G = a_G \, W_G^b \, Sc_v^{0.5}/W_L^c \qquad (19\text{-}39)$$

where W_G and W_L are the fluxes in lb/hr-ft^2, and Sc_v is the Schmidt number for the gas phase. The constants are given in Table 19-3. The expression for H_L developed by Sherwood and Holloway (1940) is

$$H_L = a_L \left(\frac{W_L}{\mu}\right)^d Sc_L^{0.5} \qquad (19\text{-}40)$$

where Sc_L is the Schmidt number for the liquid. The constants are given in Table 19-3.

These correlations are obviously easier to use than Eqs. (19-37) and (19-38) since only the Schmidt number and the viscosity need to be estimated. However, Eqs. (19-39) and (19-40) will not be as accurate; thus, they should only be used for preliminary designs.

Table 19-3. Constants for Determining H_G and H_L from Eqs. (19-39) and (19-40). Range of W_L in Eq. (19-40) Is 400 to 15,000.

Packing	a_G	b	c	Range for Eq. (19-39) W_G	W_L	a_L	d
Raschig rings							
3/8 inch	2.32	0.45	0.47	200-500	500-1500	0.0018	0.46
1	7.00	0.39	0.58	200-800	400-500	0.010	0.22
1	6.41	0.32	0.51	200-600	500-4500	--	--
2	3.82	0.41	0.45	200-800	500-4500	0.012	0.22
Berl saddles							
1/2 inch	32.4	0.30	0.74	200-700	500-1500	0.0067	0.28
1/2	0.811	0.30	0.24	200-700	1500-4500	--	--
1	1.97	0.36	0.40	200-800	400-4500	0.0059	0.28
3/2	5.05	0.32	0.45	200-1000	400-4500	0.0062	0.28

19.5. ABSORBERS AND STRIPPERS

The HTU-NTU analysis for concentrated absorbers and strippers is somewhat more complex than for distillation because total flow rates are not constant and solute A is diffusing through a stagnant film with no counterdiffusion, $N_B = 0$. We will assume that the system is isothermal. When there is appreciable mass transfer through a stagnant film the basic mass transfer equations (19-2) must be modified. With $N_B = 0$ the mass transfer flux with respect to a fixed axis system is (Bennett and Myers, 1982; McCabe *et al.*, 1984; Sherwood *et al.*, 1975)

$$N_A = J_A/(1 - y_A) \qquad (19\text{-}41)$$

where J_A is the flux with respect to an axis moving at the molar average velocity of the fluid. This leads to a transfer rate equation that is superficially similar to the previous equations.

$$N_A a = k_y' \, a(y_A - y_{AI}) \qquad (19\text{-}42)$$

Now the mass transfer coefficient is defined as

$$k_y' a = k_y \, a/(1 - y_A)_{lm} \qquad (19\text{-}43)$$

where the logarithmic mean mole fraction is defined as

$$(1 - y_A)_{lm} = \frac{(1 - y_A) - (1 - y_{AI})}{\ln\left(\dfrac{1 - y_A}{1 - y_{AI}}\right)} \qquad (19\text{-}44)$$

For very dilute systems, $(1 - y_A)_{lm} = 1$ and $k_y' a = k_y a$.

We will now repeat the analysis of a packed section using Eq. (19-41) and including the non-constant total flow rates. Figure 19-9A is a schematic diagram of an absorber. The absorber is assumed to be isothermal, and plug flow is assumed. The rate of mass transfer in a segment of the column dz is given by

$$N_A \, a \, A_c \, dz = k_y' \, a \, (y_A - y_{AI}) \, A_c \, dz \qquad (19\text{-}45)$$

Comparison of this equation with Eq. (19-7) shows that the sign on the mole fraction difference has been switched, since the direction of solute transfer in absorbers is opposite to that of transfer of the more volatile component in distillation. In addition, the modified mass transfer

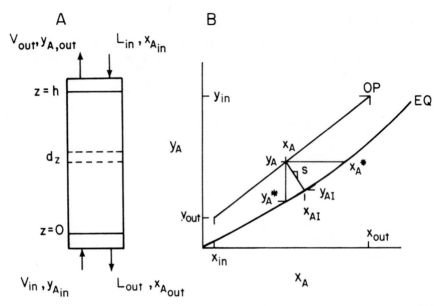

Figure 19-9.   Absorber calculation. (A) Schematic of column. (B) Calculation of interfacial mole fractions.

coefficient k_y' is used. The solute mass transfer can also be related to the change in solute flow rates in the gas or liquid streams.

$$N_A \, a \, A_c \, dz = -A_c \, d(Vy_A) = -A_c \, d(Lx_A) \qquad (19\text{-}46)$$

This equation differs from Eq. (19-8) derived for distillation since neither V nor L is constant.

The variations in V can be related to the constant flow rate of carrier gas, G.

$$V = \frac{G}{1 - y_A} \qquad (19\text{-}47)$$

which is the same as Eq. (15-11). Combining Eqs. (19-45) to (19-47), we obtain

$$-dz = \frac{d\left(\dfrac{Gy_A}{1 - y_A}\right)}{k_y' \, a \, A_c(y_A - y_{AI})}$$

After taking the derivative, substituting in Eq. (19-47), and cleaning up the algebra, we obtain

$$-dz = \frac{V \, dy_A}{k_y' \, a \, A_c(1 - y_A)(y_A - y_{AI})}$$

Integrating this equation we obtain

$$h = \int_0^h dz = - \int_{y_{A,in}}^{y_{A,out}} \frac{V}{k_y' \, a \, A_c} \frac{dy_A}{(1 - y_A)(y_A - y_{AI})} \tag{19-48}$$

This equation is difficult to integrate, because the mass transfer coefficient depends on the mole fraction in concentrated mixtures. However, we can substitute in Eq. (19-43) to obtain

$$h = - \int_{y_{A,in}}^{y_{A,out}} \left[\frac{V}{k_y \, a \, A_c} \right] \frac{(1 - y_A)_{lm} \, dy_A}{(1 - y_A)(y_A - y_{AI})} \tag{19-49}$$

The term $V/k_y a A_c$ is the height of a gas-phase transfer unit H_G defined in Eq.(19-12a).

Unfortunately, H_G is not constant. The variation in H_G can be determined from Eq. (19-37) and Figure 19-6, which are valid for both absorbers and distillation. The term that varies the most in Eq. (19-37) is the weight mass flux of liquid, W_L. H_G depends on W_L to the -0.5 to -0.6 power. In a single section of an absorber, a 20% change in liquid flow rate would be quite large. This will cause at most a 10% change in H_G. $k_y a$ is independent of concentration, since the concentration effect was included in k_y' in Eq. (19-42). Since the variation in H_G over the column section is relatively small, we will treat H_G as a constant. Then Eq. (19-49) becomes

$$h = H_G \int_{y_{A,out}}^{y_{A,in}} \frac{(1 - y_A)_{lm} \, dy_A}{(1 - y_A)(y_A - y_{AI})} \tag{19-50}$$

which is usually written as

$$h = H_G \, n_G$$

$$(19\text{-}51)$$

where

$$n_G = \int_{y_{A,out}}^{y_{A,in}} \frac{(1 - y_A)_{lm}\, dy_A}{(1 - y_A)\,(y_A - y_{AI})} \qquad (19\text{-}52)$$

and H_G is defined in Eq. (19-12a). Note that n_G for concentrated absorption is defined differently from n_G for distillation, Eq. (19-12b). The difference in the limits of integration in the two definitions for n_G occurs because the direction of transfer of component A in distillation is the negative of the direction in absorption. There are additional terms inside the integral sign in absorption because the mass transfer takes place through a stagnant film and is not equimolar mass countertransfer as in distillation.

The method for finding the interfacial compositions is similar to that used to develop Eq. (19-16) and Figure 19-3 except that Eq. (19-42) and the corresponding equation in terms of liquid mole fractions are used as the starting point. The procedure is illustrated in Figure 19-9B. The slope, $s = -k_x'/k_y'$. Use of this procedure lets us calculate the integrand in Eq. (19-52) at a series of points. The integral in Eq. (19-52) can be found either numerically or graphically.

Often, the integral in Eq. (19-52) can be simplified. The first simplification often employed is to replace the logarithmic mean with an arithmetic average.

$$(1 - y_A)_{lm} = \frac{(1 - y_A) + (1 - y_{AI})}{2} \qquad (19\text{-}53)$$

When Eq. (19-53) is substituted into Eq. (19-52), n_G can be simplified.

$$n_G = \int_{y_{A,out}}^{y_{A,in}} \frac{dy_A}{y_A - y_{AI}} + \frac{1}{2} \ln\left[\frac{1 - y_{A,out}}{1 - y_{A,in}}\right] \qquad (19\text{-}54)$$

This equation shows that n_G for absorption is essentially the n_G for distillation plus a correction factor. The interfacial mole fraction y_{AI} can be determined as shown in Figure 19-9B. The integral in Eq. (19-54) can then be determined graphically or numerically. For very dilute systems $1 - y_A$ is approximately 1.0 everywhere in the column. Then the correction factor in Eq. (19-54) will be approximately zero. Thus, n_G for *dilute* absorbers reduces to the same formula as for distillation.

For dilute absorbers and strippers, $(1 - y_A)_{lm} = 1.0$. Then $k'_y a = k_y a$ in Eq. (19-41). In this case we can use the overall gas-phase mass transfer coefficient. Following a development that parallels the analysis presented earlier for distillation, Eqs. (19-4a) and (19-20) to (19-24), we obtain for dilute absorbers

$$h = H_{OG} \, n_{OG} \tag{19-55}$$

where H_{OG} was defined in Eq. (19-24a) and

$$n_{OG} = \int_{y_{A,out}}^{y_{A,in}} \frac{dy_A}{y_A - y_A^*} \tag{19-56}$$

This n_{OG} is essentially the same as for distillation in Eq. (19-24b).

If the operating and equilibrium lines are straight, n_{OG} can be integrated analytically. The result is the Colburn equation given in Eqs. (19-31) and (19-34). An alternative integration gives an equivalent equation.

$$n_{OG} = \frac{y_{A,in} - y_{A,out}}{(y_A - y_A^*)_{in} - (y_A - y_A^*)_{out}} \ln \left[\frac{(y_A - y_A^*)_{in}}{(y_A - y_A^*)_{out}} \right] \tag{19-57}$$

The development done here in terms of gas mole fractions can obviously be done in terms of liquid mole fractions. The development is exactly analogous to that presented here. The result for liquids is

$$h = H_L \, n_L \tag{19-58}$$

where H_L was defined in Eq. (19-15a) and

$$n_L = \int_{x_{A,in}}^{x_{A,out}} \frac{(1 - x_A)_{lm} \, dx_A}{(1 - x_A)(x_{AI} - x_A)} \tag{19-59}$$

Equation (19-59) can often be simplified to

$$n_L = \int_{x_{A,in}}^{x_{A,out}} \frac{dx_A}{x_{AI} - x_A} + \frac{1}{2} \ln \left[\frac{1 - x_{A,out}}{1 - x_{A,in}} \right] \tag{19-60}$$

For dilute systems the correction factor in Eq. (19-60) becomes negligible. For dilute systems the analysis can also be done in terms of the overall transfer coefficient.

$$h = H_{OL} \, n_{OL} \tag{19-61}$$

where H_{OL} is defined in Eq. (19-26a) and

$$n_{OL} = \int_{x_{A,in}}^{x_{A,out}} \frac{dx_A}{x_A^* - x_A} \tag{19-62}$$

If the operating and equilibrium lines are both straight, n_{OL} can be integrated analytically. The result is the Colburn equation (19-34b), or the equivalent expression,

$$n_{OL} = \frac{x_{A,out} - x_{A,in}}{\left(x_A^* - x_A\right)_{out} - \left(x_A^* - x_A\right)_{in}} \ln \left[\frac{\left(x_A^* - x_A\right)_{out}}{\left(x_A^* - x_A\right)_{in}} \right] \tag{19-63}$$

The development of the equations for concentrated systems presented here is not the same as those in Cussler (1984) and Sherwood *et al.* (1975). Since the assumptions have been different, the results are slightly different. However, the differences in these equations will usually not be important, since the inaccuracies caused by assuming an isothermal system with plug flow are greater than those induced by changes in the mass transfer equations. For dilute systems all the developments reduce to the same equations.

Example 19-3. Absorption of SO_2

We are absorbing SO_2 from air with water at $20°C$ in a pilot-plant column packed with 0.5-in. metal Raschig rings. The packed section is 10 ft tall. The total pressure is 741 mmHg. The inlet water is pure. The outlet water contains 0.001 mole frac SO_2, and the inlet gas concentration is $y_{in} = 0.03082$ mole frac. $L/V = 15$. The water flux $W_L = 1000$ lb/hr-ft^2. The Henry's law constant is $H = 22,500$ mmHg/mole frac SO_2 in liquid. Estimate H_{OL} for a 10-ft high large-scale column operating at the same W_L and same fraction flooding if 2-in metal pall rings are used.

Solution

A. Define. Calculate H_{OL} for a large-scale absorber with 2-in metal pall rings.

B. Explore. We can easily determine n_{OL} for the pilot plant. Then $H_{OL} = h/n_{OL}$ for the pilot plant. Since the Henry's law constant H is large, m is probably large. This will make the liquid resistance control, and $H_L \sim H_{OL}$. Then Eq. (19-38) can be used to estimate $H_L = H_{OL}$ for the large-scale column. Since only ϕ varies, this can be estimated from Figure 19-7.

C. Plan. First calculate $m = H/P_{tot} = 22,500/741 = 30.36$. This is fairly large, and from Eq. (19-6b) the liquid resistance controls. For the pilot plant we can calculate n_{OL} from the Colburn equation (19-34b) since m is constant and L/V is approximately constant. Then $H_L = H_{OL} = h/n_{OL}$. The variation in ϕ with the change in packing can be determined from Figure 19-7, and $H_{OL} \sim H_L$ in the large column can be estimated from Eq. (19-38).

D. Do It. From Eq. (19-35b), $x^*_{out} = (y_{in} - b)/m$, so

$$x^*_{out} = \frac{0.03082 - 0.0}{30.36} = 0.001015$$

$L/mV = 15/30.36 = 0.4941$. From Eq. (19-34b) with $x_{in} = 0$ and $x_{out} = 0.001$,

$$n_{OL} = \left[\frac{1}{1 - 0.4941}\right] \ln\left[(1 - 0.4941)\left(\frac{0 - 0.001015}{0.001 - 0.001015}\right) + 0.4941\right]$$

$$n_{OL} = 7.012$$

Then $H_L \sim H_{OL} = 10 \text{ ft}/7.012 = 1.426$ ft. From Figure 19-7 at $W_L = 1000$, ϕ (0.5-in Raschig rings) = 0.32, while ϕ (2-in pall rings) = 0.62. Then from Eq. (19-38),

$$H_{OL} \sim H_L(\text{ 2-in pall rings}) = \frac{0.62}{0.32}(1.426) = 2.76 \text{ ft}$$

since all other terms in Eq. (19-38) are constant.

E. Check. These results are the correct order of magnitude. A check of n_{OL} can be made by graphically integrating n_{OL}.

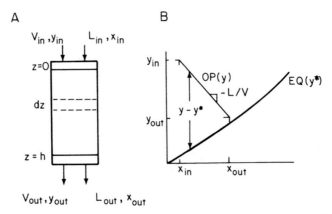

Figure 19-10.   Cocurrent absorber. (A) Schematic of column. (B) Calculation of $y - y^*$.

F. Generalization. This method of correlating H_L or H_G when packing size or type is changed can be used for other problems. The large value of m in this problem allowed the assumption of liquid-phase control. This assumption simplifies the problem since $H_{OL} \sim H_L$. If liquid-phase control is not valid, this problem becomes significantly harder.

19.6 COCURRENT ABSORBERS

In Chapter 15 we noted that cocurrent operation of absorbers was often employed when a single equilibrium stage was sufficient. Cocurrent operation has the advantage that flooding cannot occur. This means that high vapor and liquid flow rates can be used, which automatically leads to small-diameter columns.

A schematic of a cocurrent absorber is shown in Figure 19-10A. The analysis will be done for dilute systems using overall mass transfer coefficients. The system is assumed to be isothermal. The liquid and vapor are assumed to be in plug flow, and total flow rates are constant. The rate of mass transfer in segment dz is

$$N_A \, a \, A_c \, dz = K_y \, a \, (y_A - y_A^*) \, A_c \, dz \tag{19-64}$$

which can be related to the changes in solute flow rates

$$N_A \, a \, A_c \, dz = -d(Vy_A) = d(Lx_A) \tag{19-65}$$

Combining these equations we obtain

$$dz = \frac{-d(V y_A)}{k_y \, a \, A_c \, (y_A - y_A^*)} \qquad (19\text{-}66)$$

If $V/k_y a A_c$ is constant, Eq. (19-66) can be integrated to give

$$h = H_{OG} \, n_{OG} \qquad (19\text{-}67)$$

where H_{OG} is given in Eq. (19-24a) and

$$n_{OG} = \int_{y_{A,out}}^{y_{A,in}} \frac{dy_A}{y_A - y_A^*} \qquad (19\text{-}68)$$

This development obviously follows the development for counter-current systems. The analyses differ when we look at the method for calculating $y_A - y_A^*$. The operating equation is easily derived [see Eq. (15-63)]

$$y = -\frac{L}{V} x + y_{in} - \frac{L}{V} x_{in} \qquad (19\text{-}69)$$

This operating line and the calculation of $y_A - y_A^*$ are shown in Figure 19-10B. When the operating and equilibrium lines are both straight, n_{OG} can be obtained analytically. The result corresponding to the Colburn equation is (King, 1980)

$$n_{OG} = \left[\frac{1}{1 + \dfrac{mV}{L}} \right] \ln \left[\left[1 + \frac{mV}{L} \right] \left(\frac{y_{A,in} - y_{A,out}^*}{y_{A,out} - y_{A,out}^*} \right) - \frac{mV}{L} \right] \qquad (19\text{-}70)$$

where

$$y_{A,out}^* = m x_{A,out} + b \qquad (19\text{-}71)$$

For absorption with a completely irreversible reaction occurring in the liquid phase, $y_A^* = 0$ everywhere in the column. Thus, the equilibrium line is the x axis. The integration of Eq. (19-68) is now straightforward.

$$n_{OG} = \int_{y_{A,out}}^{y_{A,in}} = \frac{dy_A}{y_A} = \ln\left(\frac{y_{A,in}}{y_{A,out}}\right) \tag{19-72}$$

Exactly the same result is obtained for cocurrent and countercurrent columns. The advantage of cocurrent operation with irreversible reactions is that flooding cannot occur, so higher liquid and vapor fluxes can be used.

H_{OG} is related to the individual coefficients by Eq. (19-27a). Unfortunately, Eqs. (19-37) and (19-38) cannot be used to determine the values for H_L and H_G because the correlations are based on data in countercurrent columns at lower gas rates than those used in cocurrent columns. Reiss (1967) reviews data in cocurrent contactors and notes that the mass transfer coefficients can be considerably higher than in countercurrent systems. Harmen and Perona (1972) did an economic comparison of cocurrent and countercurrent columns. For the absorption of CO_2 in carbonate solutions where the reaction is slow they concluded that countercurrent operation is more economical. For CO_2 absorption in monoethanolamine (MEA), where the reaction is fast, they concluded that countercurrent is better at low liquid fluxes whereas cocurrent was preferable at high liquid fluxes.

19.7. MASS TRANSFER ON A STAGE

How does mass transfer affect the efficiency of a tray column? This is a question of considerable interest in the design of staged columns. We will develop a very simple model following the presentations of Cussler (1984), King (1980), and Lewis (1936).

A schematic diagram of a tray is shown in Figure 19-11. The column is operating at steady state. A mass balance will be done for the mass balance envelope indicated by the dashed outline. The vapor above the trays is assumed to be well mixed; thus, the inlet vapor mole fraction $\bar{y}_{j+1}$ does not depend on the position along the tray, ℓ. The vapor leaving the balance envelope has not yet had a chance to be mixed and its composition is a function of position ℓ. The rising vapor bubbles are assumed to perfectly mix the liquid vertically. Thus, x does not depend upon the vertical position z. The liquid mole fraction can be a function of the distance ℓ along the tray measured from the start of the active region, $\ell = 0$, to the end of the active region, $\ell = \ell_a$. At steady state a solute or more volatile component mass balance for the

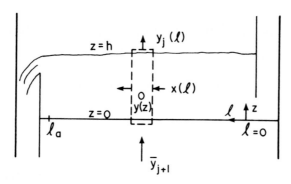

Figure 19-11.　Schematic of tray.

vapor phase is

$$(In - Out)_{convection} + (Solute\ transferred\ from\ liquid) = 0 \qquad (19\text{-}73a)$$

If we use the overall gas-phase mass transfer coefficient K_y, this equation is

$$(V/A_{active})\ [y(z) - y(z + \Delta z)] + K_y a\ \Delta z\ (y_\ell^* - y) = 0 \qquad (19\text{-}73b)$$

where y_ℓ^* is the vapor mole fraction in equilibrium with the liquid of mole fraction x_ℓ. A_{active} is the active area for vapor-liquid contact on the tray. Both A_{active} and V are assumed to be constant. Dividing Eq. (19-73b) by Δz and taking the limit as Δz goes to zero, we obtain

$$-\frac{dy}{dz} + \frac{K_y\ a\ A_{active}}{V}\ (y_\ell^* - y) = 0 \qquad (19\text{-}74)$$

This equation can now be integrated from $z=0$ to $z=h$. The boundary conditions are

$$y = \bar{y}_{j+1}, \quad z = 0 \qquad (19\text{-}75a)$$

$$y = y_\ell, \quad z = h \qquad (19\text{-}75b)$$

The solution to Eqs. (19-74) and (19-75) is

$$E_{pt} = \frac{y_\ell - \bar{y}_{j+1}}{y_\ell^* - \bar{y}_{j+1}} = 1 - \exp\ (-K_y\ a\ A_{active}\ h/V) \qquad (19\text{-}76a)$$

The point efficiency E_{pt} was defined in Chapter 12. Comparing this equation to Eqs. (19-23) and (19-24a), we obtain two alternative representations.

$$E_{pt} = 1 - \exp\left(-h/H_{OG}\right) \qquad (19\text{-}76b)$$

$$E_{pt} = 1 - \exp\left(-n_{OG}\right) \qquad (19\text{-}76c)$$

We would like to relate the point efficiency to the Murphree efficiency given by Eq. (12-2a). This relationship depends upon the liquid flow conditions on the tray. There are two limiting flow conditions that allow us to simply relate E_{pt} to E_{MV}. The first of these is a tray where the liquid is completely mixed. This means that x_ℓ is a constant and is equal to x_{out}, so that $y_\ell^* = y_{out}^*$ and $y_\ell = y_{out}$. Therefore $E_{MV} = E_{pt}$, and

$$E_{MV} = 1 - \exp\left(-K_y \, aA_{active}h/V\right) \qquad (19\text{-}77)$$

for a completely mixed stage.

The second limiting flow condition is plug flow of liquid with no mixing along the tray. If we assume that each packet of liquid has the same residence time, we can derive the relationship between E_{MV} and E_{pt} (Lewis, 1936; King, 1980):

$$E_{MV} = \frac{L}{mV}\left[\exp\left(\frac{mV}{L}\,E_{pt}\right) - 1\right] \qquad (19\text{-}78)$$

where m is the local slope of the equilibrium curve, Eq. (19-5b). Since plug flow is often closer to reality than a completely mixed tray, Eq. (19-78) is more commonly used than Eq. (19-77).

Real plates often have mixing somewhere in between these two limiting cases. These situations are discussed elsewhere (AIChE, 1958; King, 1980).

Example 19-4. Estimation of Stage Efficiency

A small distillation column separating benzene and toluene gives a Murphree vapor efficiency of 0.65 in the rectifying section where $\overline{L}/\overline{V} = 0.8$ and $x_{benz} = 0.7$. The tray is perfectly mixed and has a liquid head of 2 in. The vapor flux is 25 lb mole/hr-ft^2. (a)

Calculate K_y a. (b) Estimate E_{MV} for a large-scale column where the trays are plug flow and h becomes 2.5 in. Other parameters are constant.

Solution

a. From Eq. (19-77) we find

$$K_y a = -\frac{V}{A_c h} \ln (1 - E_{MV})$$

since $V/A_c = 25$ and $h = 2/12$ ft, this is

$$K_y a = \frac{-25}{2/12} \ln (0.35) = 157.5 \text{ lb moles/hr-ft}^3$$

b. In the large-diameter system E_{pt} is given by Eq. (19-76a). Since $h = 2.5/12$,

$$E_{pt} = 1 - \exp \left[-\frac{157.5 \ (2.5/12)}{25} \right] = 0.731$$

Increasing the liquid pool height increases the efficiency since the residence time is increased.

The Murphree vapor efficiency for plug flow is found from Eq. (19-78). The slope of the equilibrium curve, m, can be estimated. Since equilibrium is

$$y_{benz} = \frac{\alpha \ x_{benz}}{1 + (\alpha - 1) \ x_{benz}}$$

the slope is

$$m = \frac{dy_{benz}}{dx_{benz}} = \frac{\alpha}{[1 + (\alpha - 1) \ x]^2}$$

With $\alpha = 2.5$ and $x = 0.7$, $m = 0.595$. Then Eq. (19-78) is

$$E_{MV} = \frac{0.8}{0.595} \left\{ \exp \left[\frac{(0.595)(0.731)}{(0.8)} \right] - 1 \right\} = 0.971$$

The plug flow system has a significantly higher Murphree plate efficiency. Note that $K_y a$ is likely to vary throughout the column since m varies. E_{MV} is also dependent upon m and will change from stage to stage. The effect of concentration changes can be determined by calculating K_y from Eq. (19-6a).

19.8. COOLING TOWERS, HUMIDIFICATION, AND DEHUMIDIFICATION

Contacting of air and water is a very common industrial unit operation. Probably the most common application and certainly the most visible is the use of cooling towers to cool water. A cooling tower contacts air and water and cools the water by evaporating a small portion of it. Cooling towers can be either forced draft, where a fan is used to force the air through the system, or natural draft, where buoyancy effects are used. Forced draft towers are common in industrial plants and small power plants. Natural draft columns are much larger and are used at large power-generating stations. You have probably seen photographs of natural draft towers in news stories about Three Mile Island or other nuclear plants.

Cooling towers can be built as either countercurrent or cross-flow. The cross-flow cooling tower is probably more common and is the major use of cross-flow cascades. Air-water contact can also be used to either humidify or dehumidify air. The analysis procedure for these applications is essentially the same as for cooling towers.

In all the air-water contact systems heat and mass transfer occur simultaneously. This obviously will make the analysis more complicated. However, for air-water systems a major simplification can be made that is not generally valid for gas-liquid contactors and the analysis will be essentially analogous to the HTU-NTU analysis procedure for mass transfer.

To start, we need to briefly consider basic equations for heat transfer and analogies between heat and mass transfer. The equations will be derived for the countercurrent tower shown in Figure 19-12A. We will write the equations for the specific case where the liquid cools (mainly by evaporation) as it goes down the column, and the gas becomes hotter as it rises up the column. The rate of heat transfer per unit volume can be written as

$$\frac{\text{Rate heat transfer}}{\text{Volume}} = h_y \, a \, (T_I - T_y) \tag{19-79}$$

where h_y is the vapor-phase heat transfer coefficient in units such as Btu/hr-ft^2-$^\circ$F or W/m^2-K, T_I is the interfacial temperature, and T_y is the vapor-phase temperature. This equation is similar to Eq. (19-2a).

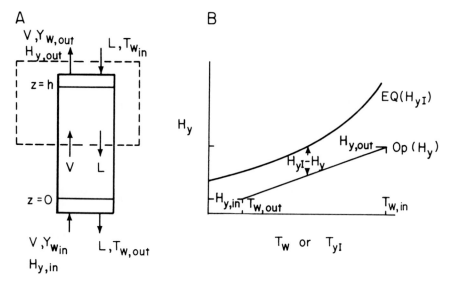

Figure 19-12.   Countercurrent tower for cooling, humidification, or dehumidification.   (A)   Schematic   of   tower.   (B) Enthalpy-temperature   diagram   for   calculation   of $H_{yI} - H_y$.

The heat transfer rate is

$$N_T \, a \, A_c \, dz = h_y \, a \, (T_I - T_y) \, A_c \, dz \qquad \text{(19-80a)}$$

which can also be related to the enthalpy change of the vapor phase

$$N_T \, a \, A_c \, dz = d[G C_{Py}(T_y - T_{ref})] \qquad \text{(19-80b)}$$

where G is the gas mass flow rate and C_{Py} is the mass heat capacity of the gas phase. In these equations N_T is the energy flux. Equating these two equations we obtain

$$dz = \frac{d[C_{Py} \, G \, (T_y - T_{ref})]}{h_y \, a \, (T_I - T_y)A_c} \qquad \text{(19-81a)}$$

For dilute solutions G and C_{Py} will be constant and Eq. (19-81a) becomes

$$dz = \frac{C_{Py} \, G d T_y}{h_y \, a \, (T_I - T_y) \, A_c} \qquad \text{(19-81b)}$$

or

$$dz = H_{Ty} \frac{dT_y}{T_I - T_y} \qquad (19\text{-}81c)$$

H_{Ty} is the gas-phase height of a transfer unit for heat transfer.

$$H_{Ty} = \frac{C_{Py} G}{h_y a A_c} \qquad (19\text{-}82)$$

If H_{Ty} is constant, Eq. (19-81c) can be integrated to give the packing height h.

$$h = H_{Ty} n_{Ty} \qquad (19\text{-}83a)$$

where n_{Ty} is the number of transfer units for heat transfer.

$$n_{Ty} = \int_{T_{y,in}}^{T_{y,out}} \frac{dT_y}{T_I - T_y} \qquad (19\text{-}83b)$$

It is evident that this heat transfer analysis parallels the mass transfer analysis for dilute systems. In comparing heat and mass transfer it will be useful to write the mass transfer expression in terms of mass units. For a dilute system the mass transfer rate in a countercurrent contactor is

$$\overline{N}_w a A_c dz = \overline{k}_y a (\overline{y}_{AI} - \overline{y}_A) A_c dz \qquad (19\text{-}84a)$$

where $\overline{N}_w$ is the mass flux of water in lb/hr-ft^2, $\overline{k}_y$ is the mass transfer coefficient in lb/hr-ft^2, and $\overline{y}_A$ is the mass fraction. This mass transfer rate can be related to the change in the water flow rate in the gas: ⁻

$$\overline{N}_w a A_c dz = G \, d\overline{y}_A \qquad (19\text{-}84b)$$

where we have assumed that the gas mass flow rate G in lb/hr is constant. Combining these equations we have

$$dz = \frac{G}{\overline{k}_y a A_c} \frac{d\overline{y}_A}{\overline{y}_{AI} - \overline{y}_A} \qquad (19\text{-}84c)$$

Integrating this equation we have the familiar form

$$h = \overline{H}_G \, \overline{n}_G \qquad (19\text{-}85a)$$

where the height of a gas-phase transfer unit in mass units, $\overline{H}_G$, is

$$\overline{H}_G = \frac{G}{\overline{k}_y \, a \, A_c}$$

and the number of transfer units is

$$\overline{n}_G = \int_{\overline{y}_{A,in}}^{\overline{y}_{A,out}} \frac{d\overline{y}_A}{\overline{y}_{AI} - \overline{y}_A} \qquad (19\text{-}85c)$$

Heat and mass transfer are analogous for dilute systems, and the transfer coefficients h_y and $\overline{k}_y$ can be related by the Chilton-Colburn analogy (Chilton and Colburn, 1934). This analogy is

$$\frac{\overline{k}_y}{v\rho} \left[\frac{\mu}{\rho \, D_{AB}} \right]^{2/3} = \frac{h_y}{\rho \, C_{Py} \, v} \left[\frac{\mu C_{Py}}{k} \right]^{2/3} = \frac{f}{2} \qquad (19\text{-}86a)$$

where k is the thermal conductivity of the gas and f is the friction factor. This correlation can be written in terms of $\overline{H}_G$ from Eq. (19-85b) and H_{Ty} from Eq. (19-82). This result is

$$\frac{\overline{H}_G}{H_{Ty}} = \frac{h_y/C_{Py}}{\overline{k}_y} = \left[\frac{k}{\rho D_{AB} C_{Py}} \right]^{2/3} = Le^{2/3} \qquad (19\text{-}86b)$$

where Le is the Lewis number. For water in air the Lewis number is about 1.15 and $Le^{2/3} = 1.1$. This is very convenient since it makes

$$H_{Ty} \sim \overline{H}_G , \quad \text{air–water} \qquad (19\text{-}87)$$

A special nomenclature for air-water systems has been developed. Instead of mass or mole fractions, it is traditional to use *absolute humidity*, which is defined as pounds (or kg) of water per pound (or kg) of *dry* air. The symbol H is often used for the absolute humidity. However, this can be confused with gas enthalpy H_y, and since humidity is a mass ratio in the gas phase we will use the symbol Y_w. The absolute humidity is related to the water mass fraction $\bar{y}_w$:

$$Y_w = \frac{\bar{y}_w}{1 - \bar{y}_w} \tag{19-88}$$

The *humid heat capacity* C_{PH} is the heat capacity of the air-water mixture per pound of dry air. The units for C_{PH} are Btu/lb (dry air)-°F or equivalent units. C_{PH} can be calculated from the heat capacities of dry air and water vapor.

$$C_{PH} = C_{P,\text{dry air}} + Y_w C_{Pw} = 0.24 + 0.46 \, Y_w \tag{19-89a}$$

The enthalpy of wet air in Btu per lb dry air is easily determined as

$$H_y = C_{PH} \left(T_y - T_{\text{ref,air}} \right) + \lambda_{\text{ref}} \, Y_w \tag{19-89b}$$

where $\lambda_{\text{ref}} = 1075.1$ Btu/lb is the latent heat at vaporization of water at the reference temperature for water, $T_{\text{ref}} = 32\,°F$. Note that it is customary to use $0\,°F$ as the reference temperature for dry air. The *wet bulb temperature* is the liquid temperature attained when a small amount of liquid is cooled by a huge volume of gas. In air-water systems the wet bulb temperature is very close to the saturation (equilibrium) temperature. Other terms are defined by Perry and Green (1984, Section 12).

A countercurrent contactor for water cooling, humidification, or dehumidification is shown in Figure 19-12A. For very dilute systems the total gas flow rate G is essentially the same as the flow rate of dry air. In addition, there will be very little evaporation or condensation of water, and the liquid flow rate $\bar{L}$ in lb/hr is approximately constant. For very dilute systems the absolute humidity is equal to the mass fraction. These assumptions are reasonable for air-water systems below about $120\,°F$, and the mass transfer expression, Eq. (19-84c), can then be written as

$$\bar{H}_G \frac{dY_w}{dz} = Y_{wI} - Y_w \tag{19-90a}$$

Equation (19-81c) can be written as

$$H_{Ty} \frac{dT_y}{dz} = T_I - T_y \tag{19-90b}$$

We multiply Eq. (19-90a) by λ_{ref}, multiply Eq. (19-90b) by C_{PH}, and add the results.

$$\overline{H}_G\lambda_{ref} \frac{dY_w}{dz} + H_{Ty} C_{PH} \frac{dT_y}{dz} = C_{PH}T_I + \lambda_{ref}Y_{wI} - (C_{PH} T_y + \lambda_{ref}Y_w)$$

The right-hand side of this equation can be recognized as the difference in the gas enthalpies, $H_{yI} - H_y$. Since $H_{Ty} = \overline{H}_G$, the left-hand side is

$$\overline{H}_G \frac{d[C_{PH}(T_y - T_{ref}) + \lambda_{ref}Y_w]}{dz}$$

The T_{ref} term can be inserted inside the derivative since the derivative of a constant is zero. Since the term inside the brackets is H_y, the final result is

$$\overline{H}_G \frac{dH_y}{dz} = H_{yI} - H_y \tag{19-91}$$

After rearrangement and integration, this becomes

$$h_p = \overline{H}_G n_{Hy} \tag{19-92a}$$

where the number of transfer units in terms of enthalpy change is

$$n_{Hy} = \int_{H_{y,in}}^{H_{y,out}} \frac{dH_y}{H_{yI} - H_y} \tag{19-92b}$$

Integration of Eq. (19-92b) to determine the height of packed bed required for humidification, dehumidification, and water cooling requires determination of $H_{yI} - H_y$. The air-water interface is assumed to be in equilibrium at temperature $T_{yI} = T_w$. This assumption is reasonable because the heat transfer coefficient in the liquid, h_x, is usually very large compared to h_y. Then the enthalpy of the wet air at the interface, H_{yI}, is equal to the equilibrium or saturation enthalpy. A short listing of the saturation data is given in Table 19-4. More detailed charts and tables are available in Perry and Green (1984, Section 12). An example of absolute humidity obtained from this table is $Y_w = 15.82 \times 10^{-3}$ at $70\,^\circ$F. The saturation enthalpy is plotted versus $T_{yI} = T_w$ in Figure 19-12B.

The air enthalpy H_y can be found from an energy balance using the balance envelope shown in Figure 19-12A.

Table 19-4. Saturation Properties of Wet Air.
Basis: Enthalpy Dry Air = 0 at $0\,°F$ and
Enthalpy of Liquid Water = 0 at $32\,°F$.

$T, °F$	$Y_{WI} \times 10^3$	H_{yI}, Btu/lb Dry Air
0	0.7872	0.835
10	1.315	3.803
20	2.152	7.106
30	3.454	10.915
40	5.213	15.230
50	7.658	20.301
60	11.08	26.46
70	15.82	34.09
80	22.33	43.69
90	31.18	55.93
100	43.19	71.73
110	59.44	92.34
120	81.49	119.54
130	111.6	155.9
140	153.4	205.7

Source: Perry and Green (1984).

$$\bar{L}h + GH_{y,out} = \bar{L}h_{in} + GH_y \tag{19-93}$$

where h is the liquid water enthalpy, which can be calculated from

$$h = C_{Pw}(T_w - T_{ref}) \tag{19-94}$$

Substituting Eq. (19-94) into Eq. (19-95) and rearranging, we obtain

$$H_y = \left[\frac{\bar{L}}{G} C_{Pw}\right]T_w + H_{y,out} - \left[\frac{\bar{L}C_{Pw}}{G}\right]T_{w,in}$$

$$= \left[\frac{\bar{L}}{G} C_{Pw}\right]T_w + H_{y,in} - \left[\frac{\bar{L}C_{Pw}}{G}\right]T_{w,out} \tag{19-95}$$

For dilute systems $\bar{L}$ and G are constant, and the operating equation
(19-95) plots as a straight line on a graph of H_y vs T_w as shown in Figure 19-12B. The value of $H_{yI} - H_y$ at any temperature can be deter-

mined as illustrated in this figure. Then Eq. (19-92) can be integrated graphically or numerically. Humidifiers have a different purpose than cooling towers, but the analysis is exactly the same, and Figure 19-12B is unchanged. In a dehumidifier a cold water stream is used and water condenses from the air. In this case the operating line will be above the equilibrium curve in Figure 19-12B.

The gas temperature can be estimated by combining Eqs. (19-90b) and (19-91),

$$\frac{dH_y}{dT_y} = \frac{H_{yI} - H_y}{T_{yI} - T_y} \qquad (19\text{-}96)$$

and integrating step by step. This is illustrated in Example 19-5. In a cooling tower, if T_y becomes equal to or greater than the saturation temperature a fog will form. Cussler (1984) has a general discussion of fog formation.

This analysis procedure is, to put it mildly, a bit strange. The procedure works for air-water systems for the following reasons:

1. The system is very dilute, and therefore:

    a. The analogy between heat and mass transfer is valid.

    b. Total gas and liquid flow rates are approximately constant.

    c. The absolute humidity equals the mass fraction.

2. The Lewis number is approximately 1.0, and $\overline{H}_G \sim H_{Ty}$.

3. The resistance to heat transfer in the liquid phase is negligible, and $T_w = T_{yI}$.

This calculation procedure cannot be used for gas-liquid contacting systems other than air-water systems, because some of the approximations become invalid. Thus, more complicated procedures are required for absorption with simultaneous heat and mass transfer (Bennett and Myers, 1982; Cussler, 1984; McCabe et al., 1984; Sherwood et al., 1975; Treybal, 1980).

In order to design a cooling tower, data for $\overline{H}_G$ are required. Typical industrial cooling towers are packed with rough wood slats arranged in such a way that the water will continually drip from one layer of slats to another. Experimental values for $\overline{H}_G$ are shown in Figure 19-13 (Sherwood et al., 1975). The value of $\overline{H}_G$ depends upon the surface area of the wood-slat packing per unit volume, a_D. Note also that the value of

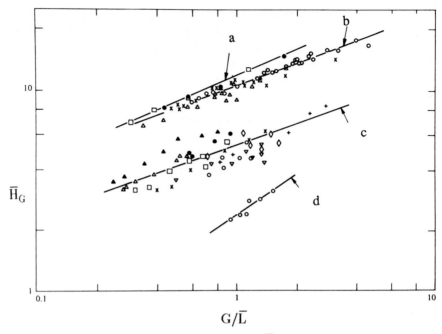

Figure 19-13.   Experimental values of $\bar{H}_G$ for slat-packed cooling towers. (a) $a_D = 0.79 \text{ ft}^{-1}$ (Kelly and Swenson, 1956). (b) $a_D = 2.7 \text{ ft}^{-1}$ (Lichtenstein,   1943).   (c) $a_D = 13.2 \text{ ft}^{-1}$ (London *et al.*, 1940). (d) $a_D = 32.9 \text{ ft}^{-1}$ (Simpson and Sherwood, 1946). From Sherwood *et al.*, *Mass Transfer* (1975). Reprinted with permission. Copyright 1975, McGraw-Hill.

$\bar{H}_G$ can be significantly greater than the values expected in distillation and absorption, where the packings are more efficient and more expensive. When packings other than wood slats are used, experimental data are required.

Cross-flow cooling towers are often used because they can be designed to have a lower pressure drop, which means that power costs for the fan can often be reduced significantly. Cross-flow towers can be designed by first designing an equivalent countercurrent tower and then applying a correction factor to compensate for differences in the driving forces for heat and mass transfer. This procedure is discussed by Sherwood *et al.* (1975).

Example 19-5. Cooling Tower Calculation

A countercurrent cooling tower with wood-slat packing with $a_D = 0.79 \text{ ft}^2/\text{ft}^3$ is being used to cool water from $112\,^\circ\text{F}$ to $79\,^\circ\text{F}$.

The entering air is at $70\,°F$ and has a wet bulb temperature of $60\,°F$. Use $\overline{L}/G = 0.5\,(\overline{L}/G)_{max}$, which is a gas flow rate twice the minimum. Determine $H_{y,in}$, $Y_{w,in}$, $(\overline{L}/G)_{max}$, $\overline{L}/G$, $H_{y,out}$, and the height of packing required, and estimate the gas temperature throughout the tower.

Solution

Since the wet bulb temperature is very close to the equilibrium temperature, the inlet gas will be saturated with water at $60\,°F$. From Table 19-4, $H_{y,in} = 26.46$ Btu/lb dry air. The water composition of the inlet air can now be determined by solving Eqs. (19-89a) and (19-89b) simultaneously.

$$Y_w = \frac{H_y - 0.24\,(T_y - T_{ref})}{0.46\,(T_y - T_{ref}) + \lambda_{ref}} \qquad (19\text{-}97)$$

which becomes

$$Y_{w,in} = \frac{26.46 - 0.24\,(70 - 0)}{0.46\,(70 - 0) + 1075.1} = 0.00872\,\frac{\text{lb water}}{\text{lb dry air}}$$

This value can also be looked up on a psychometric chart (Perry and Green, 1984, Section 12).

To determine $(C_{pw}\,\overline{L}/G)_{max}$, an enthalpy-temperature diagram is used. This is shown in Figure 19-14. Note that in this case there is a tangent minimum. $(C_{pw}\,\overline{L}/G)_{max} = 2.096$. Then

$$C_{pw}\,\overline{L}/G = 0.5\,(2.096) = 1.048$$

Since $C_{pw} = 1.0$ Btu/lb-$°F$, $\overline{L}/G = 1.048$. This is the slope of the actual operating line, which from Eq. (19-95) is

$$H_y = 1.048 T_w - 56.33$$

From Figure 19-14 or from an external balance, $H_{y,out} = 61.05$ Btu/lb dry air. To determine n_{Hy}, $H_{yI} - H_y$ is first determined from Figure 19-14. The following table is then generated.

T_w	H_y	H_{yI}	$H_y - H_{yI}$	$f = 1/(H_y - H_{yI})$
79	26.46	42.64	16.18	0.0618
95.5	43.755	64.13	20.375	0.0491
112	61.05	97.18	36.13	0.0277

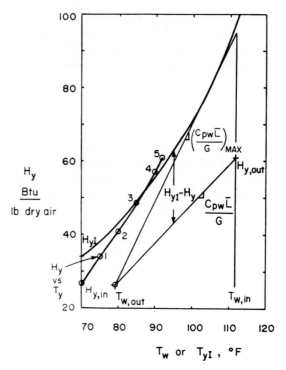

Figure 19-14.   Solution to Example 19-5.

From Simpson's rule,

$$n_{Hy} = \frac{H_{y,out} - H_{y,in}}{6} \left[f(H_{in}) + 4\, f(H_{y,avg}) + f(H_{y,out}) \right] \qquad (19\text{-}98)$$

where $H_{y,avg} = (H_{y,in} + H_{y,out})/2$. Then

$$n_{Hy} = \frac{61.05 - 26.46}{6} \left[0.0618 + 4(0.0491) + 0.0277 \right] = 1.65$$

Figure 19-13 is used to estimate $\overline{H}_G$ at $G/\overline{L} = 1/1.048 = 0.954$.

$\overline{H}_G \sim 12$ ft. Then the best estimate of the required packing height is $h_p = (1.65)(12) = 19.8$ ft.

Note: The values of H_{yI} were determined from the detailed thermodynamic charts in Perry and Green (1984, Section 12). More accuracy for the integration of n_{Hy} could be obtained by graphical integration. Since the curve of $1/(H_{yl} - H_y)$ versus H_y has no maximum or minimum, Simpson's rule should be accurate.

The gas temperature can be estimated from Eq. (19-96). This equation can be integrated step by step numerically. The modified Euler method will be used (Mickley *et al.*, 1957). At the inlet, $T_y = 70$, $T_{yl} = T_w = 79$, $H_{yl} = 42.64$, and $H_y = 26.46$. At this point Eq. (19-96) gives

$$\left(\frac{dH_y}{dT_y}\right)_0 = \frac{H_{yl} - H_y}{T_{yl} - T_y} = \frac{42.64 - 26.46}{79 - 70} = 1.80$$

Then a first guess of the gas enthalpy at $T_{yl} = T_{y0} + \Delta T_y$ is

$$H_{y,1}^{(1)} = H_{y,0} + \left(\frac{dH_y}{dT_y}\right)_0 \Delta T_y$$

If we choose $\Delta T_y = 5\,°F$, then $T_{yl} = 75\,°$ and

$$H_{y,1}^{(1)} = 26.46 + 1.8(5) = 35.5$$

At this enthalpy the water temperature $T_w = T_{yl}$ can be found from the operating line in Figure 19-14 or from the equation for the operating line, Eq. (19-95). Solving for T_w, we have

$$T_w = \left(\frac{G}{\overline{L}C_{pw}}\right) H_y - \left(\frac{G}{\overline{L}C_{pw}}\right) H_{y,in} + T_{w,out} \qquad (19\text{-}99)$$

$$T_w = 0.954\,H_y + 53.7$$

$$T_{yl,1}^{(1)} = T_{w,1}^{(1)} = (0.954)(35.5) + 53.7 = 87.6\,°F$$

At this temperature the saturation enthalpy is $H_{yl,1}^{(1)} = 52.7$. At this point we can recalculate dH_y/dT_y.

$$\left(\frac{dH_y}{dT_y}\right)_1^{(1)} = \frac{H_{yl,1}^{(1)} - H_{y,1}^{(1)}}{T_{yl,1}^{(1)} - T_{y,1}^{(1)}} = \frac{52.7 - 35.5}{87.6 - 75} = 1.37$$

Since the derivative has changed significantly, we use this derivative over the $5\,°F$ increase in gas temperature. Now the second guess for $H_{y,1}^{(2)}$ is

$$H_{y,1}^{(2)} = H_{y,0} + \left[\frac{dH_y}{dT_y}\right]_1^{(1)} (\Delta T_y) = 26.46 + (1.37)(5) = 33.3$$

Now $T_{yl,1}^{(2)} = T_{w,1}^{(2)} = 0.954 (33.31) + 53.7 = 85.5$ and $H_{yl,1}^{(2)} = 50.0$.

Then
$$\left[\frac{dH_y}{dT_y}\right]_1^{(2)} = \frac{50.0 - 33.3}{85.5 - 75} = 1.60$$

Use average slope $= 1.485$

$$H_{y,1}^{(3)} = 26.46 + (1.485)5 = 33.9$$

This is shown as point 1 in Figure 19-14. Now calculate $T_{yl,1}^{(2)}$,

$$T_{yl,1}^{(2)} = T_{w,1}^{(2)} = 0.954 (33.9) + 53.7 = 86.0\,^\circ F$$

and $H_{yl,1}^{(2)} = 50.7$. This completes the calculation for point 1. We can now calculate the derivative at point 1 to start the calculation for point 2,

$$\left[\frac{dH_y}{dT_y}\right]_1^{(1)} = \frac{50.7 - 33.9}{86.0 - 75} = 1.53$$

and proceed to calculate point 2 as follows:

$$T_{y,2} = T_{y,1} + \Delta T_y = 80\,^\circ F$$

Estimate

$$H_{y,2}^1 = H_{y,1} + \left[\frac{dH_y}{dT_y}\right]_1 \Delta T_y = 33.9 + 1.53(5) = 41.5$$

$$T_{yl,2}^{(1)} = T_{w,2}^{(1)} = (0.954)(41.5) + 53.746 = 93.3$$

$$H_{yl,2}^{(1)} = 60.8$$

$$\left[\frac{dH_y}{dT_y}\right]_2^{(1)} = \frac{60.8 - 41.5}{93.3 - 80} = 1.45$$

Then the average derivative $(dH_y/dT_y)_{avg,2} = (1.53 + 1.45)/2 = 1.49$ and

$$H_{y,2}^{(2)} = 33.9 + 1.49(5) = 41.3$$

which is close to the previous guess. This is point 2 on Figure 19-14. Proceeding with the calculation,

$$T_{yI,2}^{(2)} = T_{w,2}^{(2)} = (.954)(41.3) + 53.746 = 93.1$$

$$H_{yI,2}^{(2)} = 60.5$$

and the derivative for the next step is

$$\left.\left(\frac{dH_y}{dT_y}\right)^{(1)}\right|_2 = \frac{H_{yI,2}^{(2)} - H_{y,2}^{(2)}}{T_{yI,2}^{(2)} - T_{y,2}} = \frac{60.5 - 41.3}{93.1 - 80} = 1.46$$

We continue this procedure, making only one estimate for each derivative. The results are given in the following table.

Point	T_y	H_y	T_w
0	70	26.46	79
1	75	33.9	86.0
2	80	41.3	93.1
3	85	48.8	100.3
4	90	57.0	108.1
5	92.5	61.3	112.2

The outlet gas temperature can be estimated as the gas temperature that gives $T_w = T_{w,in} = 112°$. This is slightly less than $92.5°$ F. Note that the gas temperature curve goes through the saturated enthalpy curve. This indicates that fog will form in the tower, and the tower will not operate properly. The fog can be prevented by:

1. Not cooling the water as much (higher $T_{w,out}$)

2. Using gas with a lower $H_{y,in}$ (lower $T_{y,in}$ or lower wet bulb temperature)

3. Lowering $C_{pw} \bar{L}/G$ (increasing G)

19.9. SUMMARY - OBJECTIVES

At the end of this chapter you should be able to satisfy the following objectives:

1. Derive and use the mass transfer analysis (HTU-NTU) approach for distillation columns.

2. Use HTU-NTU analysis for dilute and concentrated absorbers and strippers.

3. Use the mass transfer correlations to determine the desired HTU.

4. Derive and use HTU-NTU analysis for cocurrent flow.

5. Use mass transfer analysis to determine tray efficiency.

6. Use the HTU-NTU procedure to analyze air-water cooling towers.

REFERENCES

AIChE, *Bubble Tray Design Manual,* AIChE, New York, 1958.

Bennett, C.O. and J.E. Myers, *Momentum, Heat and Mass Transfer,* 3rd ed., McGraw-Hill, New York, 1982.

Bolles, W.L. and J.R. Fair, "Improved Mass Transfer Model Enhances Packed-Column Design," *Chem. Eng., 89* (14), 109 (July 12, 1982).

Chilton, T.H. and A.P. Colburn, "Mass Transfer (Absorption) Coefficients. Prediction from Data on Heat Transfer and Fluid Friction," *Ind. Eng. Chem., 26,* 1183 (1934).

Cornell, D., W.G. Knapp, H.J. Close, and J.R. Fair, "Mass Transfer Efficiency-Packed Columns. Part II," *Chem. Eng. Prog., 56(8),* 48 (1960).

Cornell, D., W.G. Knapp, and J.R. Fair, "Mass Transfer Efficiency-Packed Columns. Part I," *Chem. Eng. Prog., 56* (7), 68 (1960).

Cussler, E.L., *Diffusion. Mass Transfer in Fluid Systems,* Cambridge Univ. Press, Cambridge, UK, 1984.

Fahien, R., *Transport Operations,* McGraw-Hill, New York, 1982.

Greenkorn, R.A. and D.P. Kessler, *Transfer Operations,* McGraw-Hill, New York, 1972.

Harmen, P. and J. Perona, "The Case for Co-current Operation," *Brit. Chem. Eng., 17,* 571 (1972).

Hines, A.L. and R.N. Maddox, *Mass Transfer Fundamentals and Applications,* Prentice-Hall, Englewood Cliffs, NJ, 1985.

Kelly, N.W. and L.K. Swenson, "Comparative Performance of Cooling Tower Packing Arrangements," *Chem. Eng. Prog., 52,* (7) 263 (1956).

King, C.J., *Separation Processes,* 2nd ed., McGraw-Hill, New York, 1980.

Lewis, W.K., Jr., "Rectification of Binary Mixtures. Plate Efficiency of Bubble Cap Columns," *Ind. Eng. Chem., 28,* 399 (1936).

Lichtenstein, J., "Performance and Selection of Mechanical-Draft Cooling Towers," *Trans. ASME, 65,* 779 (1943).

London, A.L., W.E. Mason, and L.M.K. Boelter, "Performance Characteristics of a Mechanically Induced Draft, Counterflow, Packed Cooling Tower," *Trans. ASME, 62,* 41 (1940).

McCabe, W.L., J.C. Smith, and P. Harriott, *Unit Operations in Chemical Engineering,* 4th ed., McGraw-Hill, New York, 1984.

Mickley, H.S., T.K. Sherwood, and C.E. Reed, *Applied Mathematics in Chemical Engineering,* 2nd ed., McGraw-Hill, New York, 1957, pp. 187-191.

Perry, R.H. and D. Green, *Perry's Chemical Engineer's Handbook,* 6th ed., McGraw-Hill, New York, 1984.

Reid, R.C., J.M. Prausnitz, and T.K. Sherwood, *The Properties of Gases and Liquids,* 3rd ed., McGraw-Hill, New York, 1977.

Reiss, L.P., "Cocurrent Gas-Liquid Contacting in Packed Columns," *Ind. Eng. Chem. Process Design Develop., 6,* 486 (1967).

Sherwood, T.K. and F.A.L. Holloway, "Performance of Packed Towers - Liquid Film Data for Several Packings," *Trans. AIChE, 36,* 39 (1940).

Sherwood, T.K., R.L. Pigford, and C.R. Wilke, *Mass Transfer*, McGraw-Hill, New York, 1975.

Simpson, W.M. and T.K. Sherwood, *Refrig. Eng.*, *52*, 535 (1946).

Treybal, R.E., *Mass-Transfer Operations*, McGraw-Hill, New York, 1955, p. 239.

Treybal, R.E., *Mass Transfer Operations*, 3rd ed., McGraw-Hill, New York, 1980.

HOMEWORK

A. *Discussion Problems*

A1.   What is a controlling resistance? How do you determine which resistance is controlling? For a cooling tower, which resistances are controlling, and which are negligible?

A2.   The mass transfer models include transfer in only the packed region. Mass transfer also occurs in the ends of the column where liquid and vapor are separated. Discuss how these "end effects" will affect a design. How could one experimentally measure the end effects?

A3.   Is a stage with a well-mixed liquid less or more efficient than a stage with plug flow of liquid across the stage (assume $K_G a$ is the same)? Explain your result with a physical argument.

A4.   a. The Bolles and Fair (1982) correlation indicates that H_G is more dependent on liquid flux than on gas flux. Explain this on the basis of a simple physical model.

b. Why do H_G and H_L depend on the packing depth?

c. Does H_G increase or decrease as μ_G increases? Does H_G increase or decrease as μ_L increases?

A5.   Why is the mass transfer analysis for concentrated absorbers considerably more complex than the analysis for binary distillation or for dilute absorbers?

A6.   Compare the advantages and disadvantages of countercurrent, cocurrent, and cross-flow cascades. Develop a decision table to help select the best cascade.

A7.   Explain what happens when a cooling tower fogs.

A8. Construct your key relations chart for this chapter.

B. *Generation of Alternatives*

B1. Develop contactor designs that combine the advantages of cocurrent, cross-flow, and countercurrent cascades.

C. *Derivations*

C1. Derive the relationships among the different NTU terms for binary distillation.

C2. Derive an equation analogous to Eq. (19-36) to relate HETP to H_{OL}.

C3. Derive Eq. (19-96).

C4. Derive the following equation to determine n_{OG} at total reflux for systems with constant relative volatility:

$$n_{OG} = \frac{1}{1 - \alpha} \ln \left[\frac{(y_{out} - 1)(y_{in})}{(y_{in} - 1)(y_{out})} \right] + \ln \left[\frac{y_{in} - 1}{y_{out} - 1} \right]$$

D. *Problems*

D1. For Examples 19-1 and 19-2, estimate an average H_{OG} in the enriching section. Then calculate n_{OG} and $h_E = H_{OG,avg} \, n_{OG}$.

D2. If 1-in metal pall rings are used instead of 2-in rings in Example 19-2:

a. Recalculate the flooding velocity and the required diameter.

b. Recalculate H_G and H_L in the enriching section.

D3. A distillation column is separating a feed that is 40 mole % methanol and 60 mole % water. The two-phase feed is 60% liquid. Distillate product should be 92 mole % methanol, and bottoms 4 mole % methanol. A total reboiler and a total condenser are used. Reflux is a saturated liquid. Operation is at 101.3 kPa. Assume constant molal overflow, and use $L/D = 0.9$. Under these conditions $H_G = 1.3$ ft and $H_L = 0.8$ ft in both the enriching and stripping sections. Determine the required heights of both the enriching and stripping sections. Equilibrium data are given in Table 3-3.

D4. A distillation column operating at total reflux is separating methanol from ethanol. The average relative volatility is 1.69. Operation is at 101.3 kPa. We obtain methanol mole fractions of $y_{out} = 0.972$ and $y_{in} = 0.016$.

    a. If there is 24.5 ft of packing, determine H_{OG} using the result of Problem 19-C4.

    b. Check your results for part a, using a McCabe-Thiele diagram.

D5. We wish to strip SO_2 from water using air at 20°C. The inlet air is pure. The outlet water contains 0.0001 mole frac SO_2, while the inlet water contains 0.0011 mole frac SO_2. Operation is at 855 mmHg, and $L/V = 0.9 (L/V)_{max}$. Assume $H_{OL} = 2.76$ ft and that the Henry's law constant is 22,500 mmHg/mole fraction SO_2. Calculate the packing height required.

D6. A packed tower is used to absorb ammonia from air using aqueous sulfuric acid. The gas enters the tower at 31 lb mole/hr-ft^2 and is 1.0 mole % ammonia. Aqueous 10 mole % sulfuric acid is fed at a rate of 24 lb mole/hr-ft^2. The equilibrium partial pressure of ammonia above a solution of sulfuric acid is zero. We desire an outlet ammonia composition of 0.01 mole % in the gas stream.

    a. Calculate n_{OG} for a countercurrent column.

    b. Calculate n_{OG} for a cocurrent column.

D7. We wish to absorb ammonia into water at 20°C. At this temperature $H = 2.7$ atm/mole frac. Pressure is 1.1 atm. Inlet gas is 0.013 mole frac NH_3, and inlet water is pure water.

    a. In a countercurrent system we wish to operate at $L/G = 15 (L/G)_{min}$. A $y_{out} = 0.00004$ is desired. If $H_{OG} = 0.25$ ft at $V/A_c = 5.7$ lb moles air/hr-ft^2, determine the height of packing required.

    b. For a cocurrent system a significantly higher V/A_c can be used, since flooding is not a problem. At $V/A_c = 22.8$, $H_{OG} = 0.12$ ft. If the same L/G is used as in part a, what is the lowest y_{out} that can be obtained? If $y_{out} = 0.00085$, determine the packing height required.

D8. We are operating a staged distillation column at total reflux to determine the Murphree efficiency. Pressure is 101.3 kPa. We are separating methanol and water. The column has a 2-in head of liquid on each well-mixed stage. The molar vapor flux is 30 lb moles/hr-ft^2. Near the top of the column, when x = 0.8 we measure $E_{MV} = 0.77$. Near the bottom, when $x = 0.16$, $E_{MV} = 0.69$. Equilibrium data are given in Table 3-3.

    a. Calculate $k_x a$ and $k_y a$.

    b. Estimate E_{MV} when $x = 0.01$.

D9.   We wish to repeat Example 19-5, but with conditions that will not cause the formation of fog in the tower. Cool the water from $112\,°F$ to $85\,°F$. Entering air is at $70\,°F$ with a $50\,°F$ wet bulb temperature. Use $\overline{L}C_{Pw}/G = 1.048$. Wood-slat packing has $a_D = 0.79$ ft^2/ft^3. Determine the height of packing required, and estimate the gas temperature throughout the tower. What is the outlet gas temperature? Does fogging occur?

D10.   We wish to dehumidify a saturated air stream at $110\,°F$ by contacting it with water at $78\,°F$. The outlet air stream should have a wet bulb temperature of $88\,°F$. Use $\overline{L}/G = 1.3\,(\overline{L}/G)_{min}$. At these conditions the packing used has $\overline{H}_G = 8$ ft. Calculate the height of packing required.

F.   *Problems Requiring Other Resources*

F1.   Repeat Problem 13-F3, except use an HTU-NTU analysis. Estimate H_G and H_L.

INDEX